LANDOLT-BÖRNSTEIN

Numerical Data and Functional Relationships in Science and Technology

New Series

Editor in Chief: K.-H. Hellwege

Group II: Atomic and Molecular Physics

Volume 14

Supplement to Volumes II/4 and II/6

Molecular Constants

mostly from Microwave, Molecular Beam and Electron Resonance Spectroscopy

Subvolume b

J. M. Brown · J. Demaison · A. Dubrulle
W. Hüttner · E. Tiemann

Editors: K.-H. Hellwege and A. M. Hellwege

Springer-Verlag Berlin · Heidelberg · New York · Tokyo 1983

LANDOLT-BÖRNSTEIN

Zahlenwerte und Funktionen
aus Naturwissenschaften und Technik

Neue Serie

Gesamtherausgabe: K.-H. Hellwege

Gruppe II: Atom- und Molekularphysik

Band 14

Ergänzung der Bände II/4 und II/6

Molekelkonstanten

vorwiegend aus Messungen der Mikrowellen-,
Molekularstrahl- und Elektronenresonanz-Spektroskopie

Teilband b

J. M. Brown · J. Demaison · A. Dubrulle
W. Hüttner · E. Tiemann

Herausgeber: K.-H. Hellwege und A. M. Hellwege

Springer-Verlag Berlin · Heidelberg · New York · Tokyo 1983

CIP-Kurztitelaufnahme der Deutschen Bibliothek

Zahlenwerte und Funktionen aus Naturwissenschaften und Technik / Landolt-Börnstein. - Berlin; Heidelberg; New York; Tokyo: Springer. Teilw. mit d. Erscheinungsorten: Berlin, Heidelberg, New York. - Parallelt.: Numerical data and functional relationships in science and technology

NE: Landolt, Hans [Begr.]; PT. N.S. / Gesamthrsg.: K.-H. Hellwege. Gruppe 2, Atom- und Molekularphysik. Bd. 14. Molekelkonstanten; vorwiegend aus Messungen d. Mikrowellen-, Molekularstrahl- u. Elektronenresonanz-Spektroskopie; Erg. d. Bd. II/4 u. II/6. Teilbd. b. J.M. Brown ··· Hrsg.: K.-H. Hellwege and A.M. Hellwege. - 1983.

ISBN 3-540-11857-8 (Berlin, Heidelberg, New York, Tokyo)
ISBN 0-387-11857-8 (New York, Heidelberg, Berlin, Tokyo)

NE: Hellwege, Karl-Heinz [Hrsg.]; Brown, John M. [Mitverf.]

Printed in Germany

Typesetting, printing and bookbinding: Universitätsdruckerei H. Stürtz AG Würzburg

2163/3020 - 543210

Contributors

John M. Brown

Department of Chemistry
University of Southampton, U.K.

Jean Demaison

Université des Sciences et Techniques de Lille, France
UER de Physique Fondamentale

Alain Dubrulle

Université des Sciences et Techniques de Lille, France
UER de Physique Fondamentale

Wolfgang Hüttner

Abteilung Chemie-Physik
Universität Ulm, Germany

Eberhard Tiemann

Institut für Atom- und Molekülphysik
Universität Hannover, Germany

Vorwort

Hiermit wird der zweite und letzte Teilband, II/14b, des Bandes II/14 „Molekelkonstanten vorwiegend aus Messungen der Mikrowellen-, Molekularstrahl- und Elektronenresonanz-Spektroskopie“ vorgelegt, der ein Ergänzungsband von II/6 und damit der zweite Ergänzungsband von II/4 ist. Diese Supplementierung erschien nötig, da in den letzten Jahren die Zahl der untersuchten Molekeln und die Menge des Datenmaterials erheblich angewachsen sind. Die drei Bände II/4, II/6 und II/14 bilden zusammen eine vollständige, kritische Sammlung aller bis 1982 veröffentlichten Werte von Molekelkonstanten, die mit Hilfe der verschiedensten klassischen und neuen spektroskopischen Methoden gewonnen wurden. Die am Anfang gewählte Art der Anordnung und der Darstellung wurde beibehalten, damit alle bis einschließlich 1982 vorliegenden Daten leicht verglichen werden können. Mit Hilfe des alle Bände übergreifenden Substanzenregisters können die zu einer Molekel gehörenden Daten schnell gefunden werden. Wir hoffen, daß den Lesern auf diese Weise ein nützlicher Datenapparat angeboten wird. Zur Ergänzung verweisen wir auf den Band II/7 „Strukturdaten freier, mehratomiger Molekeln“, der demnächst supplementiert werden soll.

Der hier vorgelegte letzte Teilband II/14b enthält die Ergänzung zu den Tabellen über freie Radikale in Band II/6, wieder einschließlich der Frequenzen und der entsprechenden Übergänge. Er bringt ferner die quantitativen Strukturzeichnungen für die in den Tabellen von II/14a behandelten diamagnetischen Molekeln sowie einen kurzen Nachtrag von neuen Daten. Den letzten Teil dieses Teilbandes II/14b bildet das Substanzenregister für alle Bände II/4, II/6 und II/14a, b.

Allen Autoren ist zu danken für ihre sorgfältige Arbeit und die Bereitwilligkeit, ihre Tabellen so weit wie irgend möglich auf den neuesten Stand zu bringen. Ebenso möchten wir wieder der „Sektion für Strukturdokumentation der Universität Ulm“ danken, deren vollständiges Dokumentationsmaterial eine wesentliche Grundlage für das Zustandekommen aller dieser Bände gewesen ist. Unser Dank gilt auch der LB-Redaktion, insbesondere Frau R. Lettmann, für ihre zuverlässige Hilfe, und dem Verlag für Geduld und Sorgfalt.

Auch dieser Band wird ohne finanzielle Hilfe von anderer Seite veröffentlicht.

Darmstadt, März 1983 **Die Herausgeber**

Preface

This is the second and last subvolume, II/14b, of volume II/14 "Molecular Constants, mostly from Microwave, Molecular Beam and Electron Resonance Spectroscopy", which is a supplement to volume II/6 and thus the second supplement to volume II/4. This supplementation will meet a need because the amount of investigated molecules has considerably increased in the last years. The three volumes II/4, II/6 and II/14 together present a complete compilation of the critically evaluated data on molecular constants from measurements with classical and modern spectroscopy which have been published until 1982. The presentation of the data in the present supplement is the same as in the previous volumes and thus all data up to 1982 can be compared easily. A comprehensive index of all substances treated in volumes II/4, II/6 and II/14a, b will help to locate the data on any individual substance in one of these volumes. Thus we hope that this compilation will be a useful source of information. – In addition, we refer to volume II/7 "Structure data of free polyatomic molecules" which will be supplemented soon.

The last subvolume, II/14b, presented herewith, contains the supplement to the tables of volume II/6 on free radicals (including again the frequencies and the respective transitions), and the qualitative structure diagrams of the diamagnetic molecules already treated in the tables of the preceding subvolume, II/14a, and a short appendix with new data of diamagnetic molecules. The last part of this subvolume, II/14b, is the index of substances for all volumes II/4, II/6 and II/14a, b.

Our thanks are due to the authors for their careful work and their willingness to supplement their tables up to the last minute. We have to thank again the "Sektion für Strukturdokumentation der Universität Ulm" who made available their documentation material to the authors. We also thank the editorial office, especially Frau R. Lettmann, for their reliable help, and Springer Verlag for their patience and care.

This subvolume is published without any outside financial support.

Darmstadt, March 1983 **The Editors**

Table of contents

Subvolume b

Survey of subvolume a

3 Figures

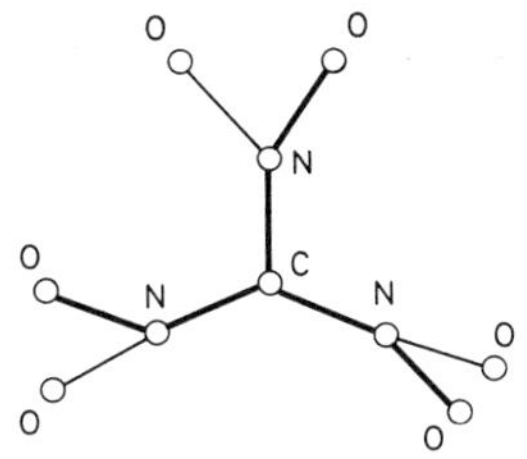

Fig. 1. $HC(NO_2)_3$.
The H is on the symmetric axis.
See subvol. a, p. 119, Nr. 87.

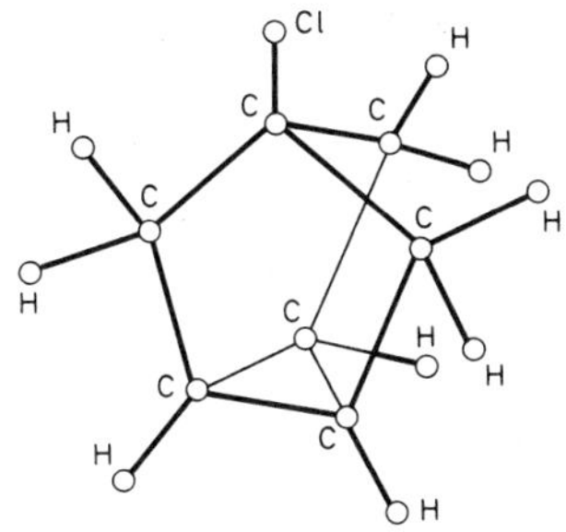

Fig. 2. C_7H_9Cl.
See subvol. a, p. 123, Nr. 95.

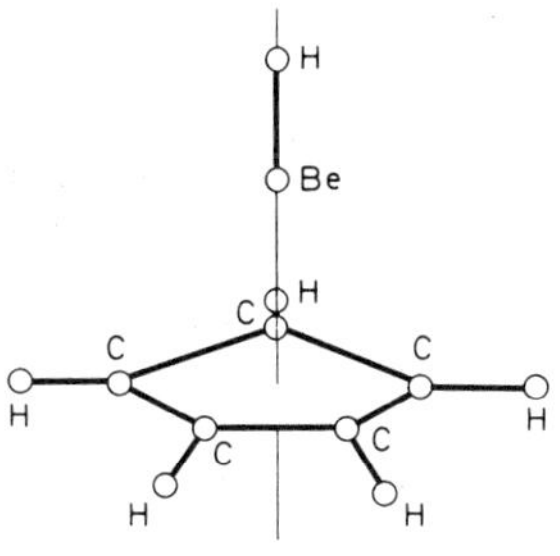

Fig. 3. C_5H_5BeH.
See subvol. a, p. 123, Nr. 96.

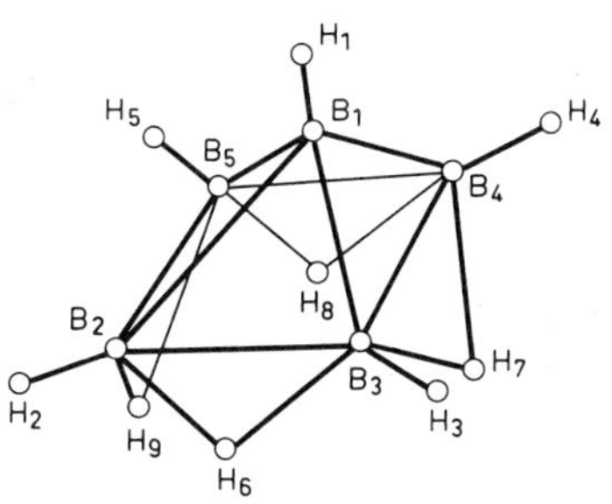

Fig. 4. B_5H_9.
See subvol. a, p. 124, Nr. 98.

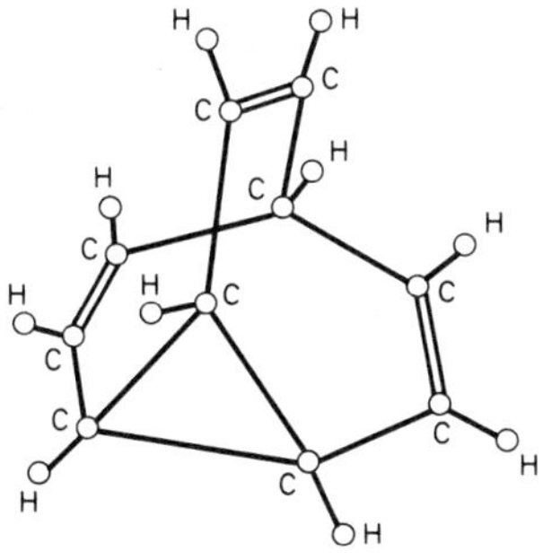

Fig. 5. $C_{10}H_{10}$.
See subvol. a, p. 125, Nr. 99.

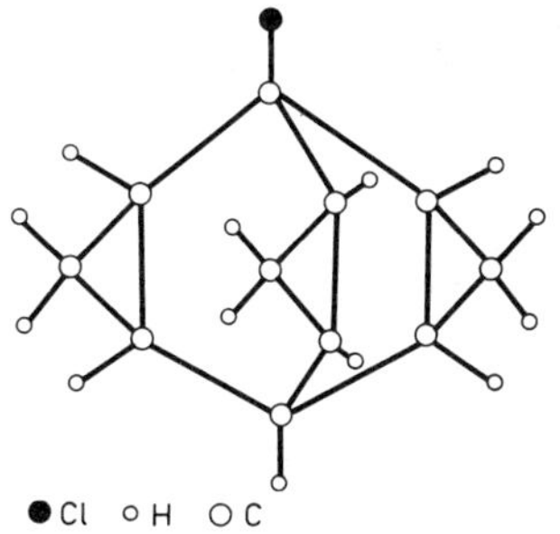

Fig. 6. $C_{11}H_{13}Cl$.
See subvol. a, p. 125, Nr. 100.

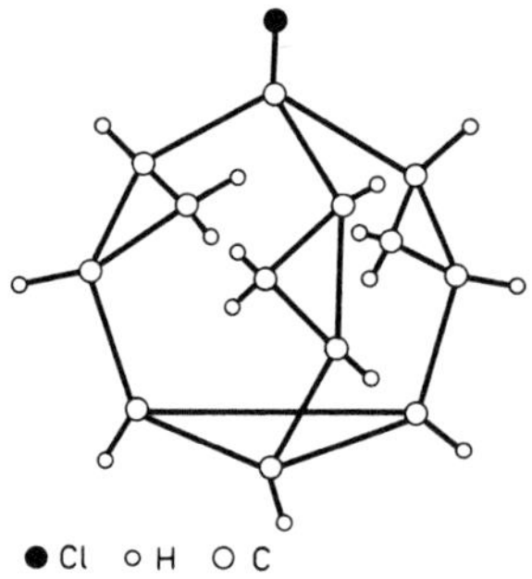

Fig. 7. $C_{13}H_{15}Cl$.
See subvol. a, p. 125, Nr. 101.

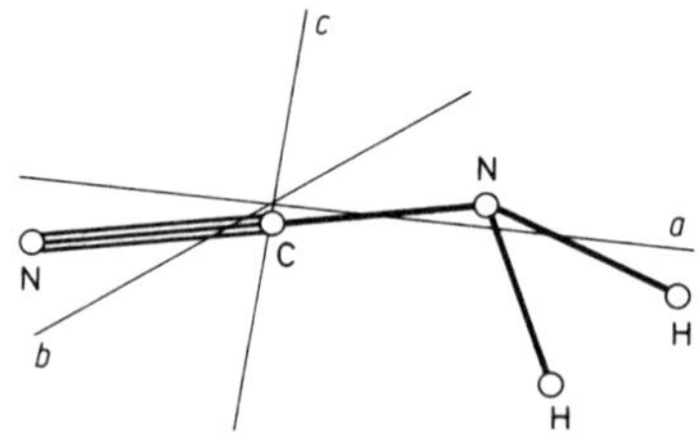

Fig. 8. NH_2CN.
See subvol. a, p. 131, Nr. 2.

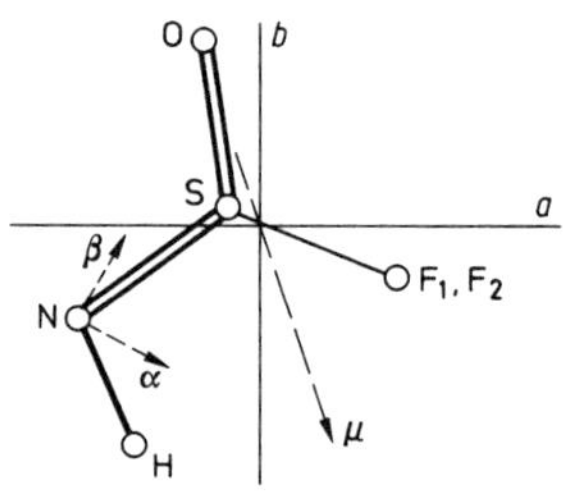

Fig. 9. $HN{=}SOF_2$.
See subvol. a, p. 132, Nr. 4.

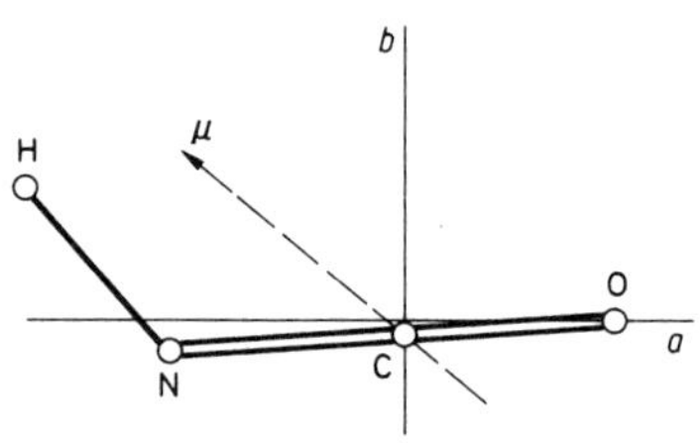

Fig. 10. HNCO.
See subvol. a, p. 133, Nr. 5.

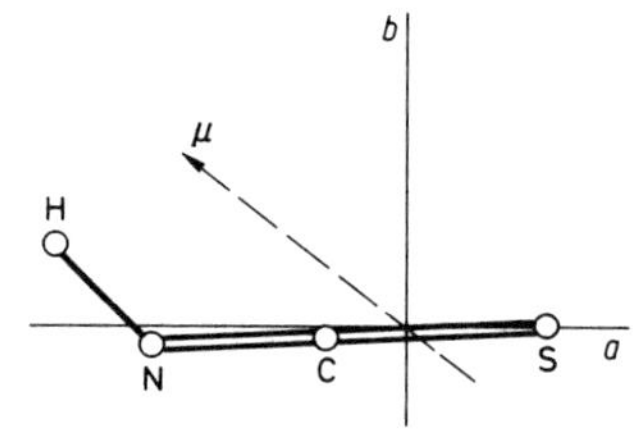

Fig. 11. HNCS.
See subvol. a, p. 135, Nr. 6.

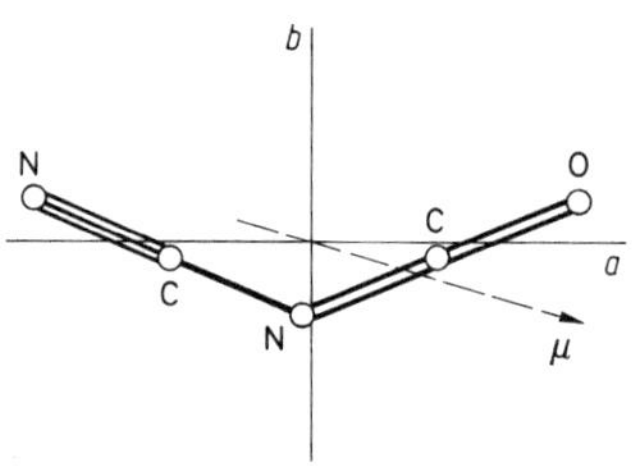

Fig. 12. NCNCO.
See subvol. a, p. 139, Nr. 9.

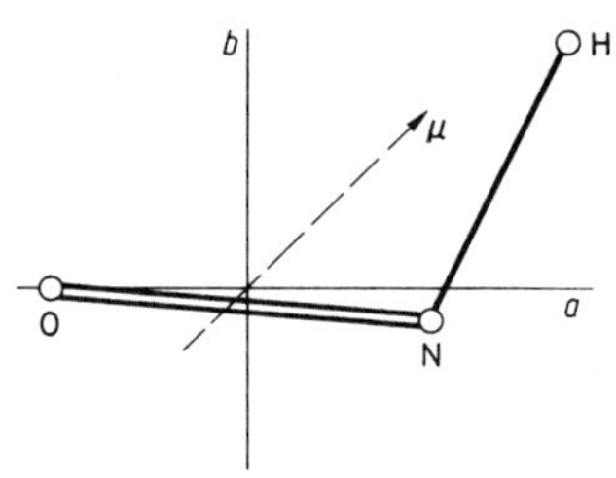

Fig. 13. HNO.
See subvol. a, p. 140, Nr. 12.

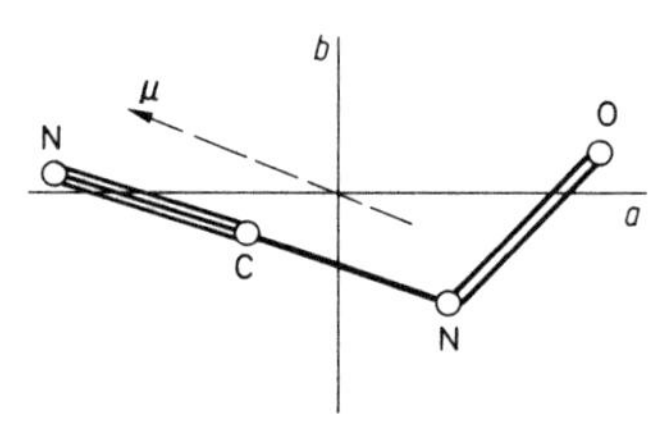

Fig. 14. NOCN.
See subvol. a, p. 142, Nr. 16.

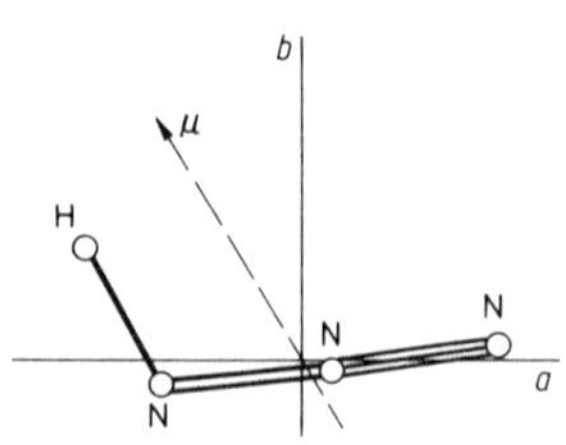

Fig. 15. HN_3.
See subvol. a, p. 145, Nr. 21.

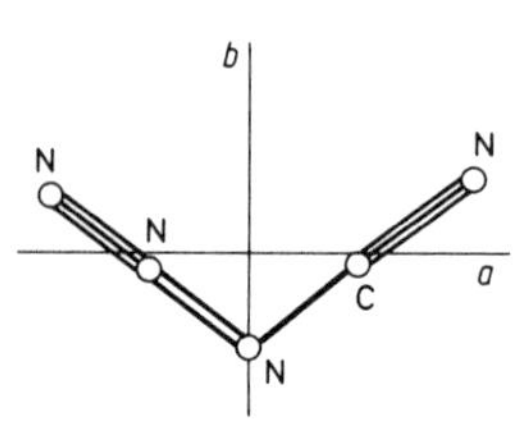

Fig. 16. NCN_3.
See subvol. a, p. 145, Nr. 22.

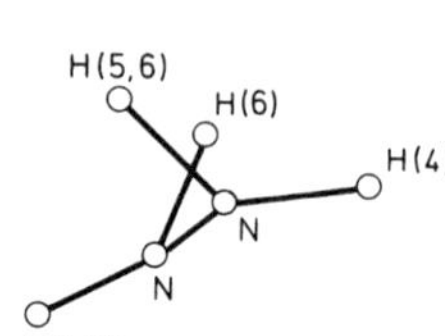

Fig. 17. N_2H_4.
See subvol. a, p. 146, Nr. 25.

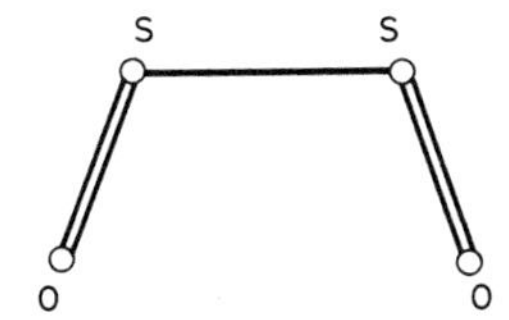

Fig. 18. $(SO)_2$.
See subvol. a, p. 151, Nr. 29.

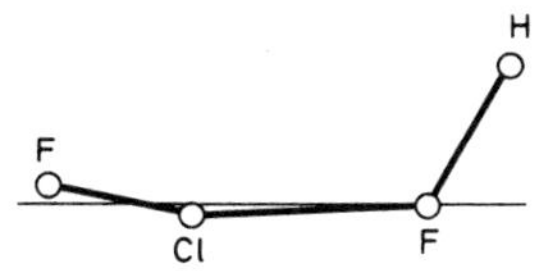

Fig. 19. HFClF.
See subvol. a, p. 156, Nr. 39.

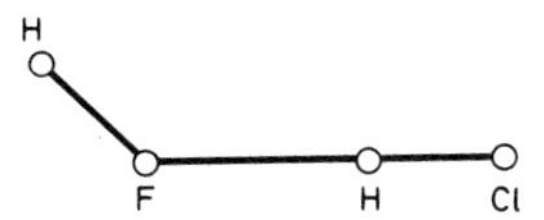

Fig. 20. HFHCl.
See subvol. a, 156, Nr. 40.

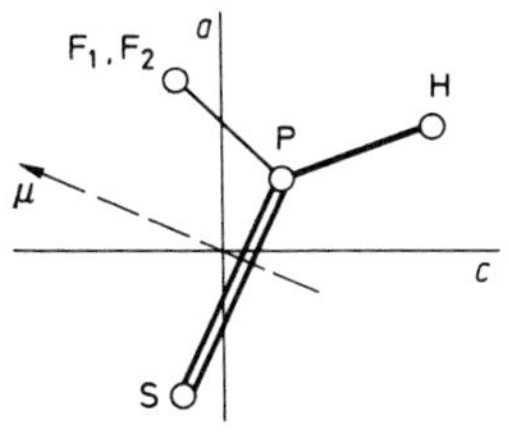

Fig. 21. $SPHF_2$.
See subvol. a, p. 157, Nr. 43.

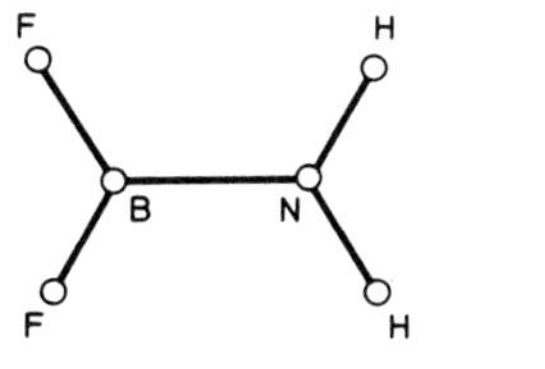

Fig. 22. BF_2NH_2.
See subvol. a, p. 158, Nr. 46.

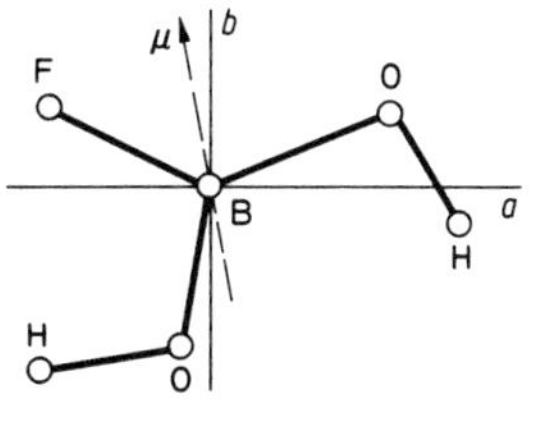

Fig. 23. $BF(OH)_2$.
See subvol. a, p. 160, Nr. 50.

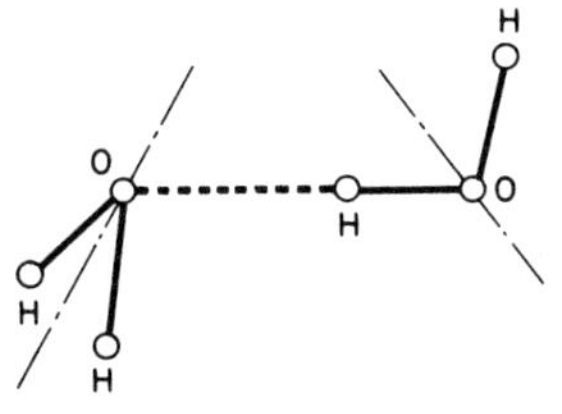

Fig. 24. $(H_2O)_2$.
See subvol. a, p. 173, Nr. 60.

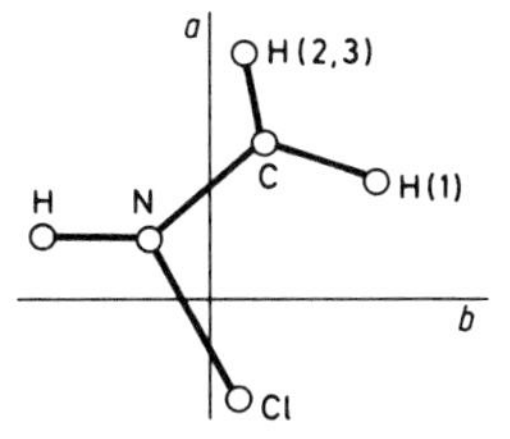

Fig. 25. CH_3NHCl.
See subvol. a, p. 177, Nr. 69.

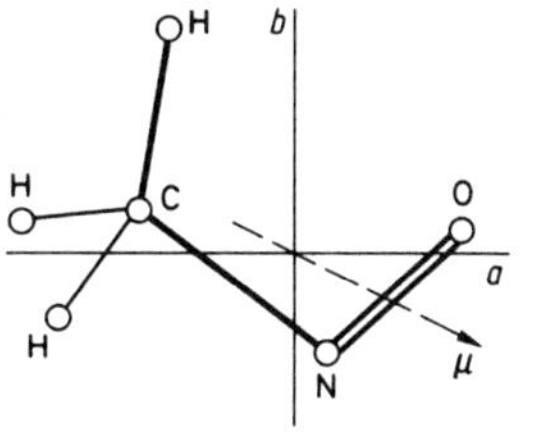

Fig. 26. CH_3NO.
See subvol. a, p. 179, Nr. 72.

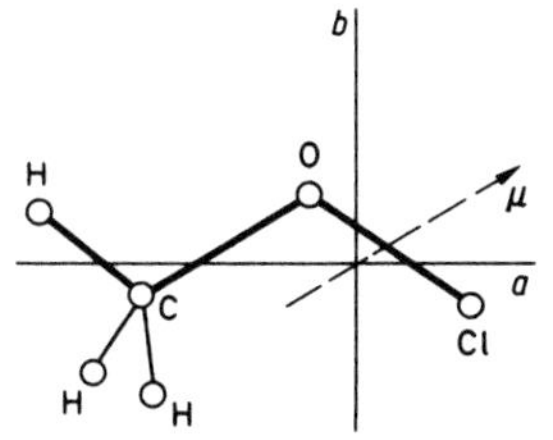

Fig. 27. CH_3OCl.
See subvol. a, p. 181, Nr. 75.

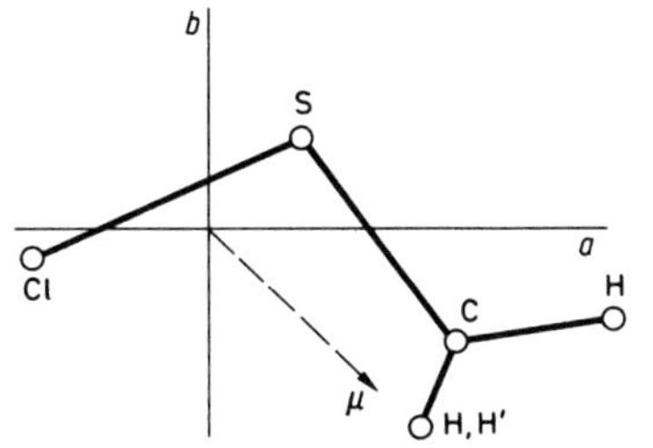

Fig. 27a. CH_3SCl.
See subvol. a, p. 181, Nr. 76.

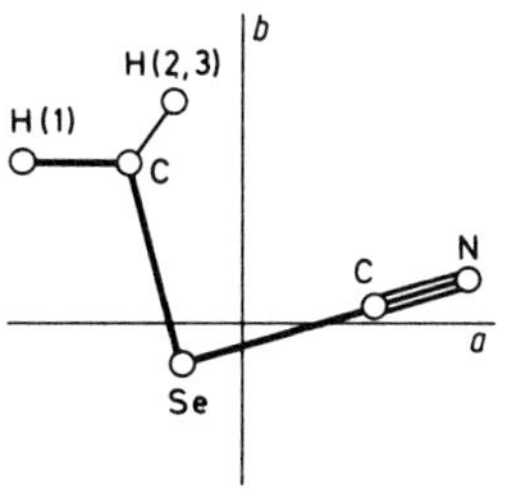

∢ $(\mu, \mathrm{CN}) = 10.2°$

Fig. 28. CH_3SeCN.
See subvol. a, p. 182, Nr. 81.

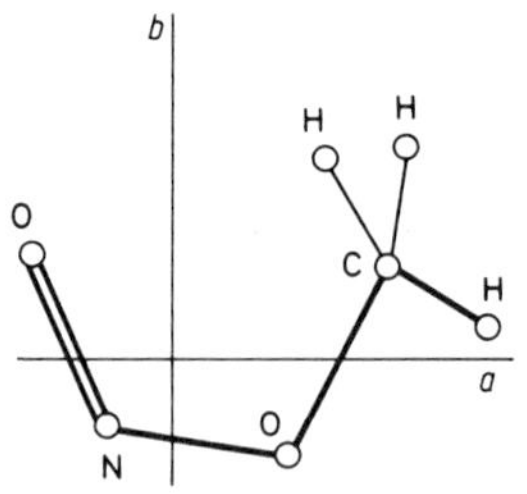

Fig. 29. CH_3ONO *cis*.
See subvol. a, p. 183, Nr. 82.

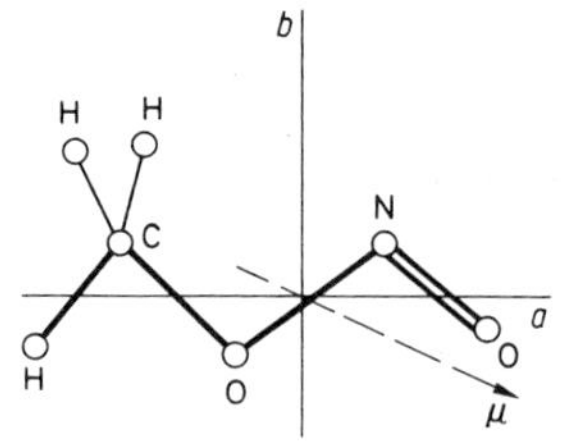

Fig. 29a. CH_3ONO *trans*.
See subvol. a, p. 183, Nr. 82.

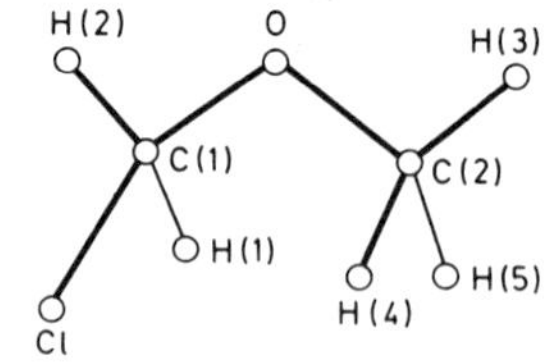

Fig. 30. CH_3OCH_2Cl.
See subvol. a, p. 187, Nr. 86.

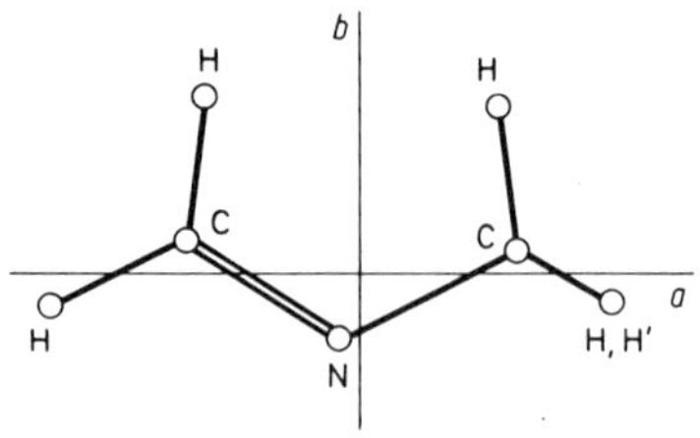

Fig. 31. $CH_3N{=}CH_2$.
See subvol. a, p. 191, Nr. 90.

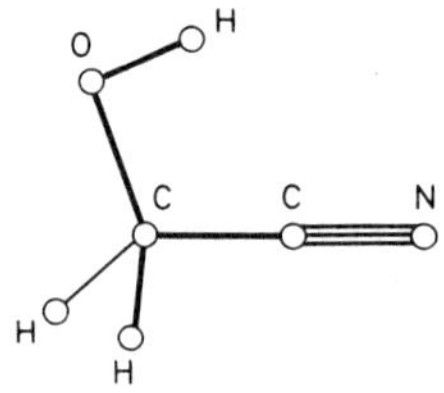

Fig. 32. CH_2OHCN.
See subvol. p. 197, Nr. 100.

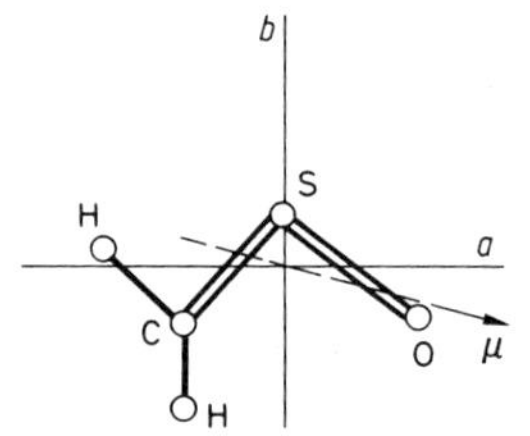

Fig. 33. $CH_2{=}SO$.
See subvol. a, p. 198, Nr. 102.

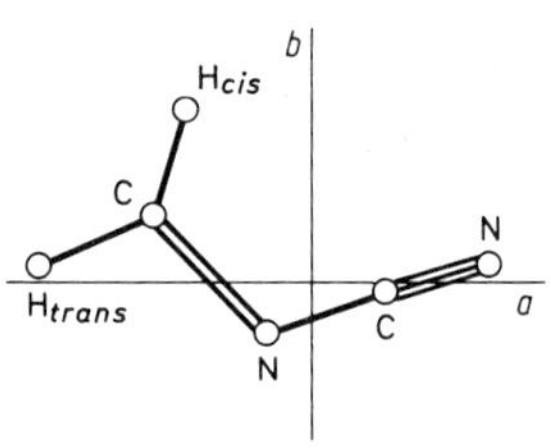

Fig. 34. $CH_2{=}N{-}CN$.
See subvol. a, p. 200, Nr. 104.

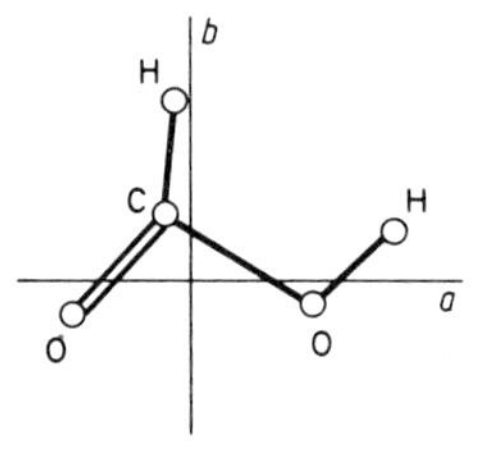

Fig. 35. $HC\genfrac{}{}{0pt}{}{{=}O}{-OH}$ *cis*.

See subvol. a, p. 209, Nr. 111.

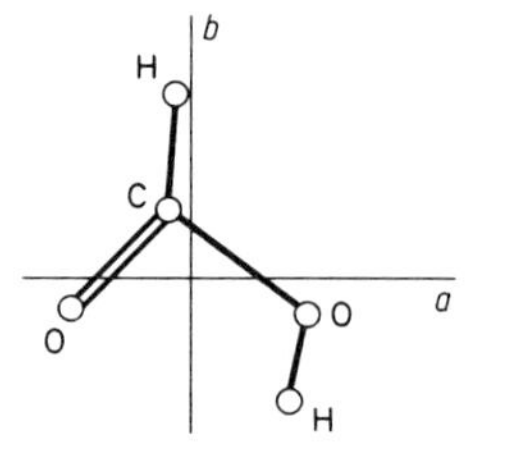

Fig. 35a. $HC\genfrac{}{}{0pt}{}{{=}O}{-OH}$ *trans*.

See subvol. a, p. 209, Nr. 111.

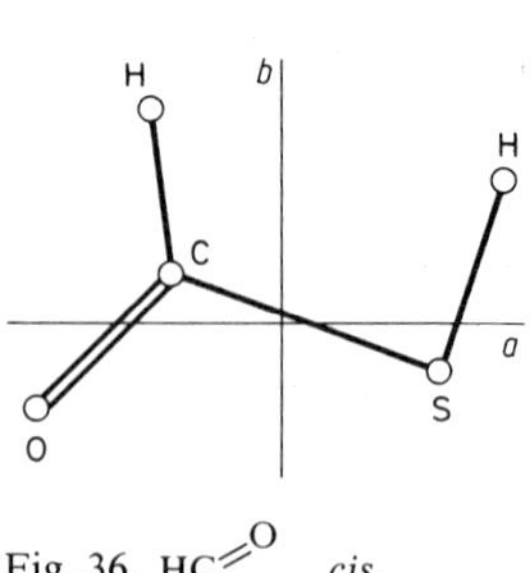

Fig. 36. $HC\genfrac{}{}{0pt}{}{{=}O}{-SH}$ *cis*.

See subvol. a, p. 214, Nr. 112.

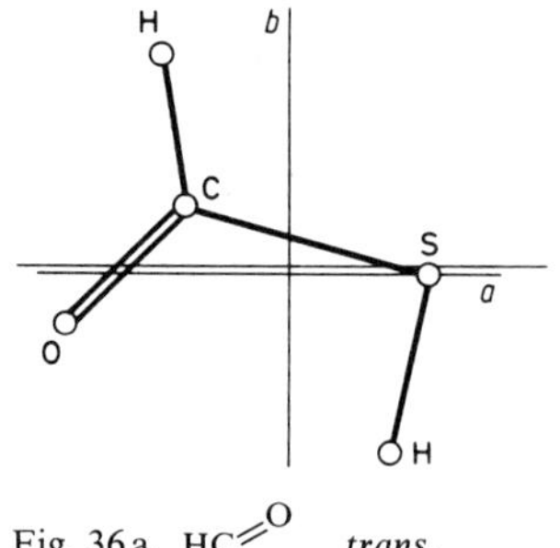

Fig. 36a. $HC\genfrac{}{}{0pt}{}{{=}O}{-SH}$ *trans*.

See subvol. a, p. 214, Nr. 112.

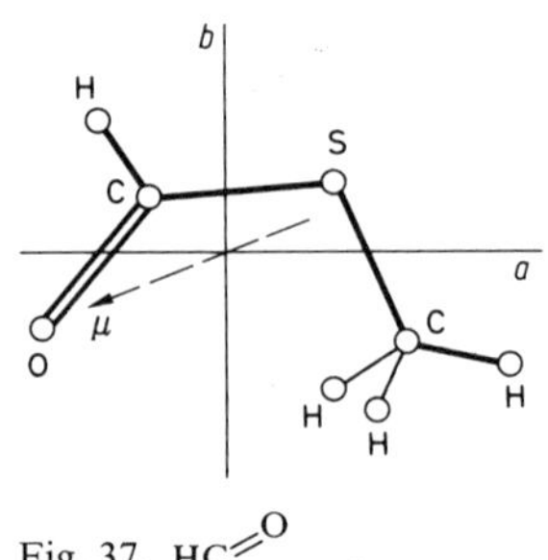

Fig. 37. $HC\genfrac{}{}{0pt}{}{{=}O}{-SCH_3}$.

See subvol. a, p. 216, Nr. 114.

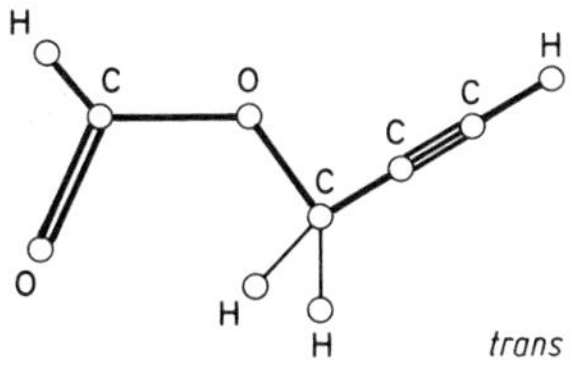

Fig. 38. $HC(=O)OCH_2C\equiv CH$.
See subvol. a, p. 216, Nr. 118.

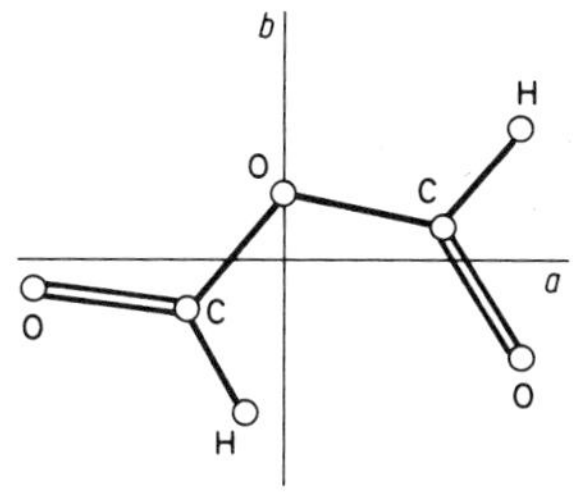

Fig. 39. $O{=}C(H){-}O{-}C({=}O){-}H$.
See subvol. a, p. 221, Nr. 123.

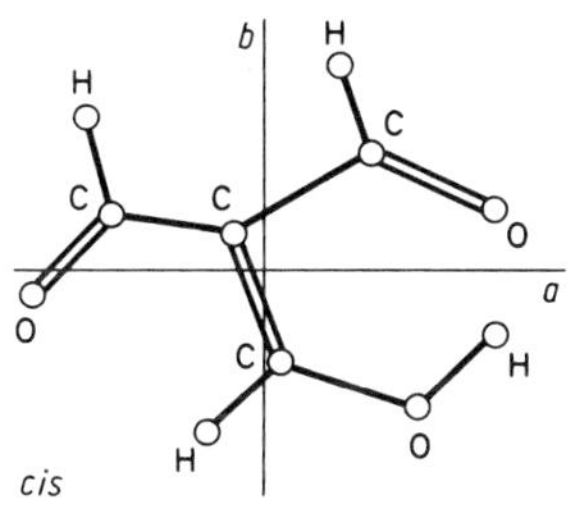

Fig. 40. $HC\left(C(=O)H\right)_3$.
"keto-enol structure".
See subvol. a, p. 223, Nr. 124.

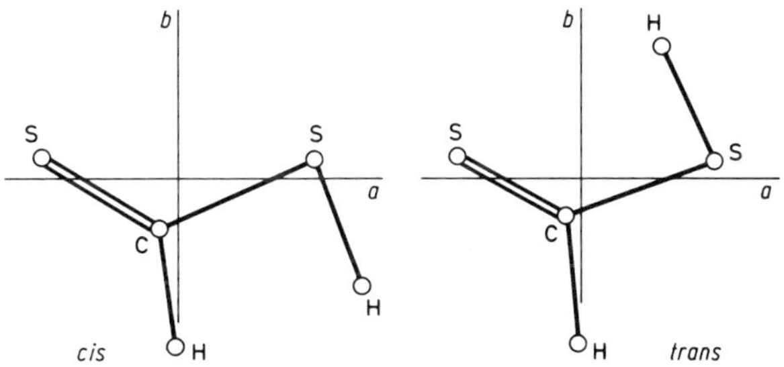

Fig. 41. $HC(=S)SH$.
See subvol. a, p. 238, Nr. 150.

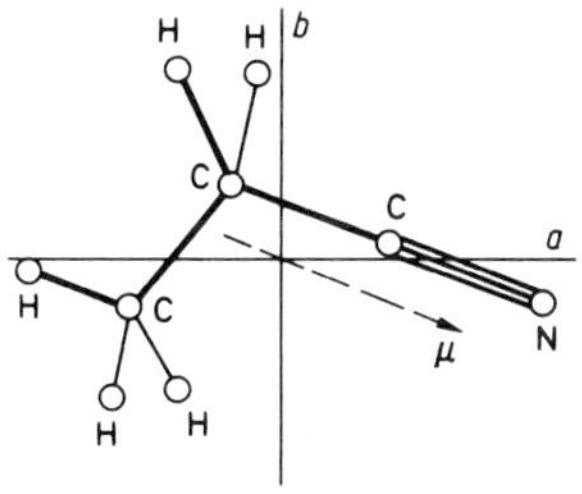

Fig. 42. CH_3CH_2CN.
See subvol. a, p. 241, Nr. 155.

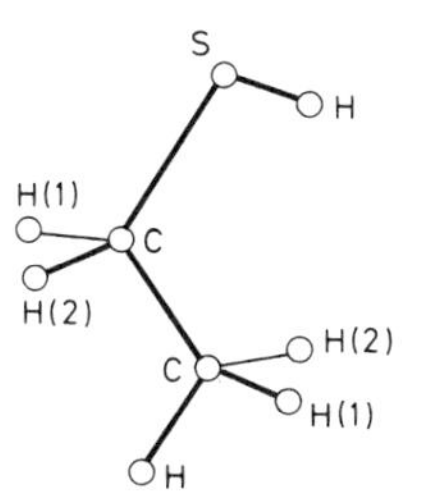

Fig. 43. CH_3CH_2SH *gauche*.
See subvol. a, p. 244, Nr. 157.

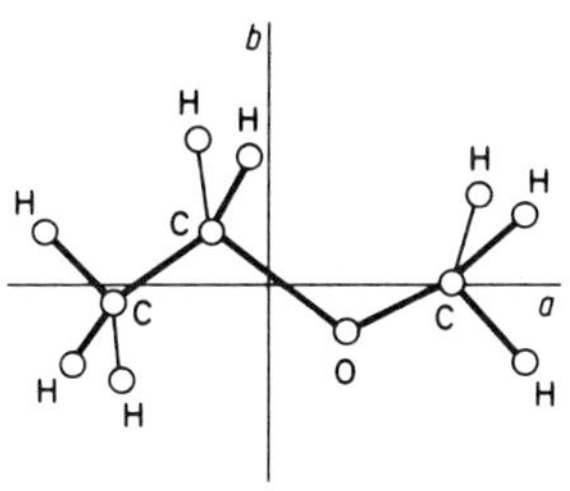

Fig. 44. $CH_3CH_2OCH_3$.
See subvol. a, p. 247, Nr. 159.

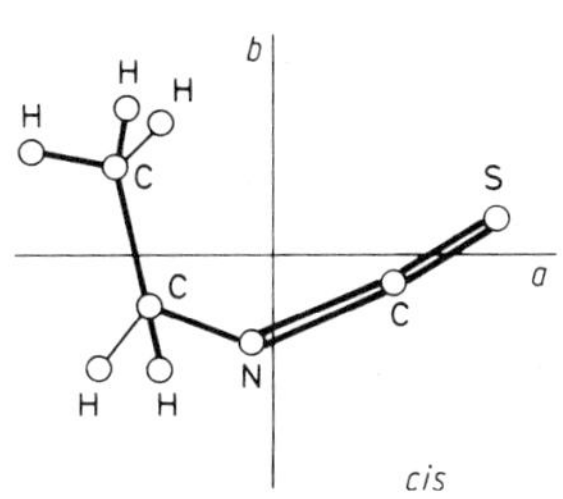

Fig. 45. CH_3CH_2NCS.
See subvol. a, p. 249, Nr. 161.

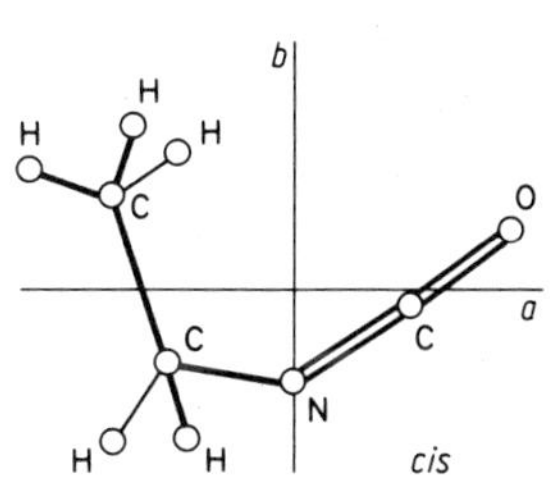

Fig. 46. CH_3CH_2NCO.
See subvol. a, p. 250, Nr. 162.

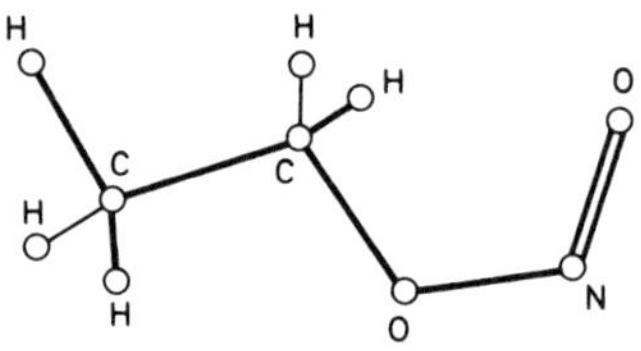

Fig. 47. CH_3CH_2ONO.
See subvol. a, p. 250, Nr. 163.
Plausible structure for cis, trans ethyl nitrite

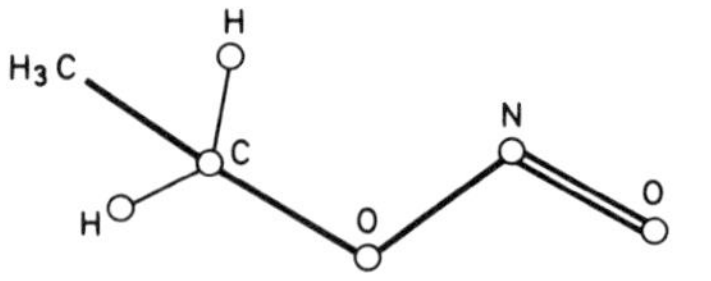

Plausible structure for trans, gauche ethyl nitrite

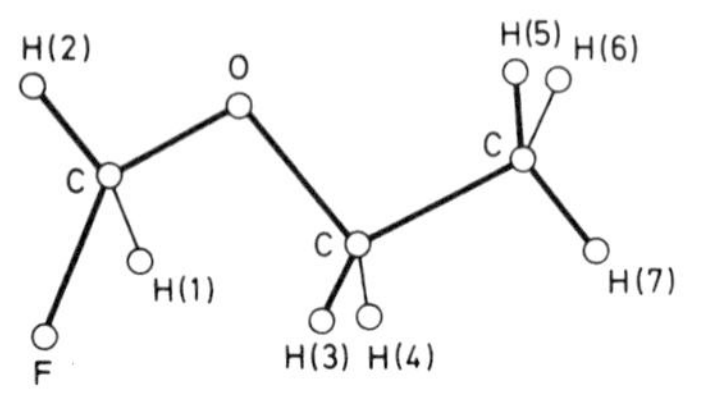

Fig. 48. $CH_3CH_2OCH_2F$.
See subvol. a, p. 254, Nr. 167.

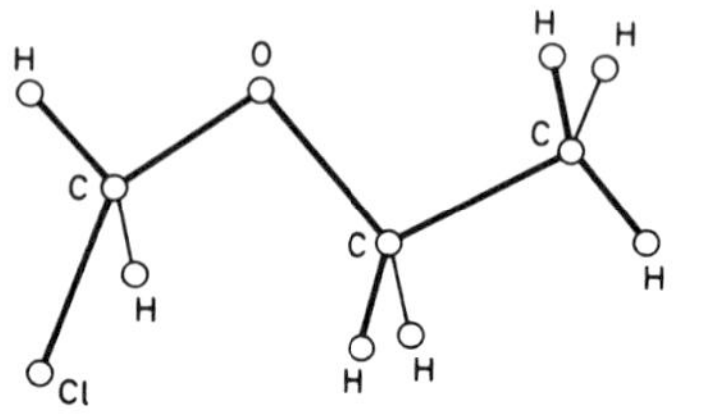

Fig. 49. $CH_3CH_2OCH_2Cl$.
For atom numbering see Fig. 48.
See subvol. a, p. 255, Nr. 168.

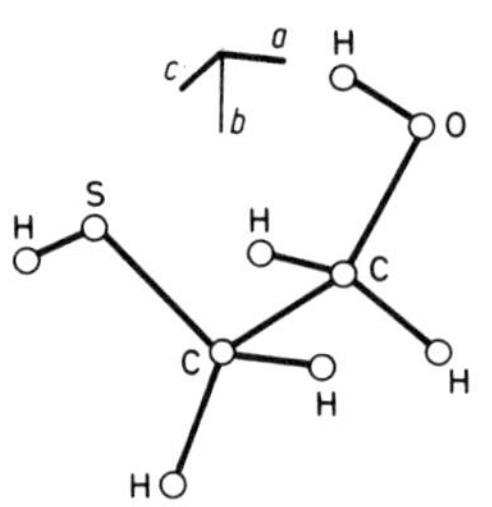

Fig. 50. $HSCH_2CH_2OH$.
Perspective drawing of 2-mercaptoethanol.
See subvol. a, p. 258, Nr. 175.

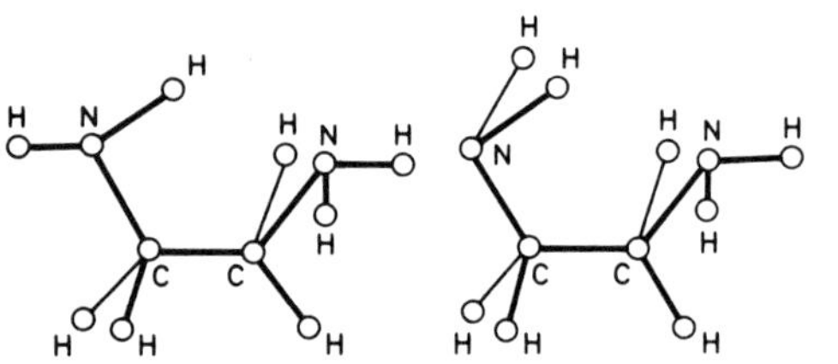

I
II

Fig. 51. $H_2NCH_2CH_2NH_2$.
The most stable conformations of ethylendiamine.
See subvol. a, p. 260, Nr. 176.

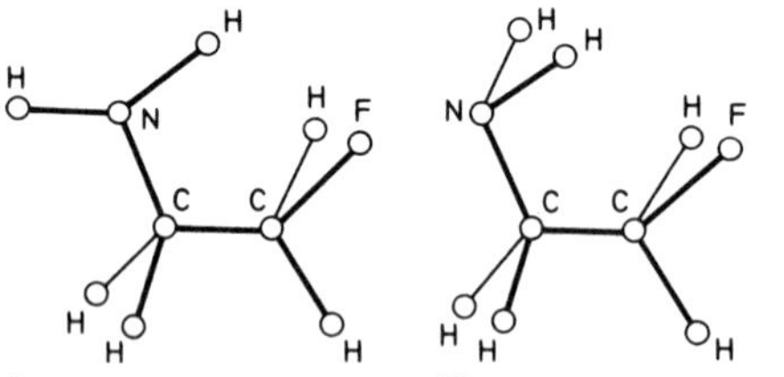

I
II

Fig. 52. $CH_2NH_2CH_2F$.
The most stable conformations of 2-fluoroethylamine.
See subvol. a, p. 261, Nr. 177.

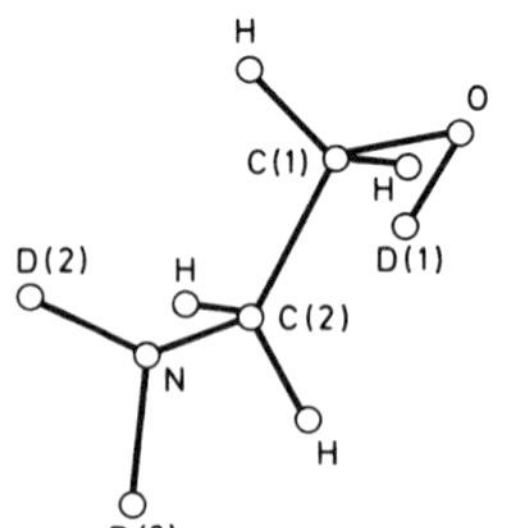

Fig. 53. $CH_2ND_2CH_2OD$ *gauche*.
View along *b*-axis.
See subvol. a, p. 262, Nr. 178.

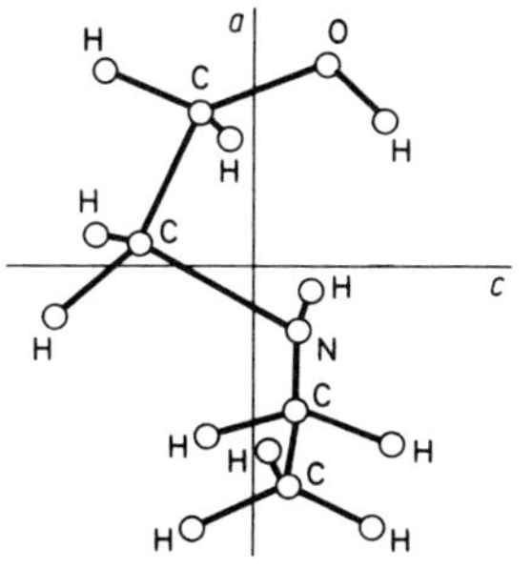

Fig. 54. $CH_2N(CH_3)_2CH_2OH$.
View down the *b*-axis.
See subvol. a, p. 263, Nr. 180.

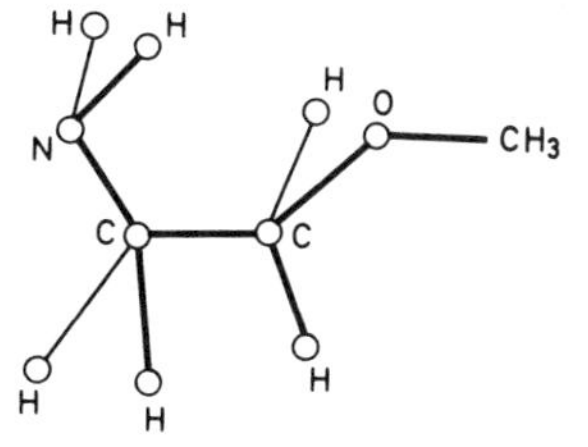

Fig. 55. $CH_3OCH_2CH_2NH_2$ (*gauche*).
See subvol. a, p. 263, Nr. 182.

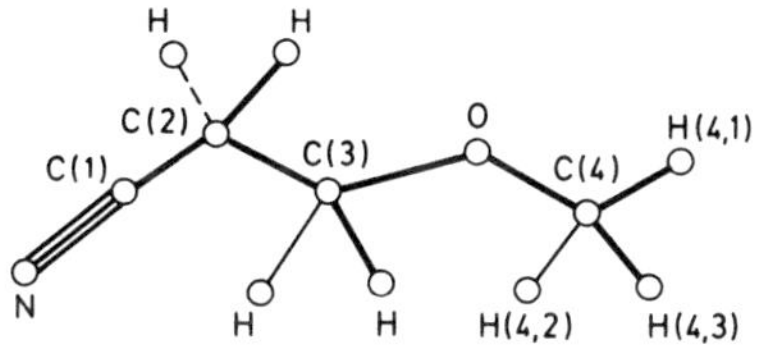

Fig. 56. $CH_3OCH_2CH_2CN$.
See subvol. a, p. 265, Nr. 183.

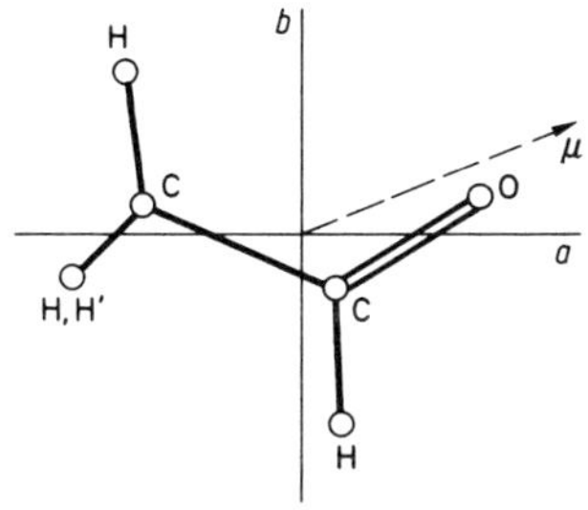

Fig. 57. $CH_3C{\lt}^{O}_{H}$.
See subvol. a, p. 266, Nr. 185.

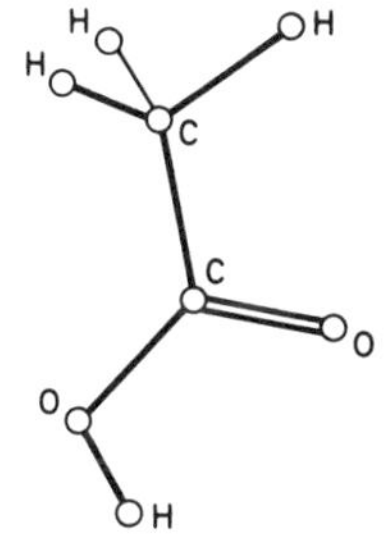

Fig. 58. $CH_3C{\lt}^{O}_{OH}$.
See subvol. a, p. 269, Nr. 188.

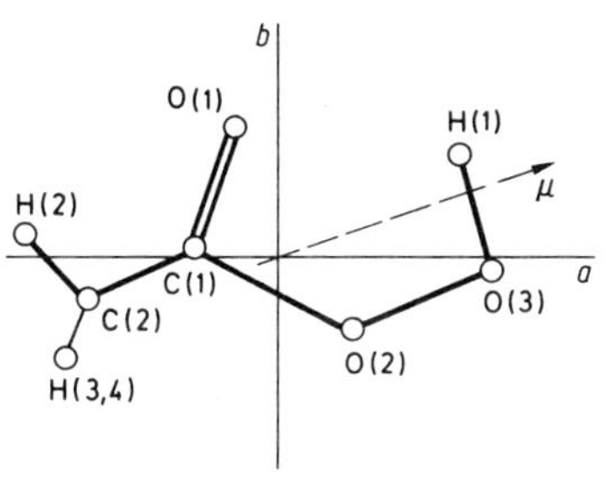

Fig. 59. $CH_3C{\lt}^{O}_{OOH}$. See subvol. a, p. 270, Nr. 189.

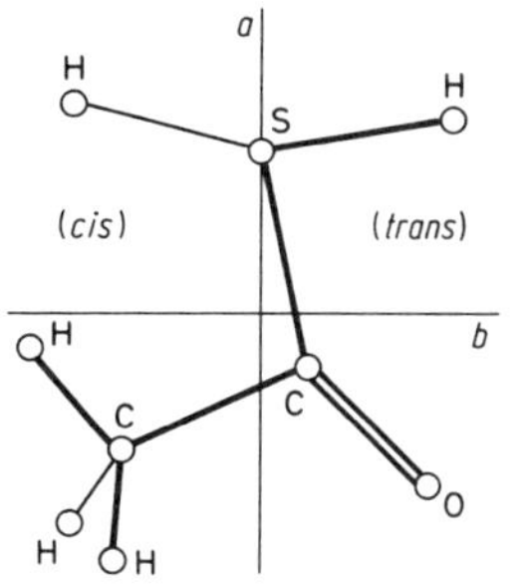

Fig. 60. $CH_3C{\lt}^{O}_{SH}$ probably *trans*.
See subvol. a, p. 270, Nr. 190.

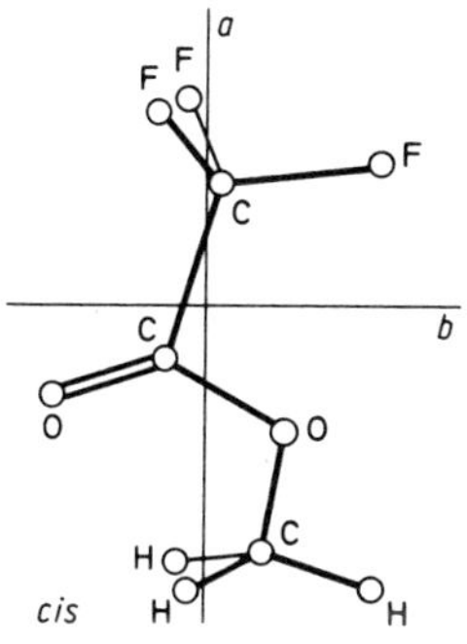

Fig. 61. $CF_3C{\lt}^{O}_{OCH_3}$.
See subvol. a, p. 270, Nr. 191.

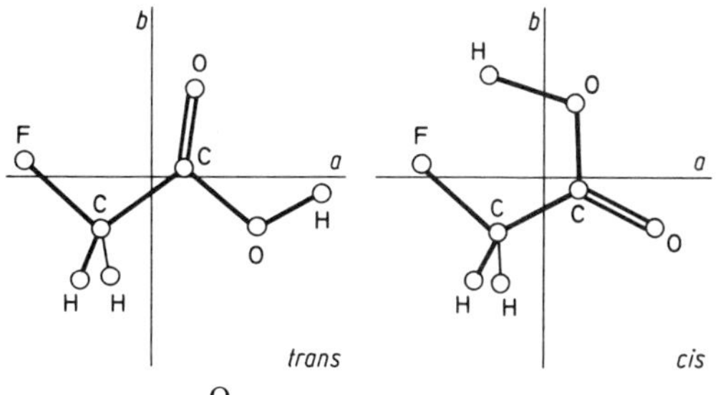

Fig. 62. $CH_2FC{\lt}^{O}_{OH}$.
See subvol. a, p. 272, Nr. 199.

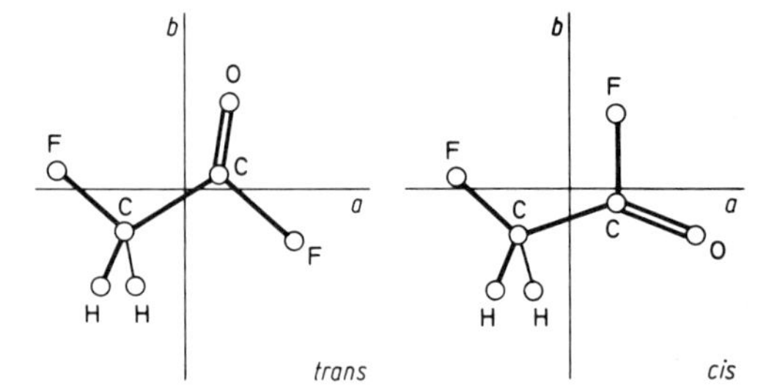

Fig. 63. $CH_2FC\begin{smallmatrix}\nearrow O\\ \searrow F\end{smallmatrix}$.

See subvol. a, p. 275, Nr. 201.

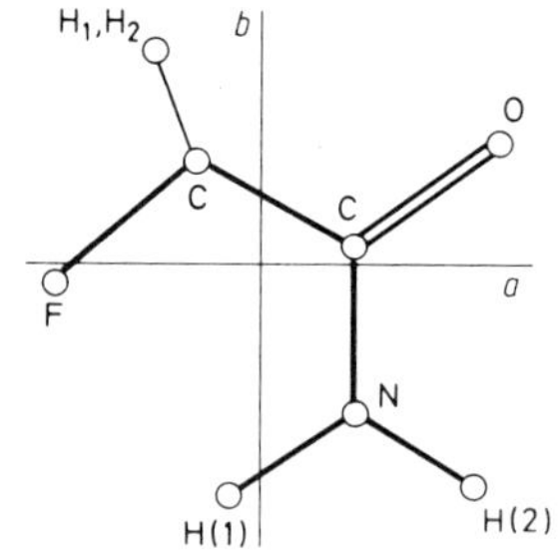

Fig. 64. $CH_2FC\begin{smallmatrix}\nearrow O\\ \searrow NH_2\end{smallmatrix}$.

Projection of stable conformer in the $a-b$ principal axes plane.

See subvol. a, p. 280, Nr. 205.

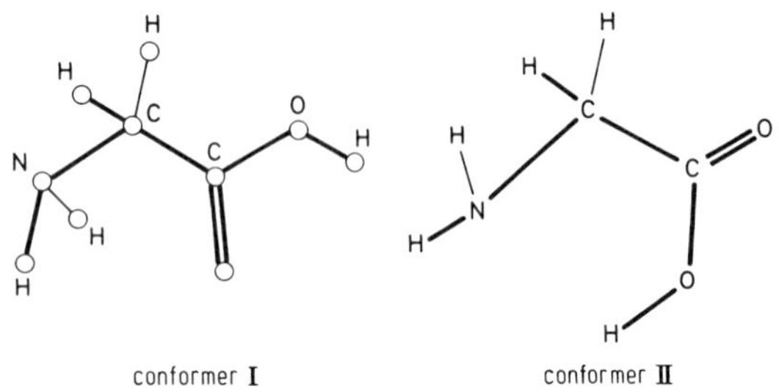

Fig. 65. $NH_2CH_2C\begin{smallmatrix}\nearrow O\\ \searrow OH\end{smallmatrix}$.

See subvol. a, p. 281, Nr. 206.

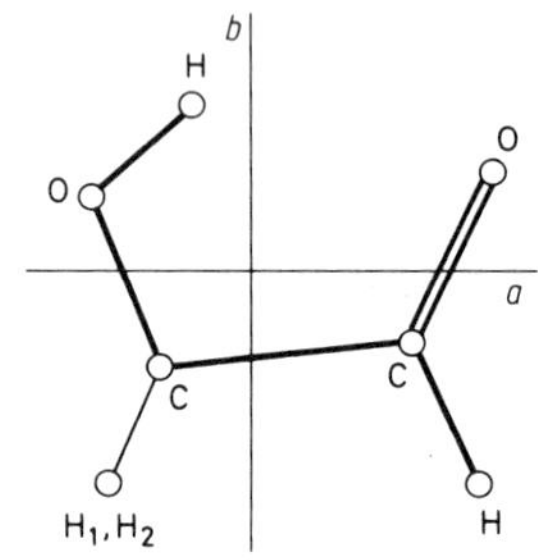

Fig. 66. $CH_2OH{-}C\begin{smallmatrix}\nearrow O\\ \searrow H\end{smallmatrix}$.

Projection of stable conformer in the principal axes plane.

See subvol. a, p. 282, Nr. 207.

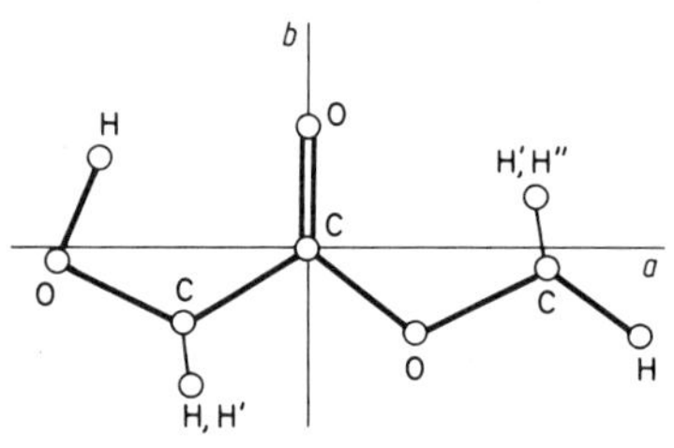

Fig. 67. $HOCH_2C\begin{smallmatrix}\nearrow O\\ \searrow OCH_3\end{smallmatrix}$.

See subvol. a, p. 283, Nr. 208.

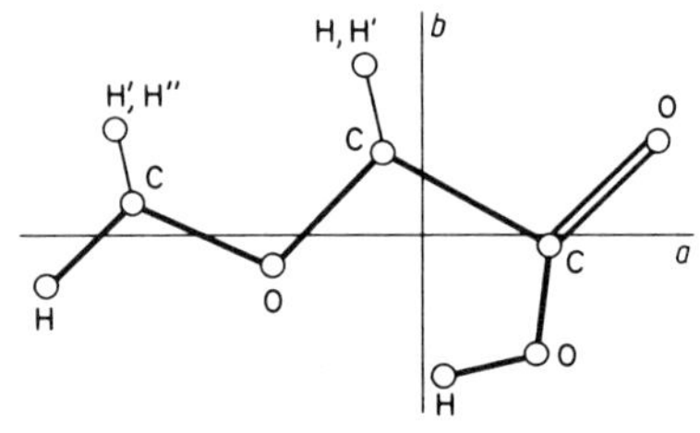

Fig. 68. $CH_3OCH_2C\begin{smallmatrix}\nearrow O\\ \searrow OH\end{smallmatrix}$.

See subvol. a, p. 283, Nr. 209.

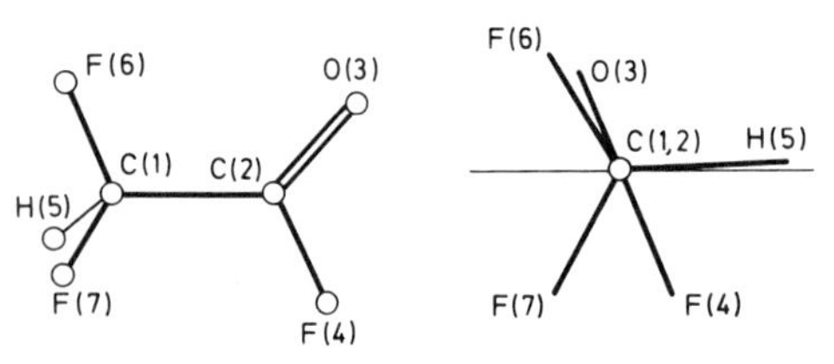

Fig. 69. $CHF_2C\begin{smallmatrix}\nearrow O\\ \searrow F\end{smallmatrix}$ *gauche.*

See subvol. a, p. 286, Nr. 211.

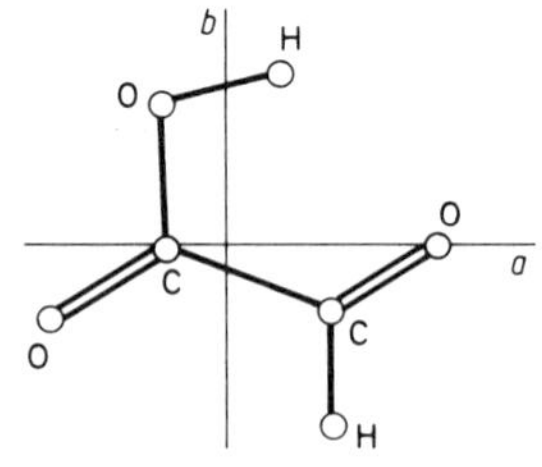

Fig. 70. $\underset{\overset{\|}{O}}{HC}{-}C\begin{smallmatrix}\nearrow O\\ \searrow OH\end{smallmatrix}$ (*trans* 1).

See subvol. a, p. 289, Nr. 213.

Trans 2 differs from trans 1 with respect to the position of the hydroxyl proton.

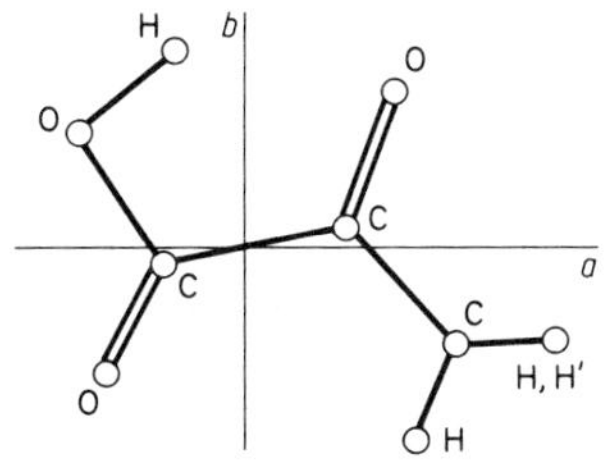

Fig. 71. $CH_3C(=O)-C(=O)-OH$.

Projection in the $a-b$ principal axes planes.
See subvol. a, p. 290, Nr. 214.

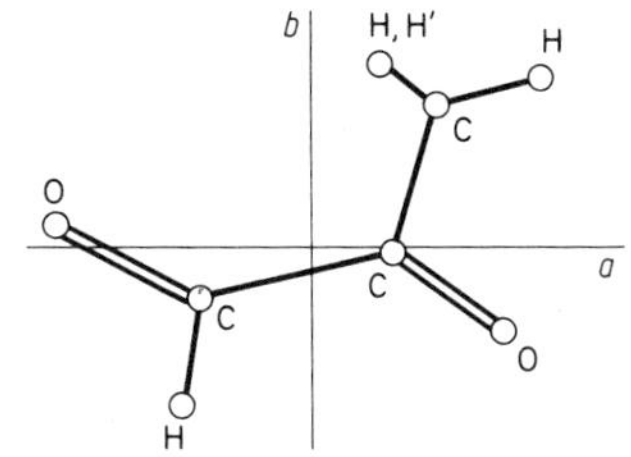

Fig. 72. $CH_3C(=O)-C(=O)H$.

Projection in the $a-b$ principal axes planes.
See subvol. a, p. 292, Nr. 215.

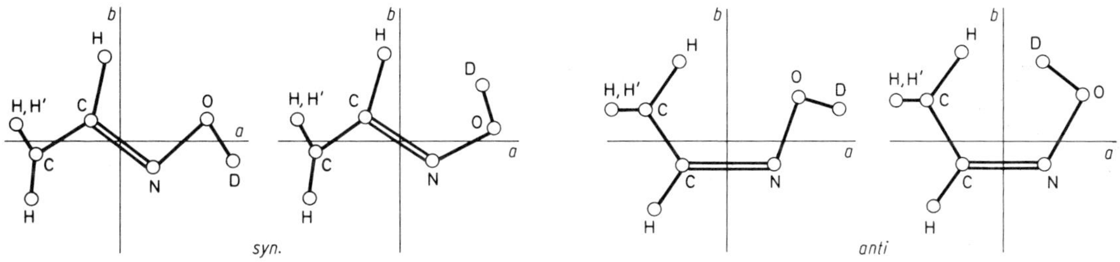

Fig. 73. $CH_3C(=NOD)H$. See subvol. a, p. 293, Nr. 216.

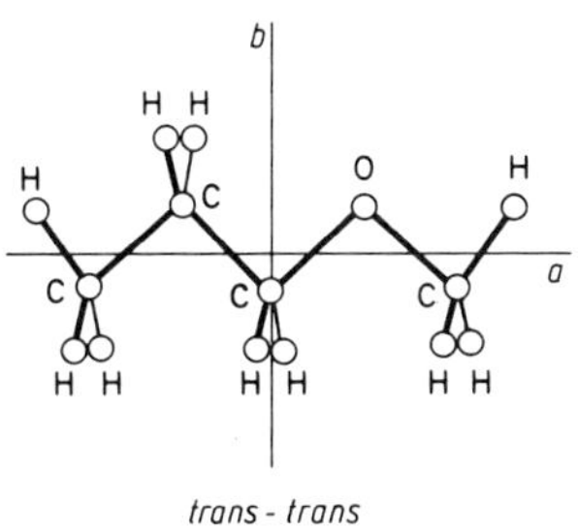

Fig. 74. $CH_3CH_2CH_2OCH_3$.
See subvol. a, p. 298, Nr. 226.

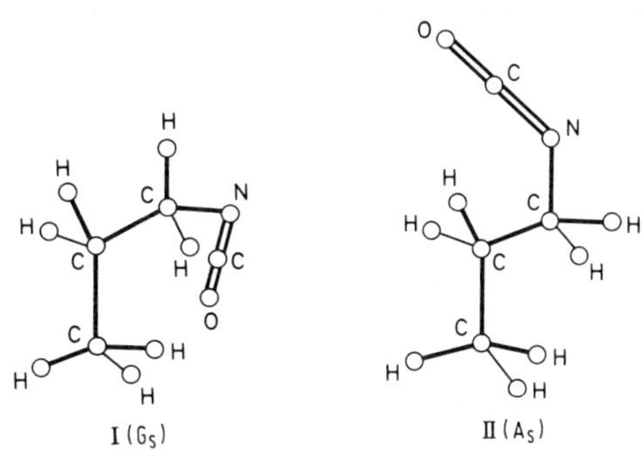

Fig. 75. $CH_3CH_2CH_2NCO$.
See subvol. a, p. 300, Nr. 228.

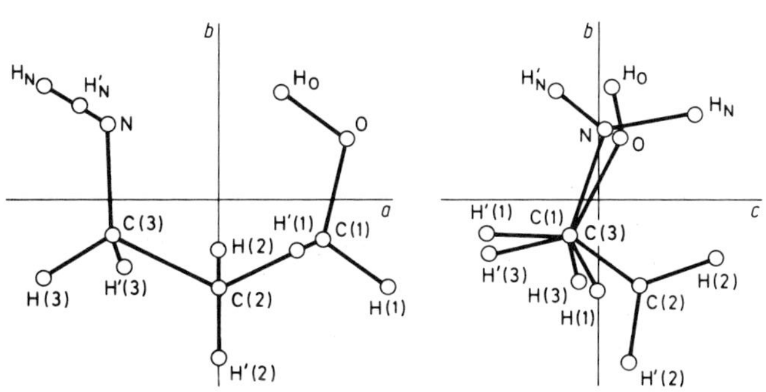

Fig. 76. $H_2NCH_2CH_2CH_2OH$.
Projection in the (a, b) and (b, c) axes plane.
See subvol. a, p. 301, Nr. 231.

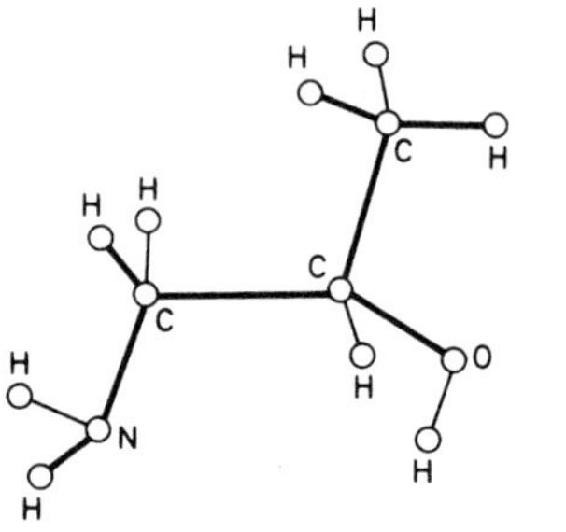

Fig. 77. $CH_3CH(OH)CH_2NH_2$.
See subvol. a, p. 303, Nr. 234.

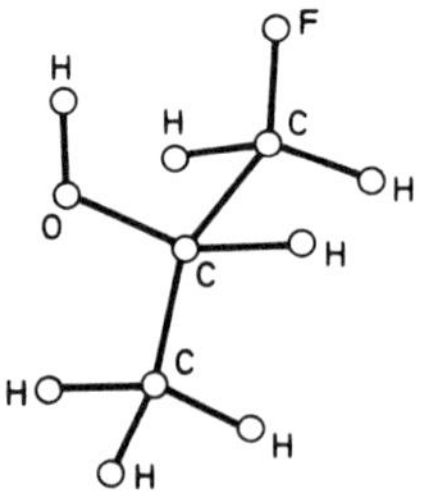

Fig. 78. $CH_3CH(OH)CH_2F$.
See subvol. a, p. 304, Nr. 235.

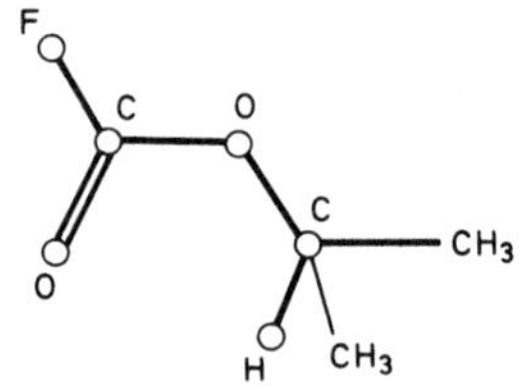

Fig. 79. $(CH_3)_2CHOC{\begin{smallmatrix}\nearrow O\\ \searrow F\end{smallmatrix}}$
See subvol. a, p. 308, Nr. 247.

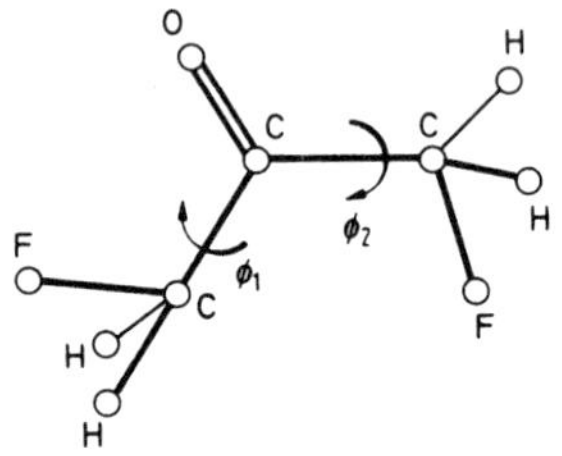

Fig. 80. $CH_2F{-}C{\begin{smallmatrix}\nearrow O\\ \searrow CH_2F\end{smallmatrix}}$.
See subvol. a, p. 309, Nr. 254.

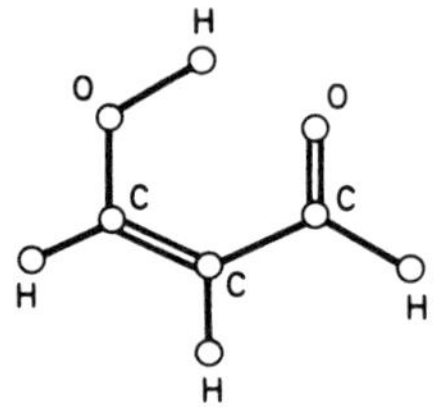

Fig. 81. $HOCH{=}CHC{\begin{smallmatrix}\nearrow O\\ \searrow H\end{smallmatrix}}$.
See subvol. a, p. 311, Nr. 256.

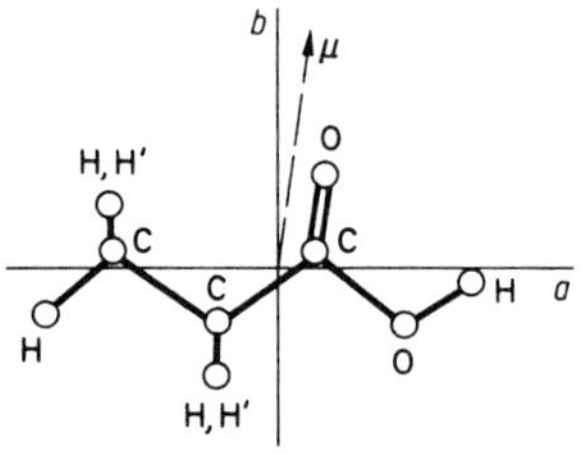

Fig. 82. $CH_3CH_2C{\begin{smallmatrix}\nearrow O\\ \searrow OH\end{smallmatrix}}$.
See subvol. a, p. 318, Nr, 260.

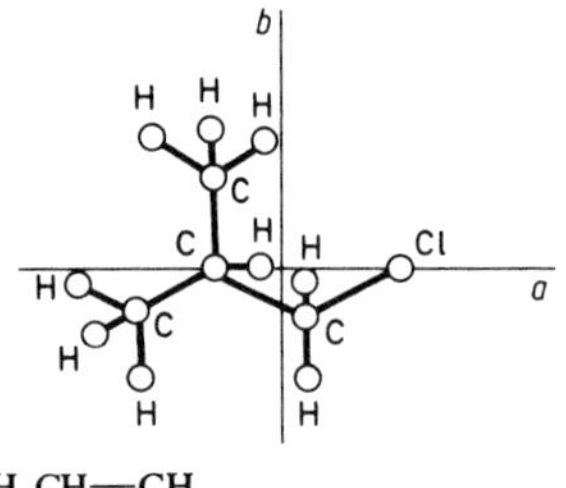

Fig. 83. $ClCH_2\underset{\displaystyle CH_3}{\underset{|}{C}}H{-}CH_3$.
See subvol. a, p. 323, Nr. 267.

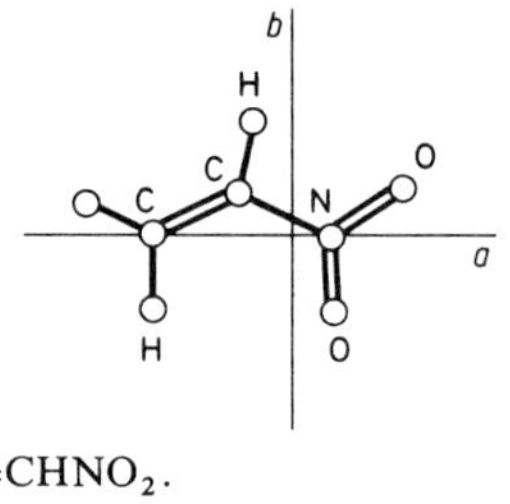

Fig. 84. $CH_2{=}CHNO_2$.
See subvol. a, p. 328, Nr. 286.

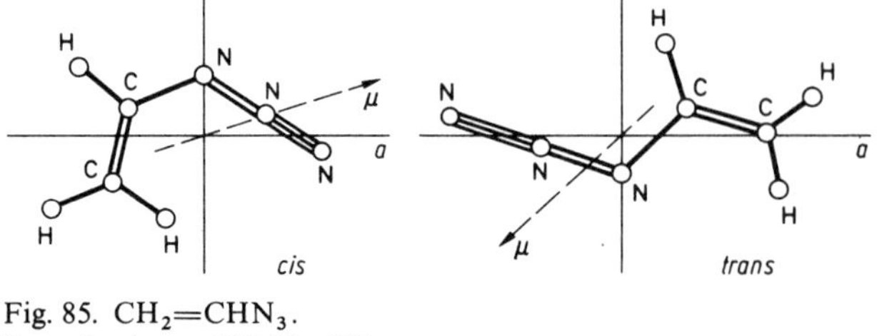

Fig. 85. $CH_2{=}CHN_3$.
See subvol. a, p. 333, Nr. 288.

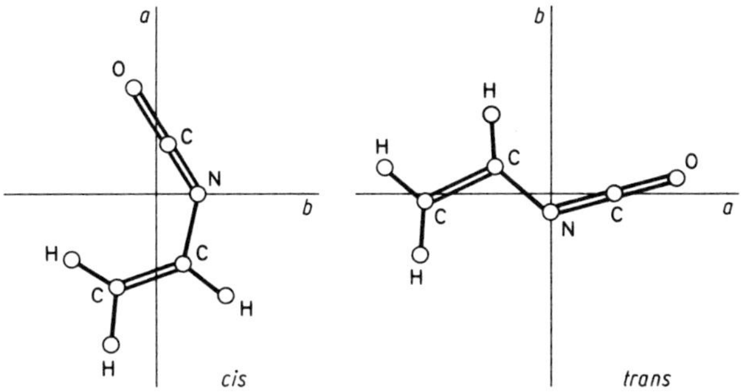

∢ (μ, CN) = 8.9°   ∢ (μ, CN) = 12.1°

Fig. 86. $CH_2{=}CHNCO$.
See subvol. a, p. 333, Nr. 289.

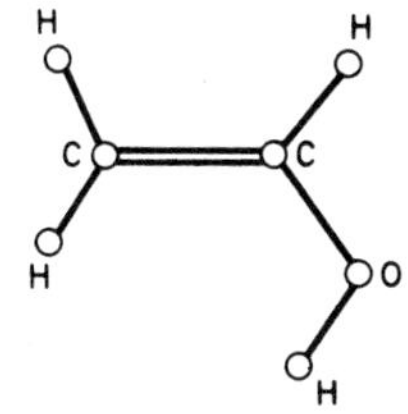

Fig. 87. $CH_2{=}CHOH$.
See subvol. a, p. 334, Nr. 290.

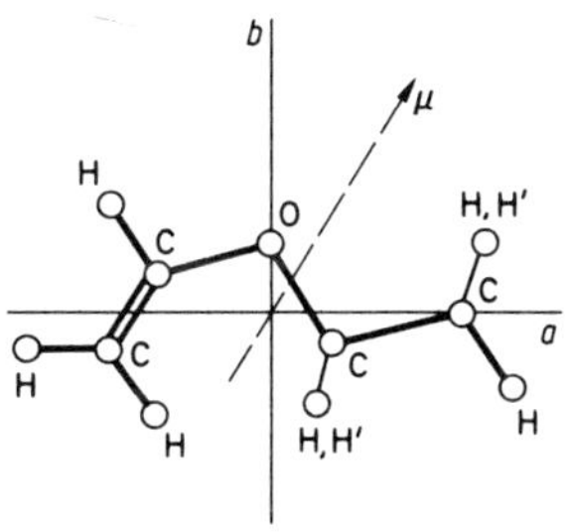

Fig. 88. $CH_2{=}CHOCH_2CH_3$.
See subvol. a, p. 334, Nr. 292.

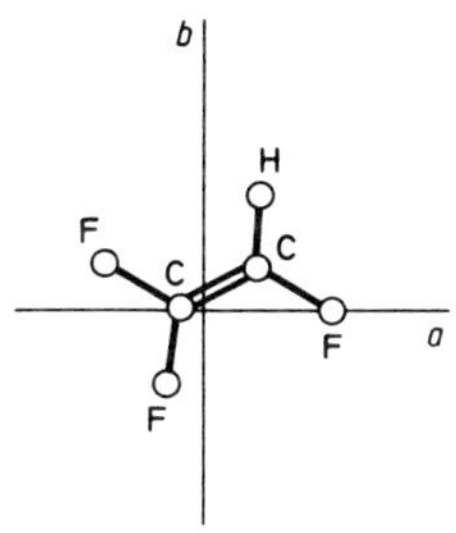

Fig. 89. $CF_2{=}CHF$.
See subvol. a, p. 342, Nr. 304.

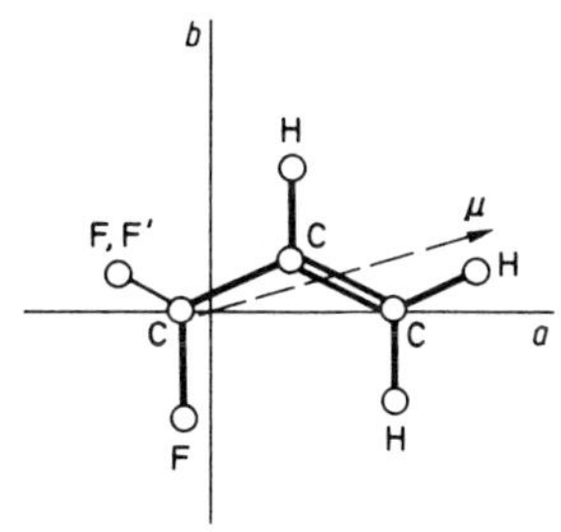

Fig. 90. $CF_3CH{=}CH_2$.
See subvol. a, p. 343, Nr. 307.

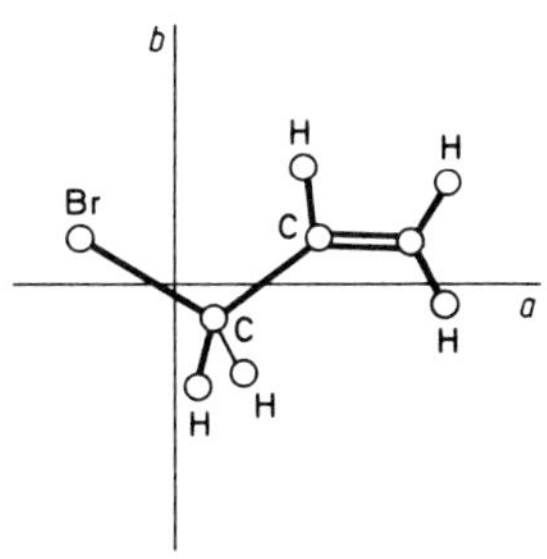

Fig. 91. $CH_2{=}CHCH_2Br$.
See subvol. a, p. 344, Nr. 309.

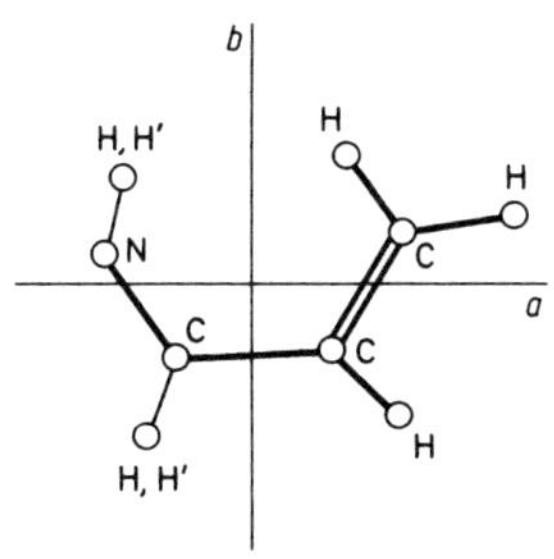

Fig. 92. $CH_2{=}CHCH_2NH_2$.
N *cis*, *N* lone pair *trans*.
See subvol. a, p. 345, Nr. 311.

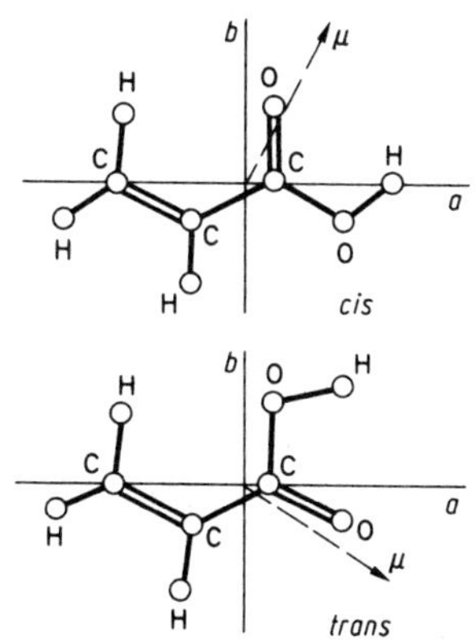

Fig. 93. $H_2C{=}CHC{\overset{O}{\underset{OH}{}}}$.
See subvol. a, p. 350, Nr. 320.

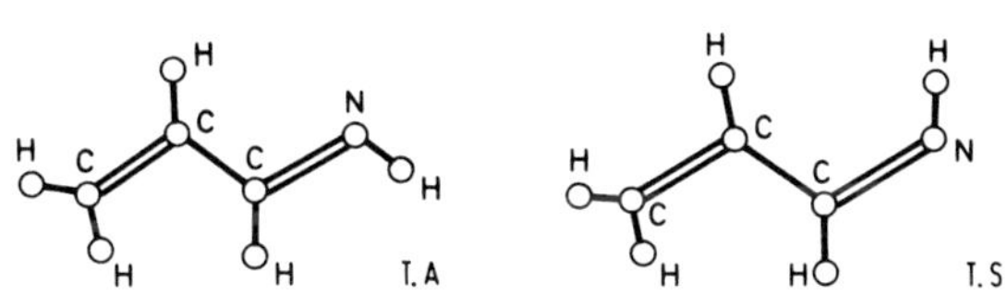

Fig. 94. $CH_2{=}CHCH{=}NH$. T.A = *trans-anti*.
See subvol. a, p. 352, Nr. 322. T.S = *trans-syn*.

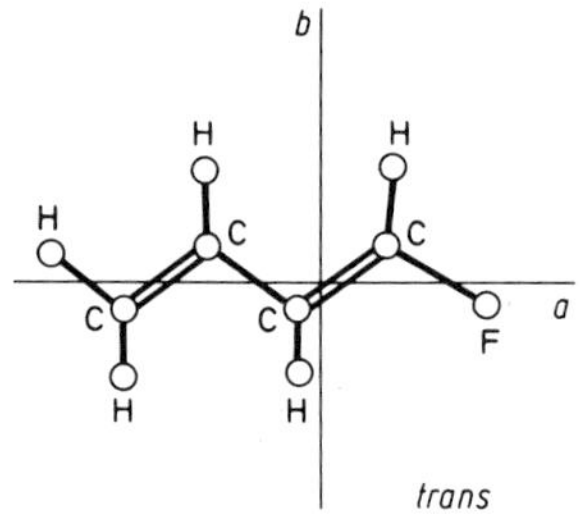

Fig. 95. $CH_2{=}CH{-}CH{=}CHF$.
See subvol. a, p. 358, Nr. 340.

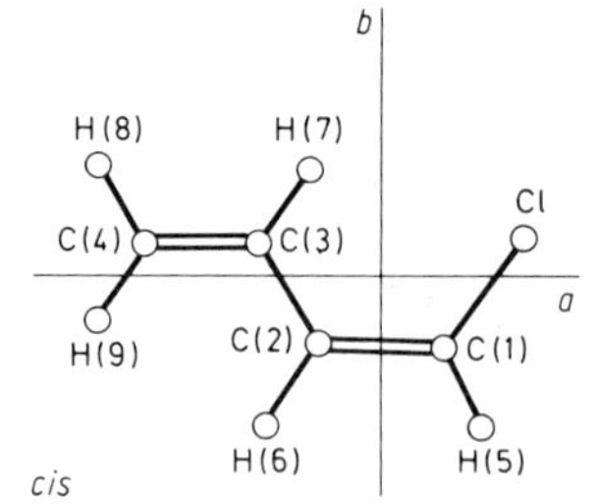

Fig. 96. $CH_2{=}CH{-}CH{=}CHCl$.
See subvol. a, p. 358, Nr. 341.

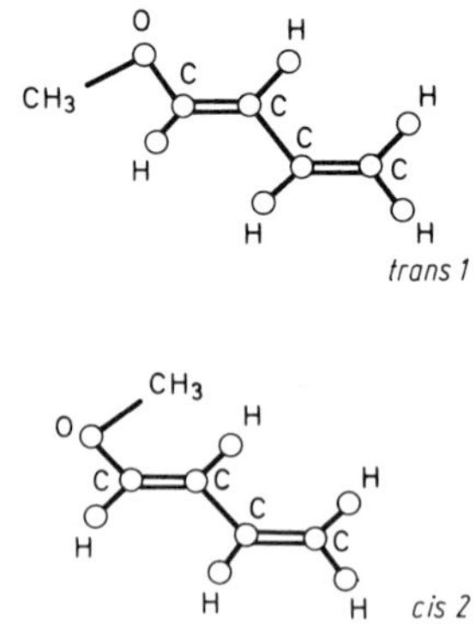

Fig. 97. $CH_2{=}CH{-}CH{=}CHOCH_3$.
See subvol. a, p. 361, Nr. 342.

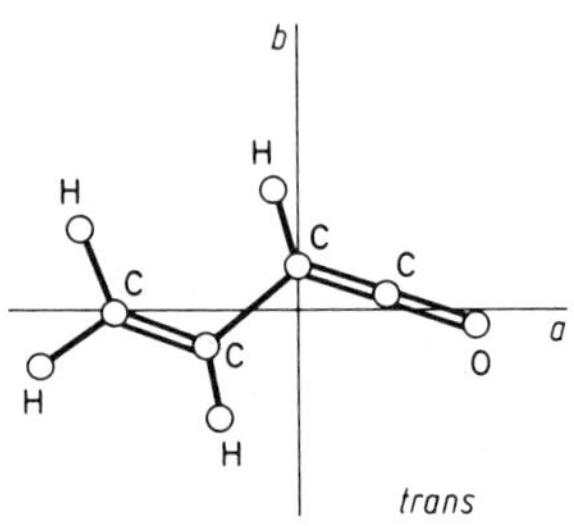

Fig. 98. $CH_2{=}CH{-}CH{=}C{=}O$.
See subvol. a, p. 362, Nr. 343.

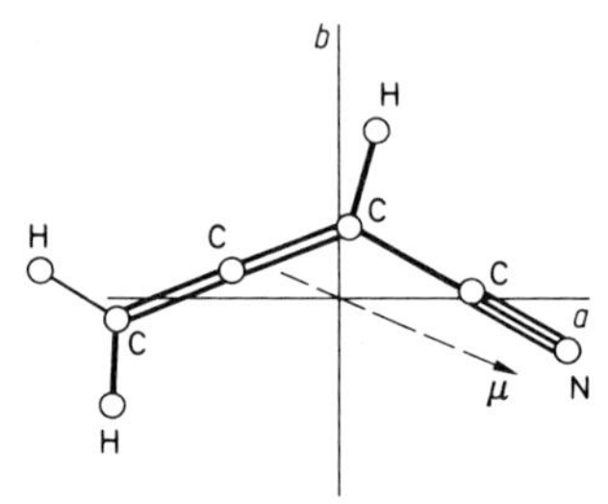

Fig. 99. $CH_2{=}C{=}CHCN$.
See subvol. a, p. 363, Nr. 347.

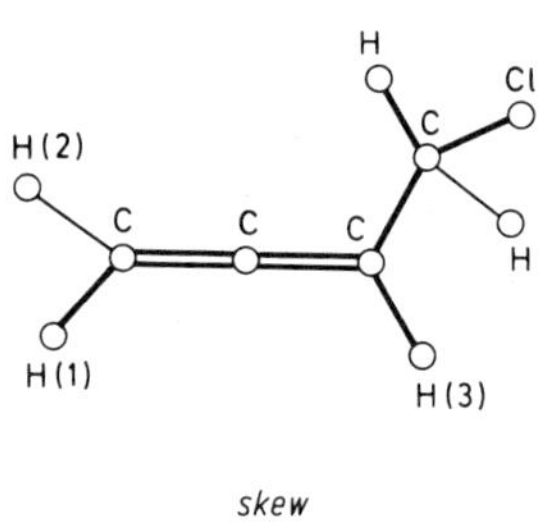

Fig. 100. $CH_2{=}C{=}CHCH_2Cl$.
See subvol. a, p. 364, Nr. 348.

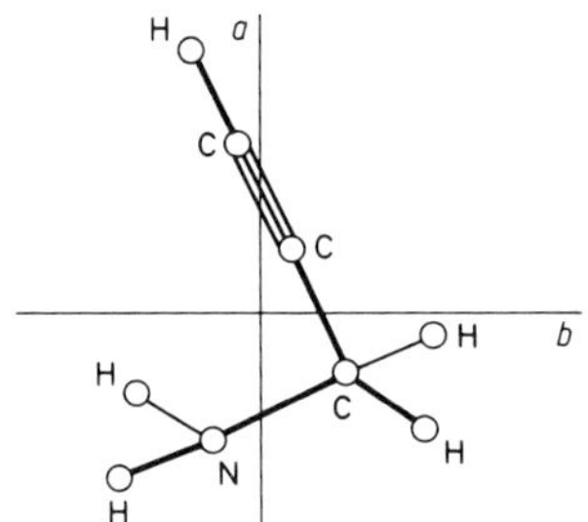

Fig. 101. $HC{\equiv}CCH_2NH_2$.
See subvol. a, p. 369, Nr. 356.

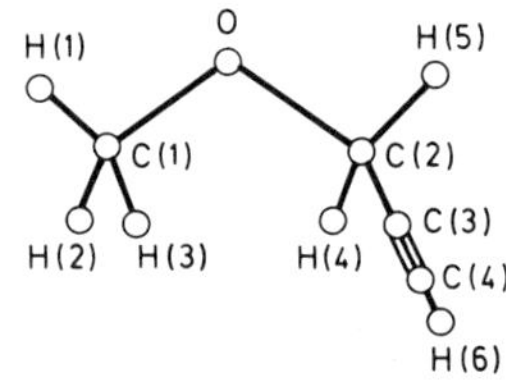

Fig. 102. $HC{\equiv}CCH_2OCH_3$.
See subvol. a, p. 370, Nr. 359.

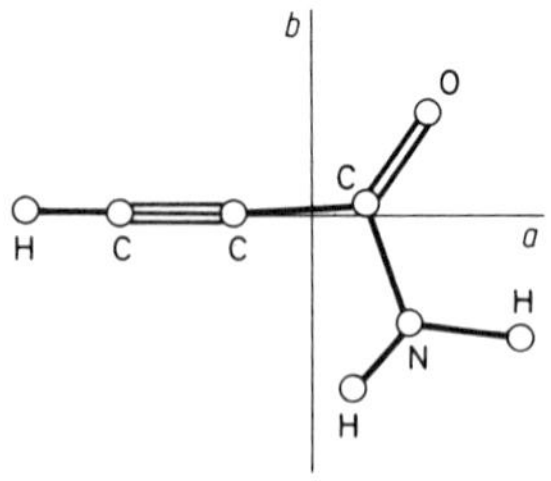

Fig. 103. $HC{\equiv}C{-}C\genfrac{}{}{0pt}{}{\diagup O}{\diagdown NH_2}$.
See subvol. a, p. 376, Nr. 368.

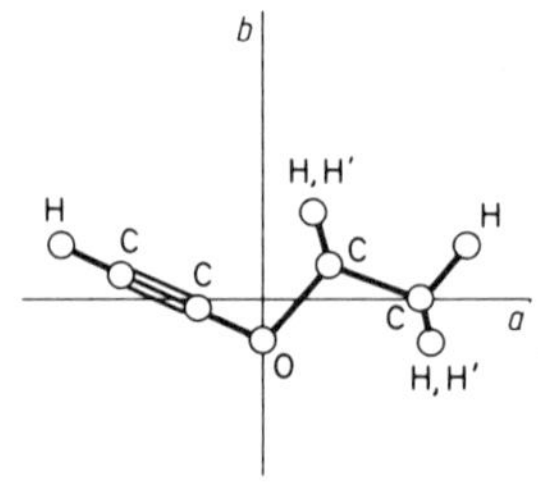

Fig. 104. $HC{\equiv}C{-}OCH_2CH_3$.
See subvol. a, p. 376, Nr. 369.

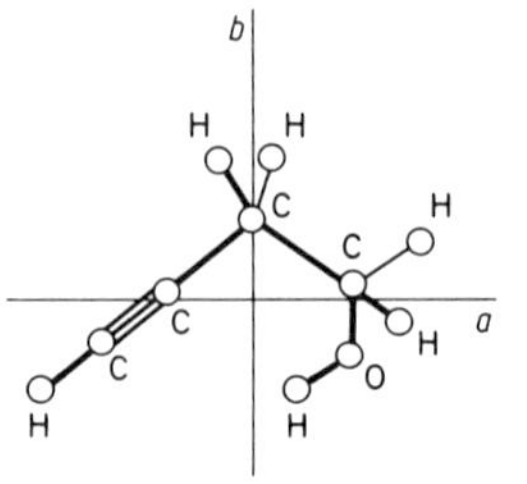

Fig. 105. $HC{\equiv}CCH_2CH_2OH$.
See subvol. a, p. 379, Nr. 372.

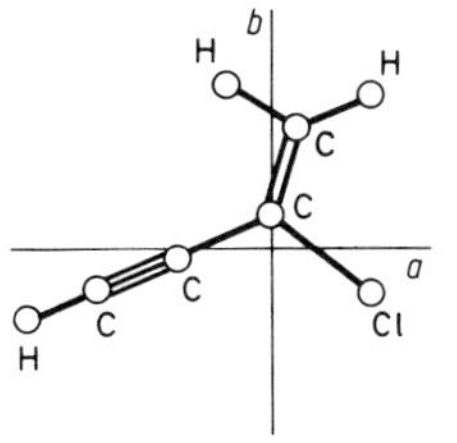

Fig. 106. HC≡C—CCl=CH$_2$.
See subvol. a, p. 380, Nr. 374.

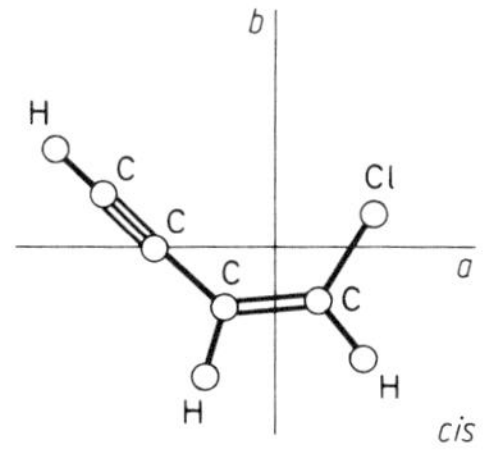

Fig. 107. HC≡C—CH=CHCl.
See subvol. a, p. 380, Nr. 375.

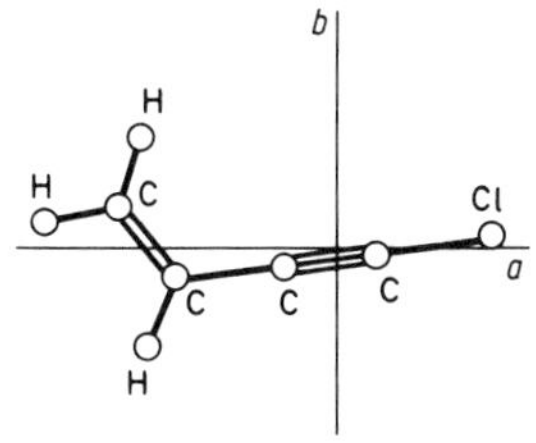

Fig. 108. ClC≡C—CH=CH$_2$.
See subvol. a, p. 381, Nr. 376.

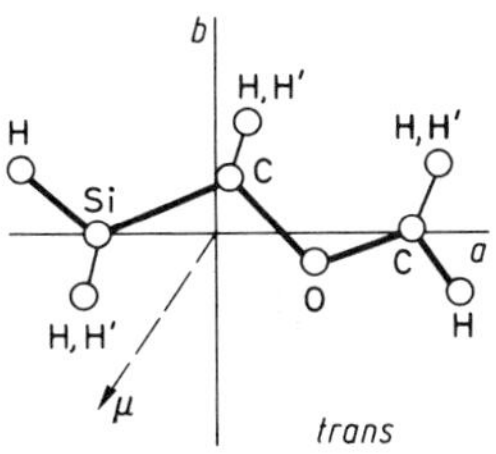

Fig. 109. SiH$_3$CH$_2$OCH$_3$.
See subvol. a, p. 387, Nr. 390.

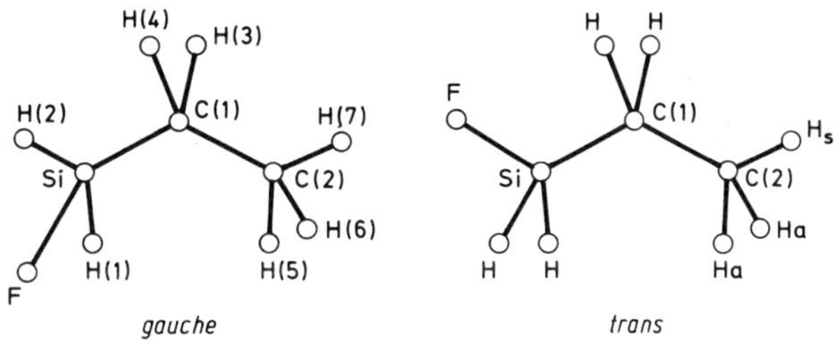

Fig. 109a. CH$_3$CH$_2$SiH$_2$F.
See subvol. a, p. 390, Nr. 395.

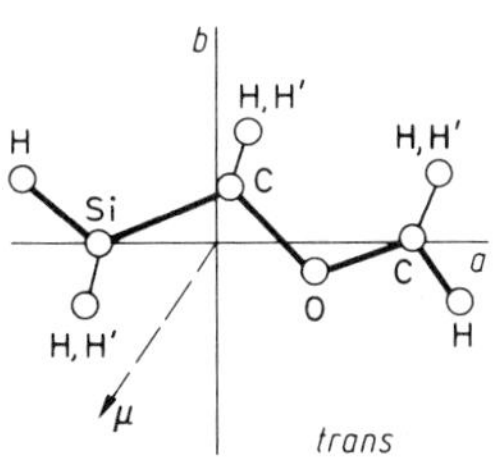

Fig. 110. CH$_3$CH$_2$SiH$_2$Cl.
See subvol. a, p. 392, Nr. 396.

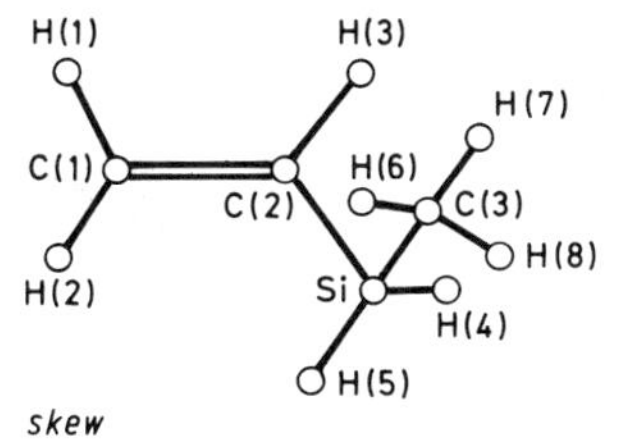

Fig. 110a. CH$_2$=CHSiH$_2$CH$_3$.
See subvol. a, p. 394, Nr. 398.

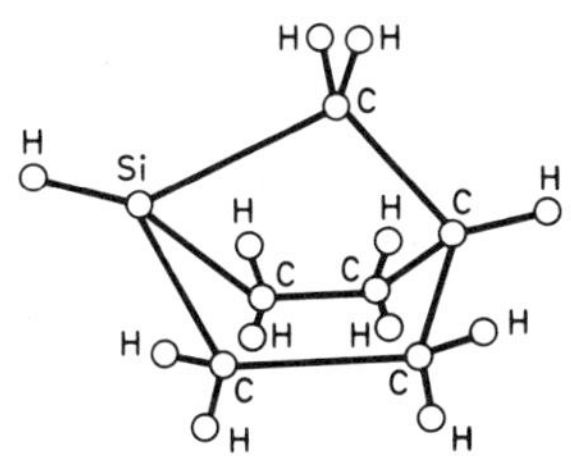

Fig. 111. 

.
See subvol. a, p. 397, Nr. 405.

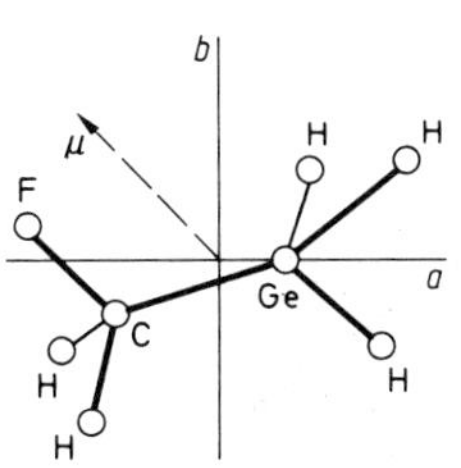

Fig. 112. GeH$_3$CH$_2$F.
See subvol. a, p. 397, Nr. 406.

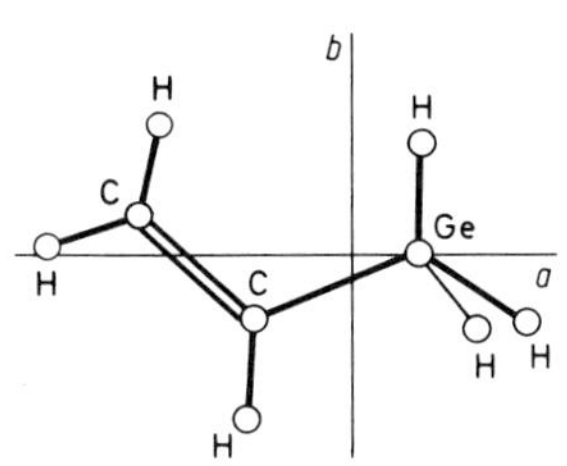

Fig. 113. GeH$_3$CH=CH$_2$.
See subvol. a, p. 399, Nr. 409.

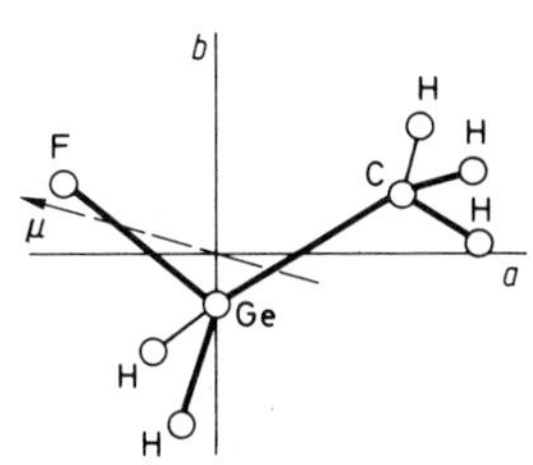

Fig. 114. CH$_3$GeH$_2$F.
See subvol. a, p. 400, Nr. 410.

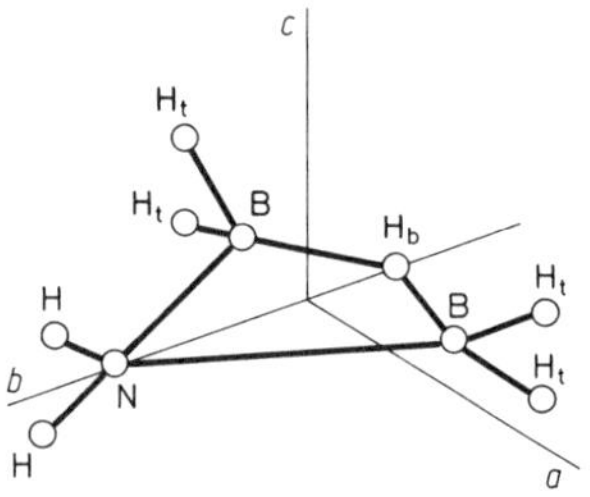

Fig. 115. $H_2NB_2H_5$.
See subvol. a, p. 404, Nr. 415.

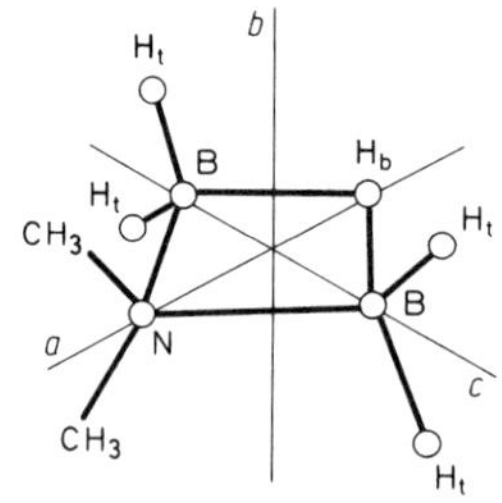

Fig. 116. $(CH_3)_2NB_2$.
See subvol. a, p. 405, Nr. 416.

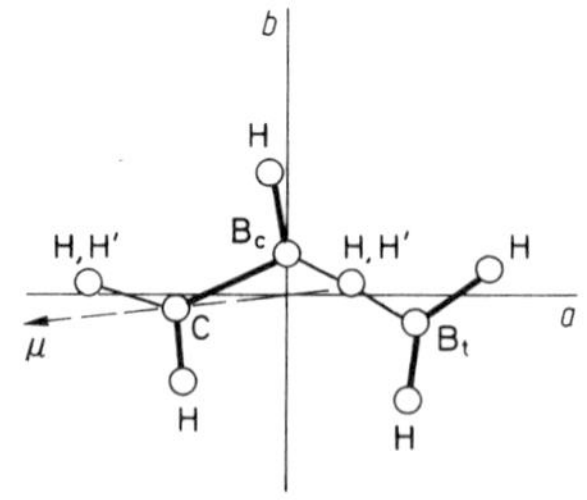

Fig. 117. $CH_3B_2H_5$.
See subvol. a, 406, Nr. 417.

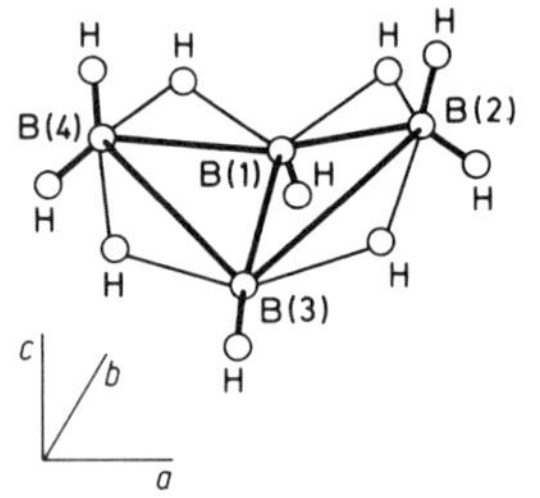

Fig. 118. B_4H_{10}.
See subvol. a, p. 406, Nr. 418.

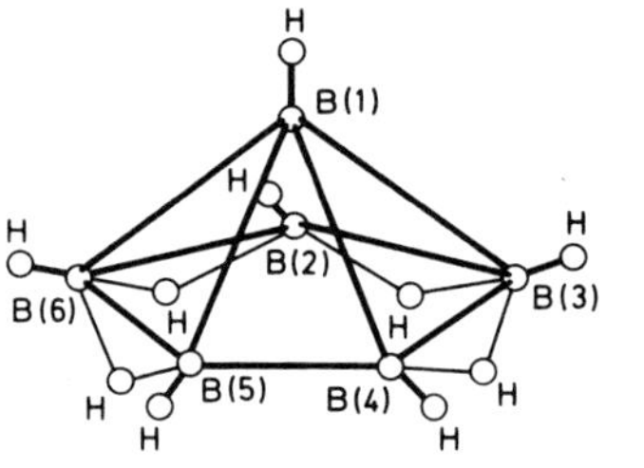

Fig. 119. B_6H_{10}.
See subvol. a, p. 407, Nr. 419.

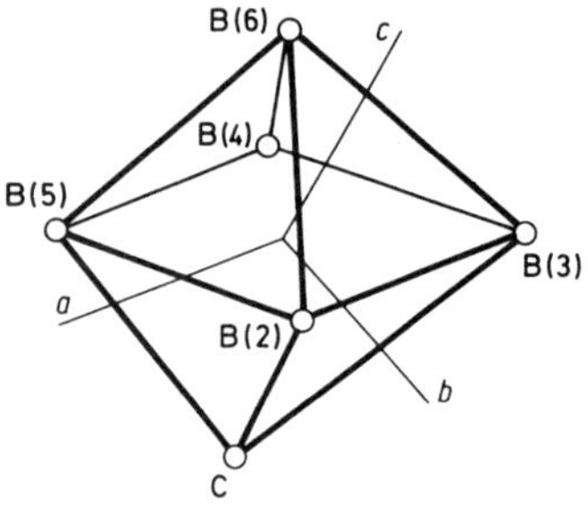

Fig. 120. CB_5H_7.
See subvol. a, p. 407, Nr. 420.

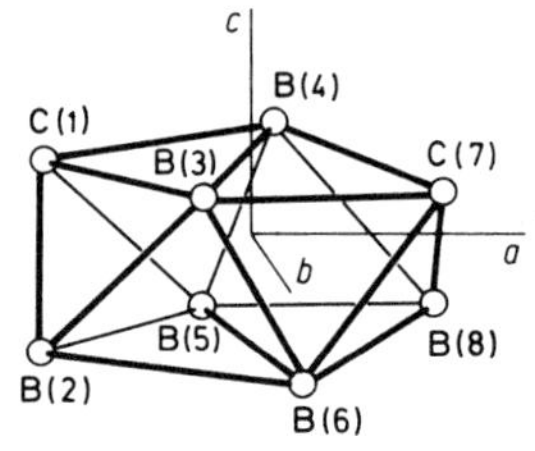

Fig. 121. 1,7—$C_2B_6H_8$.
See subvol. a, p. 408, Nr. 421.

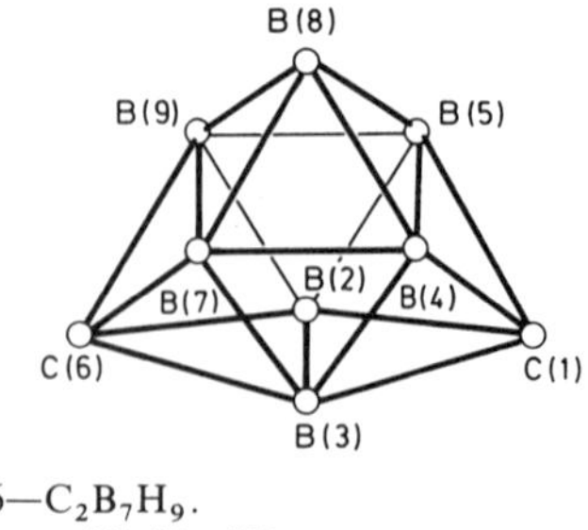

Fig. 122. 1,6—$C_2B_7H_9$.
See subvol. a, p. 409, Nr. 422.

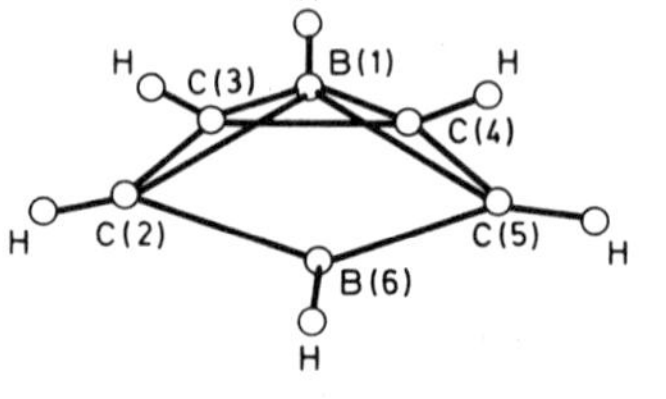

Fig. 123. 2,3,4,5 $C_4B_2H_6$.
See subvol. a, p. 410, Nr. 423.

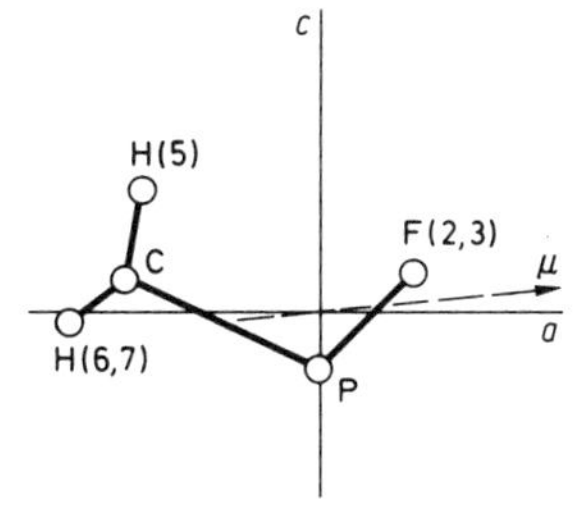

Fig. 124. CH_3PF_2.
See subvol. a, p. 410, Nr. 424.

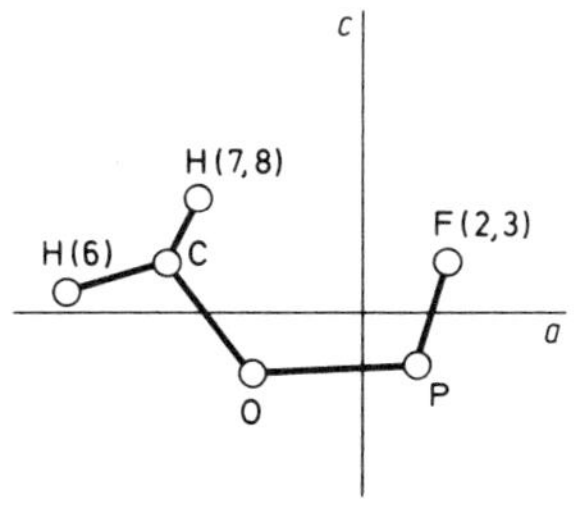

Fig. 125. CH_3OPF_2.
See subvol. a, p. 411, Nr. 426.

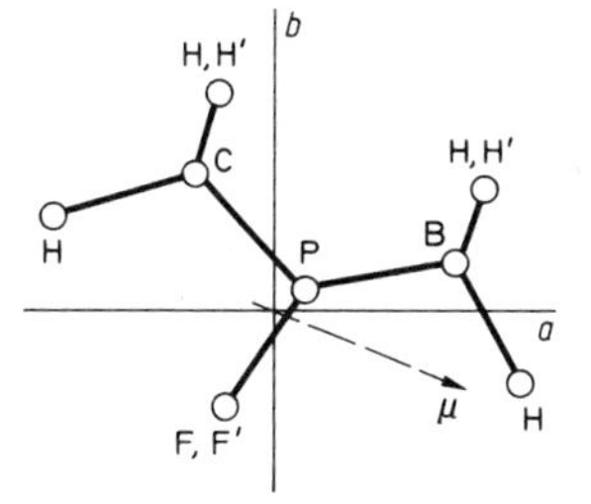

Fig. 126. $CH_3PF_2BH_3$.
See subvol. a, p. 414, Nr. 433.

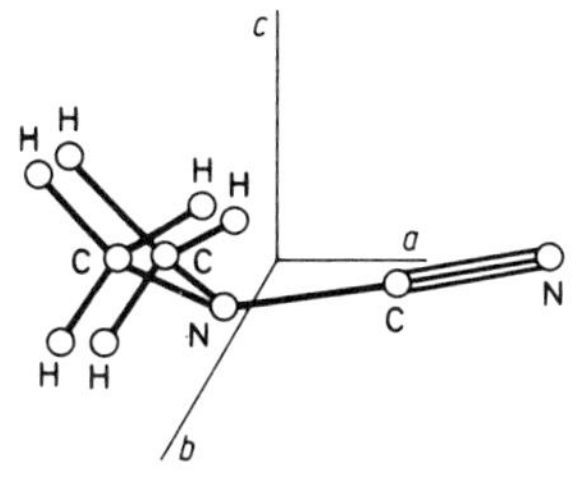

Fig. 127. $(CH_3)_2NCN$.
See subvol. a, p. 420, Nr. 438.

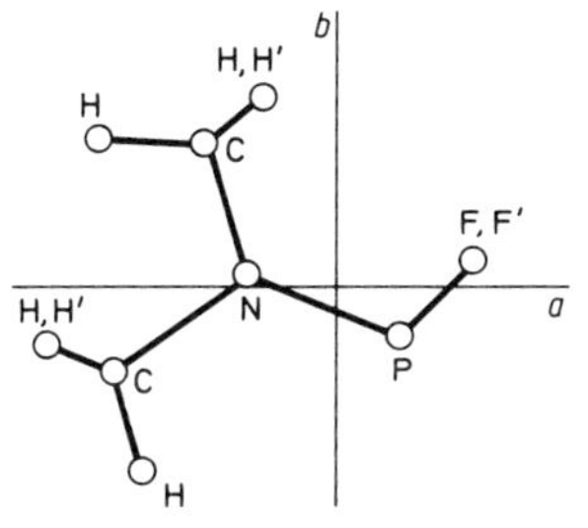

Fig. 128. $(CH_3)_2NPF_2$.
See subvol. a, p. 420, Nr. 439.

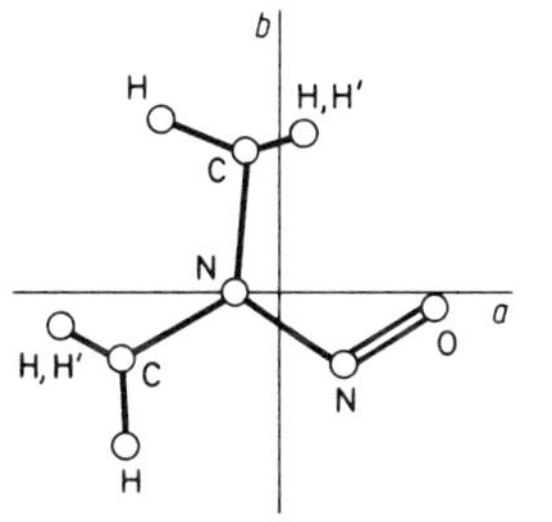

Fig. 129. $(CH_3)_2NNO$.
See subvol. a, p. 420, Nr. 440.

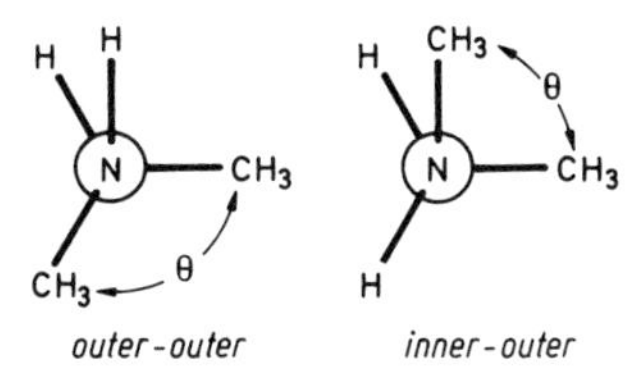

Fig. 129 a. $CH_3NHNHCH_3$.
See subvol. a, p. 422, Nr. 443 a.

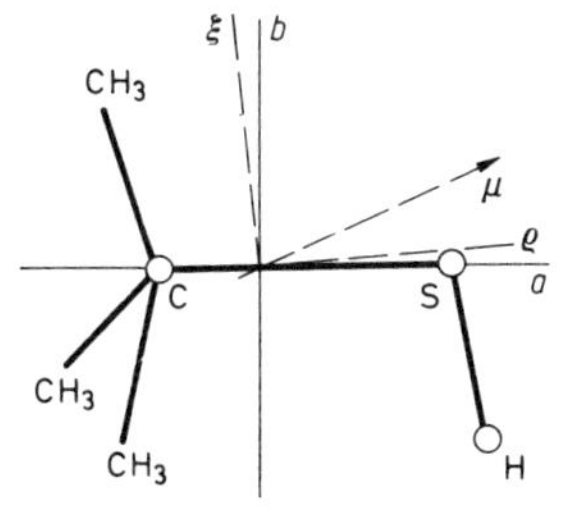

Fig. 130. $(CH_3)_3CSH$.
See subvol. a, p. 423, Nr. 446.

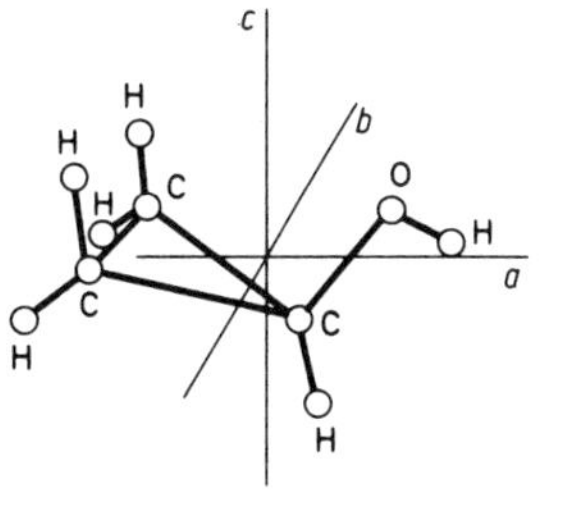

Fig. 131. H_2C——CHOH.
CH_2
See subvol. a, p. 426, Nr. 461.

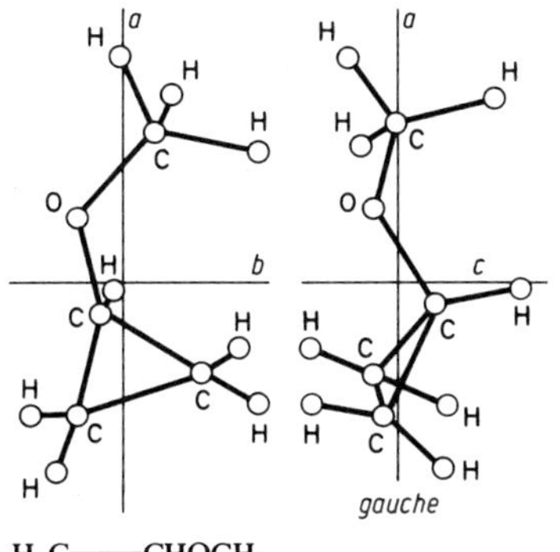

Fig. 132. H_2C——$CHOCH_3$.
CH_2
See subvol. a, p. 429, Nr. 470.

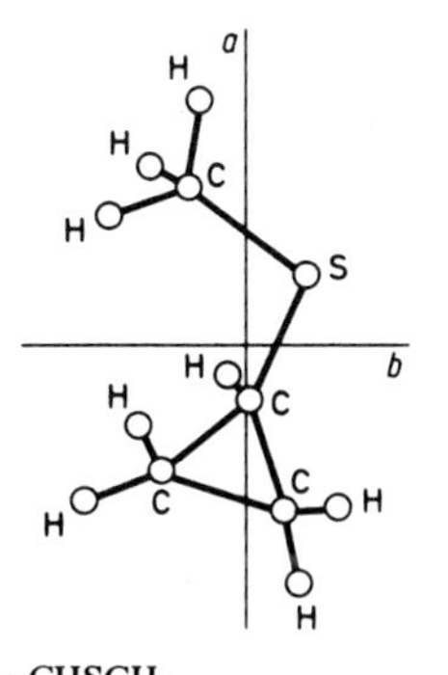

Fig. 133. H_2C——$CHSCH_3$.
CH_2
See subvol. a, p. 429, Nr. 471.

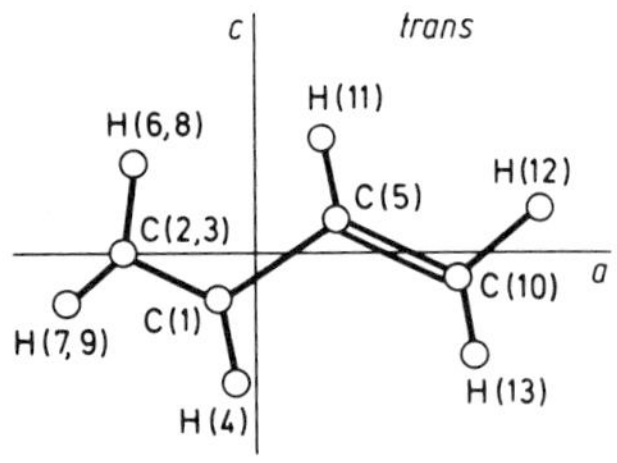

Fig. 134. H_2C——CH—$CH{=}CH_2$.
$\diagdown\diagup$
CH_2
Projection on its $a-c$ plane of symmetry.
See subvol. a, p. 430, Nr. 472.

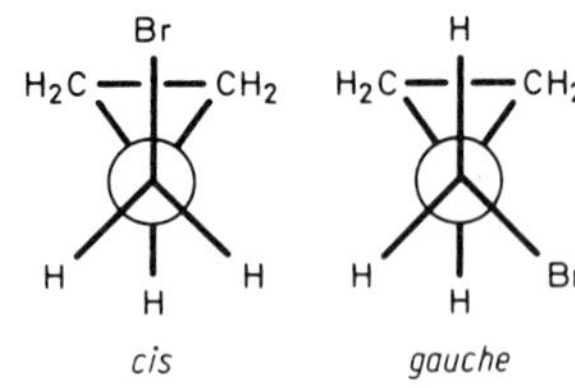

Fig. 135. H_2C——$CHCH_2Br$.
$\diagdown\diagup$
CH_2
Newman projection.
See subvol. a, p. 431, Nr. 474.

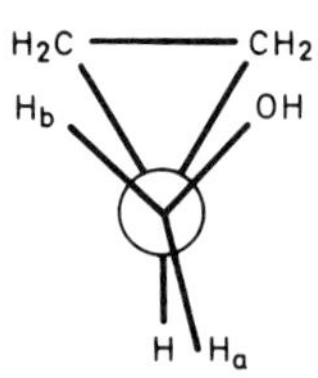

Fig. 136. H_2C——$CHCH_2OH$.
$\diagdown\diagup$
CH_2
Newman projection.
See subvol. a, p. 431, Nr. 475.

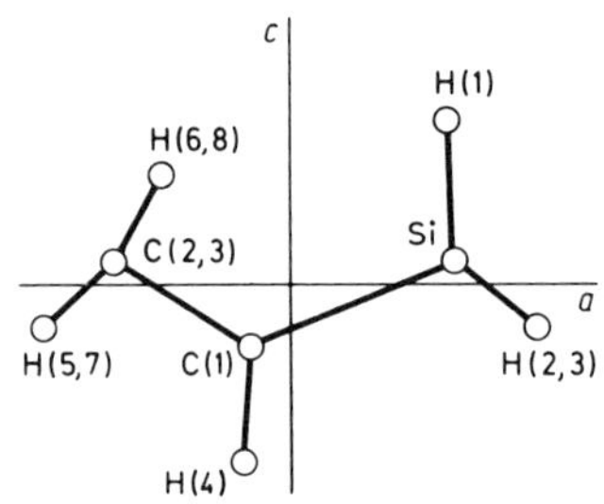

Fig. 137. H_2C——$CHSiH_3$.
$\diagdown\diagup$
CH_2
See subvol. a, p. 432, Nr. 476.

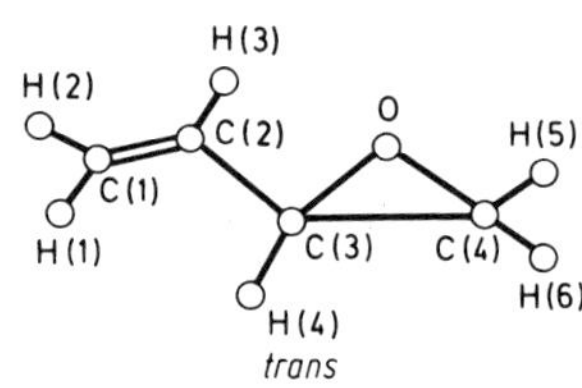

Fig. 138. H_2C——$CHCH{=}CH_2$.
$\diagdown\diagup$
O
See subvol. a, p. 436, Nr. 481.

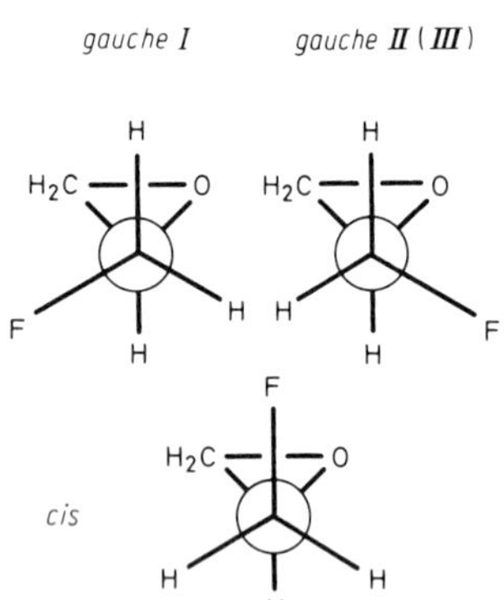

Fig. 139. H_2C——$CHCH_2F$.
$\diagdown\diagup$
O
See subvol. a, p. 437, Nr. 483.

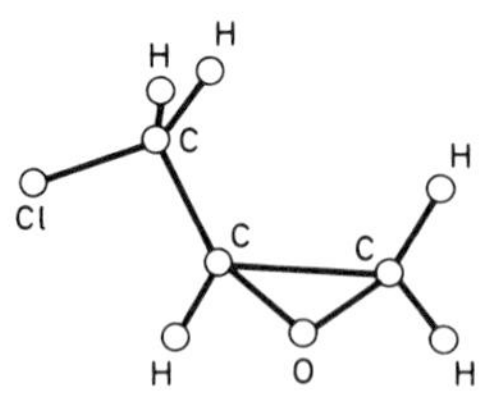

Fig. 140. H_2C——$CHCH_2Cl$.
$\diagdown\diagup$
O
See subvol. a, p. 437, Nr. 484.
See also Fig. 139.

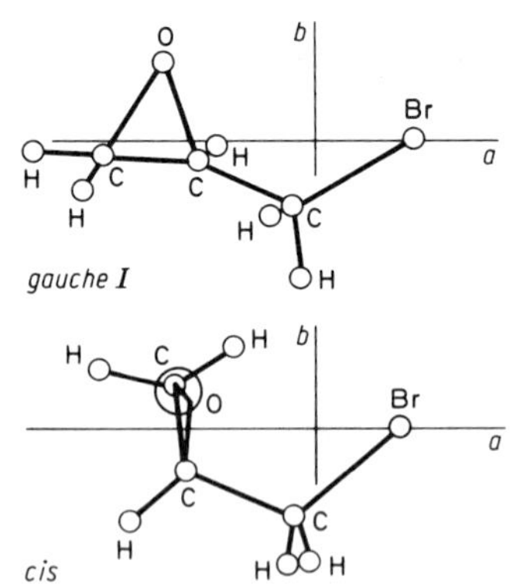

Fig. 141. H_2C——$CHCH_2Br$.
$\diagdown\diagup$
O
See subvol. a, p. 438, Nr. 485.
See also Fig. 139.

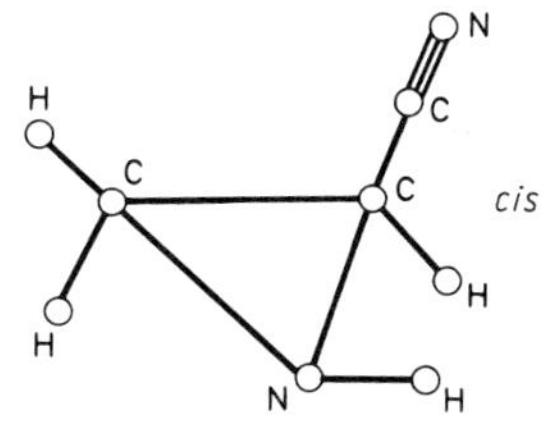

Fig. 142. H_2C—CHCN (ring closed by NH).

See subvol. a, p. 442, Nr. 492.

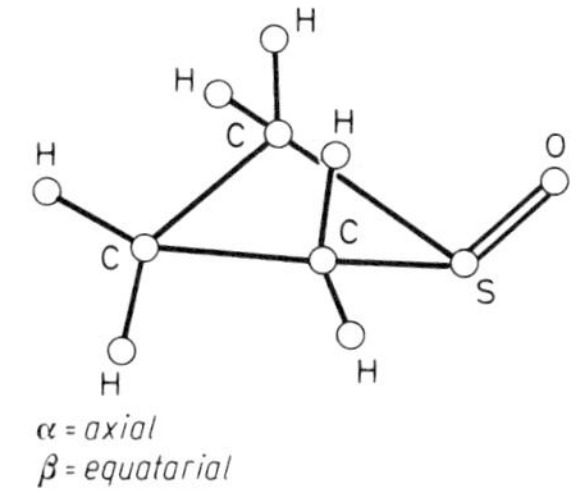

Fig. 143. H_2C—CH_2 / H_2C—SO (four-membered ring).

See subvol. a, p. 454, Nr. 507.

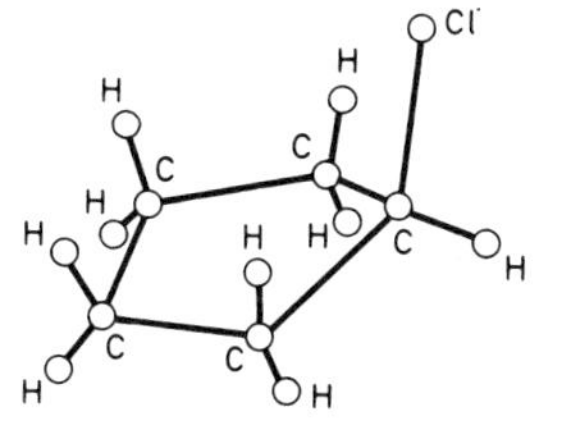

Fig. 144. 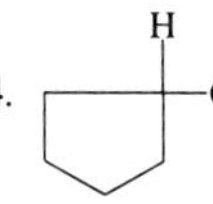 Cl.

See subvol. a, p. 456, Nr. 511.

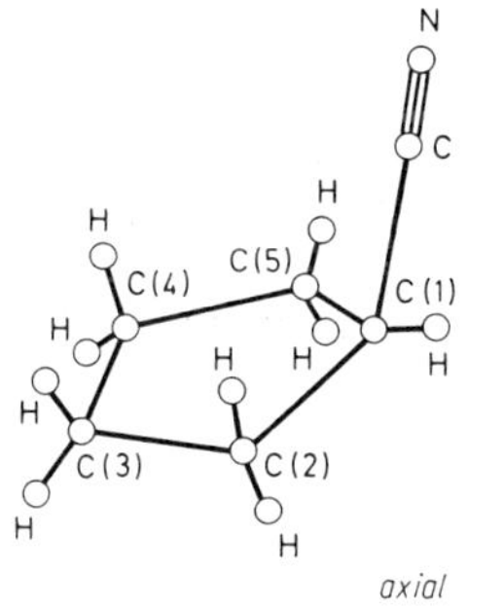

Fig. 145. 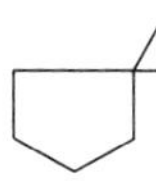 CN.

See subvol. a, p. 456, Nr. 512.

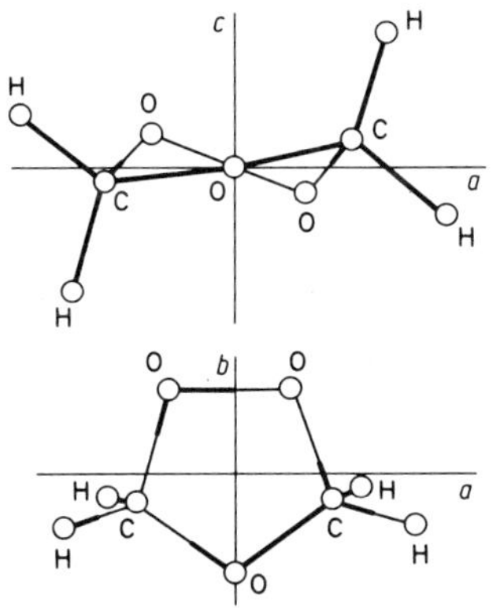

Fig. 146. O—O, H_2C O CH_2 (five-membered ring).

See subvol. a, p. 459. Nr. 515a.

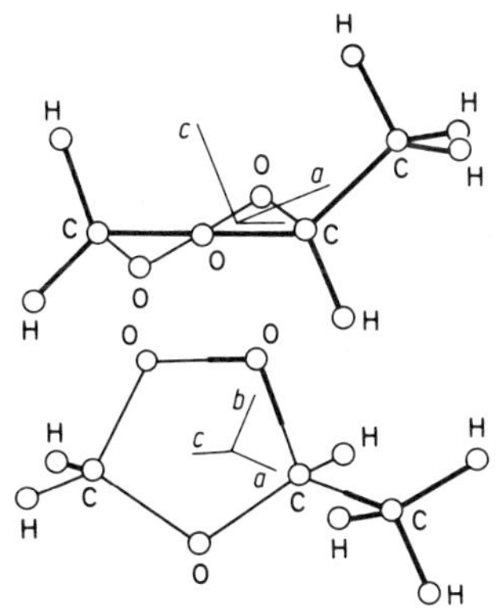

Fig. 147. O—O, H_2C O $CHCH_3$ (five-membered ring).

See subvol. a, p. 461, Nr. 518.

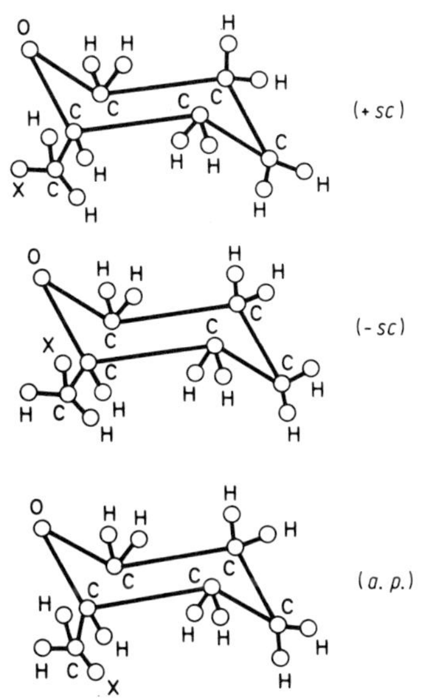

Fig. 148. 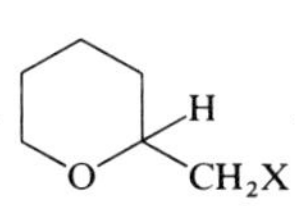

X = OH or NH_2; *sc* = synclinal, *a.p.* = antiperiplanar.

See subvol. a, p. 467, Nr. 532, 533.

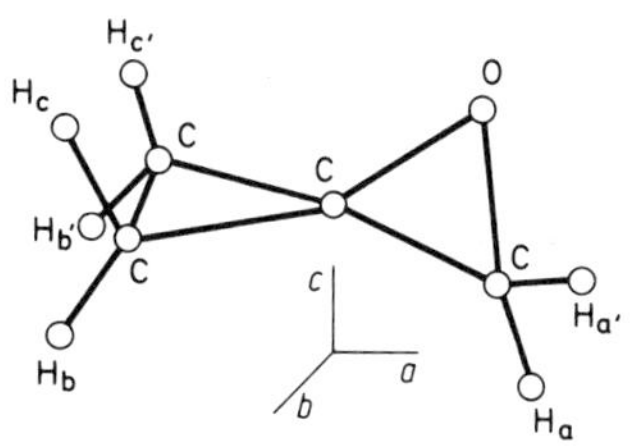

Fig. 149. H_2C, H_2C > C < O, CH_2.

See subvol. a, p. 472, Nr. 547.

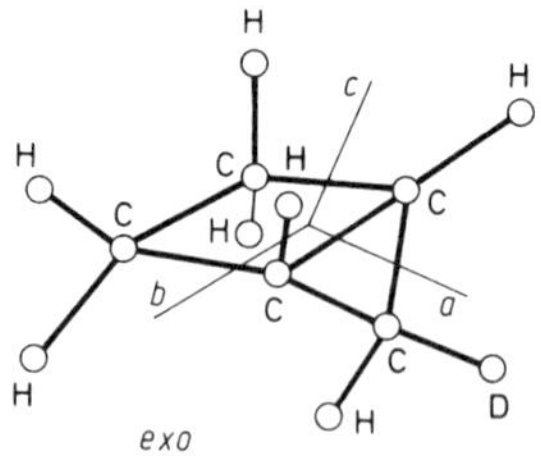

Fig. 150. 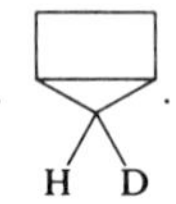.

See subvol. a, p. 473, Nr. 548.

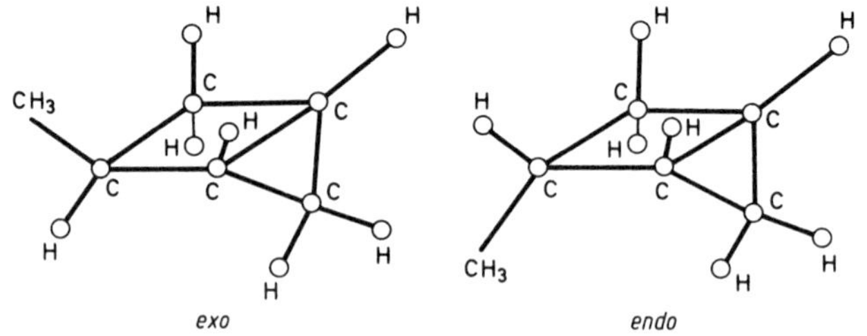

Fig. 151. 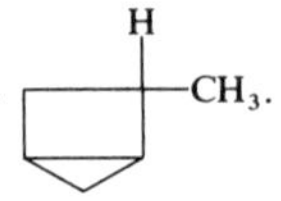.

See subvol. a, p. 474, Nr. 549.

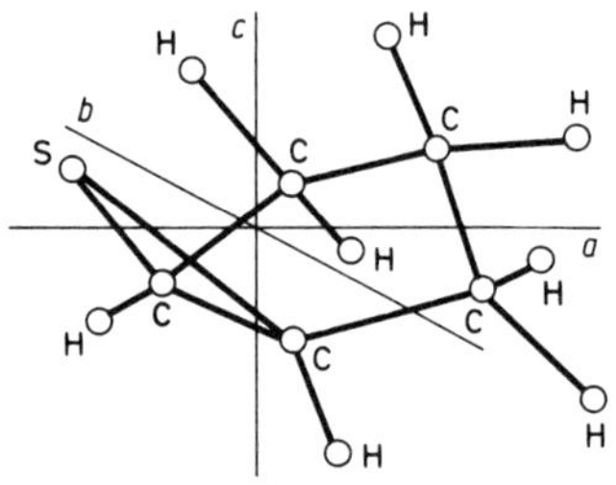

Fig. 152. 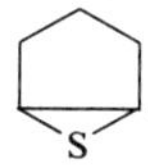.

See subvol. a, p. 474, Nr. 550.

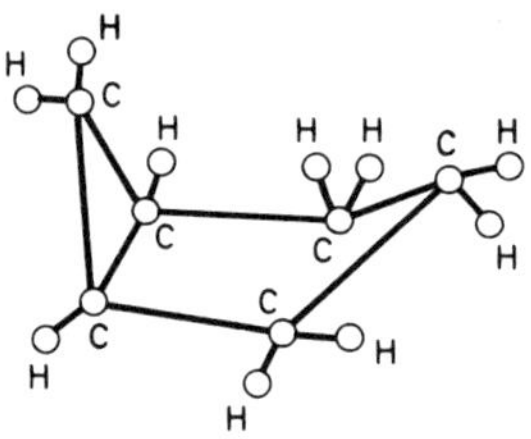

Fig. 153. 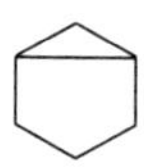.

See subvol. a, p. 474, Nr. 551.

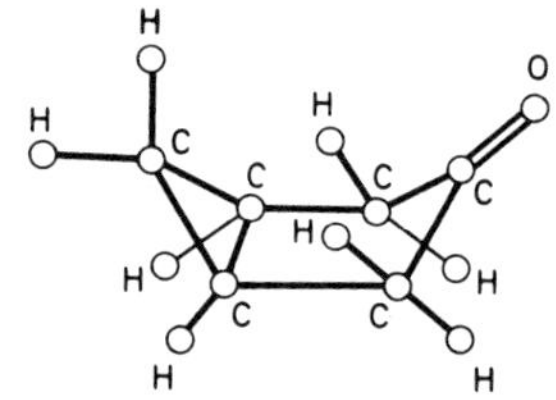

Fig. 154. 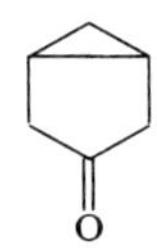.

See subvol. a, p. 475, Nr. 553.

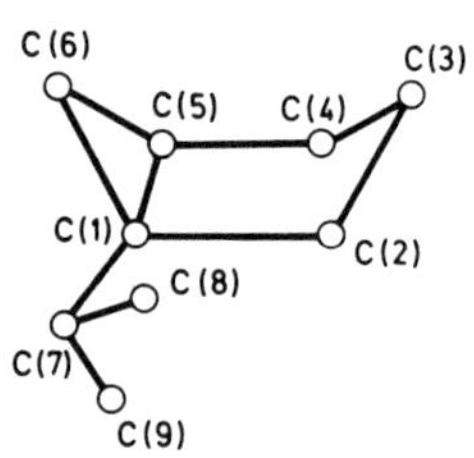

Fig. 155. The carbon atomskeleton common to the bicyclic monoterpenes 1–6.

See subvol. a, p. 476, Nr. 554···558.

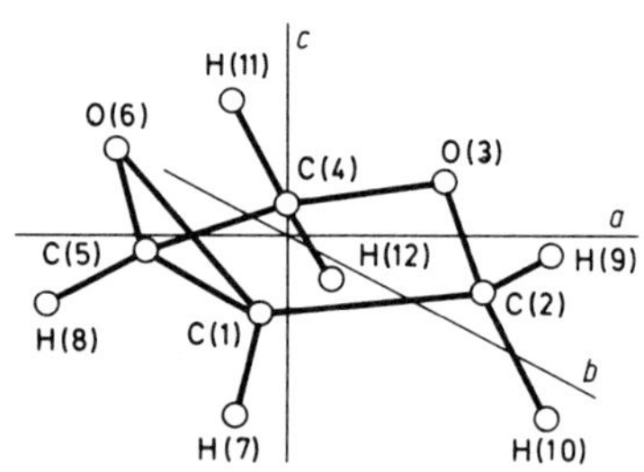

Fig. 156. 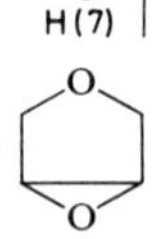.

See subvol. a, p. 476, Nr. 559.

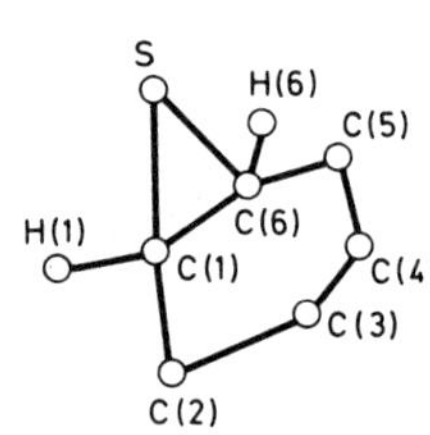

Fig. 157. 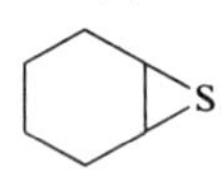.

See subvol. a, p. 477, Nr. 560.

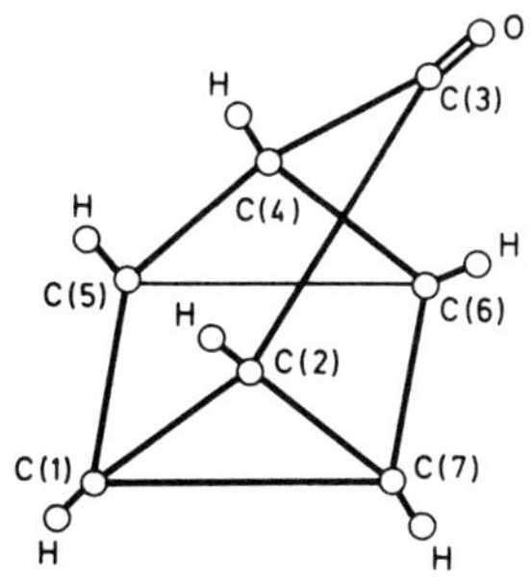

Fig. 158. 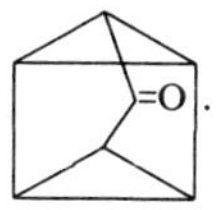.

See subvol. a, p. 477, Nr. 561.

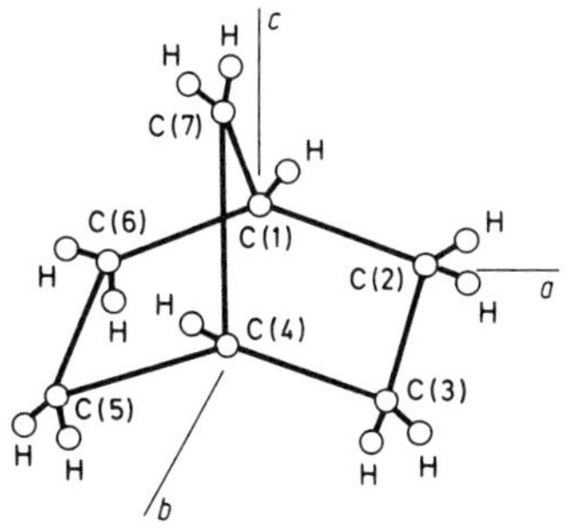

Fig. 159. .

See subvol. a, p. 478. Nr. 562.

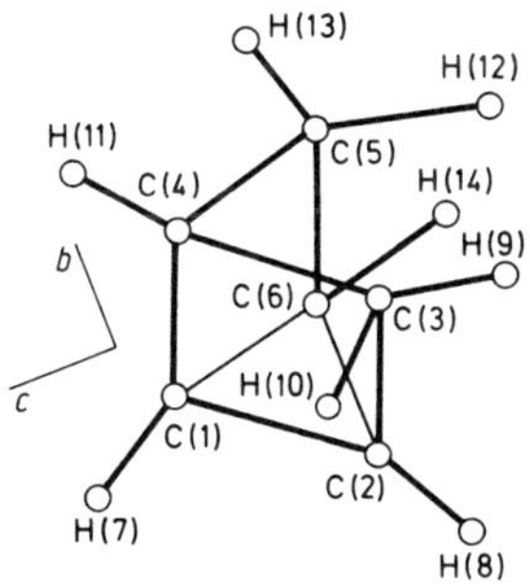

Fig. 160. 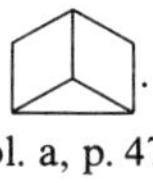.

See subvol. a, p. 479, Nr. 568.

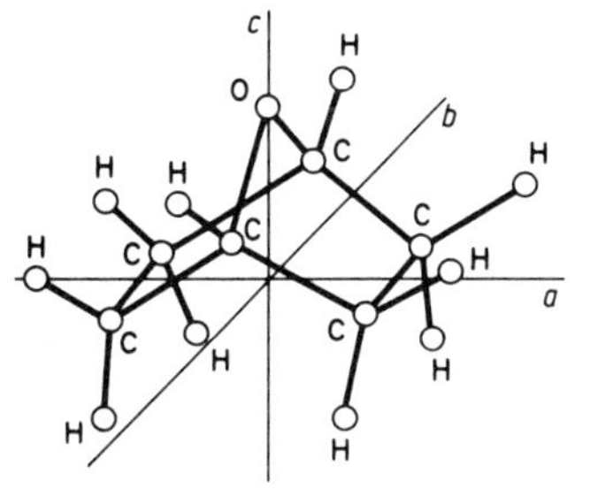

Fig. 161. 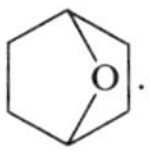.

See subvol. a, p. 480, Nr. 570.

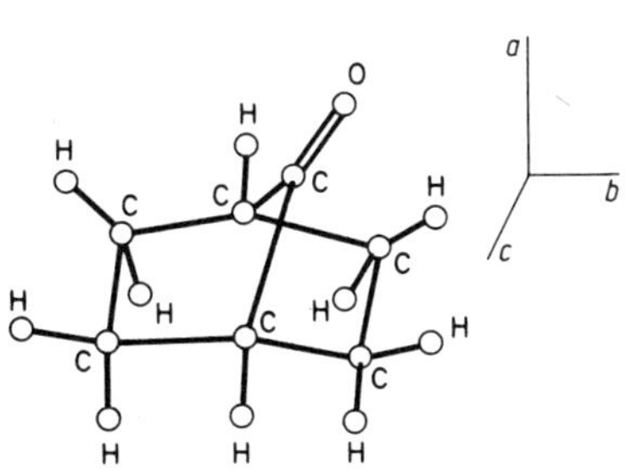

Fig. 162. 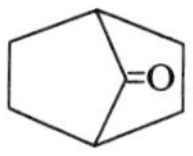.

See subvol. a, p. 481, Nr. 571.

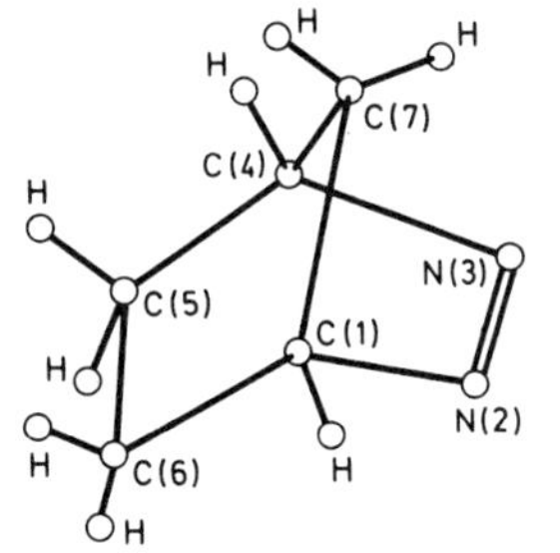

Fig. 163. .

See subvol. a, p. 481, Nr. 572.

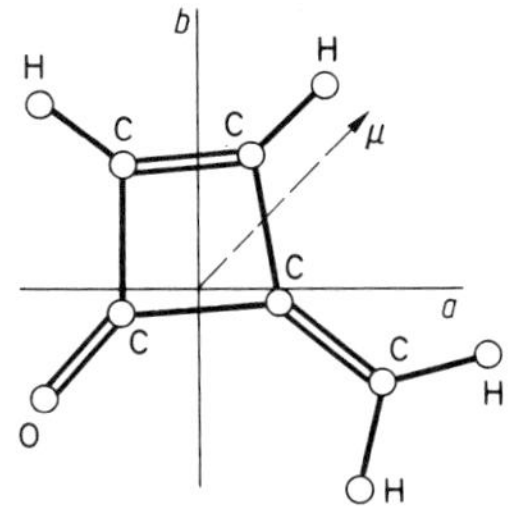

Fig. 164. HC=CH . H₂C=C—C=O

See subvol. a, p. 484, Nr. 591.

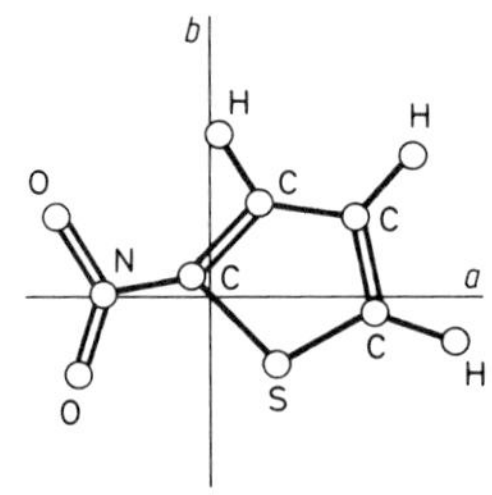

Fig. 165. 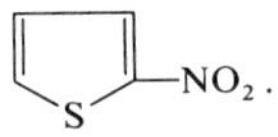

See subvol. a, p. 499, Nr. 621.

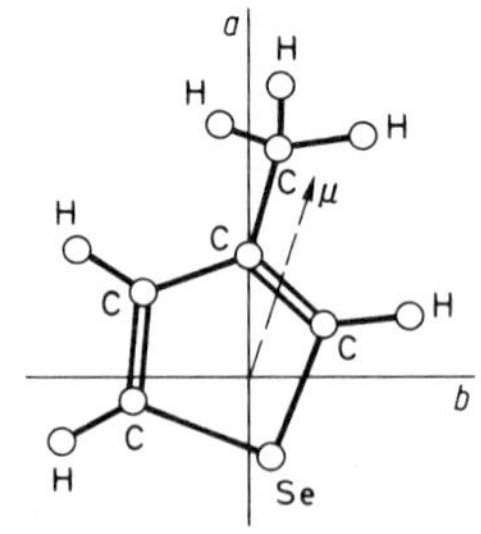

Fig. 166. 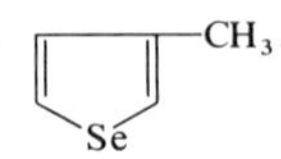

See subvol. a, p. 501, Nr. 624.

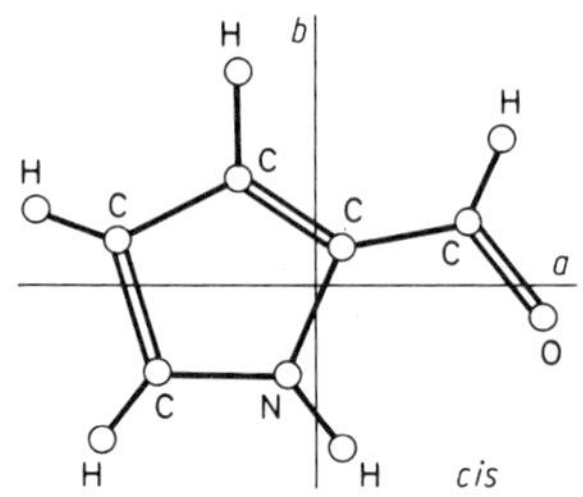

Fig. 167. 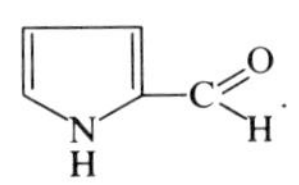

See subvol. a, p. 502, Nr. 628.

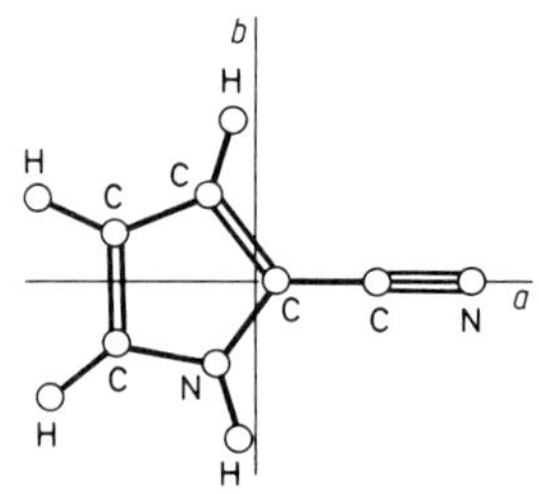

Fig. 168. 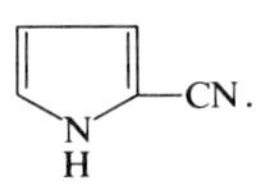

See subvol. a, p. 502, Nr. 629.

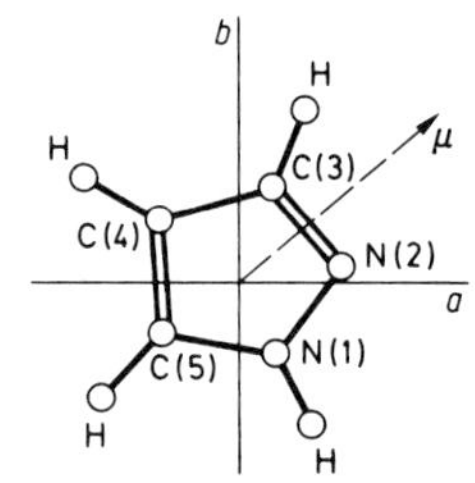

Fig. 169. 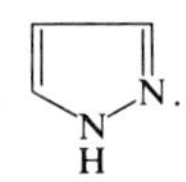

See subvol. a, p. 503, Nr. 630.

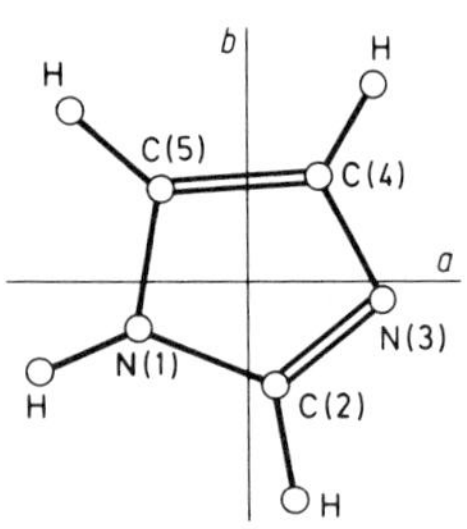

Fig. 170. 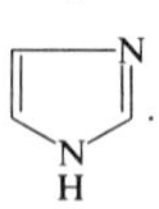

See subvol. a, p. 504, Nr. 632.

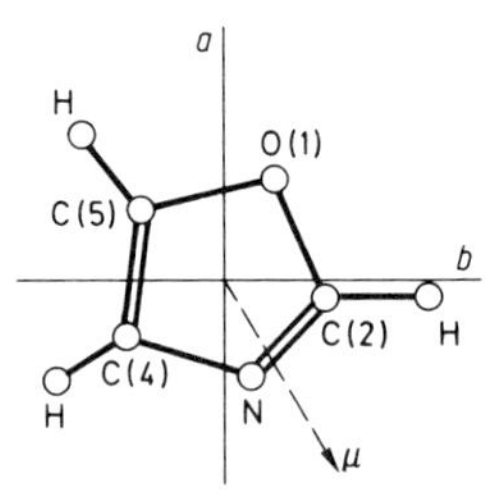

Fig. 171. 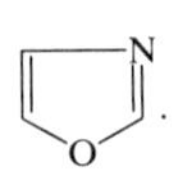

See subvol. a, p. 505, Nr. 633.

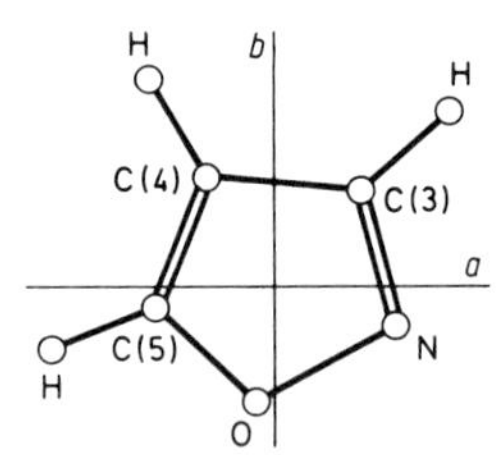

Fig. 172. 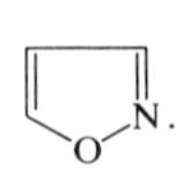

See subvol. a, p. 505, Nr. 634.

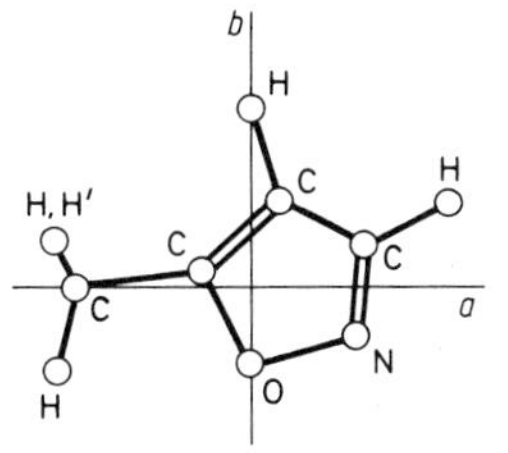

Fig. 173. CH$_3$ 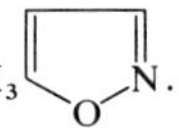 .

See subvol. a, p. 507, Nr. 635.

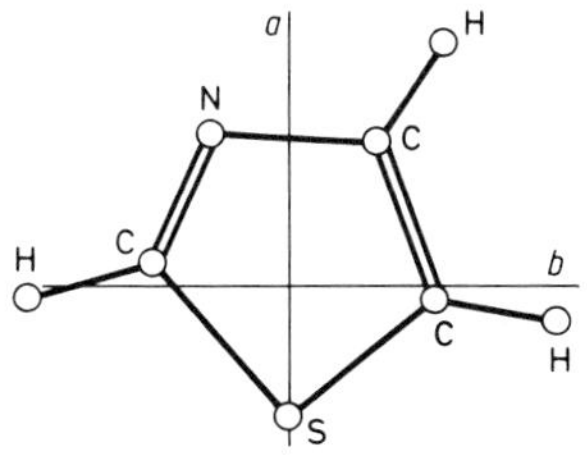

Fig. 174. 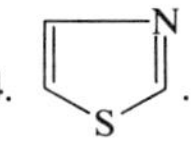 .

See subvol. a, p. 507, Nr. 636.

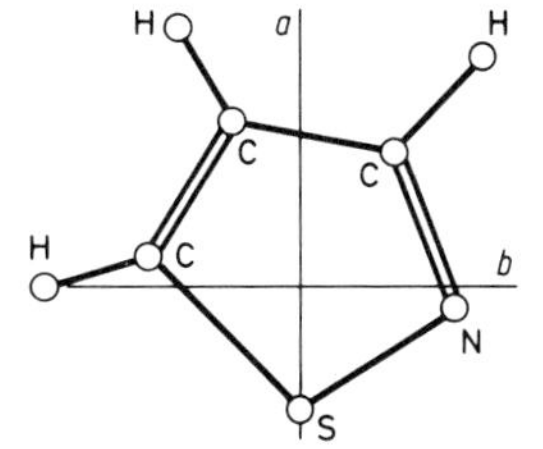

Fig. 175. 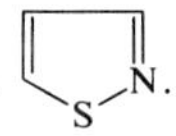 .

See subvol. a, p. 507, Nr. 637.

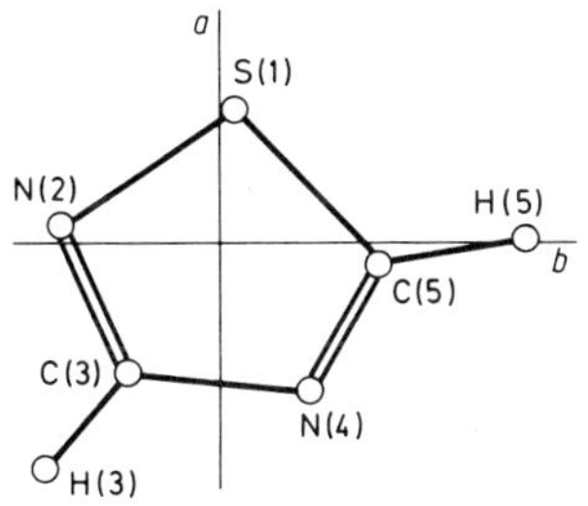

Fig. 176. 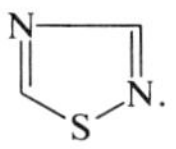 .

See subvol. a, p. 508, Nr. 639.

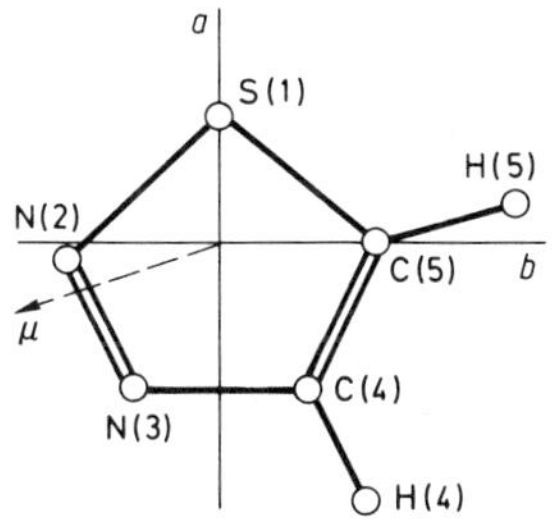

Fig. 177. 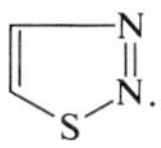 .

See subvol. a, p. 509, Nr. 640.

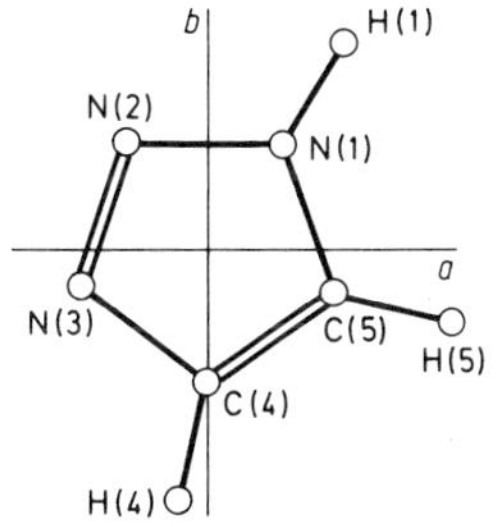

Fig. 178. 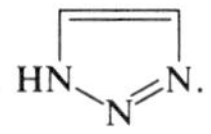 .

See subvol. a, p. 511, Nr. 642.

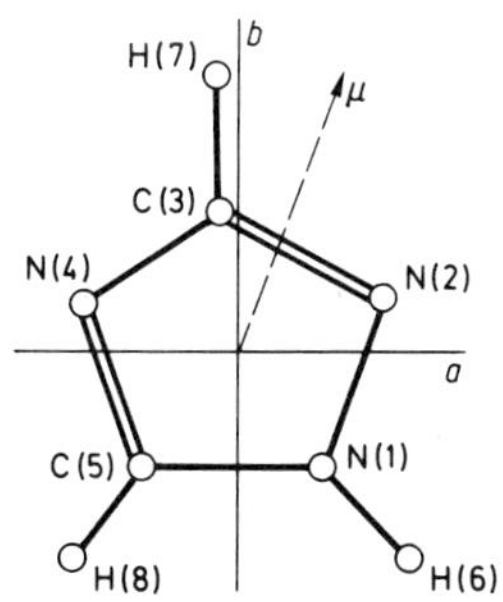

Fig. 179. 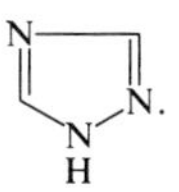 .

See subvol. a, p. 511, Nr. 643.

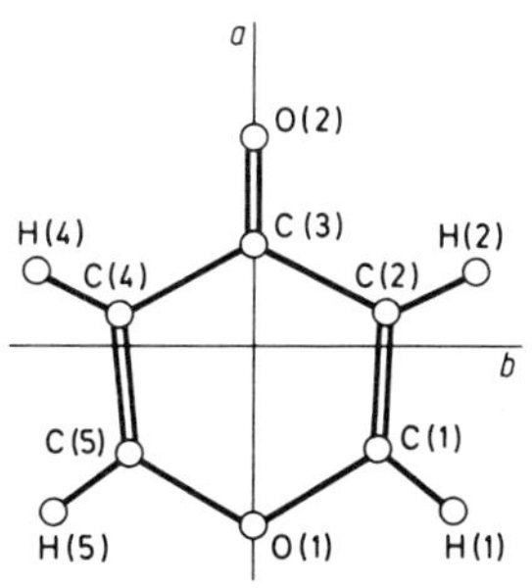

Fig. 180. 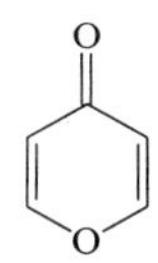 .

See subvol. a, p. 512, Nr. 646.

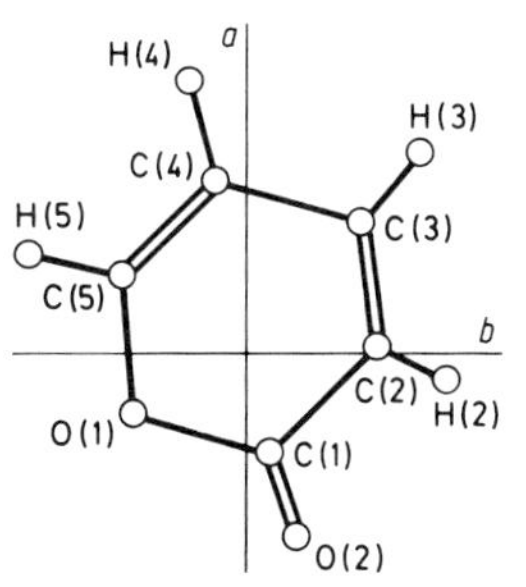

Fig. 181. 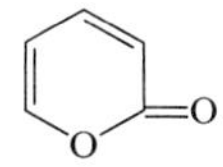 .

See subvol. a, p. 512, Nr. 647.

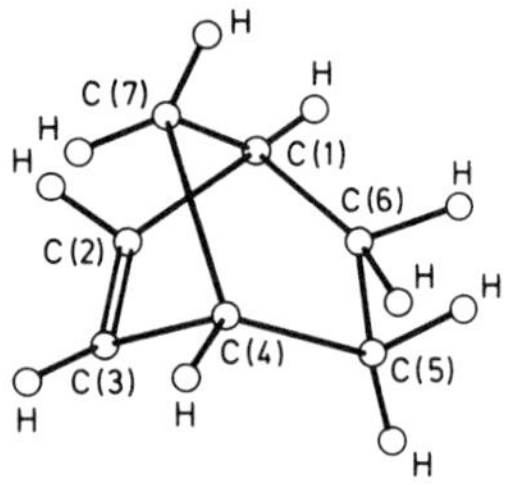

Fig. 182.  .

See subvol. a, p. 516, Nr. 663.

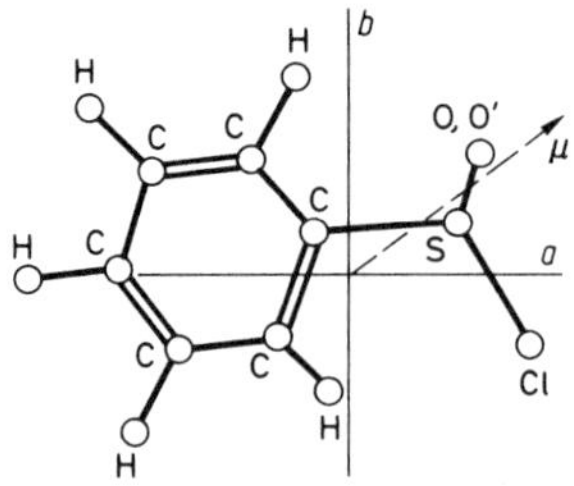

Fig. 183. 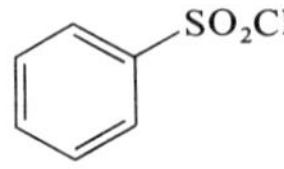

.

See subvol. a, p. 524, Nr. 678.

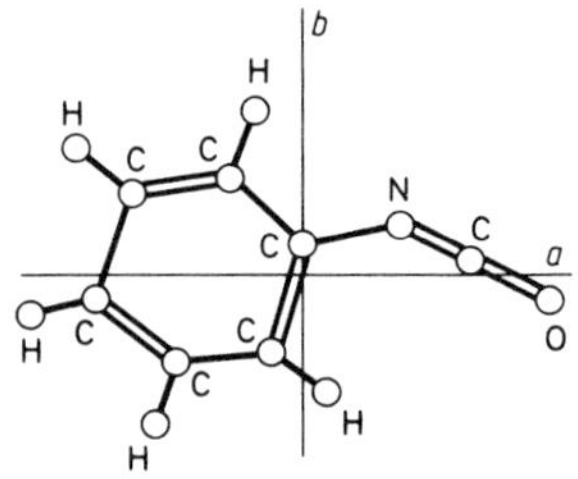

Fig. 184. 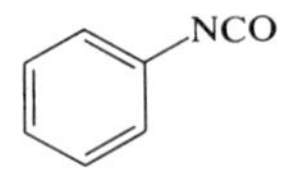

.

See subvol. a, p. 526, Nr. 681.

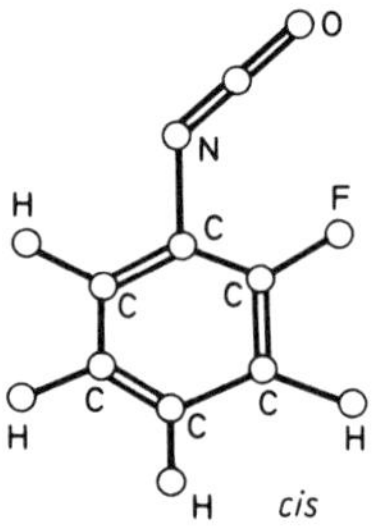

Fig. 185. 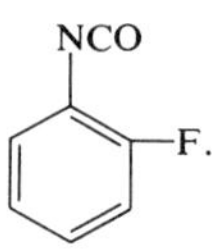

.

See subvol. a, p. 532, Nr. 697.

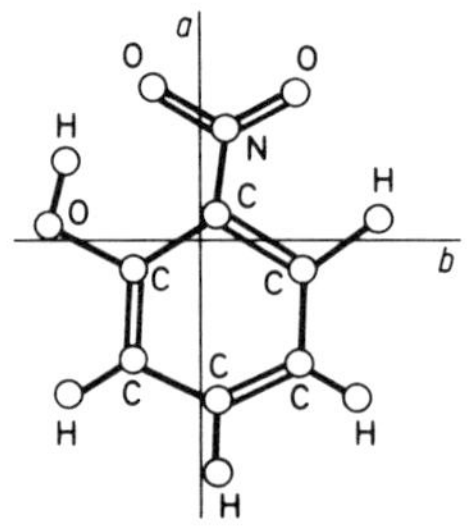

Fig. 186. 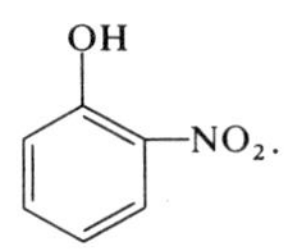

.

See subvol. a, p. 533, Nr. 700.

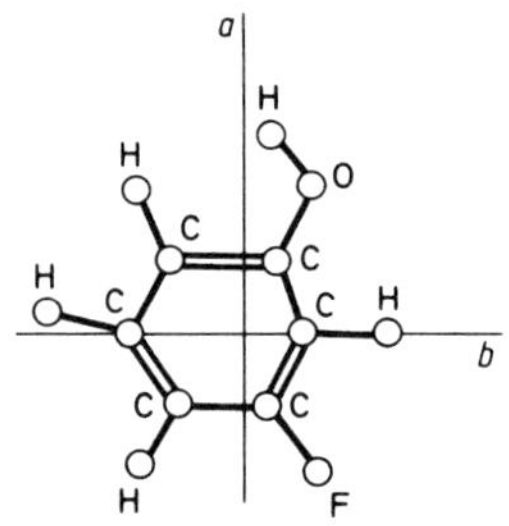

Fig. 187. 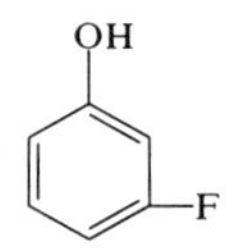

.

See subvol. a, p. 534, Nr. 703.

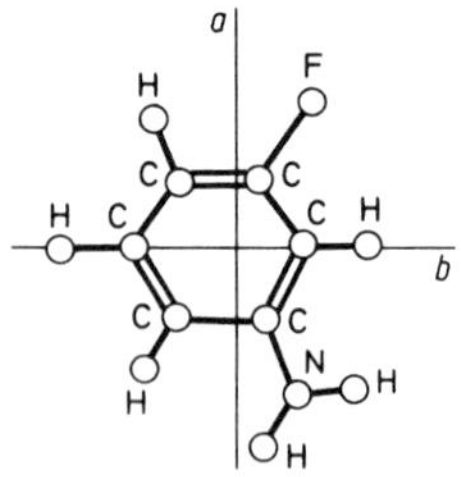

Fig. 188. 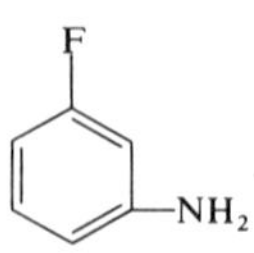

.

See subvol. a, p. 535, Nr. 704.

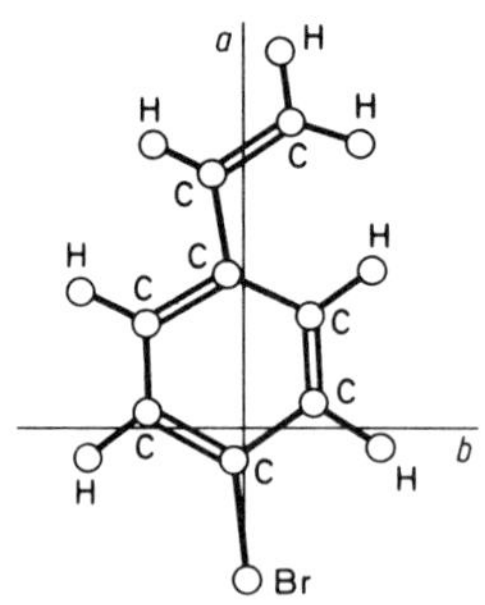

Fig. 189. 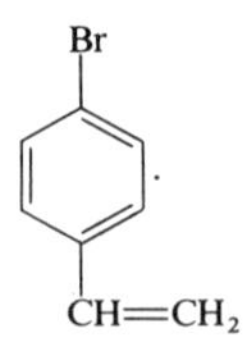

.

See subvol. a, p. 540, Nr. 719.

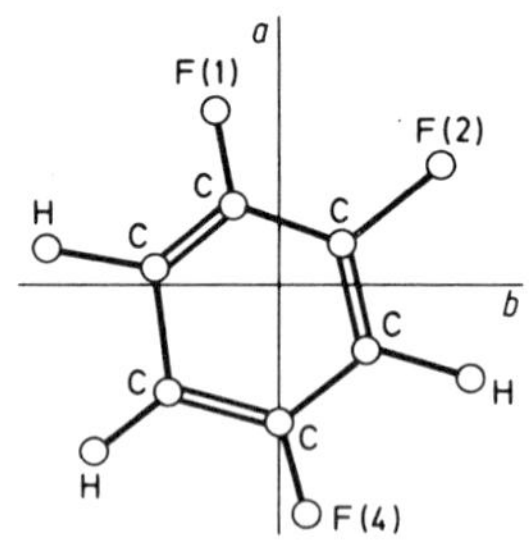

Fig. 190. 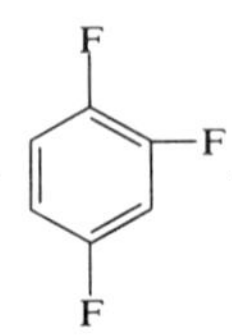

.

See subvol. a, p. 541, Nr. 724.

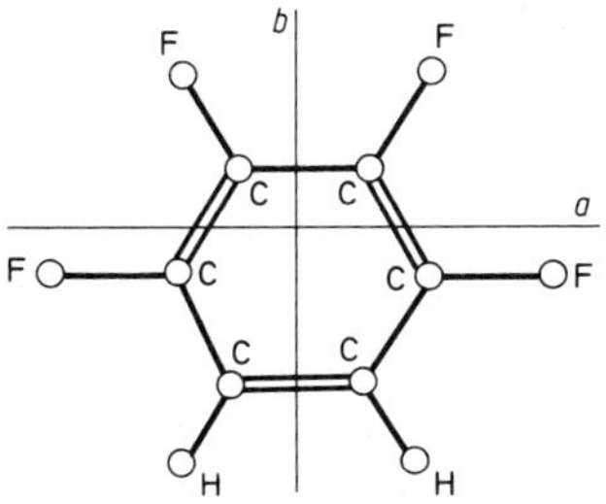

Fig. 191. 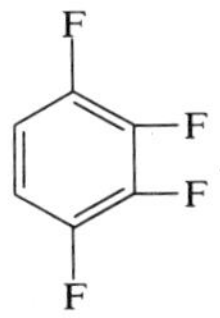.

See subvol. a, p. 542, Nr. 725.

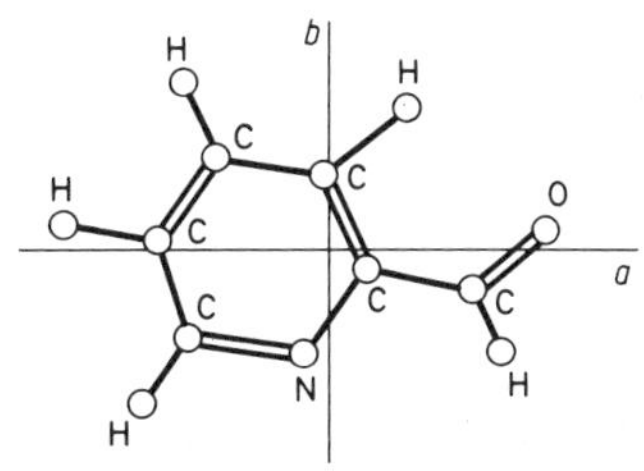

Fig. 192. 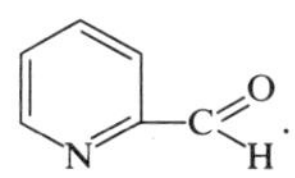.

See subvol. a, p. 544, Nr. 736.

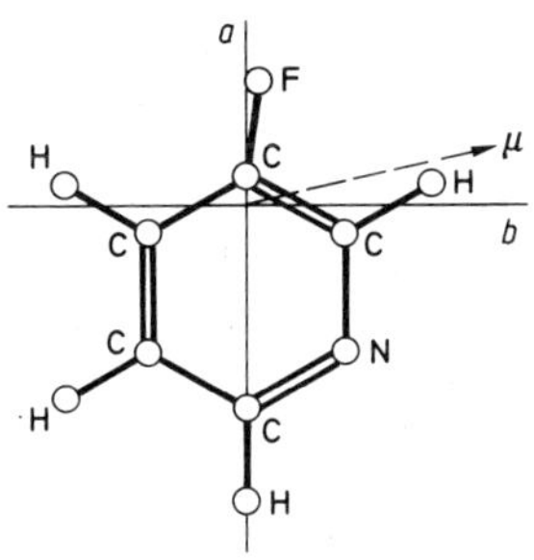

Fig. 193. 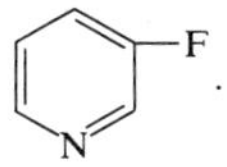.

See subvol. a, p. 545, Nr. 737.

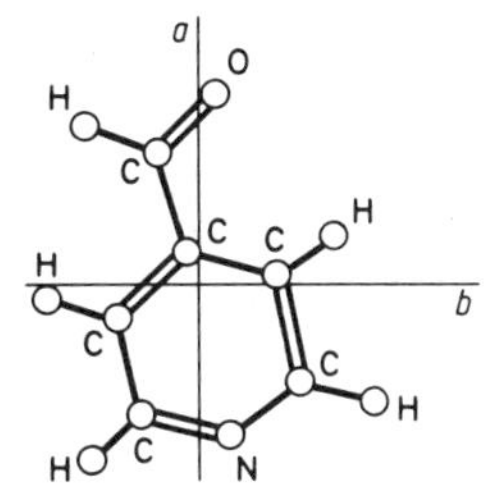

Fig. 194. 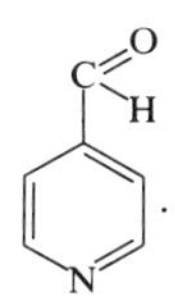.

See subvol. a, p. 548, Nr. 746.

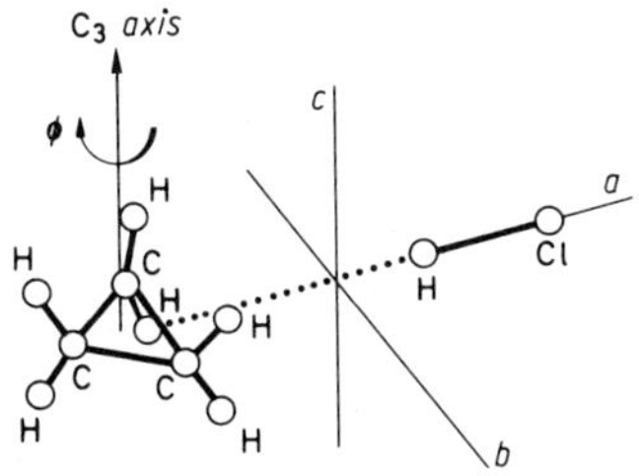

Fig. 195. $HCl \cdot C_3H_6$.
See subvol. a, p. 555, Nr. 764.

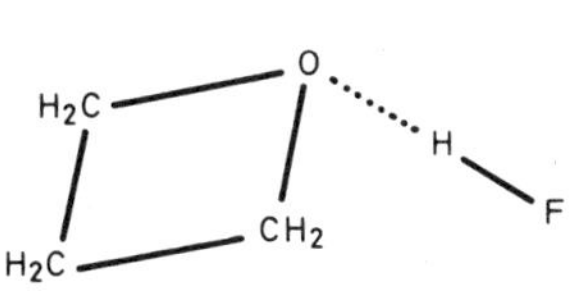

Fig. 196. $(CH_2)_3O \cdots HF$.
See subvol. a, p. 555, Nr. 765.

4 Constants of radicals

4.1 Diatomic radicals

4.1.1 Preliminary remarks

1. Introduction

All molecules in non-$^1\Sigma$ electronic states are called radicals.

Vol. II/6, Landolt-Börnstein, New Series, 5.1.1, contains the formal description of the necessary Hamiltonian to describe the rotational energy and the fine and hyperfine structure in such molecular states. Only slight modifications – in agreement with the most recent literature – will be mentioned.

The rotational angular momentum is given by the symbol $\boldsymbol{R}$. The coupling of $\boldsymbol{R}$ with the electronic orbital angular momentum $\boldsymbol{L}$ will be represented by $\boldsymbol{N}$. The elder notation was $\boldsymbol{N}$ and $\boldsymbol{K}$, respectively, but it could lead to confusion with the quantum number K in the case of symmetric top molecules and is hence replaced by $\boldsymbol{R}$ and $\boldsymbol{N}$.

In the set of the electronic hyperfine parameters a_i, b_i, c_i, d_i we will very often replace b_i by b_F, which directly describes the Fermi-contact interaction. The symbol k_i for the nuclear spin-rotation hyperfine interaction will be replaced by C_{Ii} to indicate the close similarity to the effective Hamiltonian in $^1\Sigma$-states. But it should be mentioned that the physical interpretation of C_{Ii} in case of non-$^1\Sigma$-states is not obvious.

For the Zeeman-interaction of the rotational motion g_{rot} is used instead of the separate values g^n_{rot} or g^e_{rot} for the nuclear or electronic part. Then g_{rot} is comparable to g_J in $^1\Sigma$-molecules and the specific part in the Hamiltonian thus contains the nuclear magneton μ_n and not the Bohr magneton μ_B, to obtain the right order of magnitude for this Zeeman term. The very special parts of the Zeeman-Hamiltonian for the different electronic states are briefly reviewed in the appropriate subsections. The sign of the rotational Zeeman-interaction is not consistently defined in the literature; the same definition as in part 2.9.1 is used here.

One of the main problems in compiling molecular parameters of radicals is that different effective Hamiltonians are proposed in the literature to describe the observed spectra. Therefore the definition of the molecular parameters changes from one publication to the other. The following tables try to present parameters which are as closely comparable as possible but neglect higher order terms in many cases. It was felt that including all higher order terms which resulted from a specific fitting would possibly confuse a reader who is not an expert in this field. To compensate this lack of accuracy and to simplify further calculations with a more generally accepted Hamiltonian, the spectral lines measured in the microwave and radiofrequency domain are listed here except those lines which are already contained in 5.1.1 of Vol. II/6, Landolt-Börnstein, New Series.

2. Arrangement of substances

All molecules of one specific electronic state are collected in a separate subsection. Within these subsections, the molecules are arranged according to the number of valence electrons and in increasing order of atomic number in the periodic table.

The molecule numbers introduced in part 2.2 are used again.

4.1.2 The $^2\Sigma$ electronic state

4.1.2.1 Preliminary remarks

1. Introduction

The approximate electronic quantum numbers of this molecular state are the spin $S=1/2$, the projection of the electronic angular momentum onto the molecular axis $|\Lambda|=0$, and the reflection symmetry $+/-$ in a plane containing the molecular axis. Assuming Hund's coupling case (*b*), the total molecular state is described by the additional quantum number N which is derived from the sum of the orbital angular momentum $\boldsymbol{L}$ and the rotational momentum $\boldsymbol{R}$, and by the quantum number J of the total angular momentum, by the parity $+$ or $-$ and by the vibrational state quantum number v. In the case of non-zero nuclear spins I_1 and I_2, the total angular momentum is given by quantum number F and to completely describe the molecular state the approximate intermediate quantum number $F_1(\boldsymbol{F}_1=\boldsymbol{J}+\boldsymbol{I}_1)$ or, for a homonuclear molecule, $I_T(\boldsymbol{I}_T=\boldsymbol{I}_1+\boldsymbol{I}_2)$ is used.

The appropriate phenomenological Hamiltonian is given in 5.1.2.1 of Vol. II/6, Landolt-Börnstein, New Series and will not be repeated here. There one will find the definition of the molecular parameters describing the rotational energy in such an electronic state.

For converting molecular parameters of one isotope to those of another, simple mass relations which are similar to those of a $^1\Sigma$-state in the case of Dunham-parameters Y_{lk} (compare with part 2.2) are sometimes adopted. For the spin-rotation interaction these relations are:

$\gamma_v = \gamma_e + \alpha_\gamma (v+1/2) + \cdots$

$\gamma_e \propto \frac{1}{\mu}$

$\alpha_\gamma \propto \left(\frac{1}{\mu}\right)^{3/2}$

$D_\gamma \propto \left(\frac{1}{\mu}\right)^2$ centrifugal parameter

where μ is the reduced mass of the molecule. So far the experimental results confirm these relations with a high degree of accuracy.

2. List of parameters

B_v rotational constant of the vibrational state v; when neglecting the break-down of the Born-Oppenheimer approximation this parameter is expanded in a series of Dunham parameters $B_v = \sum_l Y_{l1} (v+\frac{1}{2})^l$.

D_v centrifugal distortion constant of the vibrational state v; sometimes a similar expansion as for B_v is used: $D_v = -\sum_l Y_{l2} (v+\frac{1}{2})^l$.

In few cases higher order terms like Y_{03} are included in the centrifugal distortion correction.

ν_ϱ ϱ-type doubling frequency, which is the energy splitting of the two rotational states $J = N \pm 1/2$ for the the same vibrational state.

γ_v spin-rotation interaction constant of the vibrational state v: $\gamma_v = \gamma_e + \alpha_\gamma (v+\frac{1}{2}) + \cdots$

D_γ centrifugal distortion of the spin-rotation interaction.

γ_{lk} replacing γ_v, D_γ; some authors use the expansion of an effective spin-rotation interaction constant: $\gamma_{\text{eff}} = \sum_{l,k} \gamma_{lk}(v+\frac{1}{2})^l \cdot [N(N+1)]^k$.

b_i, c_i magnetic hyperfine structure constants for the coupling of the electronic spin with the nuclear spin of nucleus i.

b_i^{lk}, c_i^{lk} replacing b_i and c_i; some authors use the expansion of effective hyperfine interaction constants: $x_i^{\text{eff}} = \sum_{l,k} x_i^{lk}(v+\frac{1}{2})^l [N(N+1)]^k$ with $x = b, c$.

C_{Ii} magnetic hyperfine structure constant for the coupling of the molecular rotation with the nuclear spin of nucleus i.

eqQ_i electric quadrupole interaction constant of nucleus i with the electronic field gradient along the molecular axis.

μ_v molecular electric dipole moment for the vibrational state v.

g_s g-factor from electron spin including relativistic and diamagnetic corrections.

g_{rot} rotational g-factor (compare g_J for $^1\Sigma$-states, part 2.9.1).

Δg part of the g-factor which represents the influence of the external field on the spin-orbit interaction.

derived parameters

ω_e harmonic vibrational frequency.

$\omega_e x_e$ first order anharmonic correction of the vibrational energy.

$a_0, a_1, a_2, \cdots$ potential coefficients of the Dunham potential; compare part 2.2.

r_e equilibrium internuclear distance.

3. Arrangement of substances

OH
KAr
CaF, CaCl, CaBr, CaI, SrF, SrCl, BaF, BaCl
CO^+
CN
NO

4.1.2.2 Data and references

11 OH

Molecular constants of OH

Isotopic species	State		Parameter	Value	Ref.	Meth.	Remarks
	electronic	vibrational					
$^{16}O^{1}H$	$A^2\Sigma^+$	$v=0$	γ_0	6862 (43) MHz	81RAA, 80RAA	[1])	[1]) quantum beat spectroscopy
			$b_F=b+1/3c(^1H)$	774.1 (4) MHz		[1])	
			$c(^1H)$	168.9 (8) MHz		[1])	
			g_{rot}	4.52 (11)	81RAA, 80RAA, 79LEB	[1])	
			Δg	+0.0031 (35)	80RAA, 79LEB	[1])	
$^{16}O^{1}H$	$A^2\Sigma^+$	$v=0$	$(b+1/3c)(^1H)$	777.8 (20) MHz	79TER	MB/LIF	
			$c(^1H)$	165.8 (28) MHz	79TER	MB/LIF	
			γ_0	7130 (30) MHz	79TER	MB/LIF	
$^{16}O^{2}D$	$A^2\Sigma^+$	$v=0$	$b(^2D)$	109.73 (5) MHz	76GER, 73GER	LC	
			$c(^2D)$	25.73 (20) MHz	76GER, 73GER	LC	
			$eqQ(^2D)$	0.22 (3) MHz	76GER, 73GER, 76WOO	LC	
		$v=1$	$b(^2D)$	111.15 (5) MHz	76GER	LC	
			$c(^2D)$	24.40 (20) MHz	76GER	LC	
			$eqQ(^2D)$	0.05 (4) MHz	76GER	LC	
		$v=0$	μ	1.72 (10) D	73WEI	LC	

References

73GER German, K. R., Bergeman, T. H., Weinstock, E. M., Zare, R. N.: J. Chem. Phys. **58** (1973) 4304.

73WEI Weinstock, E. M., Zare, R. N.: J. Chem. Phys. **58** (1973) 4319.

76GER German, K. R.: J. Chem. Phys. **64** (1976) 4192.

76WOO Woods, R. C., Dixon, T. A.: J. Chem. Phys. **64** (1976) 5319.

79LEB Lebow, P., Raab, F., Metcalf, H.: Phys. Rev. Lett **42** (1979) 85.

79TER ter Meulen, J. J., van Mierlo, G. W. M., Dymanus, A.: Phys. Rev. Lett. **43** (1979) 29.

80RAA Raab, F., Bergeman, T., Lieberman, D., Metcalf, H.: Opt. Lett, **5** (1980) 427.

81RAA Raab, F., Bergeman, T., Lieberman, D., Metcalf, H.: Phys. Rev. A **24** (1981) 3120.

38 KAr

Molecular constants of KAr

Isotopic species	State		Parameter	Value	Ref.	Meth.	Remarks
	electronic	vibrational					
$^{39}K^{40}Ar$	$X^2\Sigma^+$	average	$\bar{\gamma}$	0.24 (1) MHz	74MAT	MB	

Reference: 74MAT Mattison, E.M., Pritchard, D.E., Kleppner, D.: Phys. Rev. Lett. **32** (1974) 507.

39 CaF

Molecular constants of CaF

Isotopic species	State	B_v MHz	D_v kHz	γ_v MHz	Ref.	Meth.	Derived constants
$^{40}Ca^{19}F$	$X^2\Sigma^+$			$\gamma_0 = 38.6\,(2)$	80BER	SLS	
		$B_0 = 10267.53\,(57)$	$D_0^{ev} = 13.97$	$\gamma_0 = 39.0\,(60)$	78NAK	MODR	
		$B_1 = 10194.4\,(27)$	$D_1^{ev} = 13.67$	$\gamma_1 = 41\,(30)$		MODR	
				$\gamma_e = \gamma_{00} = 39.65891\,(2)$	81CHI	MB	
				$D_\gamma = \gamma_{01} = 10.263\,(8)$ Hz		MB	
				$\gamma_{02} = -544.7\,(12) \cdot 10^{-6}$ Hz		MB	
				$\gamma_{03} = -845\,(68) \cdot 10^{-12}$ Hz		MB	
				$\alpha_\gamma = \gamma_{10} = -321.668\,(25)$ kHz		MB	
				$\gamma_{11} = -2.116\,(4)$ Hz		MB	
				$\gamma_{20} = -563.8\,(60)$ Hz		MB	
				$\gamma_{21} = 23.37\,(74) \cdot 10^{-3}$ Hz		MB	

Isotopic species	State		Parameter	Value	Ref.	Meth.	Remarks
	electronic	vibrational					
$^{40}Ca^{19}F$	$X^2\Sigma^+$	$v = 0$	$b(^{19}F)$	104.0 (9) MHz	80BER	SLS	
			$c(^{19}F)$	58 (6) MHz		SLS	
		equ.	$b^{00}(^{19}F)$	109.1839 (5) MHz	81CHI	MB	
			$b^{01}(^{19}F)$	−0.51120 (15) kHz		MB	
			$b^{10}(^{19}F)$	−1.41521 (30) MHz		MB	
			$c^{00}(^{19}F)$	40.1190 (26) MHz		MB	
			$c^{10}(^{19}F)$	1.0562 (15) MHz		MB	
			$C_I(^{19}F)$	28.76 (37) kHz		MB	

Line frequencies of CaF

Isotopic species	State		Transition								ν	Ref.
	electronic	vibrational v	N′	N″	J′	J″	F'_1	F''_1	F′	F″	MHz	
$^{40}Ca^{19}F$	$X^2\Sigma^+$	0	2	1	1.5	0.5					41059.1 (28)	78NAK
		0	2	1	2.5	1.5					41085.4 (11)	
		0	3	2	2.5	1.5					61586.3 (42)	
		0	3	2	3.5	2.5					61620.2 (18)	
		1	2	1	2.5	1.5					40792.5 (30)	
		1	3	2	2.5	1.5					61148.3 (43)	
		1	3	2	3.5	2.5					61185.0 (35)	
		0	1	1	1.5	0.5			2	1	147.160 (1)	81CHI
		0	2	2	2.5	1.5			3	2	177.958 (1)	
		0	2	2	2.5	1.5			2	1	71.420 (1)	
		0	2	2	2.5	2.5			3	2	32.234 (2)	
		0	2	2	1.5	1.5			1	2	74.303 (2)	
		0	3	3	3.5	2.5			4	3	211.974 (1)	
		0	6	6	6.5	5.5			7	6	322.396 (2)	
		0	6	6	6.5	5.5			6	5	214.185 (1)	
		0	8	8	8.5	7.5			9	8	398.848 (1)	
		0	8	8	8.5	7.5			8	7	290.551 (1)	
		0	9	9	9.5	8.5			10	9	437.447 (1)	
		0	9	9	9.5	8.5			9	8	329.130 (1)	
		0	10	10	10.5	9.5			11	10	476.208 (1)	
		0	10	10	10.5	9.5			10	9	367.881 (1)	
		0	11	11	11.5	10.5			12	11	515.090 (1)	
		0	11	11	11.5	10.5			11	10	406.760 (1)	
		0	12	12	12.5	11.5			13	12	554.070 (2)	
		0	12	12	12.5	11.5			12	11	445.735 (2)	
		0	14	14	14.5	13.5			15	14	632.230 (1)	
		0	14	14	14.5	13.5			14	13	523.910 (1)	
		0	16	16	16.5	15.5			17	16	710.588 (1)	
		0	16	16	16.5	15.5			16	15	602.289 (1)	
		0	19	19	19.5	18.5			20	19	828.368 (1)	
		0	19	19	19.5	18.5			19	18	720.112 (1)	
		0	25	25	25.5	24.5			26	25	1064.438 (1)	
		0	25	25	25.5	24.5			25	24	956.308 (1)	

Line frequencies of CaF (continued)

Isotopic species	State: electronic	State: vibrational v	Transition: N'	N''	J'	J''	F'_1	F''_1	F'	F''	ν MHz	Ref.
$^{40}Ca^{19}F$	$X^2\Sigma^+$	0	29	29	29.5	28.5			30	29	1222.038 (2)	81CHI
		0	29	29	29.5	28.5			29	28	1114.016 (2)	
		0	34	34	34.5	33.5			35	34	1419.191 (1)	
		0	34	34	34.5	33.5			34	33	1311.330 (1)	
		0	46	46	46.5	45.5			47	46	1892.786 (2)	
		0	46	46	46.5	45.5			46	45	1785.418 (2)	
		0	48	48	48.5	47.5			49	48	1971.761 (2)	
		0	48	48	48.5	47.5			48	47	1864.488 (2)	
		0	66	66	66.5	65.5			67	66	2682.843 (3)	
		0	66	66	66.5	65.5			66	65	2576.620 (3)	
		0	83	83	83.5	82.5			84	83	3354.533 (3)	
		0	83	83	83.5	82.5			83	82	3249.616 (3)	
		0	119	119	119.5	118.5			120	119	4773.306 (4)	
		0	119	119	119.5	118.5			119	118	4672.140 (4)	
		1	2	2	2.5	1.5			3	2	176.348 (1)	
		1	2	2	2.5	1.5			2	1	71.277 (1)	
		1	2	2	2.5	2.5			3	2	31.695 (2)	
		1	2	2	1.5	1.5			1	2	73.377 (2)	
		1	4	4	4.5	3.5			5	4	245.607 (4)	
		1	9	9	9.5	8.5			10	9	433.645 (1)	
		1	9	9	9.5	8.5			9	8	326.746 (1)	
		1	13	13	13.5	12.5			14	13	588.029 (1)	
		1	13	13	13.5	12.5			13	12	481.121 (1)	
		1	19	19	19.5	18.5			20	19	821.338 (1)	
		1	19	19	19.5	18.5			19	18	714.500 (1)	
		1	29	29	29.5	28.5			30	29	1211.746 (1)	
		1	29	29	29.5	28.5			29	28	1105.143 (1)	
		1	34	34	34.5	33.5			35	34	1407.255 (1)	
		1	34	34	34.5	33.5			34	33	1300.811 (1)	
		1	46	46	46.5	45.5			47	46	1876.855 (1)	
		1	46	46	46.5	45.5			46	45	1770.903 (1)	
		1	48	48	48.5	47.5			49	48	1995.156 (2)	
		1	48	48	48.5	47.5			48	47	1849.304 (2)	
		1	66	66	66.5	65.5			67	66	2660.054 (1)	

Line frequencies of CaF (continued)

Isotopic species	State		Transition								ν MHz	Ref.
	electronic	vibrational v	N'	N''	J'	J''	F'_1	F''_1	F'	F''		
$^{40}Ca^{19}F$	$X^2\Sigma^+$	1	66	66	66.5	65.5			66	65	2555.256 (2)	81CHI
		1	83	83	83.5	82.5			84	83	3325.668 (2)	
		1	83	83	83.5	82.5			83	82	3222.171 (1)	
		2	2	2	2.5	1.5			3	2	174.740 (1)	
		2	2	2	2.5	1.5			2	1	71.134 (1)	
		2	2	2	2.5	2.5			3	2	31.153 (3)	
		2	2	2	1.5	1.5			1	2	72.453 (2)	
		2	12	12	12.5	11.5			13	12	544.521 (1)	
		2	12	12	12.5	11.5			12	11	439.026 (1)	
		2	34	34	34.5	33.5			35	34	1395.281 (1)	
		2	34	34	34.5	33.5			34	33	1290.252 (1)	
		2	66	66	66.5	65.5			67	66	2637.207 (1)	
		2	66	66	66.5	65.5			66	65	2533.823 (1)	
		2	83	83	83.5	82.5			84	83	3296.728 (2)	
		2	83	83	83.5	82.5			83	82	3194.652 (1)	
		3	2	2	2.5	1.5			3	2	173.131 (1)	
		3	2	2	2.5	1.5			2	1	70.987 (2)	
		3	13	13	13.5	12.5			14	13	577.810 (2)	
		3	13	13	13.5	12.5			13	12	473.730 (3)	
		3	34	34	34.5	33.5			35	34	1383.275 (1)	
		3	34	34	34.5	33.5			34	33	1279.659 (1)	
		3	66	66	66.5	65.5			67	66	2614.297 (1)	
		3	66	66	66.5	65.5			66	65	2512.330 (1)	
		3	83	83	83.5	82.5			84	83	3267.727 (1)	
		3	83	83	83.5	82.5			83	82	3167.061 (2)	
		4	83	83	83.5	82.5			84	83	3238.665 (2)	
		4	83	83	83.5	82.5			83	82	3139.409 (2)	

References

78NAK Nakagawa, J., Domaille, P.J., Steimle, T.C., Harris, D.O.: J. Mol. Spectrosc. **70** (1978) 374.

80BER Bernath, P.F., Cummins, P.G.: Field, R.W.: Chem. Phys. Lett. **70** (1980) 618.

80CHI1 Childs, W.J., Goodman, L.S.: Phys. Rev. Lett. **44** (1980) 316.

80CHI2 Childs, W.J., Goodman, L.S.: Phys. Rev. A**21** (1980) 1216.

80DUL Dulick, M., Bernath, P.F., Field, R.W.: Can. J. Phys. **58** (1980) 703.

81CHI Childs, W.J., Goodman, G.L., Goodman, L.S.: J. Mol. Spectrosc. **86** (1981) 365.

40 CaCl

Molecular constants of CaCl

Isotopic species	State	B_v MHz	D_v kHz	γ_v MHz	Ref.	Meth.	Derived constants
$^{40}Ca^{35}Cl$	$X\,^2\Sigma^+$	$Y_{01}=4563.7461\,(10)$ $Y_{11}=-23.9506\,(10)$ $Y_{21}=34.46\,(12)$ kHz	$Y_{02}=-3.0860\,(4)$ $Y_{12}=-0.80\,(45)$ Hz	$\gamma_e=42.208\,(30)$ $\alpha_\gamma=-0.406\,(15)$ $D_\gamma=-55\,(8)$ Hz	81SCH, 82MOE 81SCH	MW MW MW MW	$\omega_e=370.250\,(24)\,cm^{-1}$ $r_e=2.4367869\,(84)$ Å $\omega_e x_e=1.3556\,(68)\,cm^{-1}$ $a_0=225127\,(29)\,cm^{-1}$ $a_1=-3.12735\,(8)$ $a_2=6.289\,(30)$ $a_3=-9.41\,(22)$
				$\gamma_e=\gamma_{00}=42.11784\,(5)$ $\alpha_\gamma=\gamma_{10}=-0.37152\,(6)$ $D_\gamma=\gamma_{01}=-42.32\,(2)$ Hz $\gamma_{11}=-0.31\,(2)$ Hz $\gamma_{02}=-118\,(1)\cdot 10^{-6}$ Hz	82CHI1 77DOM	MB MB MB MB MB MODR	
$^{40}Ca^{37}Cl$	$X\,^2\Sigma^+$	$Y_{01}=4432.2607\,(10)$ Y_{11}[1]$=-22.9230\,(10)$ Y_{21}[1]$=32.50\,(12)$ kHz	Y_{02}[1]$=-2.9107\,(4)$ Y_{12}[1]$=-0.73\,(45)$ Hz	$\gamma_e=40.957\,(30)$ α_γ[1]$=-0.389\,(15)$ D_γ[1]$=-52\,(8)$ Hz	81SCH, 82MOE	MW MW MW	[1]) Calculated by mass relations from the parameters obtained for $^{40}Ca^{35}Cl$.

Isotopic species	State electronic	State vibrational	Parameter	Value	Ref.	Meth.	Remarks
$^{40}Ca^{35}Cl$	$X\,^2\Sigma^+$	equ. $v=0$	$b^{00}\,(^{35}Cl)$ $b^{10}\,(^{35}Cl)$ $b^{01}\,(^{35}Cl)$ $c^{00}\,(^{35}Cl)$ $c^{10}\,(^{35}Cl)$ $eq_eQ\,(^{35}Cl)$ $eq_IQ\,(^{35}Cl)$ $C_I\,(^{35}Cl)$ μ	19.3904 (8) MHz −0.1785 (9) MHz −54.99 (8) Hz 12.387 (8) MHz 0.112 (8) MHz −0.905 (9) MHz −0.23 (1) MHz 1.6 (2) kHz 3.6 (6) D	82CHI1 82DAG	MB MB MB MB MB MB MB MB ED	Parameters were obtained by a combined fit of measurements for $^{40}Ca^{35}Cl$ and $^{40}Ca^{37}Cl$ using appropriate relations of the reduced masses and nuclear moments. But the necessary scaling factor of b^{00} is different from the expected ratio of the nuclear magnetic moments.
$^{40}Ca^{35}Cl$	$B\,^2\Sigma^+$		upper limit of the hyperfine splitting for $v=0$, $N=42$, $J=41.5$		81CHI	MB, LIF, DR	

Line frequencies of CaCl

Isotopic species	State: electronic	State: vibrational v	Transition: N'	– N''	J'	– J''	F'_1	– F''_1	F'	– F''	ν MHz	Ref.
$^{40}Ca^{35}Cl$	$X\,^2\Sigma^+$	0	5	4	4.5	3.5					45496.7 (25)	77DOM
		0	5	4	5.5	4.5					45538.7 (24)	
		0	6	5	5.5	4.5					54596.9 (38)	
		0	6	5	6.5	5.5					54639.9 (24)	
		0	7	6	6.5	5.5					63701.8 (20)	
		0	7	6	7.5	6.5					63741.0 (24)	
		1	5	4	4.5	3.5					45254.7 (29)	
		1	5	4	5.5	4.5					45297.4 (74)	
		1	6	5	5.5	4.5					54312.4 (21)	
		1	6	5	6.5	5.5					54351.7 (12)	
		1	7	6	6.5	5.5					63367.0 (43)	
		1	7	6	7.5	6.5					63406.1 (14)	
$^{40}Ca^{35}Cl$	$X\,^2\Sigma^+$	0	10	9	9.5	8.5					91002.250 (40)	82MOE, 81SCH
		0	10	9	10.5	9.5					91044.233 (40)	
		1	10	9	9.5	8.5					90524.810 (40)	
		0	30	29	29.5	28.5					272752.515 (40)	
		0	30	29	30.5	29.5					272794.363 (40)	
		1	30	29	29.5	28.5					271319.710 (40)	
		1	30	29	30.5	29.5					271361.195 (40)	
		0	31	30	30.5	29.5					281821.608 (40)	
		0	31	30	31.5	30.5					281863.453 (40)	
		1	31	30	30.5	29.5					280341.053 (40)	
		1	31	30	31.5	30.5					280382.485 (40)	
		2	31	30	30.5	29.5					278864.780 (40)	
		2	31	30	31.5	30.5					278905.793 (40)	
		0	32	31	31.5	30.5					290888.430 (40)	
		0	32	31	32.5	31.5					290930.258 (40)	
		1	32	31	31.5	30.5					289360.090 (40)	
		1	32	31	32.5	31.5					289401.525 (40)	
$^{40}Ca^{35}Cl$	$X\,^2\Sigma^+$	0	4	4	4.5	3.5			6	5	217.084 (3)	82CHI1
		0	4	4	4.5	3.5			5	4	197.116 (3)	
		0	4	4	4.5	3.5			4	4	186.379 (1)	

Line frequencies of CaCl (continued)

Isotopic species	State: electronic	State: vibrational v	Transition: N'	N''	J'	J''	F_1'	F_1''	F'	F''	ν MHz	Ref.
$^{40}Ca^{35}Cl$	$X^2\Sigma^+$	0	4	4	4.5	3.5			4	3	177.318 (2)	82CHI1
		0	4	4	4.5	3.5			3	2	159.012 (2)	
		0	4	4	4.5	3.5			5	5	206.049 (2)	
		0	4	4	4.5	3.5			3	3	167.292 (1)	
		0	4	4	3.5	3.5			4	5	8.936 (2)	
		0	7	7	7.5	6.5			6	5	285.117 (3)	
		0	7	7	7.5	6.5			9	8	343.070 (2)	
		0	7	7	7.5	6.5			8	7	323.326 (3)	
		0	7	7	7.5	6.5			7	6	303.860 (3)	
		0	22	22	22.5	21.5			24	23	971.773 (1)	
		0	22	22	22.5	21.5			23	22	952.340 (1)	
		0	22	22	22.5	21.5			22	21	933.047 (1)	
		0	22	22	22.5	21.5			21	20	913.948 (1)	
		0	43	43	43.5	42.5			45	44	1849.264 (1)	
		0	43	43	43.5	42.5			44	43	1829.982 (1)	
		0	43	43	43.5	42.5			43	42	1810.781 (1)	
		0	43	43	43.5	42.5			42	41	1791.671 (1)	
		0	58	58	58.5	57.5			60	59	2473.066 (1)	
		0	58	58	58.5	57.5			59	58	2453.891 (1)	
		0	58	58	58.5	57.5			58	57	2434.776 (1)	
		0	58	58	58.5	57.5			57	56	2415.726 (1)	
		0	100	100	100.5	99.5			102	101	4197.936 (1)	
		0	100	100	100.5	99.5			101	100	4179.155 (1)	
		0	100	100	100.5	99.5			100	99	4160.407 (2)	
		0	100	100	100.5	99.5			99	98	4141.700 (1)	
		0	101	101	101.5	100.5			103	102	4238.488 (2)	82CHI2
		0	101	101	101.5	100.5			102	101	4219.720 (2)	
		0	101	101	101.5	100.5			101	100	4200.986 (2)	
		0	101	101	101.5	100.5			100	99	4182.289 (3)	
		1	3	3	3.5	2.5			4	3	153.615 (2)	82CHI1
		1	4	4	4.5	3.5			5	4	195.413 (2)	
		1	5	5	5.5	4.5			6	5	237.146 (2)	
		1	6	6	6.5	5.5			7	6	278.835 (1)	
		1	6	6	6.5	5.5			7	7	287.749 (2)	

Line frequencies of CaCl (continued)

Isotopic species	State electronic	State vibrational v	Transition N'	N''	J'	J''	F_1'	F_1''	F'	F''	ν MHz	Ref.
$^{40}Ca^{35}Cl$	$X^2\Sigma^+$	1	6	6	6.5	5.5			6	6	268.594 (2)	82CHI1
		1	6	6	6.5	5.5			5	5	249.827 (3)	
		1	8	8	8.5	7.5			10	9	381.620 (2)	
		1	19	19	19.5	18.5			21	20	838.632 (1)	
		1	19	19	19.5	18.5			20	19	819.353 (1)	
		1	19	19	19.5	18.5			19	18	800.227 (1)	
		1	19	19	19.5	18.5			18	17	781.320 (2)	
		1	58	58	58.5	57.5			60	59	2451.004 (5)	
		0	58	58	58.5	57.5			59	58	2432.009 (2)	
		1	58	58	58.5	57.5			58	57	2413.072 (2)	
		1	65	65	65.5	64.5			66	65	2719.402 (2)	
		1	65	65	65.5	64.5			65	64	2700.511 (2)	
$^{40}Ca^{37}Cl$	$X\ ^2\Sigma^+$	0	31	30	30.5	29.5					273722.840 (40)	82MOE, 81SCH
		0	31	30	31.5	30.5					273763.470 (40)	
		0	32	31	31.5	30.5					282529.825 (40)	
		0	32	31	32.5	31.5					282570.413 (40)	
		1	32	31	31.5	30.5					281066.990 (40)	
		1	32	31	32.5	31.5					281107.203 (40)	
$^{40}Ca^{37}Cl$	$X\ ^2\Sigma^+$	0	5	5	5.5	4.5			7	6	247.460 (2)	82CHI1
		0	5	5	5.5	4.5			6	5	230.568 (1)	
		0	5	5	5.5	4.5			5	4	214.236 (1)	
		0	5	5	5.5	4.5			4	3	199.152 (2)	
		0	5	5	5.5	4.5			6	6	238.167 (1)	
		0	5	5	5.5	4.5			5	5	221.747 (1)	
		0	5	5	5.5	4.5			4	4	206.085 (2)	
		0	22	22	22.5	21.5			24	23	939.811 (1)	
		0	22	22	22.5	21.5			23	22	923.538 (2)	
		0	22	22	22.5	21.5			22	21	907.466 (1)	
		0	22	22	22.5	21.5			21	20	891.635 (1)	
		0	48	48	48.5	47.5			49	48	1978.445 (2)	
		0	58	58	58.5	57.5			60	59	2398.234 (1)	
		0	58	58	58.5	57.5			59	58	2382.222 (1)	

Line frequencies of CaCl (continued)

Isotopic species	State		Transition								ν	Ref.
	electronic	vibrational v	N'	N''	J'	J''	F_1'	F_1''	F'	F''	MHz	
$^{40}Ca^{37}Cl$	$X^2\Sigma^+$	0	58	58	58.5	57.5			58	57	2366.296 (1)	82CHI1
		0	58	58	58.5	57.5			57	56	2350.454 (1)	
		0	95	95	95.5	94.5			97	96	3877.100 (2)	
		0	95	95	95.5	94.5			96	95	3861.377 (1)	
		0	95	95	95.5	94.5			95	94	3845.703 (1)	
		0	95	95	95.5	94.5			94	93	3830.081 (1)	

References

77DOM Domaille, P.J., Steimle, T.C., Harris, D.O.: J. Mol. Spectrosc. **66** (1977) 503.

78DOM Domaille, P.J., Steimle, T.C., Harris, D.O.: J. Chem. Phys. **68** (1978) 4977.

77DOM and 78DOM contain additional data on the excited states $B^2\Sigma^+$ and $E^2\Sigma^+$ from laser induced fluorescence studies or optical-optical double resonance where the accuracy is improved by using the MODR-results on the ground state $X^2\Sigma^+$.

81CHI Childs, W.J., Cok, D.R., Goodman, L.S., Poulsen, O.: Phys. Rev. Lett. **47** (1981) 1389.

81SCH Schütze-Pahlmann, H.-U.: Dissertation Freie Universität Berlin, **1981**.

82CHI1 Childs, W.J., Cok, D.R., Goodman, L.S.: J. Chem. Phys. **76** (1982) 3993.

82CHI2 Childs, W.J., Cok, D.R., Goodman, L.S.: J. Opt. Soc. Am. **72** (1982) 717.

82DAG Dagdigian, P.J.: Chem. Phys. Lett. **88** (1982) 225.

82MOE Möller, K., Schütze-Pahlmann, H.-U., Hoeft, J., Törring, T.: Chem. Phys. **68** (1982) 399.

40a CaBr

Molecular constants of CaBr

Isotopic species	State	B_v MHz	D_v kHz	γ_v MHz	Ref.	Meth.	Derived constants
$^{40}Ca^{79}Br$	$X\,^2\Sigma^+$	$Y_{01}=2832.02600\,(63)$ $Y_{11}=-12.09966\,(50)$ $Y_{21}=\ 14.67\,(16)$ kHz	$Y_{02}=-1.23856\,(12)$	$\gamma_e=\ 91.027\,(41)$ $\alpha_\gamma=\ -0.672\,(18)$ $D_\gamma=-88.4\,(57)$ Hz	82MOE	MW MW MW	
				$\gamma_0=\ 90.72452\,(9)$	81CHI1	MB	
				$D_\gamma(v=0)=\sum_{l=0}^{\infty}\gamma_{l1}(\frac{1}{2})^l$ $=-91.11\,(3)$ Hz		MB	
				$\sum_{l=0}^{\infty}\gamma_{l2}(\frac{1}{2})^l$ $=-71.2\,(17)\cdot10^{-6}$ Hz		MB	

Molecular constants of CaBr (continued)

Isotopic species	State	B_v MHz	D_v kHz	γ_v MHz	Ref.	Meth.	Derived constants
$^{40}Ca^{79}Br$	$X\,^2\Sigma^+$			$\gamma_e = \gamma_{00} = 91.05500\,(7)$ [1]	82CHI	MB	
				$\alpha_\gamma = \gamma_{10} = -0.66105\,(5)$ [1]		MB	
				$D_\gamma = \gamma_{01} = -90.89\,(2)$ Hz [1]		MB	
				$\gamma_{11} = -0.402\,(8)$ Hz [1]		MB	
				$\gamma_{02} = -72\,(1) \cdot 10^{-6}$ Hz [1]		MB	
$^{40}Ca^{81}Br$	$X\,^2\Sigma^+$	$Y_{01} = 2808.51983\,(44)$	$Y_{02} = -1.21801\,(8)$	$\gamma_e = 90.307\,(31)$	82MOE	MW	
		$Y_{11} = -11.94898\,(52)$		$\alpha_\gamma = -0.638\,(28)$		MW	
		$Y_{21} = 14.28\,(21)$ kHz		$D_\gamma = -95.0\,(43)$ Hz		MW	
				$\gamma_0 = 89.97275\,(10)$	81CHI1	MB	
				$D_\gamma(v=0) = \sum_{l=0}^{\infty} \gamma_{l1} (\frac{1}{2})^l = -89.58\,(4)$ Hz		MB	
				$\sum_{l=0}^{\infty} \gamma_{l2} (\frac{1}{2})^l = -70.9\,(28) \cdot 10^{-6}$ Hz		MB	

[1]) Combined fit of isotopes $^{40}Ca^{79}Br$ and $^{40}Ca^{81}Br$.

Isotopic species	State		Parameter	Value	Ref.	Meth.	Remarks
	electronic	vibrational					
$^{40}Ca^{79}Br$	$X\,^2\Sigma^+$	$v=0$	$b_0\,(^{79}Br)$	95.3286 (9) MHz	81CHI1	MB	
			$\sum_{l=0}^{\infty} b^{l1} (\frac{1}{2})^l\,(^{79}Br)$	−205.4 (3) Hz		MB	
			$b^{10}\,(^{79}Br)$	−2.3 (7) MHz			derived with mass relations between $b_0\,(^{79}Br)$ and $b_0\,(^{81}Br)$.
			$c_0\,(^{79}Br)$	77.620 (7) MHz		MB	
			$C_I\,(^{79}Br)$	4.1 (2) kHz			
			$eqQ\,(^{79}Br)$	20.015 (7) MHz		MB	

Molecular constants of CaBr (continued)

Isotopic species	State electronic	State vibrational	Parameter	Value	Ref.	Meth.	Remarks
$^{40}Ca^{79}Br$	$X\,^2\Sigma^+$	equ.	b^{00} (^{79}Br)	95.7053 (9) MHz	82CHI	MB	parameters were obtained by a combined fit of isotope $^{40}Ca^{79}Br$ and $^{40}Ca^{81}Br$ using appropriate relations of the reduced masses and nuclear moments; but the necessary scaling factor of b^{00} is different to the expected ratio of the nuclear magnetic moments
			b^{10} (^{79}Br)	−0.7524 (7) MHz		MB	
			b^{01} (^{79}Br)	−205.4 (1) Hz		MB	
			c^{00} (^{79}Br)	77.546 (6) MHz		MB	
			c^{10} (^{79}Br)	0.324 (5) MHz		MB	
			eq_eQ (^{79}Br)	19.262 (6) MHz		MB	
			eq_1Q (^{79}Br)	1.482 (5) MHz		MB	
			C_I (^{79}Br)	5.19 (7) kHz		MB	
$^{40}Ca^{81}Br$	$X\,^2\Sigma^+$	$v=0$	b_0 (^{81}Br)	102.7637 (11) MHz	81CHI1	MB	
			$\sum_{l=0}^{\infty} b^{l1}(\frac{1}{2})^l$ (^{81}Br)	−219.7 (3) Hz		MB	
			c_0 (^{81}Br)	83.660 (9) MHz		MB	
			C_I (^{81}Br)	4.4 (2) kHz		MB	
			eqQ (^{81}Br)	16.714 (6) MHz		MB	
$^{40}Ca^{79}Br$	$B\,^2\Sigma^+$	$v=0$	b_0 (^{79}Br)	11.8 (4) MHz	81CHI2	LIF/MB	
			eqQ (^{79}Br)	30.4 (10) MHz		LIF/MB	
			$\|c_0\|$ (^{79}Br)	≦7 MHz		LIF/MB	
$^{40}Ca^{81}Br$	$B\,^2\Sigma^+$	$v=0$	b_0 (^{81}Br)	13.3 (9) MHz		LIF/MB	
			eqQ (^{81}Br)	21.8 (23) MHz		LIF/MB	
			$\|c_0\|$ (^{81}Br)	≦17 MHz		LIF/MB	
$^{40}Ca^{79}Br$	$B\,^2\Sigma^+$	$v=0$	γ_0	−2068.2 (11) MHz	81BER2	LIF	
$^{40}Ca^{81}Br$	$B\,^2\Sigma^+$	$v=0$	γ_0	−2052 (1) MHz	81BER1, 81BER2	LIF	

40a CaBr
Line frequencies of CaBr

Isotopic species	State		Transition								Frequency MHz	Ref.
	electronic	vibrational v	N'	N''	J'	J''	F_1'	F_1''	F'	F''		
$^{40}Ca^{79}Br$	$X\,^2\Sigma^+$	0	16	15	16.5	15.5					90456.148 (60)	82MOE
		0	16	15	15.5	14.5					90366.017 (60)	
		0	48	47	48.5	47.5					270791.213 (40)	
		0	48	47	47.5	46.5					270701.090 (40)	
		0	49	48	49.5	48.5					276408.202 (40)	
		0	49	48	48.5	47.5					276318.205 (40)	
		0	50	49	50.5	49.5					282023.725 (40)	
		0	50	49	49.5	48.5					281933.749 (40)	
		0	51	50	51.5	50.5					287637.711 (40)	
		0	51	50	50.5	49.5					287547.720 (40)	
		1	16	15	16.5	15.5					90069.524 (60)	
		1	16	15	15.5	14.5					89980.093 (60)	
		1	49	48	49.5	48.5					275224.956 (40)	
		1	49	48	48.5	47.5					275135.597 (40)	
		1	50	49	50.5	49.5					280816.355 (40)	
		1	50	49	49.5	48.5					280727.054 (40)	
		1	51	50	51.5	50.5					286406.198 (40)	
		1	51	50	50.5	49.5					286316.933 (40)	
		2	49	48	49.5	48.5					274044.595 (40)	
		2	49	48	48.5	47.5					273955.924 (40)	
		2	50	49	50.5	49.5					279611.936 (40)	
		2	50	49	49.5	48.5					279523.236 (40)	
		2	51	50	51.5	50.5					285177.721 (40)	
		2	51	50	50.5	49.5					285089.074 (40)	
$^{40}Ca^{79}Br$	$X\,^2\Sigma^+$	0	7	7	7.5	6.5			9	8	832.920 (2)	81CHI1
		0	7	7	7.5	6.5			8	7	751.380 (2)	
		0	7	7	7.5	6.5			7	6	656.616 (2)	
		0	7	7	7.5	6.5			6	5	548.224 (2)	
		0	7	7	6.5	6.5			5	6	50.442 (1)	
		0	7	7	6.5	6.5			6	7	46.948 (4)	
		0	7	7	6.5	6.5			7	8	46.114 (2)	
		0	12	12	12.5	11.5			13	12	1195.735 (2)	

Line frequencies of CaBr (continued)

Isotopic species	State: electronic	State: vibrational v	N'	N''	J'	J''	F'_1	F''_1	F'	F''	Frequency MHz	Ref.
$^{40}Ca^{79}Br$	$X^2\Sigma^+$	0	12	12	12.5	11.5			12	11	1100.665 (2)	81CHI1
		0	12	12	12.5	11.5			11	10	997.316 (2)	
		0	20	20	20.5	19.5			22	21	2005.605 (2)	
		0	20	20	20.5	19.5			21	20	1915.414 (2)	
		0	20	20	20.5	19.5			20	19	1820.253 (2)	
		0	20	20	20.5	19.5			19	18	1720.050 (2)	
		0	23	23	23.5	22.5			25	24	2276.880 (4)	
		0	64	64	64.5	63.5			66	65	5970.094 (2)	
		0	64	64	64.5	63.5			65	64	5877.212 (2)	
		0	64	64	64.5	63.5			64	63	5782.746 (2)	
		0	64	64	64.5	63.5			63	62	5686.690 (2)	
		0	100	100	100.5	99.5			100	99	8979.740 (6)	
		0	100	100	100.5	99.5			99	98	8885.482 (6)	
		0	120	120	120.5	119.5			122	121	10910.227 (5)	
$^{40}Ca^{79}Br$	$X^2\Sigma^+$	1	8	8	8.5	7.5			10	9	915.871 (1)	82CHI
		1	8	8	8.5	7.5			9	8	833.537 (1)	
		1	8	8	8.5	7.5			8	7	739.422 (1)	
		1	8	8	8.5	7.5			7	6	633.176 (1)	
		1	8	8	7.5	7.5			7	8	46.650 (1)	
		1	8	8	7.5	7.5			8	9	46.875 (1)	
		1	8	8	7.5	7.5			6	7	48.811 (1)	
		1	22	22	22.5	21.5			24	23	2170.460 (1)	
		1	22	22	22.5	21.5			23	22	2080.609 (1)	
		1	22	22	22.5	21.5			22	21	1986.206 (1)	
		1	22	22	22.5	21.5			21	20	1887.189 (1)	
		1	49	49	49.5	48.5			51	50	4589.675 (2)	
		1	49	49	49.5	48.5			50	49	4497.692 (2)	
		1	49	49	49.5	48.5			49	48	4403.633 (2)	
		1	49	49	49.5	48.5			48	47	4307.484 (2)	
		1	88	88	88.5	87.5			90	89	8047.112 (2)	
		1	88	88	88.5	87.5			89	88	7955.303 (2)	
		1	88	88	88.5	87.5			88	87	7862.339 (2)	
		1	88	88	88.5	87.5			87	86	7768.215 (2)	

Line frequencies of CaBr (continued)

Isotopic species	State: electronic	State: vibrational v	Transition: N'	N''	J'	J''	F'_1	F''_1	F'	F''	Frequency MHz	Ref.
$^{40}Ca^{81}Br$	$X\,^2\Sigma^+$	0	17	16	17.5	16.5					95307.427 (60)	82MOE
		0	17	16	16.5	15.5					95218.062 (60)	
		0	19	18	19.5	18.5					106508.175 (60)	
		0	19	18	18.5	17.5					106418.731 (60)	
		0	49	48	49.5	48.5					274121.170 (40)	
		0	49	48	48.5	47.5					274031.995 (40)	
		0	50	49	50.5	49.5					279690.529 (40)	
		0	50	49	49.5	48.5					279601.298 (40)	
		0	51	50	51.5	50.5					285258.314 (40)	
		0	51	50	50.5	49.5					285169.101 (40)	
		1	19	18	19.5	18.5					106054.865 (60)	
		1	49	48	49.5	48.5					272952.716 (40)	
		1	49	48	48.5	47.5					272864.097 (40)	
		1	50	49	50.5	49.5					278498.132 (40)	
		1	50	49	49.5	48.5					278409.567 (40)	
		2	50	49	50.5	49.5					277308.630 (40)	
$^{40}Ca^{81}Br$	$X\,^2\Sigma^+$	0	7	7	7.5	6.5			9	8	840.730 (3)	81CHI1
		0	7	7	7.5	6.5			8	7	755.172 (4)	
		0	7	7	7.5	6.5			7	6	653.570 (4)	
		0	7	7	7.5	6.5			6	5	534.380 (5)	
		0	7	7	7.5	7.5			9	8	37.856 (2)	
		0	7	7	7.5	7.5			8	7	50.500 (2)	
		0	7	7	7.5	7.5			7	6	61.492 (2)	
		0	7	7	6.5	6.5			5	6	57.694 (3)	
		0	7	7	6.5	6.5			6	7	51.102 (2)	
		0	7	7	6.5	6.5			7	8	47.705 (2)	
		0	23	23	23.5	22.5			25	24	2271.182 (3)	
		0	23	23	23.5	22.5			24	23	2174.076 (3)	
		0	23	23	23.5	22.5			23	22	2071.563 (2)	
		0	23	23	23.5	22.5			22	21	1963.461 (2)	
		0	64	64	64.5	63.5			66	65	5933.379 (3)	
		0	64	64	64.5	63.5			65	64	5833.534 (3)	
		0	64	64	64.5	63.5			64	63	5731.705 (3)	

Line frequencies of CaBr (continued)

Isotopic species	State electronic	State vibrational v	Transition N'	N''	J'	J''	F_1'	F_1''	F'	F''	Frequency MHz	Ref.
$^{40}Ca^{81}Br$	$X^2\Sigma^+$	0	64	64	64.5	63.5			63	62	5627.860 (3)	81CHI1
		0	100	100	100.5	99.5			102	101	9102.365 (6)	
		0	100	100	100.5	99.5			101	100	9003.098 (5)	
		0	100	100	100.5	99.5			100	99	8902.561 (4)	
		0	100	100	100.5	99.5			99	98	8800.745 (3)	

References

81BER1 Bernath, P.F., Pinchemel, B., Field, R.W.: J. Chem. Phys. **74** (1981) 5508.

81BER2 Bernath, P.F., Pinchemel, B., Field, R.W., Möller, K., Törring, T.: J. Mol. Spectrosc. **88** (1981) 420.

81CHI1 Childs, W.J., Cok, D.R., Goodman, G.L., Goodman, L.S.: J. Chem. Phys. **75** (1981) 501.

81CHI2 Childs, W.J., Cok, D.R., Goodman, L.S.: Can. J. Phys. **59** (1981) 1308.

82CHI Childs, W.J., Cok, D.R., Goodman, L.S.: J. Mol. Spectrosc. **95** (1982) 153.

82MOE Möller, K., Schütze-Pahlmann, H.-U., Hoeft, J., Törring, T.: Chem. Phys. **68** (1982) 399.

40b CaI

Molecular constants of CaI

Isotopic species	State electronic	State vibrational	Parameter	Value	Ref.	Meth.	Remarks
$^{40}Ca^{127}I$	$X^2\Sigma^+$	$v=0$	$b_0(^{127}I)$	116 (1) MHz	81BER	SLS	derived from the observed hyperfine structure by setting $c_0(^{127}I)=0$
			γ_0	168 (1) MHz		SLS	
	$B^2\Sigma^+$	$v=0$	$b_0(^{127}I)$	18 (1) MHz		SLS	derived from the observed hyperfine structure by setting $c_0(^{127}I)=0$
			$\Delta\, eqQ(^{127}I)$	−66 (3) MHz		SLS	$\Delta\, eqQ = eqQ(B^2\Sigma^+) - eqQ(X^2\Sigma^+)$ systematic uncertainty due to neglect of magnetic hyperfine interaction
			γ_0	−4202 (10) MHz		SLS	

Reference 81BER Bernath, P.F., Pinchemel, B., Field, R.W.: J. Chem. Phys. **74** (1981) 5508.

41 SrF

Molecular constants of SrF

Isotopic species	State	B_v MHz	D_v kHz	γ_v MHz	Ref.	Meth.	Derived constants
$^{88}Sr^{19}F$	$X\,^2\Sigma^+$	$Y_{01} = 7510.8311$ (62)	$Y_{02} = -7.4809$ (82)	$\gamma_e = 75.030$ (81)	82SCH	MW	
		$Y_{11} = -46.5014$ (60)	$Y_{12} = -9.5$ (82) Hz	$\alpha_\gamma = -0.504$ (48)		MW	
		$Y_{21} = 67.6$ (14) kHz		$D_\gamma = 0.122$ (81) kHz		MW	
				$\gamma_e = \gamma_{00} = 75.02249$ (7)	81CHI	MB	
				$\alpha_\gamma = \gamma_{10} = -0.45528$ (5)		MB	
				$D_\gamma = \gamma_{01} = 59.38$ (6) Hz		MB	
				$\gamma_{02} = -0.63$ (1) $\cdot 10^{-3}$ Hz		MB	
				$\gamma_{11} = -3.73$ (3) Hz		MB	
					81SCH	MW	$\omega_e = 501.99$ (33) cm^{-1} $r_e = 2.075391$ (7) Å
							$\omega_e x_e = 2.225$ (73) cm^{-1}
						MW	$a_0 = 251462$ (326) cm^{-1}
							$a_1 = -3.06755$ (29)
							$a_2 = 5.84$ (19)
							$a_3 = -7.6$ (14)
					77DOM	MODR	
$^{86}Sr^{19}F$	$X\,^2\Sigma^+$			$\gamma_{00} = 75.33264$ (5)	81CHI	MB	
				$\gamma_{10} = -0.45825$ (4)		MB	
				$\gamma_{01} = 59.47$ (7) Hz		MB	
				$\gamma_{02} = -0.57$ (1) $\cdot 10^{-3}$ Hz		MB	
				$\gamma_{11} = -3.63$ (2) Hz		MB	

Isotopic species	State		Parameter	Value	Ref.	Meth.	Remarks
	electronic	vibrational					
$^{88}Sr^{19}F$	$X\,^2\Sigma^+$	equ.	b^{00} (^{19}F)	97.6670 (10) MHz	81CHI	MB	
			b^{01} (^{19}F)	−330.0 (4) Hz		MB	
			b^{10} (^{19}F)	−1.1672 (8) MHz		MB	
			c^{00} (^{19}F)	29.846 (8) MHz		MB	
			c^{10} (^{19}F)	0.843 (7) MHz		MB	
			C_I (^{19}F)	2.30 (1) kHz		MB	

Molecular constants of SrF (continued)

Isotopic species	State: electronic	State: vibrational	Parameter	Value	Ref.	Meth.	Remarks
$^{86}Sr^{19}F$	$X\,^2\Sigma^+$	equ.	b^{00} (^{19}F)	97.6688 (6) MHz	81CHI	MB	
			b^{01} (^{19}F)	−330.3 (2) Hz		MB	
			b^{10} (^{19}F)	−1.1719 (5) MHz		MB	
			c^{00} (^{19}F)	29.81 (2) MHz		MB	
			c^{10} (^{19}F)	0.89 (3) MHz		MB	
			C_I (^{19}F)	2.36 (8) kHz		MB	
$^{88}Sr^{19}F$	$B\,^2\Sigma^+$		several hyperfine splittings of $v=0$ are reported in [81CHI]				

Line frequencies of SrF

Isotopic species	State: electronic	State: vibrational v	Transition: N'	N''	J'	J''	F_1'	F_1''	F'	F''	ν MHz	Ref.
$^{88}Sr^{19}F$	$X\,^2\Sigma^+$	0	2	1	1.5	0.5					29916.3 (30)	77DOM
		0	2	1	2.5	1.5					29986.2 (14)	
		0	3	2	2.5	1.5					44889.1 (26)	
		0	3	2	3.5	2.5					44961.3 (27)	
		0	4	3	3.5	2.5					59862.0 (25)	
		0	4	3	4.5	3.5					59935.5 (13)	
		1	3	2	3.5	2.5					44684.6 (53)	
		1	4	3	3.5	2.5					59491.8 (30)	
		1	4	3	4.5	3.5					59564.2 (41)	
$^{88}Sr^{19}F$	$X\,^2\Sigma^+$	0	6	5	5.5	4.5					89807.615 (40)	82SCH
		0	6	5	6.5	5.5					89881.845 (40)	
		1	6	5	5.5	4.5					89251.388 (40)	
		1	6	5	6.5	5.5					89325.245 (40)	
		0	18	17	17.5	16.5					269341.425 (40)	
		0	18	17	18.5	17.5					269416.260 (40)	
		0	19	18	18.5	17.5					284285.880 (40)	
		0	19	18	19.5	18.5					284360.819 (40)	
		1	19	18	18.5	17.5					282524.008 (40)	
		1	19	18	19.5	18.5					282598.278 (40)	
		2	19	18	18.5	17.5					280767.170 (40)	
		2	19	18	19.5	18.5					280841.035 (40)	

Line frequencies of SrF (continued)

Isotopic species	State		Transition								ν MHz	Ref.
	electronic	vibrational v	N'	N''	J'	J''	F'_1	F''_1	F'	F''		
$^{88}Sr^{19}F$	$X\,^2\Sigma^+$	0	3	3	3.5	2.5			4	3	315.115 (1)	81CHI
		0	3	3	3.5	2.5			3	2	218.709 (1)	
		0	3	3	3.5	3.5			4	3	45.510 (1)	
		0	3	3	3.5	2.5			3	3	269.605 (1)	
		0	6	6	6.5	5.5			7	6	537.371 (1)	
		0	6	6	6.5	5.5			6	5	440.485 (1)	
		0	6	6	6.5	6.5			7	6	47.040 (1)	
		0	9	9	9.5	8.5			10	9	760.960 (1)	
		0	15	15	15.5	14.5			16	15	1209.168 (1)	
		0	15	15	15.5	14.5			15	14	1112.194 (1)	
		0	30	30	30.5	29.5			31	30	2331.819 (1)	
		0	30	30	30.5	29.5			30	29	2235.051 (1)	
		0	40	40	40.5	39.5			41	40	3081.644 (1)	
		0	40	40	40.5	39.5			40	39	2985.107 (1)	
		0	60	60	60.5	59.5			61	60	4585.530 (1)	
		0	60	60	60.5	59.5			60	59	4489.656 (1)	
		1	3	3	3.5	2.5			4	3	312.928 (1)	
		1	3	3	3.5	2.5			3	2	217.707 (1)	
		1	3	3	3.5	3.5			4	3	44.981 (1)	
		1	3	3	3.5	2.5			3	3	267.947 (1)	
		1	6	6	6.5	5.5			7	6	533.823 (1)	
		1	6	6	6.5	5.5			6	5	438.109 (1)	
		1	15	15	15.5	14.5			16	15	1201.513 (1)	
		1	15	15	15.5	14.5			15	14	1105.708 (1)	
		1	30	30	30.5	29.5			31	30	2317.244 (1)	
		1	30	30	30.5	29.5			30	29	2221.642 (1)	
		1	45	45	45.5	44.5			46	45	3435.390 (1)	
		1	45	45	45.5	44.5			45	44	3340.161 (1)	
$^{86}Sr^{19}F$	$X\,^2\Sigma^+$	0	19	18	18.5	17.5					285456.963 (40)	81SCH
		0	19	18	19.5	18.5					285532.222 (40)	
$^{86}Sr^{19}F$	$X\,^2\Sigma^+$	0	6	6	6.5	5.5			7	6	539.349 (1)	81CHI
		0	6	6	6.5	5.5			6	5	442.464 (1)	
		0	6	6	6.5	6.5			7	6	47.070 (1)	

Line frequencies of SrF (continued)

Isotopic species	State electronic	State vibrational v	Transition N'	N''	J'	J''	F_1'	F_1''	F'	F''	ν MHz	Ref.
$^{86}Sr^{19}F$	$X^2\Sigma^+$	0	15	15	15.5	14.5			16	15	1213.940 (1)	81CHI
		0	15	15	15.5	14.5			15	14	1116.968 (1)	
		0	30	30	30.5	29.5			31	30	2341.234 (1)	
		0	30	30	30.5	29.5			30	29	2244.466 (1)	
		0	60	60	60.5	59.5			61	60	4604.284 (1)	
		0	60	60	60.5	59.5			60	59	4508.412 (1)	
		1	15	15	15.5	14.5			16	15	1206.238 (1)	
		1	15	15	15.5	14.5			15	14	1110.438 (1)	
		1	30	30	30.5	29.5			31	30	2326.568 (1)	
		1	30	30	30.5	29.5			30	29	2230.973 (1)	
		1	45	45	45.5	44.5			46	45	3449.327 (1)	
		1	45	45	45.5	44.5			45	44	3354.104 (1)	

77DOM, 78STE contain additional data on the excited states $B\,^2\Sigma^+$ and $A\,^2\Pi$ where the accuracy is increased using the results of the MODR-experiment.

References

77DOM Domaille, P.J., Steimle, T.C., Harris, D.O.: J. Mol. Spectrosc. **68** (1977) 146.

78STE Steimle, T.C., Domaille, P.J., Harris, D.O.: J. Mol. Spectrosc. **73** (1978) 441.

81CHI Childs, W.J., Goodman, L.S., Renhorn, I.: J. Mol. Spectrosc. **87** (1981) 522.

81SCH Schütze-Pahlmann, H.-U.: Dissertation Freie Universität Berlin, **1981**.

82SCH Schütze-Pahlmann, H.-U., Ryzlewicz, Ch., Hoeft, J., Törring, T.: Chem. Phys. Lett. **93** (1982) 74.

42 SrCl

Molecular constants of SrCl

Isotopic species	State	B_v MHz	D_v kHz	γ_v MHz	Ref.	Meth.	Derived constants
$^{88}Sr^{35}Cl$	$X\,^2\Sigma^+$	$Y_{01}=3044.65714\,(88)$	$Y_{02}=-1.37275\,(21)$	$\gamma_e=52.930\,(31)$	82SCH	MW	
		$Y_{11}=-13.5518\,(13)$	$Y_{12}=-1.68\,(28)$ Hz	$\alpha_\gamma=-0.4173\,(81)$		MW	
		$Y_{21}=13.99\,(44)$ kHz		$D_\gamma=-23.2\,(54)$ Hz		MW	
		$Y_{31}=107\,(73)$ Hz				MW	
					81SCH	MW	$\omega_e=302.630\,(55)\ cm^{-1}$ $r_e=2.5758842\,(89)$ Å
							$\omega_e x_e=0.9660\,(91)\ cm^{-1}$
						MW	$a_0=225438\,(82)\ cm^{-1}$, $a_1=-3.21090\,(25)$
							$a_2=6.546\,(60)$, $a_3=-9.75\,(46)$

Molecular constants of SrCl (continued)

Isotopic species	State	B_v MHz	D_v kHz	γ_v MHz	Ref.	Meth.	Derived constants
$^{88}Sr^{37}Cl$	$X\,^2\Sigma^+$	$Y_{01}=2926.98736\,(90)$			82SCH	MW	
$^{86}Sr^{35}Cl$	$X\,^2\Sigma^+$	$Y_{01}=3064.79184\,(90)$			82SCH	MW	

Line frequencies of SrCl

Isotopic species	State		Transition								ν MHz	Ref.
	electronic	vibrational v	N'	N''	J'	J''	F_1'	F_1''	F'	F''		
$^{88}Sr^{35}Cl$	$X\,^2\Sigma^+$	0	15	14	14.5	13.5					91091.640 (40)	82SCH
		0	15	14	15.5	14.5					91144.345 (40)	
		0	45	44	44.5	43.5					272882.663 (40)	
		0	45	44	45.5	44.4					272935.270 (40)	
		1	45	44	44.5	43.5					271665.158 (40)	
		1	45	44	45.5	44.5					271717.313 (40)	
		0	46	45	45.5	44.5					278924.333 (40)	
		0	46	45	46.5	45.5					278976.893 (40)	
		1	46	45	45.5	44.5					277679.730 (40)	
		1	46	45	46.5	45.5					277731.883 (40)	
		2	46	45	45.5	44.5					276437.753 (40)	
		2	46	45	46.5	45.5					276489.500 (40)	
		3	46	45	45.5	44.5					275198.553 (40)	
		3	46	45	46.5	45.5					275249.870 (40)	
		0	47	46	46.5	45.5					284964.415 (40)	
		0	47	46	47.5	46.5					285016.998 (40)	
		1	47	46	46.5	45.5					283692.745 (40)	
		1	47	46	47.5	46.5					283744.863 (40)	
		2	47	46	46.5	45.5					282423.788 (40)	
		2	47	46	47.5	46.5					282475.523 (40)	
		3	47	46	46.5	45.5					281157.578 (40)	
		3	47	46	47.5	46.5					281208.908 (40)	
$^{88}Sr^{37}Cl$	$X\,^2\Sigma^+$	0	47	46	46.5	45.5					273984.295 (40)	82SCH
		0	47	46	47.5	46.5					274034.830 (40)	
		1	47	46	46.5	45.5					272785.575 (40)	

Line frequencies of SrCl (continued)

Isotopic species	State: electronic	State: vibrational v	Transition N'	N''	J'	J''	F_1' – F_1''	F' – F''	ν MHz	Ref.
$^{88}Sr^{37}Cl$	$X^2\Sigma^+$	1	47	46	47.5	46.5			272835.700 (40)	82SCH
		0	48	47	47.5	46.5			279791.113 (40)	
		0	48	47	48.5	47.5			279841.658 (40)	
		1	48	47	47.5	46.5			278566.913 (40)	
		1	48	47	48.5	47.5			278616.998 (40)	
		2	48	47	47.5	46.5			277345.233 (40)	
		2	48	47	48.5	47.5			277394.960 (40)	
		0	49	48	48.5	47.5			285596.458 (40)	
		0	49	48	49.5	48.5			285646.978 (40)	
		1	49	48	48.5	47.5			284346.728 (40)	
		1	49	48	49.5	48.5			284396.843 (40)	
$^{86}Sr^{35}Cl$	$X\,^2\Sigma^+$	0	46	45	45.5	44.5			280763.238 (40)	82SCH
		0	46	45	46.5	45.5			280816.210 (40)	
		0	47	46	46.5	45.5			286842.993 (40)	
		0	47	46	47.5	46.5			286895.948 (40)	

References

81SCH Schütze-Pahlmann, H.-U.: Dissertation Freie Universität Berlin **1981**.

82SCH Schütze-Pahlmann, H.-U., Ryzlewicz, Ch., Hoeft, J., Törring, T.: Chem. Phys. Lett. **93** (1982) 74.

43 BaF

Molecular constants of BaF

Isotopic species	State	B_v MHz	D_v kHz	γ_v MHz	Ref.	Meth.	Derived constants
$^{138}Ba^{19}F$	$X\,^2\Sigma^+$	$Y_{01}=6491.3962\,(11)$ $Y_{11}=-34.8831\,(11)$ $Y_{21}=15.93\,(22)$ kHz	$Y_{02}=-5.5250\,(11)$ $Y_{12}=-9.43\,(95)$ Hz	$\gamma_e=$ 80.984 (19) $\alpha_\gamma=-58.4\,(73)$ kHz $D_\gamma=$ 112 (17) Hz	80RYZ	MW	$\omega_e=469.4\,(1)$ cm^{-1} $r_e=2.1592964\,(75)$ Å $\omega_e x_e=1.830\,(1)$ cm^{-1} $a_0=2.54404\,(50)\cdot 10^5$ cm^{-1} $a_1=-2.94159\,(53)$ $a_2=5.18267\,(50)$ $a_3=-6.060\,(12)$
$^{137}Ba^{19}F$	$X^2\Sigma^+$	$B_0=6479.6773\,(32)$	$D_0=5.544\,(3)$	$\gamma_0=81.03\,(22)$	82RYZ	MW	
$^{136}Ba^{19}F$	$X\,^2\Sigma^+$	$Y_{01}=6502.9661\,(11)$		$\gamma_e=81.152\,(43)$	80RYZ	MW	

Molecular constants of BaF (continued)

Isotopic species	State		Parameter	Value	Ref.	Meth.	Remarks
	electronic	vibrational					
$^{137}Ba^{19}F$	$X\,^2\Sigma^+$	$v=0$	$b_0\,(^{137}Ba)$	2301 (9) MHz	82RYZ	MW	
			$c_0\,(^{137}Ba)$	75 (6) MHz		MW	
			$eq_0Q\,(^{137}Ba)$	−117 (12) MHz		MW	
			$b_0\,(^{19}F)$	60 (6) MHz		MW	

Line frequencies of BaF

Isotopic species	State		Transition								ν MHz	Ref.
	electronic	vibrational v	N'	N''	J'	J''	F_1'	F_1''	F'	F''		
$^{138}Ba^{19}F$	$X\,^2\Sigma^+$	0	6	5	5.5	4.5					77642.243 (45)	80RYZ
		0	6	5	6.5	5.5					77723.196 (45)	
		1	6	5	5.5	4.5					77224.046 (45)	
		1	6	5	6.5	5.5					77304.977 (45)	
		2	6	5	5.5	4.5					76806.273 (45)	
		2	6	5	6.5	5.5					76887.125 (45)	
		0	7	6	6.5	5.5					90587.334 (45)	
		0	7	6	7.5	6.5					90668.335 (45)	
		1	7	6	6.5	5.5					90099.441 (45)	
		1	7	6	7.5	6.5					90180.360 (45)	
		2	7	6	6.5	5.5					89611.978 (45)	
		2	7	6	7.5	6.5					89692.822 (45)	
		3	7	6	6.5	5.5					89124.962 (45)	
		3	7	6	7.5	6.5					89205.744 (45)	
		4	7	6	6.5	5.5					88638.430 (45)	
		4	7	6	7.5	6.5					88719.170 (45)	
		0	8	7	7.5	6.5					103531.537 (45)	
		0	8	7	8.5	7.5					103612.483 (45)	
		0	21	20	20.5	19.5					271660.865 (45)	
		0	21	20	21.5	20.5					271741.963 (45)	
		0	22	21	21.5	20.5					284578.098 (45)	
		0	22	21	22.5	21.5					284659.205 (45)	
		1	21	20	20.5	19.5					270196.783 (45)	
		1	21	20	21.5	20.5					270277.843 (45)	

Line frequencies of BaF (continued)

Isotopic species	State: electronic	State: vibrational v	Transition: N'	N''	J'	J''	F'_1	F''_1	F'	F''	ν MHz	Ref.
$^{137}Ba^{19}F$	$X\,^2\Sigma^+$	0	8	7	[1]		6	5	6.5	5.5	103622.922 (30)	82RYZ
		0	8	7			6	5	5.5	4.5		
		0	8	7			9	8	9.5	8.5	103639.592 (30)	
		0	8	7			9	8	8.5	7.5	103639.892 (30)	
		0	8	7			7	6	6.5	5.5	103648.221 (30)	
		0	8	7			7	6	7.5	6.5	103648.521 (30)	
		0	8	7			9	8	8.5	7.5	103687.323 (30)	
		0	8	7			9	8	9.5	8.5		
		0	8	7			10	9	10.5	9.5	103703.963 (30)	
		0	8	7			10	9	9.5	8.5		
		0	22	21			20	19	20.5	19.5	284829.132 (30)	
		0	22	21			20	19	19.5	18.5		
		0	22	21			23	22	23.5	22.5	284840.730 (30)	
		0	22	21			23	22	22.5	21.5		
		0	22	21			22	21	22.5	21.5	284854.732 (30)	
		0	22	21			22	21	21.5	20.5	284855.132 (30)	
		0	22	21			21	20	20.5	19.5	284864.420 (30)	
		0	22	21			21	20	21.5	20.5	284864.920 (30)	
		0	22	21			22	21	21.5	20.5	284884.170 (30)	
		0	22	21			22	21	22.5	21.5	284884.620 (30)	
		0	22	21			23	22	22.5	21.5	284898.434 (30)	
		0	22	21			23	22	23.5	22.5	284898.814 (30)	
		0	22	21			24	23	24.5	23.5	284910.218 (30)	
		0	22	21			24	23	23.5	22.5		
$^{136}Ba^{19}F$	$X\,^2\Sigma^+$	0	7	6	6.5	5.5					90748.544 (45)	80RYZ
		0	7	6	7.5	6.5					90829.628 (45)	
		0	21	20	21.5	20.5					272225.313 (45)	
		0	22	21	21.5	20.5					285084.179 (45)	
		0	22	21	22.5	21.5					285165.499 (45)	

[1]) [82RYZ] does not specify J quantum numbers because of intermediate coupling case between Hund's case $b_{\beta S}$ and $b_{\beta J}$; but rotational states are not fully specified.

References
80RYZ Ryzlewicz, Ch., Törring, T.: Chem. Phys. **51** (1980) 329.
82RYZ Ryzlewicz, Ch., Schütze-Pahlmann, H.-U., Hoeft, J., Törring, T.: Chem. Phys. **71** (1982) 389.

44 BaCl

Molecular constants of BaCl

Isotopic species	State	B_v MHz	D_v kHz	γ_v MHz	Ref.	Meth.	Derived constants	
$^{138}Ba^{35}Cl$	$X\,^2\Sigma^+$	$Y_{01} = 2517.27236\,(21)$ $Y_{11} = -10.02186\,(18)$ $Y_{21} = 6.218\,(68)$ kHz $Y_{31} = 0.0411\,(95)$ kHz	$Y_{02} = -0.906041\,(32)$ $Y_{12} = -1.124\,(19)$ Hz $Y_{03}^{ev} = -0.69 \cdot 10^{-4}$ Hz	$\gamma_e = 49.7183\,(98)$ $\alpha_\gamma = -0.2650\,(25)$ $D_\gamma = 0.0371\,(11)$ kHz	81SCH, 82RYZ	MW	$\omega_e = 279.9179\,(49)$ cm^{-1} $\omega_e x_e = 0.8170\,(7)$ cm^{-1}	$r_e = 2.682764\,(10)$ Å
					82RYZ	MW	$a_0 = 2.332877\,(72) \cdot 10^{+5}$ cm^{-1} $a_1 = -3.212015\,(60)$ $a_2 = 6.4100\,(68)$ $a_3 = -9.114\,(64)$	
$^{138}Ba^{37}Cl$	$X\,^2\Sigma^+$	$Y_{01} = 2408.79162\,(43)$ $Y_{11} = -9.38082\,(16)$	$Y_{02} = -0.829604\,(60)$	$\gamma_e = 47.553\,(34)$ $\alpha_\gamma = -0.217\,(33)$	81SCH	MW		$r_e = 2.682762\,(10)$ Å

Isotopic species	State		Parameter	Value	Ref.	Meth.	Remarks
	electronic	vibrational					
$^{137}Ba^{35}Cl$	$X\,^2\Sigma^+$	$v=0$	$b_0\,(^{137}Ba)$ $c_0\,(^{137}Ba)$ $eq_0Q\,(^{137}Ba)$	2314 (9) MHz 96 (20) MHz −134 (42) MHz	82RYZ	MW MW MW	

Line frequencies of BaCl

Isotopic species	State		Transition								ν MHz	Ref.
	electronic	vibrational v	N'	N''	J'	J''	F_1'	F_1''	F'	F''		
$^{138}Ba^{35}Cl$	$X\,^2\Sigma^+$	0	21	20	20.5	19.5					105456.630 (30)	81SCH, 82RYZ
		0	21	20	21.5	20.5					105506.283 (30)	
		1	21	20	21.5	20.5					105085.705 (30)	
		2	21	20	20.5	19.5					104616.590 (30)	
		2	21	20	21.5	20.5					104665.685 (30)	

Line frequencies of BaCl (continued)

Isotopic species	State		Transition								ν MHz	Ref.
	electronic	vibrational v	N′	N″	J′	J″	F_1'	F_1''	F′	F″		
$^{138}Ba^{35}Cl$	$X^2\Sigma^+$	0	54	53	53.5	52.5					270728.238 (30)	81SCH, 82RYZ
		0	54	53	54.5	53.5					270778.138 (30)	
		0	55	54	54.5	53.5					275720.420 (30)	
		0	55	54	55.5	54.5					275770.340 (30)	
		1	55	54	54.5	53.5					274618.778 (30)	
		1	55	54	55.5	54.5					274668.440 (30)	
		2	55	54	54.5	53.5					273518.538 (30)	
		2	55	54	55.5	54.5					273567.930 (30)	
		0	56	55	55.5	54.5					280711.405 (30)	
		0	56	55	56.5	55.5					280761.330 (30)	
		1	56	55	55.5	54.5					279589.695 (30)	
		1	56	55	56.5	55.5					279639.390 (30)	
		2	56	55	55.5	54.5					278469.438 (30)	
		2	56	55	56.5	55.5					278518.848 (30)	
		3	56	55	55.5	54.5					277350.640 (30)	
		3	56	55	56.5	55.5					277399.773 (30)	
		0	57	56	56.5	55.5					285701.178 (30)	
		0	57	56	57.5	56.5					285751.108 (30)	
		1	57	56	56.5	55.5					284559.403 (30)	
		1	57	56	57.5	56.5					284609.095 (30)	
		2	57	56	56.5	55.5					283419.108 (30)	
		2	57	56	57.5	56.5					283468.543 (30)	
		3	57	56	56.5	55.5					282280.308 (30)	
		3	57	56	57.5	56.5					282329.458 (30)	
		4	57	56	56.5	55.5					281143.008 (30)	
		4	57	56	57.5	56.5					281191.890 (30)	
		0	58	57	57.5	56.5					290689.690 (30)	
		0	58	57	58.5	57.5					290739.648 (30)	
$^{138}Ba^{37}Cl$	$X\,^2\Sigma^+$	0	22	21	21.5	20.5					105721.405 (30)	81SCH, 82RYZ
		0	22	21	22.5	21.5					105768.943 (30)	
		1	22	21	21.5	20.5					105390.230 (30)	
		1	22	21	22.5	21.5					105356.475 (30)	
		0	57	56	56.5	55.5					273428.673 (30)	

Line frequencies of BaCl (continued)

Isotopic species	State: electronic	State: vibrational v	Transition: N'	N''	J'	J''	F'_1	F''_1	F' – F''	ν MHz	Ref.
$^{138}Ba^{37}Cl$	$X^2\Sigma^+$	0	57	56	57.5	56.5				273476.408 (30)	81SCH, 82RYZ
		0	58	57	57.5	56.5				278203.933 (30)	
		0	58	57	58.5	57.5				278251.738 (30)	
		1	58	57	57.5	56.5				277116.400 (30)	
		1	58	57	58.5	57.5				277163.948 (30)	
		0	59	58	58.5	57.5				282978.010 (30)	
		0	59	58	59.5	58.5				283025.788 (30)	
		1	59	58	58.5	57.5				281871.700 (30)	
		1	59	58	59.5	58.5				281919.335 (30)	
$^{136}Ba^{35}Cl$	$X\ ^2\Sigma^+$	0	55	54	54.5	53.5				276538.803 (30)	81SCH, 82RYZ
		0	55	54	55.5	54.5				276588.908 (30)	
		0	56	55	55.5	54.5				281544.590 (30)	
		0	56	55	56.5	55.5				281594.625 (30)	
$^{137}Ba^{35}Cl$	$X\ ^2\Sigma^+$	0	21	20	20.5[1]	19.5[1]	19	18		105612.23 (10)	81SCH, 82RYZ
		0	21	20	20.5	19.5	22	21		105621.33 (10)	
		0	21	20	20.5	19.5	20	19		105629.81 (10)	
		0	21	20	20.5	19.5	21	20		105631.60 (10)	
		0	21	20	21.5	20.5	21	20		105642.47 (10)	
		0	21	20	21.5	20.5	20	19		105644.31 (10)	
		0	21	20	21.5	20.5	22	21		105652.81 (10)	
		0	21	20	21.5	20.5	23	22		105661.84 (10)	
		0	54	53	53.5	52.5	52	51	the chlorine	271126.77 (10)	
		0	54	53	53.5	52.5	55	54	hyperfine	271132.28 (10)	
		0	54	53	53.5	52.5	54	53	structure is	271139.33 (10)	
		0	54	53	53.5	52.5	53	52	unresolved	271149.25 (10)	
		0	54	53	54.5	53.5	53	52		271154.20 (10)	
		0	54	53	54.5	53.5	54	53		271164.22 (10)	
		0	54	53	54.5	53.5	55	54		271171.21 (10)	
		0	54	53	54.5	53.5	56	55		271176.76 (10)	
		0	55	54	54.5	53.5	53	52		276126.30 (10)	
		0	55	54	54.5	53.5	56	55		276131.74 (10)	

[1]) In [82RYZ] the same transition frequencies are reported, but no J quantum numbers are specified. Other distinguishing markers must be introduced for the energy levels.

Line frequencies of BaCl (continued)

Isotopic species	State: electronic	State: vibrational v	Transition: N'	N''	J'	J''	F_1'	F_1''	F'	F''	ν MHz	Ref.
$^{137}Ba^{35}Cl$	$X^2\Sigma^+$	0	55	54	54.5[1]	53.5[1]	55	54			276138.67 (10)	81SCH, 82RYZ
		0	55	54	54.5	53.5	54	53			276148.37 (10)	
		0	55	54	55.5	54.5	54	53			276154.13 (10)	
		0	55	54	55.5	54.5	55	54			276163.67 (10)	
		0	55	54	55.5	54.5	56	55			276170.76 (10)	
		0	55	54	55.5	54.5	57	56			276176.35 (10)	
		0	56	55	55.5	54.5	54	53			281124.59 (10)	
		0	56	55	55.5	54.5	57	56			281130.01 (10)	
		0	56	55	55.5	54.5	56	55			281136.82 (10)	
		0	56	55	55.5	54.5	55	54			281146.45 (10)	
		0	56	55	56.5	55.5	55	54			281152.80 (10)	
		0	56	55	56.5	55.5	56	55			281162.45 (10)	
		0	56	55	56.5	55.5	57	56			281169.15 (10)	
		0	56	55	56.5	55.5	58	57			281174.61 (10)	

[1]) In [82RYZ] the same transition frequencies are reported, but no J quantum numbers are specified. Other distinguishing markers must be introduced for the energy levels.

References

81SCH Schütze-Pahlmann, H.-U., Dissertation Freie Universität Berlin, **1981**.

82RYZ Ryzlewicz, Ch., Schütze-Pahlmann, H.-U., Hoeft, J., Törring, T.: Chem. Phys. **71** (1982) 389.

45 CO^+

Molecular constants of CO^+

Isotopic species	State	B_v MHz	D_v kHz	γ_v MHz	Ref.	Meth.	Derived constants
$^{12}C^{16}O^+$	$X\,^2\Sigma^+$	$B_0 = 58983.040\ (12)$ $Y_{01} = 59267.60\ (9)$ $Y_{11} = -568.28\ (20)$ $Y_{21} = -1.02\ (7)$	$D_0 = 189.6\ (5)$	$\gamma_0 = 273.01\ (5)$ $\gamma_1 = 271.27\ (11)$	81SAS 82BOG	MW MW MW MW	
$^{13}C^{16}O^+$	$X\,^2\Sigma^+$	$B_0 = 56388.96\ (10)$		$\gamma_0 = 260.4\ (7)$	82PIL	MW	
$^{12}C^{18}O^+$	$X\,^2\Sigma^+$	$B_0 = 56174.61\ (6)$		$\gamma_0 = 259.8\ (2)$	82PIL	MW	

Molecular constants of CO^+ (continued)

Isotopic species	State electronic	State vibrational	Parameter	Value	Ref.	Meth.	Remarks
$^{13}C^{16}O^+$	$X\,^2\Sigma^+$	$v=0$	$b_F(^{13}C)$	1512 (12) MHz	78CAR	accelerated ion-beam and laser excitation	
			$c(^{13}C)$	92 (41) MHz			
		$v=0$	$b_F(^{13}C)$	1511.0 (100) MHz	82PIL	MW	combined analysis of MW-data and data of [78CAR]
			$c(^{13}C)$	144.6 (21) MHz		MW	

Line frequencies of CO^+

Isotopic species	State electronic	State vibrational v	Transition N'	– N''	J'	– J''	F_1'	– F_1'' [2]	F'	– F''	ν MHz	Ref.
$^{12}C^{16}O^+$	$X\,^2\Sigma^+$	0	1	0	0.5	0.5					117692.294 (103)	75DIX,
		0	1	0	1.5	0.5					118101.886 (48)	82PIL [1])
		1	1	0	0.5	0.5					116553.42 (10)	82BOG
		1	1	0	1.5	0.5					116960.32 (6)	
		2	1	0	1.5	0.5					115814.75 (12)	
		0	2	1	1.5	0.5					235789.641 (30)	81SAS
		0	2	1	2.5	1.5					236062.553 (20)	
		0	2	1	1.5	1.5					235380.046 (150)	
		0	3	2	2.5	1.5					353741.262 (100)	
		0	3	2	3.5	2.5					354014.247 (60)	
		0	4	3	3.5	2.5					471679.213 (120)	
		0	4	3	4.5	3.5					471952.343 (100)	
$^{13}C^{16}O^+$	$X\,^2\Sigma^+$	0	1	0			1	1	2	1	112902.610 (37)	82PIL [1])
		0	1	0			0	0	1	0	112753.426 (58)	
		0	1	0			1	1	1	1	112694.956 (111)	
		0	1	0			1	1	0	1	112468.609 (230)	
$^{12}C^{18}O^+$	$X\,^2\Sigma^+$	0	1	0	0.5	0.5					112088.700 (61)	82PIL [1])
		0	1	0	1.5	0.5					112478.474 (58)	

[1]) Doppler shift not corrected, compare [82BOG].
[2]) Hund's case $b_{\beta S}$: $\boldsymbol{F}_1 = \boldsymbol{I}(^{13}C) + \boldsymbol{S}$.

References

75DIX Dixon, Th.A., Woods, R.C.: Phys. Rev. Lett. **34** (1975) 61.
78CAR Carrington, A., Milverton, D.R.J., Sarre, P.J.: Mol. Phys. **35** (1978) 1505.
81BRO Brown, R.D., Godfrey, P.D., Mc Gilvery, D.C., Crofts, J.G.: Chem. Phys. Lett. **84** (1981) 437; first MODR-observation of CO^+.
81SAS Sastry, K.V.L.N., Helminger, P., Herbst, E., De Lucia, F.C.: Astrophys. J. **250** (1981) L91.
82BOG Bogey, M., Demuynck, C., Destombes, J.L.: Mol. Phys. **46** (1982) 679.
82PIL Piltch, N.D., Szanto, P.G., Anderson, T.G., Gudeman, C.S., Dixon, T.A., Woods, R.C.: J. Chem. Phys. **76** (1982) 3385.

46 CN

Molecular constants of CN

Isotopic species	State	B_v MHz	D_v kHz	γ_v MHz	Ref.	Meth.	Derived constants
$^{12}C^{14}N$	$X\,^2\Sigma^+$	$B_0=56693.096(3)$ $B_1=56170.738(15)$	$D_0^{ev}=192$ $D_1^{ev}=192$	$\gamma_0=217.488(8)$ $\gamma_1=215.072(40)$	77DIX 77DIX 74PEN, 75TUR, 82DAV	MW MW 1) 2)	$r_e=1.171821(20)$ Å
$^{12}C^{14}N$	$B\,^2\Sigma^+$			$v=0$ $\nu_\gamma=469.2(12)$	73COO	EPR	
		Magnetic resonance spectrum of the rotationally perturbed levels of $A\,^2\Pi_{3/2}$ $v=10$, $J=7/2$ and $B\,^2\Sigma^+$ $v=0$, $N=4$ has been observed in [73COO] determining different interelectronic coupling parameters between the B- and A-state.					

1) Observation by radio astronomy.
2) Infrared laser spectrum for $v=0 \rightarrow v=1$.

Isotopic species	State electronic	State vibrational	Parameter	Value	Ref.	Meth.	Remarks
$^{12}C^{14}N$	$X\,^2\Sigma^+$	$v=0$	$b(^{14}N)$	−33.968 (15) MHz	77DIX	MW	
			$c(^{14}N)$	60.322 (31) MHz	77DIX	MW	
			$eqQ(^{14}N)$	−1.287 (41) MHz	77DIX	MW	
			$C_I(^{14}N)$	7 (6) kHz	77DIX	MW	
		$v=1$	$b(^{14}N)$	−33.164 (65) MHz	77DIX	MW	
			$c(^{14}N)$	60.504 (140) MHz	77DIX	MW	
			$eqQ(^{14}N)$	−1.271 (210) MHz	77DIX	MW	

Line frequencies of CN

Isotopic species	State electronic	State vibrational v	N'	N''	J'	J''	F_1' – F_1''	F'	F''	ν MHz	Ref.
$^{12}C^{14}N$	$X\,^2\Sigma^+$	0	1	0	0.5	0.5		0.5	1.5	113144.190 (16)	77DIX
		0	1	0	0.5	0.5		1.5	0.5	113170.535 (20)	
		0	1	0	0.5	0.5		1.5	1.5	113191.325 (40)	
		0	1	0	1.5	0.5		1.5	0.5	113488.142 (13)	
		0	1	0	1.5	0.5		2.5	1.5	113490.986 (11)	
		0	1	0	1.5	0.5		0.5	0.5	113499.644 (14)	
		0	1	0	1.5	0.5		1.5	1.5	113508.934 (17)	
		1	1	0	0.5	0.5		0.5	1.5	112101.674 (43)	
		1	1	0	0.5	0.5		1.5	0.5	112128.990 (22)	
		1	1	0	0.5	0.5		1.5	1.5	112148.511 (37)	
		1	1	0	1.5	0.5		1.5	0.5	112442.814 (39)	
		1	1	0	1.5	0.5		2.5	1.5	112445.021 (28)	
		1	1	0	1.5	0.5		0.5	0.5	112453.892 (61)	
		1	1	0	1.5	0.5		1.5	1.5	112463.282 (33)	

References

73COO Cook, T.J., Levy, D.H.: J. Chem. Phys. **58** (1973) 3547.

74PEN Penzias, A.A., Wilson, R.W., Jefferts, K.B.: Phys. Rev. Lett. **32** (1974) 701.

75TUR Turner, B.E., Gammon, R.H.: Astrophys. J. **198** (1975) 71.

77DIX Dixon, T.A., Woods, R.C.: J. Chem. Phys. **67** (1977) 3956.

82DAV Davies, P.B., Hamilton, P.A.: J. Chem. Phys. **76** (1982) 2127.

80 NO

Molecular constants of NO

Isotopic species	State	B_v MHz	D_v kHz	γ_v MHz	Ref.	Meth.	Derived constants
$^{14}N^{16}O$	$A\,^2\Sigma^+$	$B_0 = 59547.45\,(45)$	$D_0 = 170.19\,(48)$	$\gamma_0 = -71\,(10)$	79AMI	FIR	
				$\gamma_0 = -86\,(6)$	78WAL	[1]	
				$\gamma_3 = -82.893\,(30)$	74BER, 76WOO	RF/DR	
$^{15}N^{16}O$	$A\,^2\Sigma^+$	$B_0 = 57441.46\,(42)$	$D_0 = 157.66\,(48)$	$\gamma_0 = -81\,(10)$	79AMI	FIR	

[1]) Two-photon laser spectroscopy

Molecular constants of NO (continued)

Isotopic species	State	B_v MHz	D_v kHz	γ_v MHz	Ref.	Meth.	Derived constants
$^{14}N^{16}O$	D $^2\Sigma^+$	$B_0 = 59685.86\,(39)$	$D_0 = 180.45\,(36)$	$\gamma_0 = -145\,(10)$	79AMI	FIR	
				$D_\gamma = 10\,(3)$ Hz	79AMI	FIR	
$^{15}N^{16}O$	D $^2\Sigma^+$	$B_0 = 57559.31\,(42)$	$D_0 = 166.86\,(48)$	$\gamma_0 = -153\,(10)$	79AMI	FIR	
				$D_\gamma = 12\,(5)$ Hz	79AMI	FIR	
$^{14}N^{16}O$	E $^2\Sigma^+$	$B_0 = 59296.43\,(36)$	$D_0 = 169.74\,(36)$	$\gamma_0 = -7\,(10)$	79AMI	FIR	
$^{15}N^{16}O$	E $^2\Sigma^+$	$B_0 = 57207.42\,(48)$	$D_0 = 156.97\,(48)$	$\gamma_0 = -5\,(10)$	79AMI	FIR	

Isotopic species	State		Parameter	Value	Ref.	Meth.	Remarks
	electronic	vibrational					
$^{14}N^{16}O$	A $^2\Sigma^+$	$v = 3$	$b(^{14}N)$	40.020 (70) MHz	74BER, 76WOO	RF/DR	
			$c(^{14}N)$	3.01 (11) MHz	74BER, 76WOO	RF/DR	
			$eqQ(^{14}N)$	−6.77 (30) MHz	74BER, 76WOO	RF/DR	
			μ	1.10 (3) D	74BER, 76WOO	RF/DR	

Line frequencies of NO

Isotopic species	State		Transition								ν MHz	Ref.
	electronic	vibrational v	N'	N''	J'	J''	F_1'	F_1''	F'	F''		
$^{14}N^{16}O$	A $^2\Sigma^+$	3	1	1	0.5	1.5			1.5	2.5	104.053 (60)	74BER
		3	1	1	0.5	1.5			1.5	1.5	146.128 (30)	
		3	1	1	0.5	1.5			1.5	0.5	159.674 (150)	
		3	1	1	0.5	1.5			0.5	1.5	160.914 (100)	

References

74BER Bergeman, T., Zare, R.N.: J. Chem. Phys. **61** (1974) 4500.
76WOO Woods, R.C., Dixon, T.A.: J. Chem. Phys. **64** (1976) 5319.
78WAL Wallenstein, R., Zacharias, H.: Opt. Comm. **25** (1978) 363.
79AMI Amiot, C., Verges, J.: Chem. Phys. Lett. **66** (1979) 570.

4.1.3 The $^3\Sigma$ electronic state

4.1.3.1 Preliminary remarks

1. Introduction

The approximate electronic quantum numbers of this molecular state are the spin $S=1$, the projection of the orbital angular momentum onto the molecular axis $|\Lambda|=0$ and the reflection symmetry $+/-$ in a plane containing the molecular axis. In the case of large spin orbit coupling the state is split into three fine structure levels which are assigned to $|\Omega|=0$ and $|\Omega|=1$ where Ω is the projection of the total electronic angular momentum on the molecular axis. The state $|\Omega|=1$ shows Ω-doubling.

Assuming Hund's coupling case (*b*), the total molecular state is described by the additional quantum number N which is the vector-sum of the orbital angular momentum $\boldsymbol{L}$ and the rotational momentum $\boldsymbol{R}$, and by the quantum number J of the total angular momentum by the parity + or −, and by the vibrational state quantum number v.

Assuming Hund's coupling case (*c*), N is no longer defined, therefore J, parity and vibrational state quantum number v are the assignments to the total molecular state.

In the case of non-zero nuclear spins I_1 and I_2, the total angular momentum is given by the quantum number F and to describe the molecular state completely the approximate intermediate quantum number F_1 ($\boldsymbol{F}_1=\boldsymbol{J}+\boldsymbol{I}_1$) or, for a homonuclear molecule, I_T ($\boldsymbol{I}_T=\boldsymbol{I}_1+\boldsymbol{I}_2$) is used.

The energy expressions for describing the rotational and the fine structure energy are given in 5.1.3.2 Vol. II/6, Landolt-Börnstein, New Series, and will not be repeated here. But the nomenclature of the molecular parameters is slightly changed in order to standardize the symbols of parameters with similar meanings in different electronic states. This is in accordance with the latest usage in the literature. The explanation of the new symbols is given in the list of the molecular parameters below.

The analysis of the phenomenological Hamiltonian used here is revised by De Santis et al. [1] including higher order corrections to the vibration-rotation interaction and to the Born-Oppenheimer approximation for the rotational energy. The derived formulae are very helpful for converting the v-dependence of the observed molecular constant to the function of the internuclear distance, which is important for the comparison with ab initio or model calculations.

Tiemann [2] uses the following correction formula for the rotational constant

$$Y_{01}=B_e^{cor}\left[1-2m_e\left(\frac{d_A}{M_A}+\frac{d_B}{M_B}\right)\right] \tag{1}$$

(compare eq. (1) of part 2.2) of the Born-Oppenheimer approximation as a phenomenological description of the isotope shift in the spin-spin coupling constant λ and the spin-rotation constant γ. The appropriate d-parameters are defined by:

$$\lambda_e=\lambda_e^{cor}\left[1-2m_e\left(\frac{d_{\lambda A}}{M_A}+\frac{d_{\lambda B}}{M_B}\right)\right], \tag{2}$$

$$\gamma_e=\gamma_e^{cor}\left[1-2m_e\left(\frac{d_{\gamma A}}{M_A}+\frac{d_{\gamma B}}{M_B}\right)\right] \tag{3}$$

where m_e is the electron mass,

M_A, M_B the atomic masses of atom A and B respectively, and

λ_e^{cor}, γ_e^{cor} the spin-spin parameter and spin-rotation parameter corrected for higher order mass-effects.

By this procedure it is possible to include measurements of all isotopic species of one molecule in the analysis of the spectrum and to derive a set of molecular parameters from which molecular constants of unmeasured isotopes can be deduced with high reliability.

Amano and Hirota [3] reexamined the spin-spin interaction in connection with the second order spin-orbit interaction and found that the Hamiltonian used so far for the rotational energy of $^3\Sigma$-states is not correct. Their proposed formulae will change the definition of the parameters B, λ, γ slightly. We have indicated these new parameters by a prime (B', λ', γ') in the tables.

Veseth and Lofthus [4] proposed to use the phenomenological Hamiltonian derived from wave functions in the Hund's case (*c*). They could fit the spectra of O_2 and SO reasonably well, but their parameter set will not be included in the tables to avoid confusion of the different definitions.

The microwave spectra of the Bi-halides which belong to the limiting Hund's coupling case (*c*) of a $^3\Sigma$-state, were fitted to the formula for the rotational energy for a $^1\Sigma$-state. Therefore the obtained Dunham parameters Y_{lk} are only effective molecular constants which describe the observed spectra very well but contain some residuals of the unknown interactions λ and γ. (Compare also the short discussion in [5]).

2. List of parameters

B_v rotational constant of the vibrational state v; this parameter is expanded in a series of Dunham parameters:

$$B_v = \sum_l Y_{l1} (v+1/2)^l$$

Compare part 2.2 for the equivalence of Y_{11} and α_e, Y_{21} and γ_e, etc. of the traditional spectroscopic notation.

d_A, d_B parameters to correct higher order mass effects which come from the break-down of the Born-Oppenheimer approximation for atom A or B of molecule AB; compare eq. (1)

B_e^{cor} rotational constant corrected for the deviations of the Born-Oppenheimer approximation

B'_v definition in [3] slightly different to B_v

D_v centrifugal distortion constant of the vibrational state v; sometimes a similar expansion as for B_v is used:

$$D_v = -\sum_l Y_{l2} (v+1/2)^l$$

In most experiments the accuracy is not sufficient to distinguish between D_v and Y_{02}. Higher order terms of the centrifugal distortion like H_v or Y_{03} are included in few cases.

λ_v spin-spin interaction constant of the vibrational state v; the phenomenological expansion is used:

$$\lambda_v = \lambda_e + \alpha_\lambda (v+1/2) + \gamma_\lambda (v+1/2)^2 + \varepsilon_\lambda (v+1/2)^3 + \cdots$$

γ_v very often contains to a large extent a contribution by the second order spin-orbit coupling which is not distinguishable from the pure spin-spin interaction in a phenomenological Hamiltonian.

$d_{\lambda A}$, $d_{\lambda B}$ parameters to correct higher order mass effects of λ caused by atom A or B of molecule AB; compare eq. (2)

λ_e^{cor} spin-spin coupling constant corrected for higher order mass-shifts

λ'_v definition in [3] slightly different to λ_v

D_λ centrifugal distortion constant for λ_v ($D_\lambda \triangleq \varrho_v$ in 5.1.3.1 of Vol. II/6 Landolt-Börnstein, New Series)

γ_v spin-rotation interaction constant of the vibrational state v; the phenomenological expansion is used:

$$\gamma_v = \gamma_e + \alpha_\gamma (v+1/2) + \gamma_\gamma (v+1/2)^2 + \cdots$$

γ'_v definition in [3], slightly different to γ_v

D_γ centrifugal distortion constant for γ_v ($D_\gamma \triangleq \bar{\gamma}_1$ in 5.1.3.1 of Vol. II/6, Landolt-Börnstein, New Series)

b_i, c_i magnetic hyperfine structure constants for the coupling of the electronic spin with the nuclear spin of nucleus i

C_{Ii} magnetic hyperfine structure constant for the coupling of the molecular rotation with the nuclear spin of nucleus i ($C_{Ii} \triangleq k_i$ in 5.1.1 of Vol. II/6, Landolt-Börnstein, New Series)

eqQ_i electric quadrupole interaction constant of nucleus i with the electronic field gradient along the molecular axis

μ_v molecular electric dipole moment for the vibrational state v

g_s g-factor from electronic spin including relativistic and diamagnetic effects

g_L^{el} net g-factor from electronic orbital motion

g_{rot} rotational g-factor (compare g_J for $^1\Sigma$-states, part 2.9.1)

Δg mainly determined by $-g_L^{el}$, but contains diamagnetic corrections

derived parameters

ω_e harmonic vibrational frequency

$\omega_e x_e$ first order anharmonic correction of the vibrational energy

r_e equilibrium internuclear distance

r_e^{cor} equilibrium internuclear distance corrected for the deviations from the Born-Oppenheimer approximation

3. Arrangement of substances

H_2
He_2
NH, PH
SH^+
N_2, BiF, BiCl, BiBr, BiI
O_2, SO, SeO, S_2

References

1 DeSantis, D., Lurio, A., Miller, T.A., Freund, R.S.: J. Chem. Phys. **58** (1973) 4625.
2 Tiemann, E.: J. Mol. Spectrosc. **91** (1982) 60.
3 Amano, T., Hirota, E.: J. Mol. Spectrosc. **53** (1974) 346.
4 Veseth, L., Lofthus, A.: Mol. Phys. **27** (1974) 511.
5 Kuijpers, P., Törring, T., Dymanus, A.: Chem. Phys. **12** (1976) 309.

4.1.3.2 Data and references

1 H_2

Molecular constants of H_2

Isotopic species	State		Remarks
	electronic	vibrational	
$^1H_2$ ortho	$p(4d)\,^3\Sigma_g^+$		data from level anticrossing spectroscopy with k $^3\Pi_u$-state, compare 4.1.6

7 He_2

Molecular constants of He_2

Isotopic species	State	λ_v MHz	γ_v MHz	Ref.	Meth.
$^4He_2$	$a\,^3\Sigma_u^+$	$\lambda_0 = -1099.167\,(5)$ $D_{\lambda 0} = 0.1970\,(7)$	$\gamma_0 = -2.422\,(3)$ $D_{\gamma 0} = 0.7\,(3)$ kHz	74LIC, 75VIE	MB [1])

[1]) The assignment to $v=0$ is not certain.

Line frequencies of He_2

Isotopic species	State		Transition				ν MHz	Ref.
	electronic	vibrational v	N' –	N''	J' –	J''		
$^4He_2$	$a\,^3\Sigma_u^+$	0	1	1	0	1	2199.968 (10)	74LIC
		0	1	1	1	2	873.668 (7)	74LIC
		0	3	3	2	3	1323.911 (6)	74LIC
		0	3	3	3	4	964.992 (6)	74LIC

References

74LIC Lichten, W., McCusker, M.V., Vierima, T.L.: J. Chem. Phys. **61** (1974) 2200.
75VIE Vierima, T.L.: J. Chem. Phys. **62** (1975) 2925.

9 NH

Molecular constants of NH

Isotopic species	State	B_v MHz	D_v kHz	λ_v MHz	γ_v MHz	Derived constants	Ref.	Meth.
$^{14}N^{1}H$	$X\,^3\Sigma^-$	$B_0=489959.26\,(68)$	$D_0^{ev}=51384\,(90)$	$\lambda_0=27581.1\,(43)$	$\gamma_0=-1642.4\,(17)$		82VAN	MW
		$B_1=470563.2\,(60)$	$D_1^{ev}=49286$	$\lambda_1=27574.9\,(60)$	$\gamma_1=-1549.9\,(60)$		76WAY	LMR
$^{14}N^{2}D$	$X\,^3\Sigma^-$	$B_0=263262.8\,(30)$	$D_0^{ev}=14489$	$\lambda_0=27533\,(12)$	$\gamma_0=-881.4\,(60)$		76WAY	LMR

Isotopic species	State electronic	State vibrational	Parameter	Value	Ref.	Meth.	Remarks
$^{14}N^{1}H$	$X\,^3\Sigma^-$	$v=0$	$b_F\,(^{14}N)$	19.22 (18) MHz	82VAN	MW	measurements with laser-sideband spectrometer using a HCN laser (MW values)
			$b_F\,(^{1}H)$	−66.23 (32) MHz	82VAN	MW	
			$c\,(^{14}N)$	−67.94 (61) MHz	82VAN	MW	
			$c\,(^{1}H)$	91.70 (160) MHz	82VAN	MW	
		$v=1$	$b_F=b+c/3\,(^{14}N)$	17.9 (16) MHz	76WAY	LMR	
			$b_F=b+c/3\,(^{1}H)$	−66.4 (20) MHz	76WAY	LMR	
			$c/3\,(^{14}N)$	−22.4 (18) MHz	76WAY	LMR	
			$c/3\,(^{1}H)$	29.4 (30) MHz	76WAY	LMR	
$^{14}N^{1}H$	$X\,^3\Sigma^-$		$g_s+g_L^{el}$ (assumed)	2.0021	76WAY	LMR	
			Δg (assumed)	−0.00169	76WAY	LMR	
			g_{rot}	−0.18	76WAY	LMR	
$^{15}N^{1}H$	$X\,^3\Sigma^-$	$v=0$	$b_F=b+c/3\,(^{15}N)$	−28.8 (34) MHz	75RAD, 76WAY	LMR	
			$b_F=b+c/3\,(^{1}H)$	−66.5 (32) MHz	75RAD, 76WAY	LMR	
			$c/3\,(^{15}N)$	31.3 (22) MHz	75RAD, 76WAY	LMR	
			$c/3\,(^{1}H)$	28.8 (22) MHz	75RAD, 76WAY	LMR	
$^{14}N^{2}D$	$X\,^3\Sigma^-$	$v=0$	$b_F=b+c/3\,(^{14}N)$	20.2 (10) MHz	76WAY	LMR	
			$b_F=b+c/3\,(^{2}D)$	−10.2 (10) MHz	76WAY	LMR	
			$c/3\,(^{14}N)$	−24.9 (20) MHz	76WAY	LMR	
			$c/3\,(^{2}D)$	5.1 (20) MHz	76WAY	LMR	

Line frequencies of NH

Isotopic species	State		Transition								ν MHz [2]	Ref.
	electronic	vibrational v	N'	N''	J'	J''	F_1'	F_1''	F' [1]	F''		
$^{14}N^1H$	$X\,^3\Sigma^-$	0	1	0	2	1			3.5	2.5	974478.32 (100)	82VAN
									2.5	1.5		
									1.5	0.5		
									2.5	1.5	974470.59 (80)	
									1.5	0.5	974461.86 (80)	
									2.5	2.5	974436.57 (100)	
									1.5	1.5		
									0.5	0.5	974450.32 (150)	
		0	1	0	1	1			2.5	1.5	1000009.00 (100)	
									1.5	1.5		
									1.5	0.5	1000026.97 (100)	
									1.5	0.5	999945.06 (80)	
									2.5	2.5	999973.41 (80)	
									1.5	1.5	1000034.30 (100)	
									1.5	0.5	1000053.97 (130)	
									0.5	0.5		
									1.5	2.5	1000000.44 (80)	
									0.5	1.5	999907.92 (100)	
									0.5	1.5	1000046.38 (130)	

[1]) The states with the same total angular momentum ***F*** may have different F_1 parentages because the hyperfine couplings of N and H nucleus are of similar magnitude.
[2]) [82VAN] gives separate values for the hyperfine splittings and the hyperfine-free rotational transition frequencies.

References

75RAD Radford, H.E., Litvak, M.M.: Chem. Phys. Lett. **34** (1975) 561.

76WAY Wayne, F.D., Radford, H.E.: Mol. Phys. **32** (1976) 1407.

82VAN Van den Heuvel, F.C., Meerts, W.L., Dymanus, A.: Chem. Phys. Lett. **92** (1982) 215.

10 PH

Molecular constants of PH

Isotopic species	State	B_v MHz	D_v kHz	λ_v MHz	γ_v MHz	Derived constants	Ref.	Meth.
$^{31}P^1H$	$X\,^3\Sigma^-$	$B_0 = 252186.0\,(8)$		$\lambda_0 = 66155\,(45)$	$\gamma_0 = -2292.7\,(14)$		75DAV, 79DAV	LMR

Molecular constants of PH (continued)

Isotopic species	State	B_v MHz	D_v kHz	λ_v MHz	γ_v MHz	Derived constants	Ref.	Meth.
$^{31}P^2D$	$X\,^3\Sigma^-$	$B_0=130799\,(9)$	$D_0^{ev}=3.481$ MHz	$\lambda_0=66197\,(39)$	$\gamma_0=-1187\,(33)$	$\nu(v=1\leftarrow0)$ $=1653.2858\,(36)\,cm^{-1}$	81UEH	LMR
		$B_1=127984\,(6)$	$D_1^{ev}=3.460$ MHz	$\lambda_1=66242\,(78)$	$\gamma_1=-1175\,(18)$		81UEH	LMR

Isotopic species	State		Parameter	Value	Ref.	Meth.	Remarks
	electronic	vibrational					
$^{31}P^1H$	$X\,^3\Sigma^-$	$v=0$	$b_F=b+c/3\,(^{31}P)$	125.4 (82) MHz	75DAV, 79DAV	LMR	
			$b_F=b+c/3\,(^1H)$	−48.9 (52) MHz	75DAV, 79DAV	LMR	
			$c/3\,(^{31}P)$	−152 (12) MHz	75DAV, 79DAV	LMR	
			$c/3\,(^1H)$	5.1 (75) MHz	75DAV, 79DAV	LMR	
			g_s (assumed)	2.00232	79DAV	LMR	
			g_L^{el} (estimated)	0.00455	79DAV	LMR	
			g_{rot}	−1.3 (7)	79DAV	LMR	
$^{31}P^2D$	$X\,^3\Sigma^-$	$v=0,1$	$b(^{31}P)$	291 (21) MHz	81UEH	LMR	
			$c(^{31}P)$	−493 (90) MHz	81UEH	LMR	

References

75DAV Davies, P.B., Russell, D.K., Thrush, B.A.: Chem. Phys. Lett. **36** (1975) 280.

79DAV Davies, P.B., Russell, D.K., Smith, D.R., Thrush, B.A.: Can. J. Phys. **57** (1979) 522.

81UEH Uehara, H., Hakuta, K.: J. Chem. Phys. **74** (1981) 4326.

10a SH^+

Molecular constants of SH^+

Isotopic species	State		Parameter	Value	Ref.	Meth.	Remarks
	electronic	vibrational					
$^{32}S^1H^+$	$X\,^3\Sigma^-$	$v=3$	$b_F=b+c/3\,(^1H)$	−85 MHz	82EDW	MB	Laser excitation, detection by daughter ions produced by predissociation.
			$c(^1H)$	12 MHz	82EDW	MB	

Reference 82EDW Edwards, C.P., MacLean, C.S., Sarre, P.J.: J. Chem. Phys. **76** (1982) 3829.

76 N_2

Molecular constants of N_2

Isotopic species	State	B_v MHz	D_v kHz	λ_v MHz	γ_v MHz	Derived constants	Ref.	Meth.
$^{14}N_2$	A $^3\Sigma_u^+$			$\lambda_e^{ev} = -39.84\,(21)$ GHz $\alpha_\lambda^{ev} = 163\,(13)$ $\gamma_\lambda^{ev} = 6.12\,(99)$			73DES	MB

Isotopic species	State		Parameter	Ref.
	electronic	vibrational		
$^{14}N_2$	A $^3\Sigma_u^+$	$0 \leqq v \leqq 12$	$b_v(^{14}N) = (25.844\,(22) - 0.0646\,(30)(v+1/2) + 0.00542\,(30)(v+1/2)^2)$ MHz	73DES, 70FRE
			$c_v(^{14}N) = (-37.980\,(27) + 0.0260\,(30)(v+1/2) + 0.00459\,(30)(v+1/2)^2)$ MHz	73DES, 70FRE
			$eqQ(^{14}N) = (-2.571\,(40) + 0.0847\,(30)(v+1/2) - 0.00012\,(20)(v+1/2)^2)$ MHz	73DES, 70FRE
			$C_I(^{14}N) = 11.9\,(50)$ kHz	73DES, 70FRE

Line frequencies of N_2

Isotopic species	State		Transition				ν MHz	Ref.
	electronic	vibrational v	N' – N''	J' – J''	I'_T – I''_T	F' – F''		
$^{14}N_2$	A $^3\Sigma_u^+$	0	3 3	2 2	2 2	4 3	40.851 (8)	73DES
		1					40.752 (8)	
		2					40.638 (8)	
		3					40.499 (8)	
		4					40.349 (8)	
		5					40.187 (8)	
		6					40.008 (8)	
		7					39.801 (8)	
		8					39.592 (8)	
		9					39.361 (8)	
		10					39.118 (8)	
		11					38.851 (8)	
		12					38.565 (13)	
		13					38.279 (13)	

Line frequencies of N_2 (continued)

Isotopic species	State: electronic	State: vibrational v	Transition: N'	N''	J'	J''	I'_T	I''_T	F'	F''	ν MHz	Ref.
$^{14}N_2$	A $^3\Sigma_u^+$	0	1	1	2	2	2	2	4	3	34.449 (8)	73DES
		1									34.279 (8)	
		2									34.084 (8)	
		3									33.873 (8)	
		4									33.633 (8)	
		5									33.381 (8)	
		6									33.109 (8)	
		7									32.819 (8)	
		8									32.511 (8)	
		9									32.191 (8)	
		10									31.855 (8)	
		11									31.512 (8)	
		12									31.152 (8)	
		13									30.774 (13)	
		0	2	2	1	1	1	1	2	1	33.358 (8)	73DES
		1									33.211 (8)	
		2									33.063 (8)	
		3									32.894 (8)	
		4									32.719 (8)	
		5									32.523 (8)	
		6									32.325 (8)	
		7									32.107 (8)	
		8									31.881 (8)	
		9									31.644 (8)	
		10									31.390 (8)	
		11									31.126 (8)	
		12									30.847 (13)	
		0	0	0	1	1	1	1	2	1	20.846 (8)	73DES
		1									20.687 (8)	
		2									20.516 (8)	
		3									20.321 (8)	
		4									20.107 (8)	
		5									19.878 (8)	

Line frequencies of N_2 (continued)

Isotopic species	State: electronic	State: vibrational v	Transition: N'	N''	J'	J''	I'_T	I''_T	F'	F''	ν MHz	Ref.
$^{14}N_2$	A $^3\Sigma_u^+$	6	0	0	1	1	1	1	2	1	19.634 (8)	73DES
		7									19.371 (8)	
		8									19.106 (8)	
		9									18.827 (8)	
		10									18.532 (8)	
		11									18.233 (8)	
		12									17.920 (13)	
		0	1	1	2	2	2	2	3	2	25.022 (8)	73DES
		1									24.918 (8)	
		2									24.800 (8)	
		3									24.664 (8)	
		4									24.516 (8)	
		5									24.354 (8)	
		6									24.177 (8)	
		7									23.985 (8)	
		8									23.780 (8)	
		9									23.570 (8)	
		10									23.350 (8)	
		11									23.113 (13)	
		12									22.864 (13)	
		0	3	3	2	2	2	2	3	2	31.720 (13)	73DES
		1									31.607 (13)	
		2									31.486 (13)	
		3									31.353 (8)	
		4									31.201 (8)	
		5	3	3	2	2	2	2	2	1	21.097 (8)	73DES
		6									20.958 (13)	
		7									20.819 (8)	

References

70FRE Freund, R.S., Miller, T.A., De Santis, D., Lurio, A.: J. Chem. Phys. **53** (1970) 2290.

73DES De Santis, D., Lurio, A., Miller, T.A., Freund, R.S.: J. Chem. Phys. **58** (1973) 4625.

83 BiF

Molecular constants of BiF

Isotopic species	State	B_v MHz	D_v kHz	λ_v MHz	γ_v MHz	Derived constants	Ref.	Meth.
$^{209}Bi^{19}F$ [1])	XO^+	$Y_{01}=6894.89594(84)$ $Y_{11}=-45.0474(11)$ $Y_{21}=81.69(45)$ kHz $Y_{31}=177(51)$ Hz	$Y_{02}=-5.5440(73)$			$r_e=2.0516(6)$ Å $a_0=2.860(7)\cdot10^5\ cm^{-1}$ $a_1=-3.429(4)$ $a_2=7.9(2)$ $a_3=-15(2)$	77KUI [1])	MW

[1]) Rotational constants shown contain small corrections due to second order spin-orbit coupling and admixture of the excited $X1^{\pm}$ state (compare preliminary remarks of this section).

Isotopic species	State		Parameter	Value	Ref.	Meth.	Remarks
	electronic	vibrational					
$^{209}Bi^{19}F$	XO^+	$v=0$	$eqQ(^{209}Bi)$	−1148.20 (12) MHz	77KUI	MW	$eq_zQ(^{209}Bi)=-1150.28(12)$ MHz
			$C_I(^{209}Bi)$	−30.0 (9) kHz	77KUI	MW	$eq_IQ(^{209}Bi)=4.20(12)$ MHz
		$v=1$	$eqQ(^{209}Bi)$	−1144.02 (16) MHz	77KUI	MW	
			$C_I(^{209}Bi)$	−30.3 (14) kHz	77KUI	MW	
		$v=2$	$eqQ(^{209}Bi)$	−1139.68 (32) MHz	77KUI	MW	
			$C_I(^{209}Bi)$	−28.2 (19) kHz	77KUI	MW	

Line frequencies of BiF

Isotopic species	State		Transition								ν MHz	Ref.
	electronic	vibrational v	N'	N''	J'	J''	F_1'	F_1''	F'	F''		
$^{209}Bi^{19}F$	$X\,^3\Sigma^-$	0	$\lvert\Omega\rvert=0$		5	4	3.5	4.5			68815.145 (30)	77KUI
							2.5	3.5			68808.145 (25)	
							1.5	2.5			68787.651 (20)	
							4.5	4.5			68752.484 (10)	
							3.5	3.5			68747.967 (15)	
							5.5	5.5			68746.363 (15)	
							2.5	2.5			68738.505 (20)	

Line frequencies of BiF (continued)

Isotopic species	State		Transition						ν	Ref.
	electronic	vibrational v	$N' - N''$	J'	J''	F_1'	F_1''	$F' - F''$	MHz	
$^{209}Bi^{19}F$	$X\,^3\Sigma^-$	0	$\|\Omega\|=0$	5	4	8.5	7.5		68735.124 (10)	77KUI
						7.5	6.5		68728.440 (20)	
						9.5	8.5		68727.063 (10)	
						6.5	6.5		68722.650 (80)	
						0.5	0.5		68721.304 (30)	
						6.5	5.5		68714.261 (15)	
						5.5	4.5		68698.380 (15)	
						4.5	3.5		68685.306 (15)	
						2.5	1.5		68679.448 (20)	
						3.5	2.5		68678.328 (20)	
						7.5	7.5		68672.925 (20)	
						8.5	8.5		68587.464 (15)	
		1	$\|\Omega\|=0$	5	4	4.5	4.5		68303.547 (20)	77KUI
						3.5	3.5		68299.048 (15)	
						5.5	5.5		68297.438 (20)	
						2.5	2.5		68289.596 (30)	
						8.5	7.5		68286.246 (15)	
						7.5	6.5		68279.573 (15)	
						9.5	8.5		68278.203 (15)	
						6.5	6.5		68273.808 (80)	
						6.5	5.5		68265.446 (20)	
						5.5	4.5		68249.621 (20)	
						4.5	3.5		68236.604 (15)	
						2.5	1.5		68230.758 (30)	
						3.5	2.5		68229.646 (20)	
						7.5	7.5		68224.275 (20)	
						8.5	8.5		68139.089 (25)	
		2	$\|\Omega\|=0$	5	4	4.5	4.5		67856.242 (20)	77KUI
						3.5	3.5		67851.749 (25)	
						5.5	5.5		67850.155 (20)	
						2.5	2.5		67842.364 (30)	
						8.5	7.5		67839.012 (20)	
						7.5	6.5		67832.386 (25)	

Line frequencies of BiF (continued)

Isotopic species	State: electronic	State: vibrational v	Transition: N' – N''	J'	J''	F_1'	F_1''	F' – F''	ν MHz	Ref.
$^{209}Bi^{19}F$	$X\ ^3\Sigma^-$	2	$\lvert\Omega\rvert=0$	5	4	9.5	8.5		67831.012 (20)	77KUI
						6.5	6.5		67826.592 (50)	
						6.5	5.5		67818.297 (20)	
						5.5	4.5		67802.522 (20)	
						4.5	3.5		67789.569 (25)	
						3.5	2.5		67782.622 (30)	
						7.5	7.5		67777.257 (30)	
		3	$\lvert\Omega\rvert=0$	5	4	8.5	7.5		67393.430 (20)	77KUI
						7.5	6.5		67386.820 (25)	
						9.5	8.5		67385.454 (15)	
						6.5	5.5		67372.785 (15)	
						5.5	4.5		67357.081 (25)	
		4	$\lvert\Omega\rvert=0$	5	4	8.5	7.5		66949.539 (30)	77KUI
						7.5	6.5		66942.950 (60)	
						9.5	8.5		66941.578 (20)	
		0	$\lvert\Omega\rvert=0$	7	6	4.5	4.5		96248.422 (20)	77KUI
						3.5	3.5		96246.858 (20)	
						5.5	5.5		96246.282 (25)	
						2.5	2.5		96243.570 (20)	
						6.5	6.5		96238.053 (20)	
						7.5	7.5		96220.842 (25)	
						10.5	9.5		96213.680 (10)	
						9.5	8.5		96211.607 (15)	
						11.5	10.5		96209.150 (15)	
						8.5	7.5		96205.854 (15)	
						2.5	1.5		96200.007 (15)	
						7.5	6.5		96198.850 (15)	
						6.5	5.5		96192.628 (15)	
						3.5	2.5		96191.962 (15)	
						8.5	8.5		96191.383 (15)	
						5.5	4.5		96188.715 (15)	
						4.5	3.5		96188.270 (15)	

Line frequencies of BiF (continued)

Isotopic species	State: electronic	State: vibrational v	Transition: N' – N''	J'	J''	F_1'	F_1''	F' – F''	ν MHz	Ref.
$^{209}Bi^{19}F$	$X\,^3\Sigma^-$	0	$\lvert\Omega\rvert=0$	7	6	9.5	9.5		96145.841 (25)	77KUI
						10.5	10.5		96079.990 (15)	
		1	$\lvert\Omega\rvert=0$	7	6	4.5	4.5		95619.881 (15)	77KUI
						3.5	3.5		95618.325 (15)	
						5.5	5.5		95617.764 (15)	
						2.5	2.5		95615.062 (35)	
						6.5	6.5		95609.562 (30)	
						7.5	7.5		95592.422 (20)	
						10.5	9.5		95585.285 (15)	
						9.5	8.5		95583.220 (15)	
						11.5	10.5		95580.760 (15)	
						8.5	7.5		95577.467 (20)	
						2.5	1.5		95571.667 (25)	
						7.5	6.5		95570.505 (25)	
						6.5	5.5		95564.306 (20)	
						3.5	2.5		95563.641 (25)	
						8.5	8.5		95563.075 (25)	
						5.5	4.5		95560.417 (30)	
						4.5	3.5		95559.988 (30)	
						9.5	9.5		95517.707 (20)	
						10.5	10.5		95452.090 (30)	
		2	$\lvert\Omega\rvert=0$	7	6	10.5	9.5		94959.210 (30)	77KUI
						9.5	8.5		94957.129 (25)	
						11.5	10.5		94954.686 (30)	
						8.5	7.5		94951.408 (40)	
						7.5	6.5		94944.485 (45)	
						6.5	5.5		94938.267 (45)	
		3	$\lvert\Omega\rvert=0$	7	6	10.5	9.5		94335.404 (35)	77KUI
						9.5	8.5		94333.368 (40)	
						11.5	10.5		94330.964 (30)	

Line frequencies of BiF (continued)

Isotopic species	State		Transition						ν MHz	Ref.
	electronic	vibrational v	$N' - N''$	J'	J''	F_1'	F_1''	$F' - F''$		
$^{209}Bi^{19}F$	$X\,^3\Sigma^-$	3	$\lvert\Omega\rvert=0$	8	7	11.5	10.5		107806.512 (30)	77KUI
						10.5	9.5		107805.246 (30)	
						12.5	11.5		107802.994 (40)	
						9.5	8.5		107801.240 (80)	
						8.5	7.5 }		107796.156	
						3.5	2.5 }			
		4	$\lvert\Omega\rvert=0$	8	7	11.5	10.5		107096.320 (25)	77KUI
						10.5	9.5		107095.060 (40)	
						12.5	11.5		107092.773 (30)	
						8.5	7.5 }		107086.003	
						3.5	2.5 }			
		5	$\lvert\Omega\rvert=0$	8	7	11.5	10.5		106388.856 (60)	77KUI
						10.5	9.5		106387.535 (60)	
						12.5	11.5		106385.321 (35)	
						8.5	7.5 }		106378.528	
						3.5	2.5 }			
		6	$\lvert\Omega\rvert\mp 0$	8	7	11.5	10.5		105684.054 (60)	77KUI
						12.5	11.5		105680.536 (40)	
						9.5	8.5		105678.640 (150)	
						8.5	7.5 }		105673.830	
						3.5	2.5 }			

References

77KUI Kuijpers, P., Dymanus, A.: Chem. Phys. **24** (1977) 97.

84 BiCl

Molecular constants of BiCl

Isotopic species	State	B_v MHz	D_v kHz	λ_v MHz	γ_v MHz	Derived constants	Ref.	Meth.
$^{209}Bi^{35}Cl$	XO^+ [1]	$Y_{01}=2761.8538(13)$ $Y_{11}=-12.0526(11)$ $Y_{21}=11.62(32)$ kHz $Y_{31}=43(25)$ Hz	$Y_{02}=-987.2(22)$ Hz			$r_e=2.47155(7)$ Å $a_0=2.577(6)\cdot 10^5$ cm^{-1} $a_1=-3.433(4)$	76KUI	MW
$^{209}Bi^{37}Cl$	XO^+ [1]	$Y_{01}=2634.0416(18)$ $Y_{11}=-11.2268(10)$ $Y_{21}=10.95(20)$ kHz	$Y_{02}=-896.8(31)$ Hz				76KUI	MW
$^{209}Bi^{35}Cl$	XO^+ [2]	$Y'_{01}=2761.996(13)$ $Y'_{11}=-12.0528(10)$ $Y'_{21}=11.67(28)$ kHz $Y'_{31}=41(23)$ Hz	$Y'_{02}=-987.1(20)$ Hz	$\lambda=3.55(35)\cdot 10^3$ cm^{-1}			77KUI	MW

[1]) Rotational constants shown contain small corrections due to second order spin-orbit coupling and admixture of the excited state $X1^{\pm}$ (compare preliminary remarks of this section).

[2]) λ is determined by isotope studies of $Bi^{35}Cl$ and $Bi^{37}Cl$. Parameters Y' are corrected for the effects mentioned in [1]).

Isotopic species	State		Parameter	Value	Ref.	Meth.	Remarks
	electronic	vibrational					
$^{209}Bi^{35}Cl$	XO^+	$v=0$	$eqQ(^{209}Bi)$ $eqQ(^{35}Cl)$ $C_I(^{209}Bi)$	$-1027(12)$ MHz $-34.5(28)$ MHz $-21(7)$ kHz	76KUI 76KUI 76KUI	MW MW MW	
$^{209}Bi^{37}Cl$	XO^+	$v=0$	$eqQ(^{209}Bi)$ $eqQ(^{37}Cl)$ $C_I(^{209}Bi)$	$-1025(28)$ MHz $-27.1(38)$ MHz $-18(22)$ kHz	76KUI 76KUI 76KUI	MW MW MW	

Line frequencies of BiCl

Isotopic species	State		Transition							ν MHz	Ref.
	electronic	vibrational v	N' – N''	J'	J''	F_1'	F_1''	F'	F''		
$^{209}Bi^{35}Cl$	$X\,^3\Sigma^-$	0	$\|\Omega\|=0$	11	10	8.5	7.5	8	7	60617.350 (70)	76KUI
		0		11	10	8.5	7.5	9	8	60617.410 (70)	
		0		11	10	9.5	8.5	9	8		
		0		11	10	8.5	7.5	10	9	60617.870 lines are	
		0		11	10	9.5	8.5	10	9	within 140 kHz	
		0		11	10	8.5	7.5	7	6		
		0		11	10	7.5	6.5	7	6		
		0		11	10	9.5	8.5	11	10	60618.375 lines are	
		0		11	10	9.5	8.5	8	7	within 150 kHz	
		0		11	10	7.5	6.5	8	7		
		0		11	10	7.5	6.5	6	5	60618.780 lines are	
		0		11	10	7.5	6.5	9	8	within 80kHz	
		0		11	10	10.5	9.5	10	9	60619.425 lines are	
		0		11	10	10.5	9.5	11	10	within 80 kHz	
		0		11	10	10.5	9.5	12	11	60619.880 lines are	
		0		11	10	10.5	9.5	9	8	within 80 kHz	
		0		11	10	6.5	5.5	6	5	60620.875 (30)	
		0		11	10	11.5	10.5	10	9		
		0		11	10	11.5	10.5	13	12	60622.130 lines are	
		0		11	10	11.5	10.5	12	11	within 200 kHz	
		0		11	10	11.5	10.5	11	10		
		0		11	10	11.5	10.5	12	11	60623.380 (50)	
		0		11	10	12.5	11.5	11	10	60623.570 (40)	
		0		11	10	12.5	11.5	13	12	60623.815 (40)	
		0		11	10	15.5	14.5	15	14		
		0		11	10	15.5	14.5	14	13		
		0		11	10	15.5	14.5	16	15	60624.260	
		0		11	10	15.5	14.5	12	11		
		0		11	10	12.5	11.5	14	13		
		0		11	10	15.5	14.5	17	16		
		0		11	10	13.5	12.5	14	13	60625.545 lines are	
		0		11	10	13.5	12.5	13	12	within 70 kHz	

Line frequencies of BiCl (continued)

Isotopic species	State		Transition							ν MHz	Ref.
	electronic	vibrational v	N' – N''	J'	– J''	F_1'	– F_1''	F'	– F''		
$^{209}Bi^{35}Cl$	$X\,^3\Sigma^-$	0	$\|\Omega\|=0$	11	10					60623.005 (25)[1]	76KUI
		0		13	12					71642.915 (30)	
		1		13	12					71330.155 (30)	
		2		13	12					71018.020 (40)	
		3		13	12					70706.485 (50)	
		4		13	12					70395.595 (50)	
		5		13	12					70085.330 (60)	
		0		17	16					93678.840 (40)	
		1		17	16					93269.830 (40)	
		2		17	16					92861.660 (40)	
		3		17	16					92454.275 (40)	
		0		18	17					99186.870 (40)	
		1		18	17					98753.815 (40)	
		2		18	17					98321.605 (40)	
		3		18	17					97890.260 (40)	
		4		18	17					97459.790 (50)	
		5		18	17					97030.205 (70)	
		6		18	17					96601.490 (70)	
		7		18	17					96173.700 (100)	
		8		18	17					95746.780 (100)	
		9		18	17					95320.790 (120)	
$^{209}Bi^{37}Cl$	$X\,^3\Sigma^-$	0	$\|\Omega\|=0$	11	10	8.5	7.5	8	7	57815.140	76KUI
		0		11	10	8.5	7.5	9	8		
		0		11	10	8.5	7.5	7	6	57815.570	
		0		11	10	8.5	7.5	10	9		
		0		11	10	9.5	8.5	9	8		
		0		11	10	9.5	8.5	10	9		
		0		11	10	9.5	8.5	11	10	57816.030	
		0		11	10	9.5	8.5	8	7		
		0		11	10	7.5	6.5	7	6		
		0		11	10	7.5	6.5	8	7		

[1]) unsplit frequency

Line frequencies of BiCl (continued)

Isotopic species	State: electronic	State: vibrational v	Transition: N' – N''	J'	J''	F_1'	F_1''	F'	F''	ν MHz	Ref.
$^{209}Bi^{37}Cl$	X $^3\Sigma^-$	0	$\lvert\Omega\rvert=0$	11	10	7.5	6.5	6	5	57816.405	76KUI
		0		11	10	7.5	6.5	9	8		
		0		11	10	10.5	9.5	10	9	57817.160	
		0		11	10	10.5	9.5	11	10		
		0		11	10	10.5	9.5	12	11	57817.545	
		0		11	10	10.5	9.5	9	8		
		0		11	10	6.5	5.5	6	5	57819.000	
		0		11	10	6.5	5.5	5	4		
		0		11	10	6.5	5.5	7	6		
		0		11	10	6.5	5.5	8	7		
		0		11	10	6.5	5.5	12	11		
		0		11	10	11.5	10.5	10	9	57819.750	
		0		11	10	11.5	10.5	13	12		
		0		11	10	11.5	10.5	12	11		
		0		11	10	11.5	10.5	11	10		
		0		11	10	11.5	10.5	12	11	57821.240	
		0		11	10	11.5	10.5	11	10		
		0		11	10	12.5	11.5	13	12	57821.530 (80)	
		0		11	10	12.5	11.5	12	11	57821.940	
		0		11	10	12.5	11.5	14	13		
		0		11	10	15.5	14.5	15	14		
		0		11	10	15.5	14.5	14	13		
		0		11	10	15.5	14.5	16	15		
		0		11	10	15.5	14.5	17	16		
		0		11	10	13.5	12.5	14	13	57823.265	
		0		11	10	13.5	12.5	13	12		
		0		11	10	13.5	12.5	12	11	57823.765	
		0		11	10	13.5	12.5	15	14		
		0		11	10	14.5	13.5	14	13		
		0		11	10	14.5	13.5	15	14		
		0		11	10	14.5	13.5	13	12		
		0		11	10	14.5	13.5	16	15		

Line frequencies of BiCl (continued)

Isotopic species	State		Transition				ν	Ref.
	electronic	vibrational v	N' – N''	J' – J''	F_1' – F_1''	F' – F''	MHz	
$^{209}Bi^{37}Cl$	$X\,^3\Sigma^-$	0	$\|\Omega\|=0$	11 10			57820.705 (40) [1]	76KUI
		0		13 12			68331.325 (50)	
		0		18 17			94602.590 (40)	
		1		18 17			94199.225 (40)	
		2		18 17			93796.620 (50)	
		3		18 17			93394.810 (70)	
		4		18 17			92993.840 (100)	
		5		18 17			92593.590 (100)	
		0		19 18			99855.765 (40)	
		1		19 18			99429.990 (40)	
		2		19 18			99005.030 (50)	
		3		19 18			98580.900 (80)	

[1]) Unsplit frequency.

References
76KUI Kuijpers, P., Törring, T., Dymanus, A.: Chem. Phys. **18** (1976) 401.
77KUI Kuijpers, P.: Doctoral thesis University Nijmegen, **1977**.

85 BiBr

Molecular constants of BiBr

Isotopic species	State	B_v MHz	D_v kHz	λ_v MHz	γ_v MHz	Derived constants	Ref.	Meth.
$^{209}Bi^{79}Br$	XO^+ [1])	$Y_{01}=1295.5609\,(12)$ $Y_{11}=-3.97809\,(18)$ $Y_{21}=2.303\,(18)$ kHz	$Y_{02}=-220.26\,(45)$ Hz			$r_e=2.60953\,(7)$ Å $a_0=2.542\,(6)\cdot 10^5$ cm^{-1} $a_1=-3.482\,(4)$	76KUI [1])	MW

[1]) Rotational constants shown contain small corrections due to second order spin-orbit coupling and admixture of the excited state $X1^{\pm}$ (compare preliminary remarks of this section).

Molecular constants of BiBr (continued)

Isotopic species	State	B_v MHz	D_v kHz	λ_v MHz	γ_v MHz	Derived constants	Ref.	Meth.
$^{209}Bi^{81}Br$	XO^+ [1])	$Y_{01}=1272.3406\,(12)$ $Y_{11}=-3.87164\,(16)$ $Y_{21}=2.225\,(14)$ kHz	$Y_{02}=-212.31\,(45)$ Hz				76KUI [1])	MW
$^{209}Bi^{79}Br$	XO^+ [2])	$Y'_{01}=1295.604\,(12)$ $Y'_{11}=-3.9778\,(3)$ $Y'_{21}=2.237\,(66)$ kHz $Y'_{31}=4\,(4)$ Hz	$Y'_{02}=-220.19\,(33)$ Hz	$\lambda=2.55\,(70)\cdot10^3\ cm^{-1}$			77KUI [2])	MW

[1]) Rotational constants shown contain small corrections due to second order spin-orbit coupling and admixture of the excited state $X1^{\pm}$ (compare preliminary remarks of this section).
[2]) λ is determined by isotope studies of $Bi^{79}Br$ and $Bi^{81}Br$. Parameters Y' are corrected for the effects mentioned in [1]).

Line frequencies of BiBr

Isotopic species	State		Transition					ν MHz	Ref.
	electronic	vibrational v	N' – N''	J'	J''	F'_1 – F''_1	F' – F''		
$^{209}Bi^{79}Br$	$X\,^3\Sigma^-$	0	$\lvert\Omega\rvert=0$	27	26			69835.510 (100)	76KUI
		1		27	26			69621.000 (100)	
		2		27	26			69406.670 (100)	
		3		27	26			69192.640 (100)	
		0		28	27			72420.665 (100)	
		1		28	27			72198.205 (100)	
		2		28	27			71975.985 (100)	
		3		28	27			71753.980 (100)	
		0		36	35			93096.090 (60)	
		1		36	35			92810.040 (60)	
		0		37	36			95679.735 (50)	
		1		37	36			95385.695 (50)	
		2		37	36			95092.000 (60)	

Line frequencies of BiBr (continued)

Isotopic species	State: electronic	State: vibrational v	Transition: N' – N''	J'	J''	F_1' – F_1''	F' – F''	ν MHz	Ref.
$^{209}Bi^{79}Br$	X $^3\Sigma^+$	3	$\|\Omega\|=0$	37	36			94798.655 (60)	76KUI
		4		37	36			94505.645 (60)	
		5		37	36			94212.940 (60)	
		0		38	37			98263.170 (50)	
		1		38	37			97961.170 (50)	
		2		38	37			97659.565 (60)	
		3		38	37			97358.245 (60)	
		4		38	37			97057.300 (60)	
		5		38	37			96756.745 (60)	
		6		38	37			96456.495 (60)	
		7		38	37			96156.623 (50)	
		8		38	37			95857.090 (60)	
		9		38	37			95557.925 (80)	
		0		39	38			100846.380 (50)	
		1		39	38			100536.450 (60)	
		2		39	38			100226.900 (60)	
		3		39	38			99917.695 (80)	
		4		39	38			99608.840 (80)	
		5		39	38			99300.315 (60)	
		6		39	38			98992.155 (60)	
		7		39	38			98684.400 (60)	
		8		39	38			98377.010 (60)	
		9		39	38			98069.930 (100)	
		8		40	39			100896.710 (80)	
		9		40	39			100581.730 (80)	
		10		40	39			100267.265 (100)	
$^{209}Bi^{81}Br$	X $^3\Sigma^-$	0	$\|\Omega\|=0$	27	26			68585.135 (100)	76KUI
		1		27	26			68376.335 (100)	
		2		27	26			68167.730 (150)	
		3		27	26			67959.390 (100)	
		0		28	27			71124.030 (100)	
		1		28	27			70907.550 (100)	
		2		28	27			70691.205 (100)	

Line frequencies of BiBr (continued)

Isotopic species	State		Transition					ν MHz	Ref.
	electronic	vibrational v	N' – N''	J'	J''	F_1' – F_1''	F' – F''		
$^{209}Bi^{81}Br$	X $^3\Sigma^-$	3	$\lvert\Omega\rvert=0$	28	27			70475.170 (150)	77KUI
		0		37	36			93966.975 (50)	
		1		37	36			93680.815 (50)	
		2		37	36			93394.955 (50)	
		3		37	36			93109.480 (80)	
		4		37	36			92824.295 (80)	
		0		38	37			96504.195 (50)	
		1		38	37			96210.305 (50)	
		2		38	37			95916.740 (60)	
		3		38	37			95623.510 (50)	
		4		38	37			95330.625 (60)	
		0		39	38			99041.240 (50)	
		1		39	38			98739.620 (50)	
		2		39	38			98438.320 (60)	
		3		39	38			98137.360 (60)	
		4		39	38			97836.770 (50)	
		5		39	38			97536.525 (50)	
		6		39	38			97236.585 (60)	
		7		39	38			96937.060 (60)	
		8		39	38			96637.800 (60)	
		9		39	38			96338.945 (60)	
		10		39	38			96040.460 (60)	
		11		39	38			95742.305 (80)	
		2		40	39			100959.695 (50)	
		3		40	39			100651.000 (60)	
		4		40	39			100342.715 (60)	
		5		40	39			100034.755 (60)	
		6		40	39			99727.150 (60)	
		7		40	39			99419.905 (60)	
		8		40	39			99113.020 (60)	
		9		40	39			98806.610 (80)	
		10		40	39			98500.340 (80)	

References: 76KUI Kuijpers, P., Dymanus, A.: Chem. Phys. Lett. **39** (1976) 217. 77KUI Kuijpers, P.: Doctoral thesis University Nijmegen, **1977**.

86 BiI

Molecular constants of BiI

Isotopic species	State	B_v MHz	D_v kHz	λ_v MHz	γ_v MHz	Derived constants	Ref.	Meth.
$^{209}Bi^{127}I$	XO^+ [1])	$Y_{01}=816.11943\,(13)$	$Y_{02}=-89.816\,(18)$ Hz			$r_e=2.80053\,(8)$ Å	76KUI	MW
		$Y_{11}=-2.09227\,(4)$	$Y_{12}=-0.106\,(6)$ Hz			$a_0=2.4736\,(6)\cdot 10^5\ cm^{-1}$		
		$Y_{21}=516\,(4)$ Hz				$a_1=-3.5760\,(6)$		
		$Y_{31}=-2.96\,(13)$ Hz				$a_2=8.12\,(4)$		
						$a_3=-14.3\,(4)$		

[1]) Rotational constants shown contain small corrections due to second order spin-orbit coupling and admixture of the excited state $X1^{\pm}$ (compare preliminary remarks of this section).

Isotopic species	State: electronic	State: vibrational	Parameter	Value	Ref.	Meth.	Remarks
$^{209}Bi^{127}I$	XO^+	$v=0$	$eqQ(^{209}Bi)$	$-909.5\,(20)$ MHz	81TIS	MW	
			$eqQ(^{127}I)$	$-995.0\,(20)$ MHz	81TIS	MW	

Line frequencies of BiI

Isotopic species	State: electronic	State: vibrational v	Transition: N' – N''	J'	J''	F_1' – F_1''	F' – F''	ν MHz	Ref.
$^{209}Bi^{127}I$	$X\,^3\Sigma^-$	0	$\lvert\Omega\rvert=0$	41	40			66811.240 (80)	76KUI
		0		42	41			68439.525 (60)	
		1		42	41			68263.850 (80)	
		2		42	41			68088.230 (80)	
		3		42	41			67912.694 (80)	
		4		42	41			67737.240 (80)	
		5		42	41			67561.851 (100)	
		6		42	41			67386.600 (100)	
		0		43	42			70067.735 (60)	

Line frequencies of BiI (continued)

Isotopic species	State		Transition					ν	Ref.
	electronic	vibrational v	N' – N''	J'	– J''	F_1' – F_1''	F' – F''	MHz	
$^{209}Bi^{127}I$	$X\ ^3\Sigma^-$	1	$\|\Omega\|=0$	43	42			69887.855 (60)	76KUI
		2		43	42			69708.075 (60)	
		3		43	42			69528.340 (100)	
		4		43	42			69348.705 (80)	
		5		43	42			69169.175 (100)	
		6		43	42			68989.720 (100)	
		0		44	43			71695.825 (60)	
		1		44	43			71511.790 (80)	
		2		44	43			71327.820 (100)	
		3		44	43			71143.880 (100)	
		4		44	43			70960.095 (100)	
		5		44	43			70776.405 (100)	
		6		44	43			70592.765 (130)	
		0		45	44			73323.855 (50)	
		1		45	44			73135.595 (50)	
		2		45	44			72947.430 (80)	
		3		45	44			72759.420 (100)	
		4		45	44			72571.400 (100)	
		5		45	44			72383.475 (100)	
		6		45	44			72195.685 (100)	
		0		59	58			96104.825 (30)	
		1		59	58			95857.995 (50)	
		0		60	59			97731.160 (30)	
		1		60	59			97480.130 (30)	
		2		60	59			97229.215 (50)	
		3		60	59			96978.400 (50)	
		4		60	59			96727.725 (50)	
		5		60	59			96477.140 (50)	
		6		60	59			96226.660 (50)	
		7		60	59			95976.320 (80)	
		0		61	60			99357.360 (30)	
		1		61	60			99102.145 (30)	
		2		61	60			98847.030 (30)	
		3		61	60			98592.030 (30)	

Line frequencies of BiI (continued)

Isotopic species	State		Transition				ν MHz	Ref.
	electronic	vibrational v	N' – N''	J' – J''	F_1' – F_1''	F' – F''		
$^{209}Bi^{127}I$	$X\ ^3\Sigma^-$	4	$\|\Omega\|=0$	61 60			98337.180 (50)	76KUI
		5		61 60			98082.420 (50)	
		6		61 60			97827.805 (50)	
		7		61 60			97573.265 (50)	
		8		61 60			97318.830 (50)	
		9		61 60			97064.515 (80)	
		10		61 60			96810.340 (150)	
		11		61 60			96556.230 (150)	
		12		61 60			96302.265 (150)	
		13		61 60			96048.380 (150)	
		0		62 61			100983.425 (30)	
		1		62 61			100724.020 (30)	
		2		62 61			100464.735 (30)	
		3		62 61			100205.555 (80)	
		4		62 61			99946.515 (50)	
		5		62 61			99687.570 (80)	
		6		62 61			99428.745 (50)	
		7		62 61			99170.080 (50)	
		8		62 61			98911.500 (80)	
		9		62 61			98653.020 (80)	
		10		62 61			98394.630 (100)	
		11		62 61			98136.410 (150)	
		12		62 61			97878.200 (150)	
		13		62 61			97620.105 (150)	
		14		62 61			97362.210 (150)	
		15		62 61			97104.330 (150)	
		16		62 61			96846.520 (100)	
		17		62 61			96588.850 (150)	
		18		62 61			96331.200 (150)	
		19		62 61			96073.740 (150)	
		20		62 61			95816.290 (150)	
		6		63 62			101029.620 (50)	
		7		63 62			100766.700 (50)	
		8		63 62			100503.965 (80)	

Line frequencies of BiI (continued)

Isotopic species	State		Transition				ν MHz	Ref.
	electronic	vibrational v	N' – N''	J' – J''	F_1' – F_1''	F' – F''		
$^{209}Bi^{127}I$	$X\,^3\Sigma^-$	9	$\|\Omega\|=0$	63 62			100241.300 (100)	76KUI
		10		63 62			99978.800 (100)	
		11		63 62			99716.330 (100)	
		12		63 62			99454.020 (150)	
		13		63 62			99191.810 (150)	

References

76KUI Kuijpers, P., Törring, T., Dymanus, A.: Chem. Phys. **12** (1976) 309.

81TIS Tischer, R., Möller, K., Törring, T.: Chem. Phys. **62** (1981) 115.

87 O_2

Molecular constants of O_2

Isotopic species	State	B_v MHz	D_v kHz	λ_v MHz	γ_v MHz	Derived constants	Ref.	Meth.
$^{16}O_2$	$X\,^3\Sigma^-$	$B_e^{cor}=43342.6\,(10)$	$Y_{02}=-145.010\,(75)$	$\lambda_e^{cor}=59424.14\,(16)$	$\gamma_e^{cor}\approx\gamma_e$ $=-252.2728\,(52)$	$d_0=0.80\,(25)$	82TIE [1])	MW
		$Y_{11}=-475.3\,(20)$		$\alpha_\lambda=145.1\,(20)$	$\alpha_\gamma=-0.627\,(10)$	$d_{\lambda 0}=-0.57\,(14)$	82TIE [1])	MW
				$D_\lambda=58.469\,(15)$ kHz	$D_\gamma=-245.6\,(10)$ Hz	$r_e^{cor}=1.20748\,(2)$ Å	82TIE [1])	MW
$^{16}O_2$	$X\,^3\Sigma^-$	$B_0=43100.430\,(47)$	$D_0=143.6\,(33)$	$\lambda_0=59501.3489\,(38)$	$\gamma_0=-252.5875\,(30)$		82END	MW
				$D_{\lambda 0}=58.305\,(19)$ kHz	$D_{\gamma 0}=-245.44\,(29)$ Hz		82END	MW
				$H_{\lambda 0}$[2]$)=0.339\,(14)$ Hz			82END	MW
		$B_1=42626.398\,(82)$	$D_1^{ev}=148.865$	$\lambda_1=59646.0785\,(66)$	$\gamma_1=-253.1925\,(11)$		82END	MW
				$D_{\lambda 1}=63.26\,(10)$ kHz	$D_{\gamma 1}=-253.4\,(36)$ Hz		82END	MW
$^{16}O^{17}O$	$X\,^3\Sigma^-$	$B_0=41830.6\,(1)$	$D_0=138.6$	$\lambda_0=59500.09\,(1)$	$\gamma_0=-245.114\,(4)$		81CAZ	MW
				$D_\lambda=57.5\,(2)$ kHz	$D_\gamma=-240\,(30)$ Hz		81CAZ	MW

[1]) This reference contains molecular parameters of $^{16}O_2$ from which the parameters of other isotopes can be derived with high reliability by simple mass relations in order to predict rotational transitions.

[2]) Third order centrifugal distortion.

Molecular constants of O_2 (continued)

Isotopic species	State	B_v MHz	D_v kHz	λ_v MHz	γ_v MHz	Derived constants	Ref.	Meth.
$^{16}O^{18}O$	$X\,^3\Sigma^-$	$B'_0=40708.0\,(12)$		$\lambda'_0=59499.17\,(11)$ $D_{\lambda 0}=52.8\,(16)$ kHz	$\gamma'_0=-238.518\,(8)$		74AMA 74AMA	MW MW
$^{17}O^{18}O$	$X\,^3\Sigma^-$	$B_0=39437.4\,(1)$	$D_0=123$	$\lambda_0=59497.80\,(4)$ $D_\lambda=54.0\,(5)$ kHz	$\gamma_0=-231.049\,(7)$ $D_\gamma=-270\,(60)$ Hz		81CAZ 81CAZ	MW MW
$^{18}O_2$	$X\,^3\Sigma^-$	$B_0=38313.721\,(10)$		$\lambda_0=59496.708\,(15)$ $D_{\lambda 0}=52.1\,(6)$ kHz	$\gamma_0=-224.438\,(6)$ $D_{\gamma 0}=-290\,(100)$ Hz		73STE 73STE	MW MW

Isotopic species	State		Parameter	Value	Ref.	Meth.	Remarks
	electronic	vibrational					
$^{16}O^{17}O$	$X\,^3\Sigma^-$	$v=0$	$b_0(^{17}O)$ $c_0(^{17}O)$ $eq_0Q(^{17}O)$	$-101.441\,(5)$ MHz $139.73\,(3)$ MHz $-8.3\,(3)$ MHz	81CAZ 81CAZ 81CAZ	MW MW MW	
$^{17}O^{18}O$	$X\,^3\Sigma^-$	$v=0$	$b_0(^{17}O)$ $c_0(^{17}O)$ $eq_0Q(^{17}O)$	$-101.46\,(1)$ MHz $139.68\,(6)$ MHz $-7.8\,(5)$ MHz	81CAZ 81CAZ 81CAZ	MW MW MW	
$^{16}O_2$	$X\,^3\Sigma^-$	$v=0$	linewidth parameters for self-broadening and foreign gas perturbers as N_2, H_2O, He, Ne, Ar, Kr, Xe		77SET 79SMI 78GIM	MW MW MW	measurements and theoretical interpretation of pressure broadening of the O_2 microwave spectrum
$^{16}O_2$	$X\,^3\Sigma^-$		linewidth parameters for self-broadening		75GAR, 81GAR	EPR	
$^{16}O_2$	$X\,^3\Sigma^-$	$v=0$	g_s g_L^{el} g_{rot}	$2.002084\,(16)$ $2.77\,(2)\cdot 10^{-3}$ $-0.213\,(18)$	78CHR 78CHR 78CHR 73COO	EPR/LMR EPR/LMR EPR/LMR EPR	measurements of earlier published works were used for a combined fit measurements on the first excited vibrational state $v=1$
$^{16}O_2$	$X\,^3\Sigma^-$	$v=0$	$\alpha_\parallel-\alpha_\perp$	$1.12\,(7)$ Å^3	74GUS	MW	anisotropy of the electric polarizability; its derivatives are determined by Raman-spectroscopy in [82HAM]

Line frequencies of O_2

Isotopic species	State: electronic	State: vibrational v	Transition: N'	N''	J'	J''	F_1'	F_1''	F'	F''	ν MHz	Ref.
$^{16}O_2$	$X\ ^3\Sigma_g^-$	0	31	31	31	30					52021.423 (9)	82END
		0	33	33	33	32					51503.348 (18)	82END
		0	35	35	35	34					50987.760 (19)	82END
		0	37	37	37	36					50474.223 (27)	82END
		0	33	33	33	34					67900.867 (11)	82END
		0	35	35	35	36					68431.005 (17)	82END
		0	37	37	37	38					68960.312 (12)	82END
		0	39	39	39	40					69489.030 (15)	82END
		0	3	1	2	2					424763.21 (10)	73STE
		0	5	3	5	3					775697.5 (70)	75TOM, 72MIZ
		0	17	15	17	15					2839400.6 (6)	75TOM
		0	21	19	21	19					3524221 (5)	75TOM
$^{16}O_2$	$X\ ^3\Sigma_g^-$	1	1	1	1	0					119039.222 (18)	82END
		1	3	3	3	2					62525.639 (19)	82END
		1	5	5	5	4					60390.862 (8)	82END, 74AMA
		1	7	7	7	6					59265.539 (13)	82END, 74AMA
		1	9	9	9	8					58433.416 (10)	82END, 74AMA
		1	11	11	11	10					57726.756 (20)	82END, 74AMA
		1	13	13	13	12					57085.381 (12)	82END
		1	15	15	15	14					56482.400 (22)	82END
		1	17	17	17	16					55904.080 (20)	82END
		1	19	19	19	18					55342.380 (34)	82END
		1	3	3	3	4					58650.837 (8)	82END
		1	5	5	5	6					59778.948 (17)	82END, 74AMA
		1	7	7	7	8					60614.812 (8)	82END, 74AMA
		1	9	9	9	10					61326.233 (9)	82END, 74AMA
		1	11	11	11	12					61973.396 (11)	82END, 74AMA
		1	13	13	13	14					62583.122 (14)	82END, 74AMA
		1	15	15	15	16					63169.240 (13)	82END
		1	17	17	17	18					63739.708 (15)	82END
		1	19	19	19	20					64299.195 (9)	82END

Line frequencies of O_2 (continued)

Isotopic species	State: electronic	State: vibrational v	Transition: N'	N''	J'	J''	F_1'	F_1''	F'	F''	ν MHz	Ref.
$^{16}O^{17}O$	$X\,^3\Sigma_g^-$	0	1	1	1	0			3.5	2.5	118804.032	81CAZ
		0	1	1	1	2			3.5	4.5	56739.737	81CAZ
		0	1	1	1	2			2.5	3.5	56502.323	81CAZ
		0	2	2	2	1			4.5	3.5	64745.263	81CAZ
		0	2	2	2	3			2.5	3.5	57665.614	81CAZ
		0	2	2	2	3			3.5	4.5	57812.656	81CAZ
		0	2	2	2	3			4.5	5.5	57993.312	81CAZ
		0	3	3	3	4			5.5	6.5	58795.503	81CAZ
		0	3	3	3	4			3.5	4.5	58502.944	81CAZ
		0	3	3	3	4			0.5	1.5	58250.104	81CAZ
		0	4	4	4	5			5.5	6.5	59248.146	81CAZ
		0	4	4	4	5			3.5	4.5	59014.377	81CAZ
		0	4	4	4	5			1.5	2.5	58859.263	81CAZ
		0	5	5	5	4			7.5	6.5	59983.058	81CAZ
		0	5	5	5	4			4.5	3.5	60348.400	81CAZ
		0	5	5	5	4			3.5	2.5	60425.307	81CAZ
		0	5	5	5	4			2.5	1.5	60480.051	81CAZ
		0	5	5	5	4			5.5	4.5	60249.049	81CAZ
		0	5	5	5	6			4.5	5.5	59518.005	81CAZ
		0	5	5	5	6			5.5	6.5	59623.861	81CAZ
		0	5	5	5	6			3.5	4.5	59428.531	81CAZ
		0	5	5	5	6			2.5	3.5	59355.446	81CAZ
		0	6	6	6	7			8.5	9.5	60309.784	81CAZ
		0	6	6	6	7			7.5	8.5	60176.562	81CAZ
		0	6	6	6	5			7.5	6.5	59517.385	81CAZ
		0	6	6	6	5			6.5	5.5	59635.081	81CAZ
		0	7	7	7	8			8.5	9.5	60563.718	81CAZ
		0	7	7	7	8			7.5	8.5	60446.847	81CAZ
		0	7	7	7	8			6.5	7.5	60342.344	81CAZ
		0	7	7	7	8			4.5	5.5	60170.202	81CAZ
		0	7	7	7	8			5.5	6.5	60250.177	81CAZ
		0	7	7	7	6			4.5	3.5	59387.677	81CAZ
		0	7	7	7	6			6.5	5.5	59234.868	81CAZ
		0	7	7	7	6			8.5	7.5	59020.323	81CAZ

Line frequencies of O_2 (continued)

Isotopic species	State		Transition								ν	Ref.
	electronic	vibrational v	N'	– N''	J'	– J''	F_1'	– F_1''	F'	– F''	MHz	
$^{16}O^{17}O$	$X\ ^3\Sigma_g^-$	0	8	8	8	7			7.5	6.5	58803.690	81CAZ
		0	8	8	8	9			6.5	7.5	60608.773	81CAZ
		0	8	8	8	9			5.5	6.5	60526.484	81CAZ
		0	9	9	9	8			9.5	8.5	58316.904	81CAZ
		0	9	9	9	8			7.5	6.5	58504.784	81CAZ
		0	10	10	10	9			11.5	10.5	57849.561	81CAZ
		0	10	10	10	9			7.5	6.5	58228.000	81CAZ
		0	10	10	10	11			9.5	10.5	61362.064	81CAZ
		0	10	10	10	11			7.5	8.5	61181.888	81CAZ
		0	10	10	10	11			8.5	9.5	61267.499	81CAZ
		0	11	11	11	12			8.5	9.5	61490.219	81CAZ
		0	11	11	11	12			10.5	11.5	61672.185	81CAZ
		0	11	11	11	12			11.5	12.5	61775.452	81CAZ
		0	11	11	11	12			12.5	13.5	61887.117	81CAZ
		0	11	11	11	12			13.5	14.5	62006.912	81CAZ
		0	11	11	11	10			9.5	8.5	57815.148	81CAZ
		0	11	11	11	10			13.5	12.5	57395.901	81CAZ
		0	12	12	12	13			14.5	15.5	62305.312	81CAZ
		0	12	12	12	13			12.5	13.5	62076.175	81CAZ
		0	12	12	12	13			11.5	12.5	61973.037	81CAZ
		0	12	12	12	13			10.5	11.5	61877.465	81CAZ
		0	12	12	12	13			9.5	10.5	61789.472	81CAZ
		0	12	12	12	11			11.5	10.5	57405.605	81CAZ
		0	12	12	12	11			9.5	8.5	57579.403	81CAZ
		0	12	12	12	11			14.5	13.5	57078.826	81CAZ
$^{16}O^{18}O$	$X\ ^3\Sigma_g^-$	0	4	4	4	3					60861.117 (114)	75STE
		0	5	5	5	4					60105.612 (77)	75STE, 74AMA
		0	6	6	6	5					59539.420 (56)	75STE, 74AMA
		0	7	7	7	6					59074.760 (67)	75STE, 74AMA
		0	8	8	8	7					58670.780 (73)	75STE
		0	9	9	9	8					58306.003 (55)	75STE, 74AMA
		0	10	10	10	9					57968.07 (15)	74AMA
		0	3	3	3	4					58656.447 (65)	75STE

Line frequencies of O_2 (continued)

Isotopic species	State		Transition								ν MHz	Ref.
	electronic	vibrational v	N′	N″	J′	J″	F_1'	F_1''	F′	F″		
$^{16}O^{18}O$	$X\,^3\Sigma_g^-$	0	4	4	4	5					59223.46 (15)	74AMA
		0	5	5	5	6					59689.56 (15)	74AMA
		0	6	6	6	7					60094.935 (17)	75STE, 74AMA
		0	7	7	7	8					60461.24 (15)	74AMA
		0	8	8	8	9					60801.121 (58)	75STE, 74AMA
		0	2	0	1	1					233946.179 (61)	75STE
$^{17}O^{18}O$	$X\,^3\Sigma_g^-$	0	1	1	1	0			3.5	2.5	118813.332	81CAZ
		0	1	1	1	2			2.5	3.5	56990.420	81CAZ
		0	1	1	1	2			3.5	4.5	57227.851	81CAZ
		0	2	2	2	1			2.5	1.5	64325.346	81CAZ
		0	2	2	2	1			3.5	2.5	64143.207	81CAZ
		0	2	2	2	1			4.5	3.5	63885.710	81CAZ
		0	2	2	2	3			2.5	3.5	57989.114	81CAZ
		0	3	3	3	4			4.5	5.5	58863.763	81CAZ
		0	3	3	3	4			5.5	6.5	59022.376	81CAZ
		0	4	4	4	3			6.5	5.5	60476.563	81CAZ
		0	4	4	4	5			5.5	6.5	59408.717	81CAZ
		0	5	5	5	6			4.5	5.5	59628.441	81CAZ
		0	5	5	5	6			7.5	8.5	59995.127	81CAZ
		0	6	6	6	5			7.5	6.5	59366.014	81CAZ
		0	6	6	6	5			8.5	7.5	59229.873	81CAZ
		0	6	6	6	7			7.5	8.5	60246.530	81CAZ
		0	7	7	7	6			7.5	6.5	59034.094	81CAZ
		0	7	7	7	6			9.5	8.5	58788.470	81CAZ
		0	7	7	7	8			7.5	8.5	60482.509	81CAZ
		0	7	7	7	8			8.5	9.5	60599.282	81CAZ
		0	8	8	8	7			6.5	5.5	58829.054	81CAZ
		0	8	8	8	7			9.5	8.5	58530.129	81CAZ
		0	8	8	8	9			5.5	6.5	60532.058	81CAZ
		0	9	9	9	8			7.5	6.5	58477.981	81CAZ
		0	9	9	9	8			9.5	8.5	58290.190	81CAZ
		0	10	10	10	9			7.5	6.5	58231.200	81CAZ
		0	10	10	10	9			11.5	10.5	57852.677	81CAZ

Line frequencies of O_2 (continued)

Isotopic species	State: electronic	State: vibrational v	Transition: N'	N''	J'	J''	F_1'	F_1''	F'	F''	ν MHz	Ref.
$^{17}O^{18}O$	$X\ ^3\Sigma_g^-$	0	11	11	11	10			9.5	8.5	57845.277	81CAZ
		0	12	12	12	13			10.5	11.5	61786.628	81CAZ
		0	12	12	12	13			11.5	12.5	61882.057	81CAZ
$^{18}O_2$	$X\ ^3\Sigma_g^-$	0	1	1	1	2					57239.907 (20)	73STE
		0	3	3	3	4					58899.771 (27)	73STE
		0	5	5	5	6					59811.414 (34)	73STE
		0	7	7	7	8					60505.782 (62)	73STE
		0	3	3	3	2					61529.854 (32)	73STE
		0	5	5	5	4					59871.473 (23)	73STE
		0	7	7	7	6					58962.067 (24)	73STE
		0	9	9	9	8					58270.727 (9)	73STE
		0	3	1	2	2					378831.51 (10)	73STE

References

72MIZ Mizushima, M., Wells, J.S., Evenson, K.M., Welch, W.M.: Phys. Rev. Lett. **29** (1972) 831.

73COO Cook, T.J., Zegarski, B.R., Breckenridge, W.H., Miller, T.A.: J. Chem. Phys. **58** (1973) 1548.

73STE Steinbach, W., Gordy, W.: Phys. Rev. **A8** (1973) 1753.

74AMA Amano, T., Hirota, E.: J. Mol. Spectrosc. **53** (1974) 346.

74GUS Gustafson, S., Gordy, W.: Phys. Lett. **49A** (1974) 161.

75GAR Gardiner, W.C., Pickett, H.M., Proffitt, M.H.: J. Chem. Phys. **63** (1975) 2149.

75STE Steinbach, W., Gordy, W.: Phys. Rev. **A11** (1975) 729.

75TOM Tomuta, L., Mizushima, M., Howard, C.J., Evenson, K.M.: Phys. Rev. **A12** (1975) 974.

77SET Setzer, B.J., Pickett, H.M.: J. Chem. Phys. **67** (1977) 340.

78CHR Christensen, H., Veseth, L.: J. Mol. Spectrosc. **72** (1978) 438.

78GIM Gimmestad.,G.G.: Ph. D. thesis, Dept. Phys. Univ. Colorado **1978**

79SMI Smith, E.W., Giraud, M.: J. Chem. Phys. **71** (1979) 4209.

81CAZ Cazzoli, G., DegliEsposti, C., Favero, P.G., Severi, G.: Il Nuovo Cimento **62B** (1981) 243.

81GAR Gardiner, W.C., Pickett, H.M., Proffitt, M.H.: J. Chem. Phys. **74** (1981) 6037.

82END Endo, Y., Mizushima, M.: Jap. J. Appl. Phys. **21** (1982) L379.

82HAM Hamaguchi, H., Buckingham, A.D., Jones, W.J.: Mol. Phys. **46** (1982) 1093.

82TIE Tiemann, E.: J. Mol. Spectrosc. **91** (1982) 60.

88 SO

Molecular constants of SO

Isotopic species	State	B_v MHz	D_v kHz	λ_v MHz	γ_v MHz	Derived constants	Ref.	Meth.
$^{32}S^{16}O$	X $^3\Sigma^-$	$Y_{01}=21609.710\,(3)$ $Y_{11}=-172.410\,(5)$ $Y_{21}=\quad 0.228\,(2)$ $Y_{31}=-2.5\,(2)$ kHz	$Y_{02}=-33.90\,(1)$	$\lambda_e=157795.7\,(2)$ $\alpha_\lambda=912.1\,(4)$ $\gamma_\lambda=\ 10.8\,(1)$ $\varepsilon_\lambda=\quad 0.38\,(1)$ $D_\lambda=\quad 0.309\,(1)$	$\gamma_e=-167.71\,(2)$ $\alpha_\gamma=-1.28\,(2)$		81BOG [1])	MW
$^{32}S^{16}O$	X $^3\Sigma^-$	$Y_{01}=21609.7067\,(24)$	$Y_{02}=-33.894\,(10)$	$\lambda_e=157795.671\,(74)$	$\gamma_e=-167.680\,(20)$	$d_S=0.915\,(28)$	82BOG, 82TIE1 [1])	MW
		$B_e^{cor}=21612.387\,(28)$	$Y_{12}^{der}=1.4$ Hz	$\lambda_e^{cor}=157787.77\,(62)$		$d_O=1.350\,(12)$	82BOG, 82TIE1	MW
		$Y_{11}=-172.4094\,(40)$	$Y_{03}^{der}=-2\cdot10^{-3}$ Hz	$\alpha_\lambda=911.92\,(19)$	$\alpha_\gamma=-1.266\,(22)$	$d_{\lambda S}=-0.098\,(80)$	82BOG, 82TIE1	MW
		$Y_{21}=\quad 0.2282\,(16)$		$\gamma_\lambda=\ 10.85\,(11)$		$d_{\lambda O}=-0.681\,(40)$	82BOG, 82TIE1	MW
		$Y_{31}=-2.57\,(17)$ kHz		$\varepsilon_\lambda=386\,(13)$ kHz			82BOG	MW
				$D_\lambda=305.3\,(12)$ $+4.3\,(12)\,(v+1/2)$ kHz	$D_\gamma=-427\,(83)$ Hz		82BOG, 82TIE1	MW
						$\omega_e=1150.710\,(11)\ \text{cm}^{-1}$ $\omega_e x_e=6.3505\,(51)\ \text{cm}^{-1}$	79KAW	LMR
SO	X $^3\Sigma^-$					$a_0=459187.2\,(88)\ \text{cm}^{-1}$ $a_1=\ -3.12222\,(5)$ $a_2=\quad 6.3127\,(47)$ $a_3=-10.284\,(38)$ $r_e^{cor}=1.4810087\,(51)$ Å	82BOG 82BOG 82BOG 82BOG 74TIE1, 82TIE1	MW MW MW MW MW
$^{32}S^{17}O$	X $^3\Sigma^-$	$B_0=20677.809\,(8)$	$D_0=31.1\,(3)$	$\lambda_0=158245.1\,(11)$ $D_\lambda=0.293\,(5)$	$\gamma_0=-161.68\,(11)$		82CAZ	MW

1) The parameter sets of both references agree very well. [81BOG] performed separate fits for each isotopic species, but [82TIE1] and [82BOG] included the data of all isotopes in a single fit taking into account the breakdown of the Born-Oppenheimer approximation. Such corrections are described by the *d*-parameters. [82TIE1] and [82BOG] contain molecular parameters of $^{32}S^{16}O$ from which the parameters of other isotopes can be derived with high reliability by simple mass relations in order to predict rotational transitions.

Molecular constants of SO (continued)

Isotopic species	State	B_v MHz	D_v kHz	λ_v MHz	γ_v MHz	Derived constants	Ref.	Meth.
$^{32}S^{18}O$	X $^3\Sigma^-$	$B_0 = 19929.251\,(18)$	$D_0 = 28.54\,(30)$	$\lambda_0 = 158233.1\,(33)$ $D_\lambda = 0.293\,(15)$	$\gamma_0 = -155.53\,(32)$		81BOG 81BOG	MW MW
$^{34}S^{16}O$	X $^3\Sigma^-$	$Y_{01} = 21186.36\,(6)$ $Y_{11} = -167.3\,(1)$ $Y_{21} = 0.21\,(7)$	$Y_{02} = -32.60\,(7)$	$\lambda_e = 157797\,(4)$ $\alpha_\lambda = 900\,(9)$ $\gamma_\lambda = 13\,(5)$ $D_\lambda = 0.299\,(3)$	$\gamma_e = -164.4\,(1)$ $\alpha_\gamma = -1.2\,(1)$		81BOG 81BOG 81BOG 81BOG	MW MW MW MW

Isotopic species	State		Parameter	Value	Ref.	Meth.	Remarks
	electronic	vibrational					
$^{33}S^{16}O$	X $^3\Sigma^-$	$v=0$	$b(^{33}S)$ $c(^{33}S)$ $eqQ(^{33}S)$	51.12 (15) MHz −96.71 (24) MHz −15.96 (65) MHz	82CAZ, 82TIE2 82CAZ, 82TIE2 82CAZ, 82TIE2	MW MW MW	reevaluation of data from [67AMA]
$^{32}S^{17}O$	X $^3\Sigma^-$	$v=0$	$b_0(^{17}O)$ $c_0(^{17}O)$ $eq_0Q(^{17}O)$	−71.68 (6) MHz 95.50 (9) MHz −3.5 (2) MHz	82CAZ, 82TIE2 82CAZ, 82TIE2 82CAZ	MW MW MW	
$^{32}S^{16}O$	X $^3\Sigma^-$	$v=0$	g_s g_L^{el} g_{rot}	2.002120 (29) $3.37\,(7)\cdot 10^{-3}$ −0.195 (73)	78CHR 78CHR 78CHR	EPR EPR EPR	measurements of earlier published works were used for a combined fit

Line frequencies of SO

Isotopic species	State		Transition								ν MHz	Ref.
	electronic	vibrational v	N'	N''	J'	J''	F_1'	F_1''	F'	F''		
$^{32}S^{16}O$	X $^3\Sigma^-$	0	1	0	1	1					286340.152 (30)	82TIE1
			2	1	3	2					99299.870 (100)	74TIE1
			2	1	2	1					86093.950 (100)	74TIE1
			2	1	1	0					329385.477 (100)	76CLA

Line frequencies of SO (continued)

Isotopic species	State		Transition								ν MHz	Ref.
	electronic	vibrational v	N'	N''	J'	J''	F_1'	F_1''	F'	F''		
$^{32}S^{16}O$	$X\,^3\Sigma^-$	0	2	1	2	2					309502.444 (100)	76CLA
			3	2	3	2					129138.850 (100)	74TIE1
			3	2	2	1					109252.100 (100)	74TIE1
			3	2	3	3					339341.459 (100)	76CLA
			3	2	2	2					336202.022 (50)	82TIE1
			4	3	5	4					178605.403 (100)	76CLA
			5	4	6	5					219949.442 (100)	76CLA
			5	4	5	4					215220.653 (100)	76CLA
			5	4	4	3					206176.005 (100)	76CLA
			5	4	4	4					100029.640 (150)	74TIE1
			6	5	7	6					261843.721 (30)	82TIE1
			6	5	5	4					251825.770 (100)	76CLA
			6	5	5	5					136634.799 (100)	76CLA
			7	6	8	7					304077.844 (100)	76CLA
			7	6	7	6					301286.124 (100)	76CLA
			7	6	6	5					296550.064 (30)	82TIE1
			7	6	6	6					174928.860 (100)	76CLA
			8	7	9	8					346528.481 (100)	76CLA
			8	7	8	7					344310.612 (100)	76CLA
			8	7	7	6					340714.155 (100)	76CLA
			8	7	7	7					214357.039 (100)	76CLA
			10	9	9	9					295355.696 (200)	82TIE1
			2	1	1	2					236452.356 (400)	82BOG
			3	2	2	3					246404.700 (500)	82BOG
			6	5	6	5					258255.948 (80)	82BOG
			9	8	8	7					384527.490 (150)	82BOG
			9	8	9	8					387328.900 (150)	82BOG
			9	8	10	9					389121.150 (200)	82BOG
			10	9	9	8					428111.024 (200)	82BOG
			10	9	10	9					430339.722 (200)	82BOG
			10	9	11	10					431808.360 (300)	82BOG
			11	10	10	9					471538.130 (200)	82BOG
			11	10	11	10					473342.280 (250)	82BOG

Line frequencies of SO (continued)

Isotopic species	State: electronic	State: vibrational v	Transition: N'	N''	J'	J''	F_1'	F_1''	F'	F''	ν MHz	Ref.
$^{32}S^{16}O$	X $^3\Sigma^-$	0	11	10	12	11					474559.700 (200)	82BOG
			12	11	11	10					514853.970 (250)	82BOG
			12	11	12	11					516336.030 (250)	82BOG
			12	11	13	12					517354.800 (300)	82BOG
		1	1	0	1	1					288261.010 (100)	82TIE1
			6	5	7	6					259857.317 (50)	82TIE1
			7	6	7	6					298878.699 (50)	82TIE1
			7	6	6	5					294047.815 (50)	82TIE1
			2	1	2	1					85406.277 (80)	82BOG
			2	1	3	2					98758.182 (80)	82BOG
			3	2	2	1					108115.605 (100)	82BOG
			3	2	3	2					128107.215 (50)	82BOG
			3	2	4	3					137284.300 (50)	82BOG
			4	3	3	2					157449.995 (50)	82BOG
			4	3	4	3					170805.752 (50)	82BOG
			4	3	5	4					177345.560 (80)	82BOG
			5	4	4	3					204319.104 (80)	82BOG
			5	4	5	4					213501.110 (80)	82BOG
			5	4	6	5					218324.553 (50)	82BOG
			6	5	5	5					134462.764 (150)	82BOG
			7	6	8	7					301733.518 (150)	82BOG
			9	8	10	9					386069.820 (500)	82BOG
		2	1	0	2	1					62452.500 (50)	82BOG
			2	1	3	2					98218.402 (80)	82BOG
			3	2	2	1					106983.370 (150)	82BOG
			3	2	3	2					127078.038 (80)	82BOG
			3	2	4	3					136394.140 (50)	82BOG
			4	3	3	2					155931.327 (80)	82BOG
			4	3	4	3					169433.544 (100)	82BOG
			4	3	5	4					176091.454 (100)	82BOG
			4	3	3	3					63581.095 (80)	82BOG
			5	4	4	3					202464.790 (100)	82BOG
			5	4	5	4					211785.880 (80)	82BOG

Line frequencies of SO (continued)

Isotopic species	State: electronic	State: vibrational v	Transition: N'	N''	J'	J''	F_1'	F_1''	F'	F''	ν MHz	Ref.
$^{32}S^{16}O$	X $^3\Sigma^-$	2	5	4	6	5					216706.173 (100)	82BOG
			7	6	7	6					296477.560 (200)	82BOG
		3	1	0	2	1					62211.158 (50)	82BOG
			2	1	3	2					97680.580 (80)	82BOG
			3	2	2	1					105855.730 (250)	82BOG
			3	2	3	2					126051.352 (80)	82BOG
			3	2	4	3					135508.200 (100)	82BOG
			4	3	3	2					154415.770 (100)	82BOG
			4	3	4	3					168064.670 (200)	82BOG
			4	3	5	4					174842.746 (100)	82BOG
			4	3	3	3					62364.772 (80)	82BOG
			5	4	4	3					200612.904 (200)	82BOG
			5	4	5	4					210074.736 (100)	82BOG
		4	1	0	2	1					61968.720 (80)	82BOG
			2	1	3	2					97144.370 (500)	82BOG
			3	2	2	1					104732.420 (250)	82BOG
			3	2	3	2					125027.212 (150)	82BOG
			3	2	4	3					134626.570 (150)	82BOG
			4	3	3	2					152903.060 (200)	82BOG
			4	3	5	4					173599.870 (100)	82BOG
			5	4	4	3					198763.056 (200)	82BOG
			5	4	5	4					208367.674 (200)	82BOG
			5	4	6	5					213488.880 (100)	82BOG
		5	1	0	2	1					61725.244 (80)	82BOG
			2	1	3	2					96610.950 (200)	82BOG
			3	2	3	2					124005.260 (150)	82BOG
			3	2	4	3					133749.200 (150)	82BOG
			4	3	5	4					172362.545 (100)	82BOG
			5	4	6	5					211889.970 (150)	82BOG
		6	1	0	2	1					61480.560 (100)	82BOG
			2	1	3	2					96079.250 (200)	82BOG
			3	2	4	3					132876.187 (200)	82BOG
			4	3	5	4					171130.780 (150)	82BOG
			5	4	6	5					210297.367 (150)	82BOG

Line frequencies of SO (continued)

Isotopic species	State: electronic	State: vibrational v	Transition: N'	N''	J'	J''	F_1'	F_1''	F'	F''	ν MHz	Ref.
$^{32}S^{16}O$	X $^3\Sigma^-$	7	1	0	2	1					61234.550 (200)	82BOG
			3	2	4	3					132007.576 (300)	82BOG
			4	3	5	4					169905.020 (400)	82BOG
			5	4	6	5					208171.368 (150)	82BOG
		8	5	4	6	5					207131.326 (250)	82BOG
$^{34}S^{16}O$	X $^3\Sigma^-$	0	0	1	1	0					29678.980 (100)	74TIE1
			1	0	2	1					62074.047 (50)	82TIE1
			1	0	1	1					286656.949 (50)	82TIE1
			2	1	3	2					97715.317 (50)	82TIE1
			2	1	2	1					84410.690 (100)	74TIE1
			3	2	3	2					126613.930 (100)	74TIE1
			3	2	2	1					106743.244 (70)	82TIE1
			3	2	2	2					34857.543 (70)	82TIE1
			5	4	4	4					96781.760 (200)	74TIE1
			6	5	7	6					256877.809 (40)	82TIE1
			7	6	6	5					290562.238 (30)	82TIE1
			7	6	7	6					295396.334 (30)	82TIE1
			7	6	8	7					298257.982 (30)	82TIE1
			10	9	9	9					287481.092 (150)	82TIE1
			3	2	4	3					135775.728 (50)	82BOG
			4	3	3	2					155506.795 (50)	82BOG
			4	3	4	3					168815.135 (80)	82BOG
			4	3	5	4					175352.766 (80)	82BOG
			5	4	4	3					201846.480 (100)	82BOG
			5	4	5	4					211013.030 (100)	82BOG
			5	4	6	5					215839.920 (100)	82BOG
			6	5	5	4					246663.470 (100)	82BOG
			6	5	5	5					132432.280 (150)	82BOG
			8	7	7	7					208292.240 (250)	82BOG
			9	8	10	9					381600.830 (350)	82BOG
		1	1	0	2	1					61837.215 (100)	82BOG
			2	1	3	2					97187.950 (100)	82BOG
			3	2	3	2					125612.565 (100)	82BOG

Line frequencies of SO (continued)

Isotopic species	State: electronic	State: vibrational v	N'	N''	J'	J''	F_1'	F_1''	F'	F''	ν MHz	Ref.
$^{34}S^{16}O$	$X\,^3\Sigma^-$	1	3	2	4	3					134908.210 (100)	82BOG
			4	3	3	2					154030.970 (80)	82BOG
			4	3	5	4					174131.590 (80)	82BOG
			5	4	5	4					209343.800 (100)	82BOG
			5	4	6	5					214264.600 (120)	82BOG
			7	6	6	5					288131.010 (200)	82BOG
		2	3	2	4	3					134044.700 (200)	82BOG
			4	3	5	4					172915.780 (200)	82BOG
		3	3	2	4	3					133185.170 (300)	82BOG
		4	3	2	4	3					132329.630 (500)	82BOG
$^{32}S^{18}O$	$X\,^3\Sigma^-$	0	1	0	2	1					59626.450 (150)	74TIE1
			1	0	1	1					287594.840 (100)	82TIE1
			2	1	3	2					93267.270 (200)	74TIE1
			2	1	2	1					79717.113 (100)	82TIE1
			3	2	2	1					99803.794 (100)	82TIE1
			7	6	8	7					282038.258 (50)	82TIE1
			7	6	7	6					278972.675 (50)	82TIE1
			7	6	6	5					273858.141 (50)	82TIE1
			3	2	4	3					129066.190 (150)	82BOG
			4	3	3	2					145874.490 (150)	82BOG
			4	3	4	3					159428.340 (150)	82BOG
			5	4	4	3					189783.250 (200)	82BOG
			5	4	5	4					199280.270 (200)	82BOG
			5	4	6	5					204388.630 (500)	82BOG
			6	5	5	4					232265.835 (200)	82BOG
			6	5	6	5					239128.682 (200)	82BOG
			6	5	7	6					243039.320 (200)	82BOG
$^{32}S^{17}O$	$X\,^3\Sigma^-$	0	1	0	2	1			4.5	3.5	61161.150	82CAZ
			1	0	2	1			3.5	2.5	61169.750	82CAZ
			1	0	2	1			2.5	1.5	61179.900	82CAZ
			1	0	2	1			1.5	1.5	61222.900	82CAZ
			1	0	2	1			2.5	2.5	61229.400	82CAZ
			1	0	2	1			3.5	3.5	61236.900	82CAZ

Line frequencies of SO (continued)

Isotopic species	State		Transition								ν MHz	Ref.
	electronic	vibrational v	N'	N''	J'	J''	F_1'	F_1''	F'	F''		
$^{32}S^{17}O$	$X\,^3\Sigma^-$	0	2	1	3	2			5.5	4.5	96085.300	82CAZ
			2	1	3	2			4.5	3.5	96089.350	82CAZ
			2	1	3	2			3.5	2.5	96095.650	82CAZ
			2	1	2	1			4.5	3.5	82701.300	82CAZ
			2	1	2	1			3.5	2.5	82724.350	82CAZ
			2	1	2	1			2.5	1.5	82740.850	82CAZ
			3	2	2	1			4.5	3.5	104247.400	82CAZ
			3	2	2	1			3.5	2.5	104262.000	82CAZ
			3	2	4	3			6.5	5.5	133331.300	82CAZ
			3	2	4	3			5.5	4.5	133332.900	82CAZ
			3	2	4	3			4.5	3.5	133336.420	82CAZ
			3	2	4	3			3.5	2.5	133341.870	82CAZ
			3	2	4	3			2.5	1.5	133349.500	82CAZ
			3	2	4	3			1.5	0.5	133359.250	82CAZ
			4	3	3	2			5.5	4.5	152033.400	82CAZ
			4	3	3	2			4.5	3.5	152036.250	82CAZ
			4	3	3	2			3.5	2.5	152034.800	82CAZ
			4	3	3	2			2.5	1.5	152029.000	82CAZ
			3	2	3	2			5.5	4.5	124059.770	82CAZ
			3	2	3	2			4.5	3.5	124066.650	82CAZ
			3	2	3	2			3.5	2.5	124071.500	82CAZ
			3	2	3	2			2.5	1.5	124074.600	82CAZ
			4	3	4	3			6.5	5.5	165413.200	82CAZ
			4	3	4	3			5.5	4.5	165416.250	82CAZ
			4	3	4	3			4.5	3.5	165418.500	82CAZ
			4	3	4	3			2.5	1.5	165421.000	82CAZ
			4	3	4	3			1.5	0.5	165421.000	82CAZ
			4	3	5	4			7.5	6.5	172058.050	82CAZ
			4	3	5	4			6.5	5.5	172058.300	82CAZ
			4	3	5	4			5.5	4.5	172060.350	82CAZ
			4	3	5	4			4.5	3.5	172063.850	82CAZ

For References see next page.

References to No. 88 (SO)

67AMA Amano, T., Hirota, E., Morino, Y.: J. Phys. Soc. Jpn. **22** (1967) 399.
74DAV Davies, P.B., Wayne, F.D., Stone, A.J.: Mol. Phys. **28** (1974) 1409.
74TIE1 Tiemann, E.: J. Mol. Spectrosc. **51** (1974) 316.
74TIE2 Tiemann, E.: J. Phys. Chem. Ref. Data **3** (1974) 259.
76CLA Clark, W.W., De Lucia, F.C.: J. Mol. Spectrosc. **60** (1976) 332.
78CHR Christensen, H., Veseth, L.: J. Mol. Spectrosc. **72** (1978) 438.
79KAW Kawaguchi, K., Yamada, C., Hirota, E.: J. Chem. Phys. **71** (1979) 3338.
81BOG Bogey, M., Demuynck, C., Destombes, J.L.: priv. com. **1981**.
82BOG Bogey, M., Demuynck, C., Destombes, J.L.: Chem. Phys. **66** (1982) 99.
82CAZ Cazzoli, G., Degli Esposti, C., Favero, P.G.: Il Nuovo Cimento **1D** (1982) 648.
82TIE1 Tiemann, E.: J. Mol. Spectrosc. **91** (1982) 60.
82TIE2 Tiemann, E.: private communication, report of preliminary results in: Verhandlungen DPG **(VI) 13** (1978) 542.

89 SeO

Molecular constants of SeO

Isotopic species	State	B_v MHz	D_v kHz	λ_v MHz	γ_v MHz	Derived constants	Ref.	Meth.
$^{80}Se^{16}O$	$X\,^3\Sigma^-$	$Y_{01}=14107.731\,(17)$ $Y_{11}=-98.272\,(13)$ $Y_{21}=51.9\,(35)$ kHz	$Y_{02}=-15.19\,(56)$	$\lambda_e^{ev}=2516.9\,(8)$ GHz $\alpha_\lambda^{ev}=9.9\,(10)$ GHz $D_\lambda=2.126\,(16)$	$\gamma_e=-204.3\,(15)$ [1] $\alpha_\gamma=-7.3\,(10)$ [1]	$r_e=1.639499\,(6)$ Å	79PAR 79PAR 79PAR	MW MW MW
$^{78}Se^{16}O$	$X\,^3\Sigma^-$	$Y_{01}=14168.101\,(23)$ $Y_{11}=-98.906\,(17)$	$Y_{02}=-15.48\,(71)$	$D_\lambda=2.136\,(26)$	$\gamma_e=-204.7\,(27)$ [1] $\alpha_\gamma=-8.0\,(41)$ [1]		79PAR 79PAR	MW MW
$^{76}Se^{16}O$	$X\,^3\Sigma^-$	$Y_{01}=14231.590\,(53)$ $Y_{11}=-99.570\,(15)$	$Y_{02}=-15.3\,(19)$		$\gamma_e=-205.9\,(20)$ [1]		79PAR 79PAR	MW MW
$^{82}Se^{16}O$	$X\,^3\Sigma^-$	$Y_{01}=14050.276\,(61)$ $Y_{11}=-97.675\,(16)$			$\gamma_e=-203.9\,(80)$ [1]		79PAR 79PAR	MW MW
SeO	$X\,^3\Sigma^-$	accurate vibrational constants for several isotopes obtained by the LMR-method [81HAK]						

[1]) Strong correlation of the λ- and γ-parameters will result in an underestimation of the errors shown here.

(Molecular constants of SeO)

Isotopic species	State: electronic	State: vibrational	Parameter	Value	Ref.	Meth.	Remarks
$^{77}Se^{16}O$	X $^3\Sigma^-$	$v=0$	b (^{77}Se)	169.5 (96) MHz	79PAR	MW	
			c (^{77}Se)	−506.9 (104) MHz	79PAR	MW	
			$\|b+c\|$ (^{77}Se)	337 (14) MHz	81HAK	LMR	
$^{80}Se^{16}O$	X $^3\Sigma^-$	$v=0$	g_s	2.00049 (7)	78BRO	EPR	
			g_L^{el}	−0.011 (4)	78BRO	EPR	
$^{78}Se^{16}O$	X $^3\Sigma^-$	$v=0$	g_s	2.0006 (1)	78BRO	EPR	
			g_L^{el}	−0.007 (7)	78BRO	EPR	

Line frequencies of SeO

Isotopic species	State: electronic	State: vibrational v	Transition: $\|\Omega\|$	parity	J' − J''	F_1' − F_1''	F' − F''	ν MHz	Ref.
$^{74}Se^{16}O$	X $^3\Sigma^-$	0	0	+ −	4 3			112645.755 (80)	79PAR
$^{76}Se^{16}O$	X $^3\Sigma^-$	0	0	+ −	4 3			112125.915 (80)	79PAR
		0	1	− +	4 3			113462.225 (150)	
		0	1	+ −	3 2			85097.925 (130)	
		0	1	− +	3 2			86073.475 (220)	
		1	0	+ −	4 3			111352.225 (100)	
$^{77}Se^{16}O$	X $^3\Sigma^-$	0	0	+ −	4 3		4.5 3.5	111876.450 (80)	79PAR
		0	0	+ −	4 3		3.5 2.5	111874.540 (80)	
		0	1	− +	4 3		3.5 2.5	113214.275 (100)	
		0	1	− +	4 3		4.5 3.5	113191.660 (150)	
		0	1	+ −	3 2		2.5 1.5	84919.840 (400)	
		0	1	+ −	3 2		3.5 2.5	84877.690 (600)	
		0	1	+ −	4 3		3.5 2.5	114507.645 (100)	
		0	1	+ −	4 3		4.5 3.5	114486.955 (150)	
		1	0	+ −	4 3		4.5 3.5	111105.185 (200)	
		1	0	+ −	4 3		3.5 2.5	111103.310 (200)	

Line frequencies of SeO (continued)

Isotopic species	State		Transition										ν MHz	Ref.
	electronic	vibrational v	$\|\Omega\|$	parity		J' –	J''	F_1' –	F_1''	F' –	F''			
$^{78}Se^{16}O$	$X\,^3\Sigma^-$	0	0	+	–	4	3						111632.305 (80)	79PAR
		0	0	+	–	2	1						55817.015 (100)	
		0	1	–	+	4	3						112956.765 (100)	
		0	1	+	–	3	2						84719.025 (120)	
		0	1	–	+	2	1						56479.930 (220)	
		0	1	+	–	4	3						114245.410 (100)	
		0	1	–	+	3	2						85685.810 (90)	
		0	1	+	–	2	1						57124.760 (170)	
		1	0	+	–	4	3						110863.640 (150)	
		1	1	–	+	4	3						112166.620 (250)	
		1	1	+	–	4	3						113433.015 (250)	
$^{80}Se^{16}O$	$X\,^3\Sigma^-$	0	0	+	–	4	3						111163.020 (80)	79PAR
		0	0	+	–	2	1						55582.395 (60)	
		0	1	–	+	4	3						112476.425 (100)	
		0	1	–	+	2	1						56239.670 (130)	
		0	1	+	–	4	3						113754.035 (100)	
		0	1	–	+	3	2						85317.280 (70)	
		0	1	+	–	2	1						56879.100 (130)	
		1	0	+	–	4	3						110399.105 (100)	
		1	1	–	+	4	3						111691.230 (200)	
		1	1	+	–	4	3						112947.050 (150)	
		2	0	+	–	4	3						109635.910 (120)	
		2	1	–	+	4	3						110907.050 (150)	
		2	1	+	–	4	3						112141.075 (150)	
		3	0	+	–	4	3						108873.195 (100)	
		3	1	–	+	4	3						110123.490 (350)	
		3	1	+	–	4	3						111335.850 (250)	
		4	0	+	–	4	3						108110.970 (150)	
$^{82}Se^{16}O$	$X\,^3\Sigma^-$	0	0	+	–	4	3						110716.330 (80)	79PAR
		0	1	–	+	4	3						112019.270 (100)	
		0	1	+	–	4	3						113286.555 (100)	
		0	1	–	+	3	2						84966.610 (170)	
		1	0	+	–	4	3						109957.080 (100)	

References

78BRO Brown, J. M., Dumper, K., Parent, C. R.: Mol. Phys. **36** (1978) 1149.
79PAR Parent, C. R., Kuijpers, P. J. M.: Chem. Phys. **40** (1979) 425.
81HAK Hakuta, K., Uehara, H.: J. Mol. Spectrosc. **85** (1981) 97.

90 S_2

Molecular constants of S_2

Isotopic species	State	B_v MHz	D_v kHz	λ_v MHz	γ_v MHz	Derived constants	Ref.	Meth.
$^{32}S_2$	$X\ ^3\Sigma_g^-$	$B_0=8831.868\ (10)$	$D_0=5.9\ (10)$	$\lambda_0=353040\ (27)$	$\gamma_0=-200.9\ (10)$		79PIC	MW

Isotopic species	State electronic	State vibrational	Parameter	Value	Ref.	Meth.	Remarks
$^{32}S_2$	$X\ ^3\Sigma_g^-$	$v=0$	$g_S+g_L^{el}$	2.0020 (14)	79PIC, 74WAY	MW/EPR	
			Δg	−0.0089 (17)	79PIC, 74WAY	MW/EPR	
			g_{rot}	−0.057 (13)	74WAY	EPR	

Line frequencies of S_2

Isotopic species	State electronic	State vibrational v	Transition N'	N''	J'	J''	I'_T	I''_T	F'	F''	ν MHz	Ref.
$^{32}S_2$	$X\ ^3\Sigma_g^-$	0	5	3	4	3					79805.30 (4)	79PIC
		0	5	5	5	4					79163.83 (4)	
		0	7	5	6	5					124928.77 (4)	
		0	7	7	7	6					104686.60 (4)	
		0	9	9	9	8					127054.98 (4)	
		0	11	11	11	10					146581.90 (4)	

References

74WAY Wayne, F. D., Davies, P. B., Thrush, B. A.: Mol. Phys. **28** (1974) 989.
79PIC Pickett, H. M., Boyd, T. L.: J. Mol. Spectrosc. **75** (1979) 53.

4.1.5 The $^2\Pi$ electronic state

4.1.5.1 Preliminary remarks

1. Introduction

The approximate electronic quantum numbers of this molecular state are the spin $S=\frac{1}{2}$, the projection of the orbital angular momentum onto the molecular axis $|\Lambda|=1$ and, assuming Hund's coupling case (*a*), the projection $|\Sigma|=\frac{1}{2}$ of the spin to give the projection of the total electronic angular momentum onto the molecular axis $|\Omega|=\frac{1}{2},\frac{3}{2}$ for the two fine structure states $^2\Pi_{\frac{1}{2}}$ and $^2\Pi_{\frac{3}{2}}$.

The molecular levels of the vibrational state v are further described by the quantum number J of the total angular momentum, coupling Ω with the rotational momentum $\boldsymbol{R}$. For each J-state two levels called Λ-doublet exist with different parity + or −.

For Hund's coupling case (*b*), Λ couples with the nuclear rotation $\boldsymbol{R}$ to give the angular momentum $\boldsymbol{N}$ for which two Λ-doublet levels are given with different parity + or −. $\boldsymbol{N}$ then couples with the spin $\boldsymbol{S}$ to the total angular momentum $\boldsymbol{J}$.

In the case of non-zero nuclear spins I_1 and I_2, they couple with $\boldsymbol{J}$ to the total angular momentum F, and the approximate quantum number F_1 ($\boldsymbol{F}_1=\boldsymbol{J}+\boldsymbol{I}_1$) or I_T ($\boldsymbol{I}_T=\boldsymbol{I}_1+\boldsymbol{I}_2$) are used for a complete description of the molecular level.

The phenomenological Hamiltonian is principally discussed in 5.1.1 and 5.1.5.1 of Vol. II/6 Landolt-Börnstein, New Series, and will therefore not be repeated here. Since that review, different approaches to an effective Hamiltonian have been published and should be briefly mentioned to show the difficulty of tabulating molecular parameters which are generally accepted.

Meerts [1] refined his perturbation treatment, but did not introduce new molecular parameters. Only the physical interpretation has changed slightly. When neglecting higher order terms, the parameters of Meerts are linear combinations of the traditional molecular parameters which are used for the following table. Therefore the latter are approximately calculable only. A very similar approach is given by Brown et al. [2]; but they used the traditional spectroscopic constants and gave a table to compare the different notations used in the literature.

Beaudet and Poynter [3] used numerical diagonalization of 3×3 matrices containing the $^2\Pi_{\frac{3}{2}}$, $^2\Pi_{\frac{1}{2}}$ and the perturbing $^2\Sigma_{\frac{1}{2}}$ states. This is in contrast to the other scientists who made a contact-transformation for the $^2\Sigma$ state into the $^2\Pi$ state in order to get reduced 2×2 matrices for the $^2\Pi$ state. Therefore the molecular parameters of Beaudet and Poynter are not directly comparable to others because of the influence of higher order correction terms. In the first procedure one also directly obtains the interaction matrix-elements of the $^2\Pi$ and $^2\Sigma$ states instead of the usual Λ-doubling parameters p, q.

The value of the molecular parameter and its centrifugal distortion constant depend on the power expansion used for the effective parameter. The traditional power series in $\boldsymbol{R}^2$ is used by [4] and [5] for example. Brown et al. [8] propose the series in $\boldsymbol{N}^2$ and in [3] the phenomenological ansatz with $J(J+1)$ is applied.

In order to reduce high correlations between molecular parameters several authors [4], [5], [6] made combined fits of microwave spectral lines with optical and/or infrared measurements. This seems to be the best way to get the most consistent set of molecular parameters, but introduces slightly different definitions of the molecular parameters to compensate for changes in the perturbation order.

The goal of this compilation is to present data which is closely comparable. In order to achieve this for the most extensively studied molecules like OH or NO and for molecules with a small set of data, the molecular parameters which mainly determine the rotational energy levels are selected but a large amount of their higher order terms are excluded, and in this sense the values shown for the parameters are only approximately correct. On the other side it was felt that including all existing fitting parameters would make the tables useful for the expert only who precisely knows the effective Hamiltonian which was used in deriving these special parameters. Remarks will be found in the tables in occasions where higher order terms are calculated or discussed. Parameter sets obtained by very different calculations are chosen to indicate the status of research on the molecule under discussion. Again, to simplify later calculations the primarily measured transition frequencies are given.

The effective Hamiltonian for the Zeeman-effect is reviewed by Brown et al. [2] and by Veseth [7]. The g-factors as defined in [2] are preferred for the tables.

The following list of parameters is a short description only. All parameters are precisely defined in the references cited and follow as closely as possible the definitions used in 5.1.1 and 5.1.5.1 of Vol. II/6, Landolt-Börnstein, New Series.

2. List of parameters

$W_\Sigma - W_\Pi$	energy difference of the Σ-state and Π-state if only one $^2\Sigma$-state interacts with the $^2\Pi$-state under consideration
$B_{v\,\mathrm{eff}}$	effective rotational constant for one fine structure component $\|\Omega\|$ for the vibrational state v; this includes parity-independent contributions of the Λ-doubling interaction
B_v	rotational constant for the vibrational state v, which has the same meaning as for $^1\Sigma$-molecules (compare 2.2) $B_v = \sum_l Y_{l1}(v+\frac{1}{2})^l$; $Y_{11} = -\alpha_e$, $Y_{21} = \gamma_e$ etc.
$D_{v\,\mathrm{eff}}$, D_v	contrifugal distortion constant corresponding to the rotational parameter $B_{v\,\mathrm{eff}}$ and B_v respectively
A_v	spin-orbit interaction constant for the vibrational state v
$A_v/B_v = \lambda$	the parameter determines the intermediate coupling case between Hund's case (*a*) and (*b*)
γ_v	spin-rotation interaction constant for the vibrational state v
ν_Λ	Λ-doubling frequency
α_p, β_p	Λ-doubling parameters for a specific vibrational state; compare the definition by perturbation theory in 5.1.5.1 of Vol. II/6, Landolt-Börnstein, New Series
p, q	$p = -2(\alpha_p - 2\beta_p)$, $q = -2\beta_p$
$\langle\Sigma\|BL_y\|\Pi\rangle$, $\langle\Sigma\|AL_y\|\Pi\rangle$, $\langle\Sigma\|(2B+A)L_y\|\Pi\rangle$	matrix elements which describe the Λ-doubling for the interaction of the Π-state with the $^2\Sigma$-state. α_p, β_p are in general the sums of products of such matrix elements divided by the appropriate energy-difference
	Sometimes effective centrifugal distortion parameters for the Λ-doubling or the specific matrix elements are given in the literature. They are not listed in the tables (compare the discussion in the introduction)
a_i, b_i, c_i, d_i	magnetic hyperfine coupling constants for nucleus i due to the electronic orbital (a_i) and spin (b_i, c_i, d_i)
$b_F = b + c/3$	coupling constant of Fermi-contact term
C_{Ii}	effective magnetic spin-rotation hyperfine coupling constant of nucleus i
$eq^{(a)}Q_i$	quadrupole coupling constant of nucleus i according to the electric field gradient in the direction of the molecular axis *)
$eq^{(b)}Q_i$	quadrupole coupling constant of nucleus i according to the electric field gradient perpendicular to the molecular axis *)
μ	electric dipole moment
g_s	g-factor of the electron spin including relativistic corrections
g'_L	orbital g-factor including the non-adiabatic contribution Δg_L from the admixture of $^2\Sigma$ and $^2\Delta$ states: $g'_L = g_L + \Delta g_L$
g_l	anisotropic contribution to the electron spin Zeeman-interaction arising from the admixture of $^2\Delta$ and $^2\Sigma$ states
g_{rot}	rotational g-factor; the rotational magnetic moment is defined in nuclear magnetons (compare part 2.9)
g'_l, $g^{e'}_{\mathrm{rot}}$	contributions which represent lambda-doubling analogues of g_l and the electronic part of g_{rot}, respectively. $g^{e'}_{\mathrm{rot}}$ is given in the nuclear magneton unit

*) The parameters $eq^{(a)}Q$ and $eq^{(b)}Q$ are identical with eq_0Q and eq_2Q respectively as defined in 4.2.1 for polyatomic radicals.

3. Arrangement of substances

CH
OH, SH, SeH
HCl^+, HBr^+
CaCl, CaBr, CaI
CO^+
CN
CF, CCl
NO, NS, AsO
OF, OCl, OBr, OI, SF

4. References

1 Meerts, W.L.: Chem. Phys. **14** (1976) 421 (and literature cited therein).
2 Brown, J.M., Kaise, M., Kerr, C.M.L., Milton, D.J.: Mol. Phys. **36** (1978) 553.
3 Beaudet, R.A., Poynter, R.L.: J. Phys. Chem. Ref. Data **7** (1978) 311.
4 Destombes, J.L., Marlière, C., Rohart, F.: J. Mol. Spectrosc. **67** (1977) 93.
5 Coxon, J.A.: Can. J. Phys. **58** (1980) 933.
6 Van den Heuvel, F.C., Meerts, W.L., Dymanus, A.: J. Mol. Spectrosc. **84** (1980) 162.
7 Veseth, L.: J. Mol. Spectrosc. **63** (1976) 180.
Veseth, L.: J. Mol. Spectrosc. **66** (1977) 259.
Veseth, L.: J. Mol. Spectrosc. **77** (1979) 195.
8 Brown, J.M., Colbourn, E.A., Watson, J.K.G., Wayne, F.D.: J. Mol. Spectrosc. **74** (1979) 294.

4.1.5.2 Data and references

8 CH

Molecular constants of $^{16}C^1H$

Electronic state: X $^2\Pi$
Vibrational state: $v=0$

Parameter	Value	Ref.	Meth.	Remarks
		78HOU	LMR	rotational energy levels and Λ-doubling splittings are calculated in [78HOU] transition frequencies of [74RYD] are used for fitting routine
$a(^1H)$	52 (2) MHz	78HOU	LMR	hyperfine structure calculations in [75LEV]
$b(^1H)$	−74 (3) MHz	78HOU	LMR	
$c(^1H)$	52 (7) MHz	78HOU	LMR	
$d(^1H)$	43.6 (4) MHz	78HOU	LMR	

Line frequencies of CH

Isotopic species	State: electronic	State: vibrational	Transition: parity	rotational J'	J''	hyperfine F_1'	F_1''	F'	F''	ν MHz	Ref.	Remarks
$^{16}C^1H$	$X^2\Pi_{1/2}$ or $X^2\Pi, N=1$	$v=0$	− +	1/2	1/2			1	0	3349.193 (3)	74RYD	the electronic state is more closely related to Hund's coupling case *b*
			− +	1/2	1/2			1	1	3335.481 (2)	74RYD	
			− +	1/2	1/2			0	1	3263.794 (3)	74RYD	

References

74RYD Rydbeck, O.E.H., Elldér, J., Irvine, W.M., Sume, A., Hjalmarson, Å.: Astron. Astrophys. **34** (1974) 479.

75LEV Levy, D.H., Hinze, J.: Astrophys. J. **200** (1975) 236.

76RYD Rydbeck, O.E.H., Kollberg, E., Hjalmarson, Å., Sume, A., Elldér, J., Irvine, W.M.: Astrophys. J. Suppl. **31** (1976) 333.

78HOU Hougen, J.T., Mucha, J.A., Jennings, D.A., Evenson, K.M.: J. Mol. Spectrosc. **72** (1978) 463.

11 OH

Molecular constants of OH

Isotopic species	State: electronic	State: vibrational	Parameter	Value	Ref.	Meth.	Remarks
$^{16}O^1H$	$X^2\Pi$	$v=0$					parameters were obtained by fitting the interaction matrix
			$W_\Sigma - W_\Pi$	32683.477 cm^{-1}	78BEA	MW/MB	values from the analysis of the optical spectrum; rotational constants of the perturbing $^2\Sigma$-state $B_0=507471.7$ MHz, $D_0=61.4491$ MHz
			B_0	556140.9 MHz	78BEA	MW/MB	
			D_0	58.3519 MHz	78BEA	MW/MB	
			A_0	−4157.663 (450) GHz	78BEA	MW/MB	centrifugal distortion parameters in [78BEA]
			$\langle BL_y \rangle$	376.6535 (240) GHz	78BEA	MW/MB	
			$\langle (A+2B) L_y \rangle$	−1534.2919 (970) GHz	78BEA	MW/MB	
			$a(^1H)$	86.1118 (33) MHz	78BEA	MW/MB	
			$b(^1H)$	−116.281 (16) MHz	78BEA	MW/MB	
			$c(^1H)$	130.221 (19) MHz	78BEA	MW/MB	
			$d(^1H)$	56.6550 (36) MHz	78BEA	MW/MB	centrifugal distortion parameter in [78BEA]
			$C_I(^1H)$	−99.65 (79) kHz	78BEA	MW/MB	effective parameter only

Molecular constants of OH (continued)

Isotopic species	State: electronic	State: vibrational	Parameter	Value	Ref.	Meth.	Remarks
$^{16}O^{1}H$	$X\ ^2\Pi$						similar fits using all existing MW- and MB-data are presented in [77MEE] and [78BRO]. [79COX] is prefered because of adequate comparison with the results in vibrational excited states
		$v=0$	B_0^{ev}	556127 MHz	79COX	MW/MB	higher order corrections and centrifugal distortion parameters in [79COX]
			A_0^{ev}	−139.277 cm^{-1}	79COX	MW/MB	
			p	7056.4480 (11) MHz	79COX	MW/MB	
			q	−1162.4160 (1) MHz	79COX	MW/MB	
			$a(^1H)$	86.0123 (19) MHz	79COX	MW/MB	
			$b_F(^1H)$	−73.1258 (39) MHz	79COX	MW/MB	
			$c(^1H)$	130.749 (10) MHz	79COX	MW/MB	
			$d(^1H)$	56.6320 (37) MHz	79COX	MW/MB	
			$C_I(^1H)$	−99.48 (54) kHz	79COX	MW/MB	
		$v=1$	B_1^{ev}	534788 MHz	79COX, 79MEE, 79DES	MW/MB	[79MEE] presents a slightly different fit, higher order corrections in [79COX]
			A_1^{ev}	−139.549 cm^{-1}	79COX		
			p	6739.0955 (40) MHz	79COX, 79MEE, 79DES	MW/MB	
			q	−1109.8926 (10) MHz	79COX, 79MEE, 79DES	MW/MB	
			$a(^1H)$	81.9763 (44) MHz	79COX, 79MEE, 79DES	MW/MB	
			$b_F(^1H)$	−77.074 (12) MHz	79COX, 79MEE, 79DES	MW/MB	higher order corrections in [79COX]
			$c(^1H)$	125.438 (27) MHz	79COX, 79MEE, 79DES	MW/MB	

Molecular constants of OH (continued)

Isotopic species	State electronic	State vibrational	Parameter	Value	Ref.	Meth.	Remarks
$^{16}O^{1}H$	X $^{2}\Pi$	$v=1$	$d(^{1}H)$	53.7743 (85) MHz	79COX, 79MEE, 79DES	MW/MB	
			$C_I(^{1}H)$	−98.2 (19) kHz	79COX, 79MEE, 79DES	MW/MB	
		$v=2$	B_2^{ev}	513733 MHz	79COX, 79MEE	MW/MB	higher order corrections in [79COX], [79MEE] presents a slightly different fit
			A_2^{ev}	−139.813 cm^{-1}	79COX, 79MEE		
			p	6416.829 (47) MHz	79COX, 79MEE	MW/MB	
			q	−1057.0045 (15) MHz	79COX, 79MEE	MW/MB	
			$a(^{1}H)$	78.0353 (49) MHz	79COX, 79MEE	MW/MB	
			$b_F(^{1}H)$	−81.914 (13) MHz	79COX, 79MEE	MW/MB	
			$c(^{1}H)$	120.284 (31) MHz	79COX, 79MEE	MW/MB	
			$d(^{1}H)$	50.9953 (90) MHz	79COX, 79MEE	MW/MB	
			$C_I(^{1}H)$	−96.2 (20) kHz	79COX, 79MEE	MW/MB	
		$v=3$	B_3^{ev}	492886 MHz	79COX	MW	higher order corrections in [79COX]
			A_3^{ev}	−140.069 cm^{-1}	79COX	MW	
			p^{der}	6080.2 MHz	79COX	MW	
			q	−1003.752 (31) MHz	79COX	MW	
			$a^{der}(^{1}H)$	74.20 MHz	79COX	MW	
			$b_F^{der}(^{1}H)$	−87.64 MHz	79COX	MW	
			$c^{der}(^{1}H)$	115.29 MHz	79COX	MW	
			$d(^{1}H)$	48.5 (25) MHz	79COX	MW	
			$C_I^{der}(^{1}H)$	−94.9 kHz	79COX	MW	

Molecular constants of OH (continued)

Isotopic species	State: electronic	State: vibrational	Parameter	Value	Ref.	Meth.	Remarks
$^{16}O^1H$	$X\ ^2\Pi$	$v=4$	q	−949.965 (90) MHz	79COX	MW	the parameters B_4, A_4, p, a, b_F, c and C_I are derived by extrapolation of the known constants of $0 \leqq v \leqq 2$ in [79COX]
			$d(^1H)$	46.6 (40) MHz	79COX	MW	
		$v=5$	q	−893.860 (23) MHz	79COX	MW	the parameters B_5, A_5, p, a, b_F, c and C_I are derived by extrapolation of the known constants of $0 \leqq v \leqq 2$ in [79COX]
			$d(^1H)$	44.0 (44) MHz	79COX	MW	
		$v=0$	A_0 [1])	−4168639.13 (78) MHz	81BRO	LMR	λ-doubling, hyperfine interaction parameters and molecular g-factors were constrained in the least-squares fit at values of [78BRO], but they were reevaluated with a Hamiltonian expanded in powers of $\boldsymbol{N}^2$
			B_0	555660.97 (11) MHz	81BRO	LMR	
			D_0	57.1785 (86) MHz	81BRO	LMR	
			γ_0 [1])	−3574.88 (49) MHz	81BRO	LMR	
			p	7053.09846 MHz	81BRO, 78BRO	EPR/MB/MW	centrifugal distortion parameters in [81BRO]; no errors are given in [81BRO], the figures shown are not all physically significant but used to avoid round off errors in recalculating the spectra
			q	−1159.991650 MHz	81BRO, 78BRO	EPR/MB/MW	
			$a(^1H)$	86.1116 MHz	81BRO, 78BRO	EPR/MB/MW	
			$b_F(^1H)$	−73.2537 MHz	81BRO, 78BRO	EPR/MB/MW	
			$c(^1H)$	130.641 MHz	81BRO, 78BRO	EPR/MB/MW	
			$d(^1H)$	56.6838 MHz	81BRO, 78BRO	EPR/MB/MW	[1]) Effectiv parameters setting the centrifugal distortion A_D of the spin-orbit interaction A zero.
			$C_I(^1H)$	−99.71 kHz	81BRO, 78BRO	EPR/MB/MW	
			g_s	2.00152 (36)	78BRO	EPR	
			g_L'	1.00107 (15)	78BRO	EPR	
			g_l	0.00400 (56)	78BRO	EPR	
			g_l'	0.006386 (30)	78BRO	EPR	
			g_{rot}	−1.162 (35)	78BRO	EPR	
			$g_{rot}^{e'}$	3.7539 (42)	78BRO	EPR	

Molecular constants of OH (continued)

Isotopic species	State electronic	State vibrational	Parameter	Value	Ref.	Meth.	Remarks
$^{16}O^{1}H$	X $^2\Pi_{1/2}$	$v=0$	μ	1.6676 (9) D	73MEE1	MB	
	X $^2\Pi_{1/2}$	$v=1$	$\nu_\Lambda (J=1/2)$	4495 (134) MHz	79DAV	LMR	
		$v=2$	$\nu_\Lambda (J=1/2)$	4290 (129) MHz	79DAV	LMR	
		$v=3$	$\nu_\Lambda (J=1/2)$	4087 (123) MHz	79DAV	LMR	
	X $^2\Pi_{3/2}$	$v=5$	$\nu_\Lambda (J=3/2)$	1061.45 MHz	74LEE	EPR	molecular g-factors and a linear combination of the hyperfine parameters a, b, c are also contained in [74LEE] for the excited vibrational states
		$v=6$	$\nu_\Lambda (J=3/2)$	948.38 MHz	74LEE	EPR	
		$v=7$	$\nu_\Lambda (J=3/2)$	838.93 MHz	74LEE	EPR	
		$v=8$	$\nu_\Lambda (J=3/2)$	727.82 MHz	74LEE	EPR	
		$v=9$	$\nu_\Lambda (J=3/2)$	642.70 MHz	74LEE	EPR	
	X $^2\Pi$	$0 \leqq v \leqq 4$			73HIN		interpretation of the vibrational dependence of the g-factors and the Λ-doubling parameters studied by [71CLO]
$^{18}O^{1}H$	X $^2\Pi$	$v=0$	$W_\Sigma - W_\Pi$	32683.477 cm^{-1}	78BEA	MW/MB	values from the analysis of the optical spectrum; rotational constants of the perturbing $^2\Sigma$-state: $B_0=504122.4$ MHz, $D_0=60.641$ MHz
			B_0	552470.4 MHz	78BEA	MW/MB	
			D_0	57.584 MHz	78BEA	MW/MB	
			A_0	−4157.0715 (260) GHz	78BEA	MW/MB	centrifugal distortion parameters in [78BEA]
			$\langle BL_y \rangle$	374.1645 (11) GHz	78BEA	MW/MB	
			$\langle (A+2B) L_y \rangle$	−1539.4127 (46) GHz	78BEA	MW/MB	
			$a(^1H)$	86.1248 (20) MHz	78BEA	MW/MB	
			$b(^1H)$	−116.281 MHz	78BEA		
			$c(^1H)$	130.22 MHz	78BEA		
			$d(^1H)$	56.6664 (75) MHz	78BEA	MW/MB	centrifugal distortion parameter in [78BEA]
			$C_I(^1H)$	−99.29 kHz	78BEA		effective parameter only
	Several parameters of $^{18}O^{1}H$ were obtained by isotopic scaling, which probably results in an underestimation of errors of the fitted parameters.						
$^{16}O^{2}D$	X $^2\Pi$	$v=0$	$W_\Sigma - W_\Pi$	32680.856 cm^{-1}	78BEA	MW/MB	values from the analysis of the optical spectrum; rotational constants of the perturbing $^2\Sigma$-state: $B_0=270817.0$ MHz, $D_0=17.2636$ MHz
			B_0	296312.0 MHz	78BEA	MW/MB	
			D_0	16.1417 MHz	78BEA	MW/MB	
			A_0	−4165.28 (230) GHz	78BEA	MW/MB	centrifugal distortion parameters in [78BEA]
			$\langle BL_y \rangle$	200.4377 (790) GHz	78BEA	MW/MB	
			$\langle (A+2B) L_y \rangle$	−1892.097 (751) GHz	78BEA	MW/MB	
			$a(^2D)$	13.3051 (5) MHz	78BEA	MW/MB	

Molecular constants of OH (continued)

Isotopic species	State electronic	State vibrational	Parameter	Value	Ref.	Meth.	Remarks
$^{16}O^{2}D$	X $^2\Pi$	$v=0$	$b(^2D)$	−17.865 (11) MHz	78BEA	MW/MB	
			$c(^2D)$	20.138 (11) MHz	78BEA	MW/MB	
			$d(^2D)$	8.7698 (5) MHz	78BEA	MW/MB	centrifugal distortion parameter in [78BEA]
			$C_I(^2D)$	−8.07 (22) kHz	78BEA	MW/MB	effective parameter only
			$eq^{(a)}Q(^2D)$	0.2850 (11) MHz	78BEA	MW/MB	
			$eq^{(b)}Q(^2D)$	−0.1189 (42) MHz	78BEA	MW/MB	
	X $^2\Pi$	$v=0$	A_0	−139.22 (2) cm^{-1}	73MEE2, 75MEE	MB	parameters were obtained from a fit including third order perturbation theory, higher order parameters in [75MEE]
			α_p	−1553.176 (4) MHz	73MEE2, 75MEE	MB	
			β_p	164.69 (2) MHz	73MEE2, 75MEE	MB	
			$a-1/2(b+c)(^2D)$	12.156 (2) MHz	73MEE2, 75MEE	MB	$a^{der}(^2D)=13.296$ (3) MHz
			$a+1/2(b+c)(^2D)$	14.436 (5) MHz	73MEE2, 75MEE	MB	$c^{der}(^2D)=20.35$ (3) MHz
			$b(^2D)$	−18.068 (20) MHz	73MEE2, 75MEE	MB	
			$d(^2D)$	8.768 (2) MHz	73MEE2, 75MEE	MB	
			$eq^{(a)}Q(^2D)$	285 (7) kHz	73MEE2, 75MEE	MB	
			$eq^{(b)}Q(^2D)$	−124 (12) kHz	73MEE2, 75MEE	MB	
	X $^2\Pi$	$v=0$	A_0 [1])	−4167594.57 (63) MHz	82BRO		parameters were obtained by fitting all existing measurements of Λ-doubling frequencies by MB, of pure rotational transitions, and of EPR and far-infrared LMR observations; higher order centrifugal corrections are used for the calculation
			B_0	296158.573 (61) MHz	82BRO		
			D_0	16.114 MHz	82BRO		
			γ_0 [1])	−1929.40 (11) MHz	82BRO		
			p_0	3761.84628 (96) MHz	82BRO		
			q_0	−327.97256 (10) MHz	82BRO		
			$a(^2D)$	13.30486 (54) MHz	82BRO		
			$b_F(^2D)$	−11.1803 (24) MHz	82BRO		

[1]) Effectiv parameters, setting the centrifugal distortion A_D of the spin-orbit interaction A zero

Molecular constants of OH (continued)

Isotopic species	State: electronic	State: vibrational	Parameter	Value	Ref.	Meth.	Remarks
$^{16}O^{2}D$	$X\,^2\Pi$	$v=0$	$c(^2D)$	20.1673 (47) MHz	82BRO		
			$d(^2D)$	8.77197 (60) MHz	82BRO		
			$C_I(^2D)$	−8.00 (23) kHz	82BRO		
			$eq^{(a)}Q(^2D)$	285.1 (12) kHz	82BRO		
			$eq^{(b)}Q(^2D)$	−123.3 (10) kHz	82BRO		
			A_0	−4167662.7 (27) MHz	82BRO		deduced from the effective parameters by using the results of $^{16}O^{1}H$ and $^{16}O^{2}D$ (for A_0, γ_0, A_D)
			γ_0	−1851.5 (21) MHz	82BRO		
			A_D	−11.34 (27) MHz	82BRO		
			g'_L	1.000610 (24)	82BRO	EPR	
			g_l	0.00391 (23)	82BRO	EPR	
			g'_l	0.00663 (16)	82BRO	EPR	
			g_{rot}	−0.676 (35)	82BRO	EPR	
			$g^{e'}_{rot}$	1.988 (16)	82BRO	EPR	
	$X\,^2\Pi_{1/2}$	$v=0$	μ_0	1.65312 (14) D	73MEE1	MB	
$^{16}O^{2}D$	[75COX1] and [75COX2] contain analysis of optical data which are necessary for the interpretation of the microwave measurements.						
$^{17}O^{1}H$	[73VAL] presents predictions of Λ-doubling transition frequencies for the ground vibrational state. [78BEA] reports first measurements of Λ-doubling transitions by Gottlieb, C. A., Radford, H. E., and Smith, B. P.						
$^{16}O^{1}H$	combined fit of microwave and optical data in [80COX, 77DES1, 77DES2] radial dependence of the spin-orbit and Λ-doubling parameters evaluated from the vibrational dependence of these parameters [82COX]						
$^{16}O^{1}H$	$X\,^2\Pi_{3/2}$	$v=0$ $J=11/2$			79BAS	MW	pressure broadening parameter for the Λ-doubling transition: $a(OH, NO_2)=4.5\,(3)$ MHz/Torr $a(OH, H_2O)=6.6\,(5)$ MHz/Torr $a(OH, N_2)=3.4\,(2)$ MHz/Torr $a(OH, H_2)=2.3\,(2)$ MHz/Torr $a(OH, He)=1.4\,(1)$ MHz/Torr $a(OH, CH_3Cl)=4.6\,(3)$ MHz/Torr
$^{16}O^{1}H$	$X\,^2\Pi_{3/2}$	$v=0$ $J=9/2$			76BUS	MW	pressure broadening parameter for the Λ-doubling transition: $a(OH, NO_2)=3.1\,(4)$ MHz/Torr $a(OH, H_2O)=9.2\,(6)$ MHz/Torr

Line frequencies of OH

Isotopic species	State: electronic	State: vibrational	Transition: parity	rotational J′	rotational J″	hyperfine F_1'	hyperfine F_1''	F′	F″	ν MHz	Ref.	Remarks
$^{16}O^{1}H$	X $^2\Pi_{3/2}$	$v=0$	− +	5/2	5/2			2	3	6016.746 (5)	75MEE	
			− +	5/2	5/2			2	2	6030.74850 (20)	76TER1	
			− +	5/2	5/2			3	3	6035.09321 (20)	76TER1	
			− −	7/2	7/2			3	4	7.435 (15)	75DES2	
			+ −	7/2	7/2			4	3	13433.56 (5)	80SAS	
			+ −	7/2	7/2			3	3	13434.63740 (20)	76TER1	
			+ −	7/2	7/2			4	4	13441.41727 (20)	76TER1	
			+ −	7/2	7/2			3	4	13442.129 (5)	80SAS	
			− −	9/2	9/2			4	5	12.315 (15)	75DES2	
			+ +	9/2	9/2			4	5	21.320 (15)	75DES2	
			− +	9/2	9/2			5	4	23805.297 (10)	75DES1	
			− +	9/2	9/2			4	4	23817.6153 (20)	76TER2	
			− +	9/2	9/2			5	5	23826.6211 (30)	76TER2	
			− +	9/2	9/2			4	5	23838.933 (10)	75DES1	
			+ −	11/2	11/2			6	5	36963.480 (30)	75DES1	
			+ −	11/2	11/2			5	5	36983.470 (20)	75DES1	
			+ −	11/2	11/2			6	6	36994.430 (20)	75DES1	
			+ −	11/2	11/2			5	6	37014.420 (30)	75DES1	
			− +	13/2	13/2			7	6	52696.720 (30)	75DES1	
			− +	13/2	13/2			6	6	52722.040 (20)	75DES1	
			− +	13/2	13/2			7	7	52734.560 (20)	75DES1	
			− +	13/2	13/2			6	7	52759.890 (30)	75DES1	
			+ −	15/2	15/2			7	7	70845.081 (20)	75DES1	
			+ −	15/2	15/2			8	8	70858.930 (20)	75DES1	
	X $^2\Pi_{1/2}$	$v=0$	− +	5/2	5/2			2	3	8118.051 (5)	75MEE	
			− +	5/2	5/2			2	2	8135.870 (5)	75MEE	
			− +	5/2	5/2			3	3	8189.587 (5)	75MEE	
			− +	5/2	5/2			3	2	8207.402 (5)	75MEE	
			+ −	7/2	7/2			3	4	5449.436 (5)	75MEE	
			+ −	7/2	7/2			3	3	5473.045 (5)	75MEE	
			+ −	7/2	7/2			4	4	5523.438 (5)	75MEE	
			+ −	7/2	7/2			4	3	5547.042 (5)	75MEE	
			+ −	9/2	9/2			4	5	88.9504 (5)	73MEE1	

Line frequencies of OH (continued)

Isotopic species	State: electronic	State: vibrational	Transition: parity		Transition: rotational J'	J''	hyperfine F_1'	F_1''	F'	F''	ν MHz	Ref.	Remarks
$^{16}O^{1}H$	$X\ ^2\Pi_{1/2}$	$v=0$	+	−	9/2	9/2			5	5	117.1495 (5)	73MEE1	
			+	−	9/2	9/2			4	4	164.7960 (5)	73MEE1	
			+	−	9/2	9/2			5	4	192.9957 (5)	73MEE1	
			−	+	11/2	11/2			6	6	8534.860 (20)	75DES1	
			−	+	11/2	11/2			5	5	8580.170 (20)	75DES1	
			+	−	13/2	13/2			7	7	19518.612 (20)	75DES1	
			+	−	13/2	13/2			6	6	19561.897 (20)	75DES1	
			−	+	15/2	15/2			8	8	32879.80 (5)	78DES	
			−	+	15/2	15/2			7	7	32921.32 (5)	78DES	
			+	−	17/2	17/2			9	9	48454.04 (5)	79COX	
			+	−	17/2	17/2			8	8	48493.98 (5)	79COX	
			−	+	19/2	19/2			10	10	66094.76 (5)	79COX	
			−	+	19/2	19/2			9	9	66133.35 (5)	79COX	
	$X\ ^2\Pi_{3/2}$	$v=1$	+	−	3/2	3/2			1	2	1489.438 (3)	78BEK	
			+	−	3/2	3/2			1	1	1536.944 (3)	78BEK	
			+	−	3/2	3/2			2	2	1538.702 (3)	78BEK	
			+	−	3/2	3/2			2	1	1586.213 (3)	78BEK	
			−	+	5/2	5/2			2	3	5583.476 (7)	79MEE	
			−	+	5/2	5/2			2	2	5594.246 (5)	79MEE	
			−	+	5/2	5/2			3	3	5598.168 (5)	79MEE	
			−	+	5/2	5/2			3	2	5608.935 (6)	79MEE	
			+	−	7/2	7/2			4	3	12520.8 (3)	79COX	
			+	−	7/2	7/2			3	3	12524.1 (3)	79COX	
			+	−	7/2	7/2			4	4	12530.1 (3)	79COX	
			+	−	7/2	7/2			3	4	12533.8 (3)	79COX	
			−	+	9/2	9/2			4	4	22296.57 (14)	79COX, 79DES	
			−	+	9/2	9/2			5	5	22304.86 (14)	79COX, 79DES	
			+	−	11/2	11/2			5	5	34738.215 (50)	79COX, 79DES	

Line frequencies of OH (continued)

Isotopic species	State		Transition							ν MHz	Ref.	Remarks
	electronic	vibrational	parity	rotational		hyperfine						
				J'	J''	F_1'	F_1''	F'	F''			
$^{16}O^{1}H$	$X\ ^2\Pi_{3/2}$	$v=1$	+ −	11/2	11/2			6	6	34748.275 (50)	79COX, 79DES	
			− +	13/2	13/2			6	6	49650.59 (5)	79COX, 79DES	
			− +	13/2	13/2			7	7	49662.09 (5)	79COX, 79DES	
			+ −	15/2	15/2			7	7	66851.04 (10)	79DES	
			+ −	15/2	15/2			8	8	66863.88 (10)	79DES	
	$X\ ^2\Pi_{1/2}$	$v=1$	− +	1/2	1/2			0	1	4449.064 (5)	79MEE	
			− +	1/2	1/2			1	1	4537.381 (5)	79MEE	
			− +	1/2	1/2			1	0	4553.998 (5)	79MEE	
			+ −	3/2	3/2			1	2	7454.592 (6)	79MEE	
			+ −	3/2	3/2			1	1	7467.323 (8)	79MEE	
			+ −	3/2	3/2			2	2	7522.842 (6)	79MEE	
			+ −	3/2	3/2			2	1	7535.563 (5)	79MEE	
			− +	5/2	5/2			2	2	7937.2 (5)	79COX	
			− +	5/2	5/2			3	3	7988.5 (5)	79COX	
	$X\ ^2\Pi_{3/2}$	$v=2$	+ −	3/2	3/2			1	2	1371.377 (3)	78BEK	
			+ −	3/2	3/2			1	1	1412.850 (3)	78BEK	
			+ −	3/2	3/2			2	2	1414.424 (3)	78BEK	
			+ −	3/2	3/2			2	1	1455.896 (3)	78BEK	
			− +	5/2	5/2			2	3	5161.489 (7)	79MEE	
			− +	5/2	5/2			2	2	5168.657 (6)	79MEE	
			− +	5/2	5/2			3	3	5172.192 (6)	79MEE	
			− +	5/2	5/2			3	2	5179.376 (8)	79MEE	
			+ −	7/2	7/2			3	3	11628.9 (3)	79COX	
			+ −	7/2	7/2			4	4	11634.0 (3)	79COX	
			− +	9/2	9/2			4	4	20791.0 (3)	79COX	
			− +	9/2	9/2			5	5	20798.7 (3)	79COX	
			+ −	11/2	11/2			5	5	32505.22 (5)	79COX	
			+ −	11/2	11/2			6	6	32514.36 (5)	79COX	
			− +	13/2	13/2			6	6	46584.86 (5)	79COX	
			− +	13/2	13/2			7	7	46595.38 (5)	79COX	

Line frequencies of OH (continued)

Isotopic species	State: electronic	State: vibrational	Transition: parity	Transition: rotational J'	J''	hyperfine F_1'	F_1''	F'	F''	ν MHz	Ref.	Remarks
$^{16}O^{1}H$	$X\ ^2\Pi_{1/2}$	$v=2$	− +	1/2	1/2			0	1	4233.346 (5)	79MEE	
			− +	1/2	1/2			1	1	4319.938 (5)	79MEE	
			− +	1/2	1/2			1	0	4338.543 (6)	79MEE	
			+ −	3/2	3/2			1	2	7146.208 (6)	79MEE	
			+ −	3/2	3/2			1	1	7160.046 (7)	79MEE	
			+ −	3/2	3/2			2	2	7212.791 (6)	79MEE	
			+ −	3/2	3/2			2	1	7226.627 (5)	79MEE	
	$X\ ^2\Pi_{3/2}$	$v=3$	+ −	7/2	7/2			3	3	10745.5 (3)	79COX	
			+ −	7/2	7/2			4	4	10750.5 (3)	79COX	
			− +	9/2	9/2			4	4	19296.4 (6)	79COX	
			− +	9/2	9/2			5	5	19303.2 (6)	79COX	
			− +	13/2	13/2			6	6	43511.93 (5)	79COX	
			− +	13/2	13/2			7	7	43521.72 (5)	79COX	
		$v=4$	+ −	7/2	7/2			3	3	9871.6 (3)	79COX	
			+ −	7/2	7/2			4	4	9876.2 (3)	79COX	
			− +	9/2	9/2			4	4	17805.2 (7)	79COX	
			− +	9/2	9/2			5	5	17811.5 (9)	79COX	
		$v=5$	+ −	7/2	7/2			3	3	9001.3 (3)	79COX	
			+ −	7/2	7/2			4	4	9005.5 (3)	79COX	
			− +	9/2	9/2			4	4	16311.0 (10)	79COX	
			− +	9/2	9/2			5	5	16316.0 (10)	79COX	
		$v=6$	− +	7/2	5/2					2779526 (60)	81DAV	extrapolation from LMR-measurements
			+ −	7/2	5/2					2784292 (60)	81DAV	
$^{17}O^{1}H$	$X\ ^2\Pi_{3/2}$	$v=0$		3/2	3/2	3	2	5/2	3/2	1302.119 (5)	78BEA	the following coupling scheme was used: $\boldsymbol{F}_1=\boldsymbol{J}+\boldsymbol{I}_{^{17}O}$, $\boldsymbol{F}=\boldsymbol{F}_1+\boldsymbol{I}_{^{1}H}$
				3/2	3/2	3	2	7/2	5/2	1322.464 (5)	78BEA	
				3/2	3/2	2	1	3/2	1/2	1418.202 (5)	78BEA	
				3/2	3/2	2	1	5/2	3/2	1455.729 (5)	78BEA	
				3/2	3/2	4	4	7/2	7/2	1624.518 (5)	78BEA	
				3/2	3/2	4	4	9/2	9/2	1626.161 (5)	78BEA	
				3/2	3/2	3	3	7/2	7/2	1656.542 (5)	78BEA	

Line frequencies of OH (continued)

Isotopic species	State: electronic	State: vibrational	Transition: parity	rotational J'	rotational J''	hyperfine F_1'	hyperfine F_1''	hyperfine F'	hyperfine F''	ν MHz	Ref.	Remarks
$^{17}O^1H$	$X\ ^2\Pi_{3/2}$	$v=0$		3/2	3/2	3	3	5/2	5/2	1655.499 (5)	78BEA	the following
				3/2	3/2	1	1	3/2	3/2	1683.540 (5)	78BEA	coupling
				3/2	3/2	1	1	1/2	1/2	1684.542 (5)	78BEA	scheme
				3/2	3/2	1	2	3/2	5/2	1902.093 (5)	78BEA	was used:
				3/2	3/2	1	2	3/2	3/2	1912.439 (5)	78BEA	$\boldsymbol{F}_1 = \boldsymbol{J} + \boldsymbol{I}_{^{17}O}$,
				3/2	3/2	1	2	1/2	3/2	1940.287 (5)	78BEA	$\boldsymbol{F} = \boldsymbol{F}_1 + \boldsymbol{I}_{^1H}$
				3/2	3/2	2	3	5/2	7/2	2008.354 (5)	78BEA	
				3/2	3/2	2	3	3/2	5/2	2027.323 (5)	78BEA	
				3/2	3/2	3	4	7/2	9/2	2102.781 (5)	78BEA	
				3/2	3/2	3	4	5/2	7/2	2117.800 (5)	78BEA	
				5/2	5/2	4	3	7/2	5/2	5752.916 (5)	78BEA	
				5/2	5/2	4	3	9/2	7/2	5760.223 (5)	78BEA	
				5/2	5/2	3	2	5/2	3/2	5851.057 (5)	78BEA	
				5/2	5/2	3	2	7/2	5/2	5857.578 (5)	78BEA	
				5/2	5/2	5	5	9/2	9/2	5914.960 (5)	78BEA	
				5/2	5/2	5	5	11/2	11/2	5918.956 (5)	78BEA	
				5/2	5/2	2	1	3/2	1/2	5931.908 (5)	78BEA	
				5/2	5/2	2	1	5/2	3/2	5938.801 (5)	78BEA	
				5/2	5/2	4	4	7/2	7/2	5971.048 (5)	78BEA	
				5/2	5/2	4	4	9/2	9/2	5974.374 (5)	78BEA	
				5/2	5/2	1	0	3/2	1/2	6005.626 (5)	78BEA	
				5/2	5/2	3	3	7/2	7/2	6013.646 (5)	78BEA	
				5/2	5/2	0	1	1/2	3/2	6114.648 (5)	78BEA	
				5/2	5/2	1	2	3/2	5/2	6157.620 (5)	78BEA	
				5/2	5/2	2	3	5/2	7/2	6200.038 (5)	78BEA	
				5/2	5/2	2	3	3/2	5/2	6202.176 (5)	78BEA	
	$X\ ^2\Pi_{1/2}$	$v=0$		1/2	1/2	3	3	5/2	5/2	3980.229 (5)	78BEA	the following
				1/2	1/2	3	3	7/2	7/2	4025.914 (5)	78BEA	coupling
				1/2	1/2	2	3	5/2	7/2	5473.032 (5)	78BEA	scheme
				1/2	1/2	2	3	3/2	5/2	5523.435 (5)	78BEA	was used:
				1/2	1/2	2	2	5/2	5/2	5716.505 (5)	78BEA	$\boldsymbol{F}_1 = \boldsymbol{J} + \boldsymbol{I}_{^{17}O}$,
				1/2	1/2	2	2	3/2	3/2	5746.333 (5)	78BEA	$\boldsymbol{F} = \boldsymbol{F}_1 + \boldsymbol{I}_{^1H}$

Line frequencies of OH (continued)

Isotopic species	State: electronic	State: vibrational	Transition: parity	rotational J'	– J''	hyperfine F_1' – F_1''	F'	– F''	ν MHz	Ref.	Remarks
$^{18}O^1H$	$X\ ^2\Pi_{3/2}$	$v=0$	+ −	3/2	3/2		1	2	1584.274 (2)	78BEA	
			+ −	3/2	3/2		1	1	1637.564 (2)	78BEA	
			+ −	3/2	3/2		2	2	1639.503 (2)	78BEA	
			+ −	3/2	3/2		2	1	1692.795 (2)	78BEA	
			− +	5/2	5/2		2	3	5920.505 (5)	78BEA	
			− +	5/2	5/2		2	2	5934.644 (6)	78BEA	
			− +	5/2	5/2		3	3	5938.967 (5)	78BEA	
			− +	5/2	5/2		3	2	5953.116 (4)	78BEA	
	$X\ ^2\Pi_{1/2}$	$v=0$	− +	1/2	1/2		0	1	4644.650 (6)	78BEA	
			− +	1/2	1/2		1	1	4735.073 (6)	78BEA	
			− +	1/2	1/2		1	0	4749.971 (5)	78BEA	
$^{16}O^2D$	$X\ ^2\Pi_{3/2}$	$v=0$	+ −	3/2	3/2		3/2	1/2	317.3290 (40)	73MEE2	
			+ −	3/2	3/2		5/2	3/2	322.4800 (20)	73MEE2	
			+ −	3/2	3/2		1/2	1/2	310.1445 (5)	73MEE2	
			+ −	3/2	3/2		3/2	3/2	310.2147 (5)	73MEE2	
			+ −	3/2	3/2		5/2	5/2	310.3627 (10)	73MEE2	
			+ −	3/2	3/2		3/2	5/2	298.0970 (10)	73MEE2	
			+ −	3/2	3/2		1/2	3/2	303.0320 (20)	73MEE2	
			− +	5/2	5/2		5/2	3/2	1194.3390 (10)	73MEE2	
			− +	5/2	5/2		7/2	5/2	1196.0060 (30)	73MEE2	
			− +	5/2	5/2		3/2	3/2	1190.5659 (2)	73MEE2	
			− +	5/2	5/2		5/2	5/2	1190.7741 (2)	73MEE2	
			− +	5/2	5/2		7/2	7/2	1191.1047 (2)	73MEE2	
			− +	5/2	5/2		5/2	7/2	1185.8712 (10)	73MEE2	
			− +	5/2	5/2		3/2	5/2	1186.9986 (30)	73MEE2	
			+ −	7/2	7/2		7/2	5/2	2823.5328 (20)	73MEE2	
			+ −	7/2	7/2		9/2	7/2	2824.2276 (20)	73MEE2	
			+ −	7/2	7/2		5/2	5/2	2822.0007 (20)	73MEE2	
			+ −	7/2	7/2		7/2	7/2	2822.3883 (20)	73MEE2	
			+ −	7/2	7/2		9/2	9/2	2822.9250 (20)	73MEE2	
			+ −	7/2	7/2		7/2	9/2	2821.0857 (20)	73MEE2	
			+ −	7/2	7/2		5/2	7/2	2820.8514 (20)	73MEE2	

Line frequencies of OH (continued)

Isotopic species	State: electronic	State: vibrational	Transition: parity	rotational J'	rotational J''	hyperfine F_1'	hyperfine F_1''	F'	F''	ν MHz	Ref.	Remarks
$^{16}O^{2}D$	$X\ ^2\Pi_{3/2}$	$v=0$	− +	9/2	9/2			9/2	7/2	5303.934 (5)	75MEE	
			− +	9/2	9/2			11/2	9/2	5304.359 (5)	75MEE	
			− +	9/2	9/2			7/2	7/2	5304.015 (5)	75MEE	
			− +	9/2	9/2			9/2	9/2	5304.600 (5)	75MEE	
			− +	9/2	9/2			11/2	11/2	5305.372 (5)	75MEE	
			− +	9/2	9/2			9/2	11/2	5305.619 (5)	75MEE	
			− +	9/2	9/2			7/2	9/2	5304.681 (5)	75MEE	
			+ −	11/2	11/2			9/2	9/2	8671.60 (8)	69POY	
			+ −	11/2	11/2			11/2	11/2	8672.37 (8)	69POY	
			+ −	11/2	11/2			13/2	13/2	8673.35 (8)	69POY	
			− +	13/2	13/2			11/2	11/2	12917.08 (8)	69POY	
			− +	13/2	13/2			13/2	13/2	12918.13 (8)	69POY	
			− +	13/2	13/2			15/2	15/2	12919.33 (8)	69POY	
			+ −	15/2	15/2			13/2	13/2	18009.46 (8)	69POY	
			+ −	15/2	15/2			15/2	15/2	18010.63 (8)	69POY	
			+ −	15/2	15/2			17/2	17/2	18012.13 (8)	69POY	
			− +	17/2	17/2			15/2	15/2	23907.08 (8)	69POY	
			− +	17/2	17/2			17/2	17/2	23908.44 (8)	69POY	
			− +	17/2	17/2			19/2	19/2	23910.04 (8)	69POY	
			+ −	19/2	19/2			17/2	17/2	30565.42 (8)	69POY	
			+ −	19/2	19/2			19/2	19/2	30566.95 (8)	69POY	
			+ −	19/2	19/2			21/2	21/2	30568.72 (8)	69POY	
			− +	21/2	21/2			19/2	19/2	37940.73 (8)	69POY	
			− +	21/2	21/2			21/2	21/2	37942.42 (8)	69POY	
			− +	21/2	21/2			23/2	23/2	37944.35 (8)	69POY	
	$X\ ^2\Pi_{1/2}$	$v=0$	− +	1/2	1/2			3/2	1/2	3114.5294 (10)	73MEE2	
			− +	1/2	1/2			1/2	1/2	3093.6057 (10)	73MEE2	
			− +	1/2	1/2			3/2	3/2	3111.1414 (10)	73MEE2	
			− +	1/2	1/2			1/2	3/2	3090.2163 (10)	73MEE2	
			+ −	3/2	3/2			3/2	1/2	5895.135 (5)	75MEE	
			+ −	3/2	3/2			5/2	3/2	5906.712 (5)	75MEE	
			+ −	3/2	3/2			1/2	1/2	5887.741 (5)	75MEE	
			+ −	3/2	3/2			3/2	3/2	5894.680 (5)	75MEE	

Line frequencies of OH (continued)

Isotopic species	State: electronic	State: vibrational	Transition: parity	rotational J'	J''	hyperfine F_1'	F_1''	F'	F''	ν MHz	Ref.	Remarks
$^{16}O^2D$	$X\ ^2\Pi_{1/2}$	$v=0$	+ −	3/2	3/2			5/2	5/2	5906.215 (5)	75MEE	
			+ −	3/2	3/2			3/2	5/2	5894.179 (5)	75MEE	
			+ −	3/2	3/2			1/2	3/2	5887.282 (5)	75MEE	
			− +	5/2	5/2			5/2	3/2	8118.733 (5)	75MEE	
			− +	5/2	5/2			7/2	5/2	8128.961 (5)	75MEE	
			− +	5/2	5/2			3/2	3/2	8110.717 (5)	75MEE	
			− +	5/2	5/2			5/2	5/2	8118.013 (5)	75MEE	
			− +	5/2	5/2			7/2	7/2	8128.181 (5)	75MEE	
			− +	5/2	5/2			5/2	7/2	8117.228 (5)	75MEE	
			− +	5/2	5/2			3/2	5/2	8109.994 (5)	75MEE	
			+ −	7/2	7/2			5/2	5/2	9578.90 (8)	69POY	
			+ −	7/2	7/2			7/2	7/2	9586.32 (8)	69POY	
			+ −	7/2	7/2			9/2	9/2	9595.74 (8)	69POY	
			− +	9/2	9/2			7/2	7/2	10192.21 (8)	69POY	
			− +	9/2	9/2			9/2	9/2	10199.56 (8)	69POY	
			− +	9/2	9/2			11/2	11/2	10208.50 (8)	69POY	
			+ −	11/2	11/2			9/2	9/2	9914.70 (8)	69POY	
			+ −	11/2	11/2			11/2	11/2	9921.95 (8)	69POY	
			+ −	11/2	11/2			13/2	13/2	9930.44 (8)	69POY	
		$v=0$	+ −	3/2	1/2			3/2	3/2	944412.0	82BRO	private communication of van den Heuvel, E., and Dymanus, A.
			+ −	3/2	1/2			5/2	3/2 }	944424.4	82BRO	
			+ −	3/2	1/2			1/2	1/2 }			
			+ −	3/2	1/2			3/2	1/2	944433.3	82BRO	

References

69POY Poynter, R.L., Beaudet, R.A.: 24th Symp. Mol. Spectrosc. Columbus, Ohio 1969, Abstract 78R2 (frequency listing in [78BEA]).

71CLO Clough, P.N., Curran, A.H., Thrush, B.A.: Proc. Roy. Soc. London **A 323** (1971) 451.

73HIN Hinkley, R.K., Walker, T.E.H., Richards, W.G.: Proc. Roy. Soc. London **A 331** (1973) 553.

73MEE1 Meerts, W.L., Dymanus, A.: Chem. Phys. Lett **23** (1973) 45.

73MEE2 Meerts, W.L., Dymanus, A.: Astrophys. J. **180** (1973) L93.

continued

References to No. 11 (OH), continued

73VAL Valtz, I.E., Soglasnova, V.A.: Astrophys. Lett. **13** (1973) 23.
74DES Destombes, J.-L., Marlière, C., Rohart, F., Burie, J.: C.R. Acad. Sci. Paris **B 278** (1974) 275.
74LEE Lee, K.P., Tam, W.G.: Chem. Phys. **4** (1974) 434.
75COX1 Coxon, J.A.: J. Mol. Spectrosc. **58** (1975) 1.
75COX2 Coxon, J.A., Hammersley, R.E.: J. Mol. Spectrosc. **58** (1975) 29.
75DES1 Destombes, J.-L., Journel, G., Marlière, C., Rohart, F.: C.R. Acad. Sci. Paris **B 280** (1975) 809.
75DES2 Destombes, J.-L., Marlière, C.: Chem. Phys. Lett. **34** (1975) 532.
75MEE Meerts, W.L., Dymanus, A.: Can. J. Phys. **53** (1975) 2123.
76BUS Bustreel, R., Destombes, J.-L., Marlière, C.: Chem. Phys. Lett. **42** (1976) 154.
76TER1 TerMeulen. J.J., Meerts, W.L., van Mierlo, G.W.M., Dymanus, A.: Phys. Rev. Lett. **36** (1976) 1031.
76TER2 TerMeulen, J.J.: Ph.D. Thesis, Katholieke Universiteit Nijmegen, **1976**.
77DES1 Destombes, J.-L., Marlière, C., Rohart, F.: J. Mol. Spectrosc. **67** (1977) 93.
77DES2 Destombes, J.-L., Marlière, C., Baudry, A., Brillet, J.: Astron. Astrophys. **60** (1977) 55.
77MEE Meerts, W.L.: Chem. Phys. Lett. **46** (1977) 24.
78BEA Beaudet, R.A., Poynter, R.L.: J. Phys. Chem. Ref. Data **7** (1978) 311.
78BEK Bekooy, J.P., Meerts, W.L., Dymanus, A.: Astrophys. J. **224** (1978) L77.
78BRO Brown, J.M., Kaise, M., Kerr, C.M.L., Milton, D.J.: Mol. Phys. **36** (1978) 553.
78DES Destombes, J.-L.: Thèse L'Université des Science et Techniques de Lille, France, **1978**.
79BAS Bastard, D., Bretenoux, A., Charru, A., Picherit, F.: J. de Phys. **40** (1979) L533.
79COX Coxon. J.A., Sastry, K.V.L.N., Austin, J.A., Levy, D.H.: Can. J. Phys. **57** (1979) 619.
79DAV Davies, P.B., Hack, W., Preuss, A.W., Temps, F.: Chem. Phys. Lett. **64** (1979) 94.
79DES Destombes, J.-L., Lemoine, B., Marlière-Demuynck, C.: Chem. Phys. Lett. **60** (1979) 493.
79MEE Meerts, W.L., Bekooy, J.P., Dymanus, A.: Mol. Phys. **37** (1979) 425.
80COX Coxon, J.A.: Can. J. Phys. **58** (1980) 933.
80SAS Sastry, K.V.L.N., Van der Linde, J.: J. Mol. Spectrosc. **83** (1980) 332.
81BRO Brown, J.M., Kerr, C.M.L., Wayne, F.D., Evenson, K.M., Radford, H.E.: J. Mol. Spectrosc. **86** (1981) 544.
81DAV Davies, P.B., Hack, W., Wagner, H.Gg.: Farad. Disc. Chem. Soc. **71** (1981) 15.
82BRO Brown, J.M., Schubert, J.E.: J. Mol. Spectrosc. **95** (1982) 194.
82COX Coxon, J.A., Foster, S.C.: J. Mol. Spectrosc. **91** (1982) 243.

12 SH

Molecular constants of SH

Isotopic species	State		Parameter	Value	Ref.	Meth.	Remarks
	electronic	vibrational					
$^{32}S^1H$	$X^2\Pi$	$v=0$	α_p	−4216.0 (4) MHz	75MEE	MB	parameters were obtained from a fit including third order perturbation theory
			β_p	141.9 (1) MHz	75MEE	MB	
			$a-1/2(b+c)\,(^1H)$	48.08 (12) MHz	75MEE	MB }	$a^{der}(^1H)=32.58\,(7)$ MHz
			$a+1/2(b+c)\,(^1H)$	17.078 (5) MHz	75MEE	MB }	$c^{der}(^1H)=32.44\,(14)$ MHz
			$b(^1H)$	−63.44 (4) MHz	75MEE	MB	
			$d(^1H)$	27.36 (12) MHz	75MEE	MB	

Molecular constants of SH (continued)

Isotopic species	State: electronic	State: vibrational	Parameter	Value	Ref.	Meth.	Remarks
$^{32}S^{1}H$	X $^2\Pi$	$v=0$	B_0	283577.0 (9) MHz	78DAV, 75MEE	LMR/MB	combined fit
			p	9007.28 (15) MHz	78DAV, 75MEE	LMR/MB	centrifugal distortion parameters in [78DAV]
			q	−284.404 (33) MHz	78DAV, 75MEE	LMR/MB	
			g_s^{ev}	2.00213	78DAV, 73TAN	LMR/EPR	
			g'_L	1.0008 (9)	78DAV, 73TAN	LMR/EPR	
	X $^2\Pi_{3/2}$	$v=0$	μ	0.7580 (1) D	75MEE	MB	
$^{34}S^{1}H$	X $^2\Pi$	$v=0$	B_0	283072.4 (9) MHz	78DAV	LMR	extrapolated values of other parameters from the results on $^{32}S^{1}H$ were used to fit the data
$^{32}S^{2}D$	X $^2\Pi$	$v=0$	B_0	146886 (3) MHz	78DAV, 75MEE	LMR/MB	combined fit
			p	4669.3 (10) MHz	78DAV, 75MEE	LMR/MB	centrifugal distortion parameters in [78DAV]
			q	−76.1 (3) MHz	78DAV, 75MEE	LMR/MB	
			α_p	−2254.6 MHz	75MEE	MB	obtained from the results of SH by isotopic substitution
			β_p	38.07 MHz	75MEE	MB	
	X $^2\Pi_{3/2}$	$v=0$	$a+1/2(b+c)\,(^2D)$	2.6813 (33) MHz	75MEE	MB	
			$b(^2D)$	−9.648 (30) MHz	75MEE	MB	
			$d(^2D)$	4.20 (16) MHz	75MEE	MB	
			$eq^{(a)}Q(^2D)$	149 (4) kHz	75MEE	MB	
			$eq^{(b)}Q(^2D)$	−50 (24) kHz	75MEE	MB	
			μ	0.7571 (1) D	75MEE	MB	

Line frequencies of SH

Isotopic species	State: electronic	State: vibrational	Transition: parity	Transition: rotational J' – J''	Transition: hyperfine F_1' – F_1''	Transition: hyperfine F' – F''	ν MHz	Ref.	Remarks
$^{32}S^{1}H$	X $^{2}\Pi_{3/2}$	$v=0$	+ –	3/2 3/2		1 1	111.4862 (5)	74MEE	
			+ –	3/2 3/2		2 2	111.5452 (5)	74MEE	
			+ –	3/2 3/2		1 2	100.293 (3)	74MEE	
			+ –	3/2 3/2		2 1	122.737 (3)	74MEE	
			– +	5/2 5/2		2 2	442.4781 (5)	74MEE	
			– +	5/2 5/2		3 3	442.6277 (5)	74MEE	
			– +	5/2 5/2		2 3	437.9154 (20)	74MEE	
			– +	5/2 5/2		3 2	447.1876 (20)	74MEE	
			+ –	7/2 7/2		3 3	1094.1871 (5)	74MEE	
			+ –	7/2 7/2		4 4	1094.4596 (10)	74MEE	
			+ –	7/2 7/2		3 4	1093.5326 (20)	74MEE	
			+ –	7/2 7/2		4 3	1095.1179 (20)	74MEE	
			– +	9/2 9/2		4 4	2158.2986 (10)	74MEE	
			– +	9/2 9/2		5 5	2158.7239 (10)	74MEE	
			– +	9/2 9/2		4 5	2160.5318 (10)	74MEE	
			– +	9/2 9/2		5 4	2156.4902 (10)	74MEE	
			+ –	11/2 11/2		5 5	3714.8864 (20)	74MEE	
			+ –	11/2 11/2		6 6	3715.4892 (20)	74MEE	
	X $^{2}\Pi_{1/2}$	$v=0$	– +	1/2 1/2		1 1	8445.211 (5)	75MEE	
			– +	1/2 1/2		1 0	8459.034 (5)	75MEE	
$^{32}S^{2}D$	X $^{2}\Pi_{3/2}$	$v=0$	+ –	3/2 3/2		5/2 5/2	16.120 (3)	75MEE	
			+ –	3/2 3/2		1/2 3/2	14.719 (5)	75MEE	
			+ –	3/2 3/2		3/2 5/2	13.643 (3)	75MEE	
			+ –	3/2 3/2		5/2 3/2	18.591 (3)	75MEE	
			– +	5/2 5/2		3/2 3/2	64.295 (1)	75MEE	
			– +	5/2 5/2		5/2 5/2	64.299 (1)	75MEE	
			– +	5/2 5/2		7/2 7/2	64.307 (1)	75MEE	
			– +	5/2 5/2		3/2 5/2	63.410 (5)	75MEE	
			– +	5/2 5/2		5/2 3/2	65.181 (1)	75MEE	
			– +	5/2 5/2		5/2 7/2	63.096 (3)	75MEE	
			– +	5/2 5/2		7/2 5/2	65.505 (2)	75MEE	
			+ –	7/2 7/2		5/2 5/2	160.1546 (10)	75MEE	
			+ –	7/2 7/2		7/2 7/2	160.1617 (10)	75MEE	

Line frequencies of SH (continued)

Isotopic species	State: electronic	State: vibrational	Transition: parity	Transition: rotational J'	J''	Transition: hyperfine F_1'	F_1''	F'	F''	ν MHz	Ref.	Remarks
$^{32}S^{2}D$	X $^2\Pi_{3/2}$	$v=0$	+ −	7/2	7/2			9/2	9/2	160.1767 (7)	75MEE	
			+ −	7/2	7/2			5/2	7/2	159.635 (2)	75MEE	
			+ −	7/2	7/2			7/2	5/2	160.6807 (10)	75MEE	
			+ −	7/2	7/2			7/2	9/2	159.5752 (10)	75MEE	
			+ −	7/2	7/2			9/2	7/2	160.7660 (10)	75MEE	
			− +	9/2	9/2			7/2	7/2	318.8188 (7)	75MEE	
			− +	9/2	9/2			9/2	9/2	318.8334 (7)	75MEE	
			− +	9/2	9/2			11/2	11/2	318.8568 (7)	75MEE	
			− +	9/2	9/2			7/2	9/2	318.583 (2)	75MEE	
			− +	9/2	9/2			9/2	7/2	319.071 (1)	75MEE	
			− +	9/2	9/2			9/2	11/2	318.640 (1)	75MEE	
			− +	9/2	9/2			11/2	9/2	319.052 (1)	75MEE	
			+ −	11/2	11/2			13/2	13/2	554.8335 (10)	75MEE	
			− +	13/2	13/2			11/2	11/2	881.757 (2)	75MEE	
			− +	13/2	13/2			13/2	13/2	881.788 (2)	75MEE	
			− +	13/2	13/2			15/2	15/2	881.831 (1)	75MEE	
			− +	13/2	13/2			11/2	13/2	881.959 (5)	75MEE	

References

73TAN Tanimoto, M., Uehara, H.: Mol. Phys. **25** (1973) 1193.
74MEE Meerts, W.L., Dymanus, A.: Astrophys. J. **187** (1974) L45.
75MEE Meerts, W.L., Dymanus, A.: Can. J. Phys. **53** (1975) 2123.
78DAV Davies, P.B., Handy, B.J., Murray, Lloyd E.K., Russel, D.K.: Mol. Phys. **36** (1978) 1005.
80LOW Lowe, R.S.: Mol. Phys. **41** (1980) 929.
80ROH Rohrbeck, W., Hinz, A., Urban, W.: Mol. Phys. **41** (1980) 925.

13 SeH

Molecular constants of SeH

Isotopic species	State: electronic	State: vibrational	Parameter	Value	Ref.	Meth.	Remarks
$^{80}Se^{1}H$	X $^2\Pi$	$v=0$	$A_0+\gamma_0$	−1764.01895 (16) cm^{-1}	79BRO	LMR	according to the authors the errors shown are certainly underestimated because of neglected systematic errors
			γ_0^{ev}	−22.78 GHz	79BRO	LMR	
			B_0	233458.5 (19) MHz	79BRO	LMR	
			D_0^{ev}	9.68 MHz	79BRO	LMR	

Molecular constants of SeH (continued)

Isotopic species	State		Parameter	Value	Ref.	Meth.	Remarks
	electronic	vibrational					
$^{80}Se^{1}H$	$X\ ^2\Pi$	$v=0$	α_p	−16980.1 (16) MHz	79BRO	LMR	centrifugal distortion parameters are used in the analysis
			β_p	89.55 (61) MHz	79BRO	LMR	
			g_s^{ev}	2.00122	79BRO	LMR	
			g'_L	1.00062 (11)	79BRO	LMR	
			g_l^{ev}	0.0349	79BRO	LMR	
			g'_l	0.0693 (12)	79BRO	LMR	
			g_{rot}	−0.538 (57)	79BRO	LMR	
			$g_{rot}^{e'ev}$	1.452	79BRO	LMR	
$^{80}Se^{2}D$	$X\ ^2\Pi$	$v=0$	$A_0+\gamma_0$	−1763.2859 (10) cm^{-1}	80BRO	LMR	parameter contains some centrifugal distortion
			γ_0^{ev}	−15.0 GHz	80BRO	LMR	
			B_0+2D_0	118808.06 (26) MHz	80BRO	LMR	
			D_0^{ev}	2.61 MHz	80BRO	LMR	
			α_p	−8843 (14) MHz	80BRO	LMR	centrifugal distortion parameters are used in the analysis
			β_p^{der}	22.9 MHz	80BRO	LMR	
			g'_L	1.00054 (20)	80BRO	LMR	other g-factors derived from [79BRO]
SeH	$X\ ^2\Pi$	$v=0$			82BRO		reanalysis of the data of [79BRO] and [80BRO] taking account of the mixing between the spin components of the $v=0$ and 1 levels, deriving a set of so-called deperturbed molecular parameters. Therefore the parameters shown in the present tables are the effective parameters describing the spectra with the familiar formulas
SeD	$X\ ^2\Pi$	$v=0$					
$^{82}Se^{2}D$	$X\ ^2\Pi$	$v=0$	B_0	118731.9 (12) MHz	80CLI	LMR	
$^{80}Se^{2}D$	$X\ ^2\Pi$	$v=0$	B_0	118803.0 (12) MHz	80CLI	LMR	
$^{78}Se^{2}D$	$X\ ^2\Pi$	$v=0$	B_0	118877.3 (12) MHz	80CLI	LMR	
$^{77}Se^{2}D$	$X\ ^2\Pi$	$v=0$	B_0	118914.8 (12) MHz	80CLI	LMR	
$^{76}Se^{2}D$	$X\ ^2\Pi$	$v=0$	B_0	118955.6 (12) MHz	80CLI	LMR	
$^{74}Se^{2}D$	$X\ ^2\Pi$	$v=0$	B_0	119037.4 (12) MHz	80CLI	LMR	

Molecular constants of SeH (continued)

Isotopic species	State electronic	State vibrational	Parameter	Value	Ref.	Meth.	Remarks
$^{80}Se^{2}D$	$X\,^2\Pi$	equ.	$Y_{01} \approx B_e$	120010.07 (33) MHz	82BRO	LMR	
			$Y_{11} = -\alpha_e$	−2394.08 (60) MHz	82BRO	LMR	
			r_e	1.4640715 (25) Å	82BRO	LMR	

References

78DAV Davies, P.B., Handy, B.J., Murray Lloyd, E.K., Russell, D.K.: J. Chem. Phys. **68** (1978) 3377.

79BRO Brown, J.M., Carrington, A., Sears, T.J.: Mol. Phys. **37** (1979) 1837.

80BRO Brown, J.M., Carrington, A., Fackerell, A.D.: Chem. Phys. Lett. **75** (1980) 13.

80CLI Cliff, D.I., Davies, P.B., Handy, B.J., Thrush, B.A., Murray Lloyd, E.K.: Chem. Phys. Lett. **75** (1980) 9.

82BRO Brown, J.M., Fackerell, A.D.: Physica Scripta **25** (1982) 351.

13a HCl^+

Molecular constants of HCl^+

Isotopic species	State electronic	State vibrational	Parameter	Value	Ref.	Meth.	Remarks
$^1H^{35}Cl^+$	$X\,^2\Pi_{3/2}$	$v=0$	B_0	293438.14 (8) MHz	82RAY	LMR	
			q	−336.83 (97) MHz	82RAY	LMR	
			$a+1/2(b+c)(^{35}Cl)$	365.2 (20) MHz	82RAY	LMR	
			$eq^{(a)}Q\,(^{35}Cl)$	7.65 (182) MHz	82RAY	LMR	
			$eq^{(b)}Q\,(^{35}Cl)$	−131.8 (564) MHz	82RAY	LMR	
			D_0	16.2095 MHz	82RAY		parameters from [79BRO] used for fitting the LMR-spectra
			A_0	−643.396 (21) cm^{-1}	82RAY		
			p	18.467 (39) GHz	82RAY		
			γ	−8.99 (16) GHz	82RAY		
			g_s	2.00129	82RAY		estimated g-factors
			g_l	0.002226	82RAY		
			g_{rot}	−0.29	82RAY		

References

79BRO Brown, J.M., Colbourn, E.A., Watson, J.K.G., Wayne, F.D.: J. Mol. Spectrosc. **74** (1979) 294.

82RAY Ray, D., Lubic, K.G., Saykally, R.J.: Mol. Phys. **46** (1982) 217.

14 HBr^+

Molecular constants of HBr^+

Isotopic species	State electronic	State vibrational	Parameter	Value	Ref.	Meth.	Remarks
$^1H^{79}Br^+$	$X\,^2\Pi_{3/2}$	$v=0$	$a+1/2(b+c)\,(^{79}Br)$	1460 (30) MHz	79SAY	LMR	
			$E_{RES}(J=5/2\leftarrow J=3/2)$	1188700 (15) MHz	79SAY	LMR	rotational fine-structure difference
			$\Delta\nu_\Lambda=\nu_\Lambda(5/2)-\nu(3/2)$	39 (5) MHz	79SAY	LMR	difference between Λ-doubling frequencies
$^1H^{81}Br^+$	$X\,^2\Pi_{3/2}$	$v=0$	$a+1/2(b+c)\,(^{81}Br)$	1580 (30) MHz	79SAY	LMR	
			$E_{RFS}(J=5/2\leftarrow J=3/2)$	1188660 (15) MHz	79SAY	LMR	rotational fine-structure difference
			$\Delta\nu_\Lambda=\nu_\Lambda(5/2)-\nu_\Lambda(3/2)$	36 (5) MHz	79SAY	LMR	difference between Λ-doubling frequencies
			Nuclear quadrupole structure and second-order magnetic hyperfine interactions are not well analyzed in [79SAY]				

Reference: 79SAY Saykally, R.J., Evenson, K.M.: Phys. Rev. Lett. **43** (1979) 515.

40 CaCl

Molecular constants of CaCl

Isotopic species	State electronic	State vibrational	Parameter	Value	Ref.	Meth.	Remarks
$^{40}Ca\,^{35}Cl$	$A\,^2\Pi_{3/2}$	$v=0$			82CHI	MB/LIF	upper limit of hyperfine structure is determined to be 1 MHz for $J=101.5$

Reference: 82CHI Childs, W.J., Cok, D.R., Goodman, L.S.: J. Opt. Soc. Am. **72** (1982) 717.

40a CaBr

Molecular constants of CaBr

Isotopic species	State electronic	State vibrational	Parameter	Value	Ref.	Meth.	Remarks
$^{40}CaBr$	$A\,^2\Pi$	$v=0$	$\Delta eqQ(^{79}Br)$	31 (1) MHz	81BER	SLS	$\Delta eqQ=eqQ(A\,^2\Pi)-eqQ(X\,^2\Sigma^+)$
			$\Delta eqQ(^{81}Br)$	28 (2) MHz	81BER	SLS	systematic uncertainty due to neglect of magnetic hyperfine interactions

Reference: 81BER Bernath, P.F., Pinchemel, B., Field, R.W.: J. Chem. Phys. **74** (1981) 5508.

40b CaI
Molecular constants of CaI

Isotopic species	State		Parameter	Value	Ref.	Meth.	Remarks
	electronic	vibrational					
$^{40}Ca^{127}I$	A $^2\Pi$	$v=0$	$\Delta\, eqQ\,(^{127}I)$	−55 (1) MHz	81BER	SLS	$\Delta\, eqQ = eqQ(A\,^2\Pi) - eqQ(X\,^2\Sigma^+)$, systematic uncertainty due to neglect of magnetic hyperfine interactions

Reference: 81BER Bernath, P.F., Pinchemel, B., Field, R.W.: J. Chem. Phys. **74** (1981) 5508.

45 CO^+
Molecular constants of CO^+

Isotopic species	State		Parameter	Value	Ref.	Meth.	Remarks
	electronic	vibrational					
$^{13}C^{16}O^+$	A $^2\Pi$	$v=1$	$[a+1/2\, b_F+1/3\, c]$ (^{13}C)	67.8 (198) MHz	78CAR		accelerated ion-beam and laser excitation spectroscopy
			$[a-1/2\, b_F-1/3\, c]^{ev}$ (^{13}C)	11.5 MHz	78CAR		

Reference: 78CAR Carrington, A., Milverton, D.R.J., Sarre, P.J.: Mol. Phys. **35** (1978) 1505.

46 CN
Molecular constants of CN

Isotopic species	State		Parameter	Value	Ref.	Meth.	Remarks
	electronic	vibrational					
$^{12}C^{14}N$	A $^2\Pi_{3/2}$	$v=10$	μ_{10}	0.56 (3) D	73COO		level anticrossing spectroscopy
			Magnetic resonance spectrum of the rotational perturbed levels of A $^2\Pi_{3/2}$ $v=10$, $J=7/2$ and B $^2\Sigma^+$ $v=0$, $N=4$ has been observed in [73COO] determining different interelectronic coupling parameters between the *B*- and *A*-state.				

Reference: 73COO Cook, T.J., Levy, D.H.: J. Chem. Phys. **59** (1973) 2387.

78 CF

Molecular constants of CF

Isotopic species	State electronic	State vibrational	Parameter	Value	Ref.	Meth.	Remarks
$^{12}C^{19}F$	X $^2\Pi_{1/2}$	$v=0$	$\nu_0(J=3/2\leftarrow1/2)$	124201.973 (24) MHz	80SAI	MW	transition frequency excluding Λ-doubling and hyperfine energies
			p_{eff}	256.679 (12) MHz	80SAI	MW	the parameter includes a contribution of the Λ-doubling constant q
			$a-1/2(b+c)(^{19}F)$	747.61 (3) MHz	80SAI	MW	
			$b(^{19}F)$	270 (1) MHz	80SAI	MW	
			$d(^{19}F)$	792.16 (3) MHz	80SAI	MW	
			$a(^{19}F)$	705.3 (15) MHz	80SAI		adopting $a+1/2(b+c)=662.9\ (30)$ MHz from measurements by ESR [C1] of Landolt-Börnstein N.S. II/6, p. 5-51
			$c(^{19}F)$	−354.7 (40) MHz	80SAI		
			μ	0.645 (5) D	80SAI	MW	
	X $^2\Pi$	$v=0$	B_0	42196.45 (63) MHz	81KAW	IR/MW	
			D_0	199.1 (45) kHz	81KAW	IR/MW	
			p	257.22 (63) MHz	81KAW	IR/MW	
			A_0	77.11 cm^{-1}	81KAW		adopted from optical spectroscopy
		$v=1$	B_1	41650.86 (66) MHz	81KAW	IR/MW	
			D_1	196.7 (36) kHz	81KAW	IR/MW	
			p	255.4 (33) MHz	81KAW	IR/MW	
			A_1-A_0	−19.487 (27) GHz = −0.6500 (9) cm^{-1}	81KAW	IR/MW	
			$\nu_0(v=0\rightarrow v=1)$	1286.1281 (5) cm^{-1}	81KAW	IR/MW	
		equ.	Y_{01}	42471.7 (11) MHz	81KAW	IR/MW	
			Y_{11}	−552.2 (15) MHz	81KAW	IR/MW	
			Y_{21}	3.30 (60) MHz	81KAW		adopted from optical spectroscopy
			r_e	1.271977 (17) Å	81KAW	IR/MW	
		$v=0$	B_0	42197.031 (59) MHz	82VAN	MW	the values are obtained by a combined analysis with data of [81KAW]; but the effective parameters were expanded as a power series in $\boldsymbol{N}^2$ and not in $\boldsymbol{R}^2$ as in [81KAW]. This results in slight deviations for the values shown in both cases
			D_0	198.70 (23) kHz	82VAN	MW	
			p	255.60 (35) MHz	82VAN	MW	
			q	0.760 (14) MHz	82VAN	MW	

Molecular constants of CF (continued)

Isotopic species	State: electronic	State: vibrational	Parameter	Value	Ref.	Meth.	Remarks
$^{12}C^{19}F$	X $^2\Pi$	$v=0$	B_0	42196.35 (3) MHz	82SAY	LMR	
			$h=a+1/2(b+c)(^{19}F)$	665.7 (83) MHz	82SAY	LMR	
	X $^2\Pi_{3/2}$		g_s	2.00196	82SAY		the estimated g-factors were used to fit the observations of the LMR-experiment
			g_L'	0.99989	82SAY		
			g_l	−0.00181	82SAY		
			g_{rot}	0.37	82SAY		

Line frequencies of CF

Isotopic species	State: electronic	State: vibrational	Transition: parity	rotational J'	rotational J''	hyperfine F_1'	hyperfine F_1''	F'	F''	ν MHz	Ref.	Remarks
$^{12}C^{19}F$	X $^2\Pi_{1/2}$	$v=0$	+ −	3/2	1/2			2	1	124309.999 (6)	80SAI	
			+ −	3/2	1/2			1	0	124708.812 (12)	80SAI	
			+ −	3/2	1/2			1	1	123682.520 (9)	80SAI	
			− +	3/2	1/2			2	1	124001.566 (18)	80SAI	
			− +	3/2	1/2			1	0	124185.447 (10)	80SAI	
			− +	3/2	1/2			1	1	124217.490 (10)	80SAI	
	X $^2\Pi_{1/2}$	$v=0$	− +	19/2	17/2			9	8	786742.7 (8)	82VAN	
			− +	19/2	17/2			10	9	786745.8 (8)		
			+ −	19/2	17/2			10	9	786974.1 (8)		
			+ −	19/2	17/2			9	8	786976.6 (8)		
			+ −	21/2	19/2			10	9	869612.7 (8)		
			+ −	21/2	19/2			11	10	869616.7 (8)		
			− +	21/2	19/2			10	9 }	869840.0 (8)		
			− +	21/2	19/2			11	10 }			
			− +	23/2	21/2			11	10	952484.6 (8)		
			− +	23/2	21/2			12	11	952488.7 (8)		
			+ −	23/2	21/2			11	10 }	952705.8 (8)		
			+ −	23/2	21/2			12	11 }			
			+ −	25/2	23/2			12	11	1035354.2 (8)		
			+ −	25/2	23/2			13	12	1035359.0 (8)		
			− +	25/2	23/2			12	11 }	1035569.9 (8)		
			− +	25/2	23/2			13	12 }			

Line frequencies of CF (continued)

Isotopic species	State		Transition							v	Ref.	Remarks
	electronic	vibrational	parity		rotational		hyperfine			MHz		
					J'	J''	F_1' – F_1''	F'	F''			
$^{12}C^{19}F$	X $^2\Pi_{3/2}$	$v=0$	–	+	19/2	17/2		10	9	815230.5 (8)	82VAN	
			–	+	19/2	17/2		9	8	815248.4 (8)		
			+	–	19/2	17/2		10	9	815261.0 (8)		
			+	–	19/2	17/2		9	8	815274.3 (8)		
			+	–	21/2	19/2		11	10	900671.8 (8)		
			+	–	21/2	19/2		10	9	900687.3 (8)		
			–	+	21/2	19/2		11	10	900708.2 (8)		
			–	+	21/2	19/2		10	9	900719.1 (8)		
			+	–	25/2	23/2		13	12	1071239.7 (8)		
			+	–	25/2	23/2		12	11	1071252.2 (8)		
			–	+	25/2	23/2		13	12	1071287.9 (8)		
			–	+	25/2	23/2		12	11	1071294.9 (8)		

References

80SAI Saito, S., Endo, Y., Hirota, E.: Ann. Rev. of the Institute for Mol. Science, Okazaki, Japan, **1980**.

81KAW Kawaguchi, K., Yamada, Ch., Hamada, Y., Hirota, E.: J. Mol. Spectrosc. **86** (1981) 136.

82SAY Saykally, R.J., Lubic, K.G., Scalabrin, A., Evenson, K.M.: J. Chem. Phys. **77** (1982) 58.

82VAN Van den Heuvel, F.C., Meerts, W.L., Dymanus, A.: Chem. Phys. Lett. **88** (1982) 59.

79 CCl

Molecular constants of CCl

Isotopic species	State		Parameter	Value	Ref.	Meth.	Remarks
	electronic	vibrational					
$^{12}C^{35}Cl$	X $^2\Pi$	$v=0$	B_0	20797.1725 (38) MHz	82END	MW	
			D_0	52.71 (12) kHz	82END	MW	
			A_0	4045.94 GHz	82END		adopted from optical spectroscopy
			p	401.696 (25) MHz	82END	MW	
			q	1.176 (48) MHz	82END	MW	
			$a+1/2(b+c)(^{35}Cl)$	66.46 (11) MHz	82END	MW }	derived constants:
			$a-1/2(b+c)(^{35}Cl)$	93.936 (37) MHz	82END	MW }	$a(^{35}Cl)=$ 80.199 (54) MHz
			$b(^{35}Cl)$	12.0 (15) MHz	82END	MW }	$c(^{35}Cl)=-39.5$ (16) MHz
			$d(^{35}Cl)$	82.212 (37) MHz	82END	MW	

Molecular constants of CCl (continued)

Isotopic species	State		Parameter	Value	Ref.	Meth.	Remarks
	electronic	vibrational					
$^{12}C^{35}Cl$	X $^2\Pi$	$v=0$	$eq^{(a)}Q(^{35}Cl)$	−34.26 (13) MHz	82END	MW	
			$eq^{(b)}Q(^{35}Cl)$	84.6 (95) MHz	82END	MW	
			$A_1 - A_0$	−37.334 (22) GHz	82END, 81YAM	IR/MW	
			$B_1 - B_0 \approx Y_{11}$	−202.91 (14) MHz	82END, 81YAM	IR/MW	
		equ.	$Y_{01} \approx B_e$	20898.63 (10) MHz	82END, 81YAM	IR/MW	
			r_e	1.645218 (16) Å	82END, 81YAM	IR/MW	
			$Y_{10} \approx \omega_e$	876.7442 (14) cm^{-1}	82END, 81YAM	IR/MW	
			$Y_{20} = -\omega_e x_e$	−5.32968 (76) cm^{-1}	82END, 81YAM	IR/MW	

Line frequencies of CCl

Isotopic species	State		Transition								ν MHz	Ref.	Remarks
	electronic	vibrational	parity		rotational J'	J''	hyperfine F_1'	F_1''	F'	F''			
$^{12}C^{35}Cl$	X $^2\Pi_{1/2}$	$v=0$	+	−	3/2	1/2			2	1	62397.197 (50)	82END	
			−	+	3/2	1/2			2	1	61891.581 (50)		
			+	−	3/2	1/2			3	2	62272.573 (50)		
			−	+	3/2	1/2			3	2	61854.665 (50)		
			+	−	3/2	1/2			1	1	62335.147 (50)		
			−	+	3/2	1/2			1	1	61917.047 (50)		
			+	−	3/2	1/2			2	2	62162.317 (50)		
			−	+	3/2	1/2			2	2	61876.027 (50)		
			−	+	5/2	3/2			1	0	103691.504 (50)		
			+	−	5/2	3/2			1	0	103273.581 (50)		
			−	+	5/2	3/2			2	1	103694.817 (50)		
			+	−	5/2	3/2			2	1	103264.710 (50)		
			−	+	5/2	3/2			3	2	103686.760 (50)		

Line frequencies of CCl (continued)

Isotopic species	State: electronic	State: vibrational	Transition: parity	Transition: rotational J' – J''	Transition: hyperfine F_1' – F_1''	Transition: hyperfine F' – F''	ν MHz	Ref.	Remarks
$^{12}C^{35}Cl$	$X\ ^2\Pi_{1/2}$	$v=0$	+ −	5/2 3/2		3 2	103259.899 (50)	82END	
			− +	5/2 3/2		4 3	103662.532 (50)		
			+ −	5/2 3/2		4 3	103254.174 (50)		
			− +	5/2 3/2		2 2	103632.775 (50)		
			− +	5/2 3/2		3 3	103576.526 (50)		
			+ −	7/2 5/2		2 1	145058.877 (50)		
			− +	7/2 5/2		2 1	144650.832 (50)		
			+ −	7/2 5/2		3 2	145059.808 (50)		
			− +	7/2 5/2		3 2	144645.782 (50)		
			+ −	7/2 5/2		4 3	145056.321 (50)		
			− +	7/2 5/2		4 3	144643.364 (50)		
			+ −	7/2 5/2		5 4	145046.119 (50)		
			− +	7/2 5/2		5 4	144641.297 (50)		
			+ −	9/2 7/2		3 2	186027.407 (50)		
			+ −	9/2 7/2		4 3	186024.201 (50)		
			− +	9/2 7/2		5 4	186430.191 (50)		
			+ −	9/2 7/2		5 4	186022.793 (50)		
			− +	9/2 7/2		6 5	186424.619 (50)		
			+ −	9/2 7/2		6 5	186021.715 (50)		
	$X\ ^2\Pi_{3/2}$	$v=0$		5/2 3/2		3 2	104510.501 (50)	82END	unresolved Λ-doublets
				5/2 3/2		4 3	104484.971 (50)		unresolved Λ-doublets
			+ −	7/2 5/2		2 1	146314.570 (50)		
			− +	7/2 5/2		2 1	146313.817 (50)		
			+ −	7/2 5/2		3 2	146307.942 (50)		
			− +	7/2 5/2		3 2	146306.897 (50)		
			+ −	7/2 5/2		4 3	146298.116 (50)		
			− +	7/2 5/2		4 3	146297.130 (50)		
			+ −	7/2 5/2		5 4	146286.956 (50)		
			− +	7/2 5/2		5 4	146286.199 (50)		
			− +	9/2 7/2		3 2	188096.318 (50)		
			+ −	9/2 7/2		3 2	188095.078 (50)		

Line frequencies of CCl (continued)

Isotopic species	State: electronic	State: vibrational	Transition: parity	Transition: rotational J' – J''		Transition: hyperfine F_1' – F_1''	Transition: hyperfine F' – F''		ν MHz	Ref.	Remarks
$^{12}C^{35}Cl$	X $^2\Pi_{3/2}$	$v=0$	− +	9/2	7/2		4	3	188092.168 (50)	82END	
			+ −	9/2	7/2		4	3	188090.671 (50)		
			− +	9/2	7/2		5	4	188086.952 (50)		
			+ −	9/2	7/2		5	4	188085.448 (50)		
			− +	9/2	7/2		6	5	188080.657 (50)		
			+ −	9/2	7/2		6	5	188079.369 (50)		

References

81YAM Yamada, Ch., Nagai, K., Hirota, E.: J. Mol. Spectrosc. **85** (1981) 416.

82END Endo, Y., Saito, S., Hirota, E.: J. Mol. Spectrosc. **94** (1982) 199.

80 NO

Molecular constants of NO

Isotopic species	State: electronic	State: vibrational	Parameter	Value	Ref.	Meth.	Remarks
$^{14}N^{16}O$	X $^2\Pi$	$v=0$	A_0	3691.619 (3) GHz	80VAN	MB/MW/IR	the analysis includes IR, MB, and MW measurements from other sources; the parameters A_0, B_0 contain parity-independent contributions of the interaction which produces the Λ-doubling of the rotational levels. Centrifugal distortion parameters are used for A, p, and q in [80VAN]
			B_0	50847.801 (1) MHz	80VAN	MB/MW/IR	
			D_0	0.16416 (3) MHz	80VAN	MB/MW/IR	
			p	350.37543 (7) MHz	81LOW	MB/MW/IR/DR	combined analysis of data from different sources, extension of [80VAN]
			q	2.83712 (5) MHz	81LOW	MB/MW/IR/DR	
			$a(^{14}N)$	84.2163 (29) MHz	76MEE	MB/MW	the analysis includes MW-measurements from other sources; the hyperfine structure parameters shown are calculated from differently defined parameters reported in [76MEE]. [76MEE] contains also fitted values of the rotational and Λ-doubling parameters
			$b_F(^{14}N)$	21.863 (27) MHz	76MEE	MB/MW	
			$c(^{14}N)$	−58.125 (46) MHz	76MEE	MB/MW	
			$d(^{14}N)$	112.5974 (4) MHz	76MEE	MB/MW	
			$C_I(^{14}N)$	12.3 (3) kHz	76MEE	MB/MW	
			$eq^{(a)}Q(^{14}N)$	−1.857 (1) MHz	76MEE	MB/MW	
			$eq^{(b)}Q(^{14}N)$	23.160 (24) MHz	76MEE	MB/MW	

Molecular constants of NO (continued)

Isotopic species	State electronic	State vibrational	Parameter	Value	Ref.	Meth.	Remarks
$^{14}N^{16}O$	$X\,^2\Pi$	$v=1$	A_1	3684.290 (4) GHz	80VAN	MB/MW/IR	combined fit of IR, MW, and MB-data; centrifugal distortion parameters are also used for A, p, and q
			B_1	50321.03 (1) MHz	80VAN	MB/MW/IR	
			D_1	0.16463 (3) MHz	80VAN	MB/MW/IR	
			p	349.972 (2) MHz	81LOW	MB/MW/IR/DR	combined analysis of data from different sources extension of [80VAN]
			q	2.7920 (2) MHz	81LOW	MB/MW/IR/DR	
			$a-1/2(b+c)$ (^{14}N)	92.46 (8) MHz	81LOW	MB/MW/IR/DR	
			$d(^{14}N)$	112.568 (9) MHz	81LOW	MB/MW/IR/DR	
			$eq^{(a)}Q\,(^{14}N)$	−1.78 (3) MHz	81LOW	MB/MW/IR/DR	
			$eq^{(b)}Q\,(^{14}N)$	23.0 (1) MHz	81LOW	MB/MW/IR/DR	
	From pure IR-measurements spin-orbit parameters A, rotational constants B and Λ-doubling parameters are derived for vibrational states up to $v=15$ [78AMI, 80AMI, 79HAL, 79PIN].						
$^{14}N^{17}O$	$X\,^2\Pi$	$v=0$	B_0	49453.70 (87) MHz	77DAL	LMR	the parameters A and D were calculated from the results of $^{14}N^{16}O$ by isotopic relations to analyse the data, therefore the error shown for B is underestimated
		$v=1$	B_1	48930.7 (16) MHz	77DAL	LMR	
$^{14}N^{18}O$	$X\,^2\Pi$	$v=0$	A_0	123.02 (23) cm^{-1}	77DAL	LMR	centrifugal distortion parameters are used for the spin-orbit interaction
			B_0	48210.6 (34) MHz	77DAL	LMR	
			D_0	0.147 (5) MHz	77DAL	LMR	
		$v=1$	A_1	122.78 (23) cm^{-1}	77DAL	LMR	
			B_1	47723.7 (34) MHz	77DAL	LMR	
			D_1	0.146 (4) MHz	77DAL	LMR	
$^{15}N^{16}O$	$X\,^2\Pi$	$v=0$	A_0	123.1419 (7) cm^{-1}	77DAL	LMR	centrifugal distortion parameters are used for the spin-orbit interaction
			B_0	49051.77 (45) MHz	77DAL	LMR	
			D_0	0.1527 (2) MHz	77DAL	LMR	
		$v=1$	A_1	122.8986 (8) cm^{-1}	77DAL	LMR	
			B_1	48552.65 (45) MHz	77DAL	LMR	
			D_1	0.1531 (3) MHz	77DAL	LMR	

Molecular constants of NO (continued)

Isotopic species	State electronic	State vibrational	Parameter	Value	Ref.	Meth.	Remarks
$^{15}N^{18}O$	$X\,^2\Pi$	$v=0$	A_0	123.56 (27) cm^{-1}	77DAL	LMR	centrifugal distortion parameters are used for the spin-orbit interaction
			B_0	46417.0 (19) MHz	77DAL	LMR	
			D_0	0.140 (1) MHz	77DAL	LMR	
		$v=1$	A_1	123.32 (27) cm^{-1}	77DAL	LMR	
			B_1	45957.6 (19) MHz	77DAL	LMR	
			D_1	0.140 (1) MHz	77DAL	LMR	
$^{14}N^{16}O$	$X\,^2\Pi$	equ.	A_e	123.2614 (12) cm^{-1}	77DAL	LMR	$A_v=A_e-\alpha_A(v+1/2)+\dots$ $\alpha_A=0.2443\,(10)\,cm^{-1}$
			$B_e=Y_{01}$	51111.71 (12) MHz	77DAL	LMR	both parameters contain contributions of higher order Y_{11}
			$\alpha_e=-Y_{11}$	527.305 (30) MHz	77DAL	LMR	
	[78AMI] also reports some results on the isotopes $^{14}N^{17}O$, $^{14}N^{18}O$, and $^{15}N^{16}O$ from Fourierspectroscopy of vibrational bands						
	$X\,^2\Pi$	$v=0$	g_s	2.002095 (40)	80MEE	MB	
			g'_L	0.999906 (20)	80MEE	MB	
			g_l	0.002401 (70)	80MEE	MB	
			g_{rot}	−0.2679 (55)	80MEE	MB	
			g'_l	0.0035245 (31)	80MEE	MB	
			$g^{e'}_{rot}$	−0.1113 (44)	80MEE	MB	
			A_D	−1.43 (20) MHz	80MEE	MB	pure centrifugal distortion of the spin-orbit interaction.
			γ_0	−244.1 (70) MHz	80MEE	MB	γ_0 is calculated using a relation with g_l and B_0; the fitting of the spectroscopic data yields only an effective centrifugal distortion constant A_D which contains the spin-rotation interaction
		$v=0$	$3\xi_T=\xi_\perp-\xi_\parallel$	81 (71) Hz/kG^2	80MEE	MB	tensor part of the magnetic susceptibility constant (compare part 2.9.1)
		$v=0$	Θ^{el}/e	−7.50 (7) $Å^2$	80MEE	MB	electronic part of the molecular quadrupole moment (compare part 2.9.1)
	$X\,^2\Pi_{1/2}$	$v=1$	μ_1	0.1416 (4) D	75HOY		Laser Stark-spectroscopy of the vibrational transition $v'=1-v''=0$
			$\mu_1-\mu_0$	−0.01703 (40) D	75HOY		
	$X\,^2\Pi_{3/2}$		$\mu_1-\mu_0$	−0.01735 (8) D	75HOY		

Line frequencies of NO

Isotopic species	State: electronic	State: vibrational	Transition: parity	Transition: rotational J'	J''	hyperfine F_1' – F_1''	F'	F''	ν MHz	Ref.	Remarks
$^{14}N^{16}O$	$X\,^2\Pi_{1/2}$	$v=0$	− +	9/2	9/2		11/2	11/2	1863.878 (1)	76MEE	parities for the
			+ −	11/2	11/2		11/2	11/2	2087.217 (1)		transition fre-
			+ −	11/2	11/2		13/2	13/2	2207.877 (2)		quencies given
			+ −	11/2	11/2		9/2	9/2	1982.393 (20)	77DAL	in Landolt-
			+ −	11/2	11/2		9/2	11/2	2025.125 (25)		Börnstein N.S. II/6
			+ −	11/2	11/2		11/2	9/2	2044.477 (25)		must be reversed
			+ −	11/2	11/2		11/2	13/2	2134.594 (25)		
			+ −	11/2	11/2		13/2	11/2	2160.487 (25)		
			− +	25/2	25/2		23/2	23/2	4207.85 (3)	81LOW	
			− +	25/2	25/2		25/2	25/2	4316.13 (3)		
			− +	25/2	25/2		27/2	27/2	4426.82 (3)		
			+ −	35/2	35/2		33/2	33/2	5533.92 (3)		
			+ −	35/2	35/2		35/2	35/2	5641.70 (3)		
			+ −	35/2	35/2		37/2	37/2	5747.48 (3)		
			+ −	39/2	39/2		37/2	37/2	5993.12 (3)		
			+ −	39/2	39/2		39/2	39/2	6100.37 (3)		
			+ −	39/2	39/2		41/2	41/2	6204.34 (3)		
			− +	41/2	41/2		39/2	39/2	6206.90 (3)		
			− +	41/2	41/2		41/2	41/2	6313.85 (3)		
			− +	41/2	41/2		43/2	43/2	6416.93 (3)		
			+ −	17/2	15/2		15/2	13/2	851913.54 (100)	80VAN	
			− +	17/2	15/2		15/2	13/2	852239.91 (100)		
			− +	19/2	17/2		17/2	15/2	952145.08 (30)		
			+ −	19/2	17/2		17/2	15/2	952464.10 (30)		
			+ −	21/2	19/2		21/2	19/2	1052369.91 (30)		
			− +	21/2	19/2		21/2	19/2	1052680.83 (30)		
	$X\,^2\Pi_{1/2}$	$v=1$	− +	5/2	5/2		3/2	5/2	962.66 (3)	81LOW	
			− +	5/2	5/2		5/2	3/2	990.35 (3)		
			− +	5/2	5/2		5/2	7/2	1071.27 (3)		
			− +	5/2	5/2		7/2	5/2	1113.19 (3)		
			− +	5/2	5/2		3/2	3/2	927.89 (3)		
			− +	5/2	5/2		5/2	5/2	1025.11 (3)		
			− +	5/2	5/2		7/2	7/2	1159.35 (3)		

Line frequencies of NO (continued)

Isotopic species	State		Transition								ν	Ref.	Remarks
	electronic	vibrational	parity		rotational		hyperfine				MHz		
					J'	J''	F_1'	F_1''	F'	F''			
$^{14}N^{16}O$	$X\ ^2\Pi_{1/2}$	$v=1$	–	+	9/2	9/2			7/2	7/2	1635.180 (20)	77DAL	
			–	+	9/2	9/2			9/2	9/2	1738.375 (20)		
			–	+	9/2	9/2			11/2	11/2	1861.787 (20)		
			–	+	9/2	9/2			7/2	9/2	1676.436 (25)		
			–	+	9/2	9/2			9/2	7/2	1697.118 (25)		
			–	+	9/2	9/2			9/2	11/2	1785.967 (25)		
			–	+	9/2	9/2			11/2	9/2	1814.180 (25)		
			+	–	27/2	27/2			25/2	25/2	4493.34 (3)	81LOW	
			+	–	27/2	27/2			27/2	27/2	4601.62 (3)		
			+	–	27/2	27/2			29/2	29/2	4711.34 (3)		
			–	+	33/2	33/2			31/2	31/2	5293.20 (3)		
			–	+	33/2	33/2			33/2	33/2	5401.18 (3)		
			–	+	33/2	33/2			35/2	35/2	5508.01 (3)		
			–	+	37/2	37/2			35/2	35/2	5776.50 (3)		
			–	+	37/2	37/2			37/2	37/2	5884.03 (3)		
			–	+	37/2	37/2			39/2	39/2	5989.02 (3)		
			+	–	39/2	39/2			37/2	37/2	6002.74 (3)		
			+	–	39/2	39/2			39/2	39/2	6110.00 (3)		
			+	–	39/2	39/2			41/2	41/2	6214.12 (3)		
	$X\ ^2\Pi_{3/2}$	$v=0$	+	–	17/2	15/2			17/2	15/2	875930.36 (30)	80VAN	
			–	+	17/2	15/2			17/2	15/2	875961.25 (30)		
			–	+	19/2	17/2			19/2	17/2	978770.21 (30)		
			+	–	19/2	17/2			19/2	17/2	978808.04 (30)		

References

75HOY Hoy, A. R., Johns, J.W.C., McKellar, A.R.W.: Can. J. Phys. **53** (1975) 2029.

76MEE Meerts, W.L.: Chem. Phys. **14** (1976) 421.

77DAL Dale, R.M., Johns, J.W.C., McKellar, A.R.W., Riggin, M.: J. Mol. Spectrosc. **67** (1977) 440.

78AMI Amiot, C., Bacis, R., Guelachvili, G.: Can. J. Phys. **56** (1978) 251.

79HAL Hallin, K.-E.J., Johns, J.W.C., Lepard, D.W., Mantz, A.W., Wall, D.L., Rao, K.N.: J. Mol. Spectrosc. **74** (1979) 26.

79PIN Pine, A.S., Johns, J.W.C., Robiette, A.G.: J. Mol. Spectrosc. **74** (1979) 52.

80AMI Amiot, C., Verges, J.: J. Mol. Spectrosc. **81** (1980) 424.

80MEE Meerts, W.L., Veseth, L.: J. Mol. Spectrosc. **82** (1980) 202.

80VAN van den Heuvel, F.C., Meerts, W.L., Dymanus, A.: J. Mol. Spectrosc. **84** (1980) 162.

81LOW Lowe, R.S., McKellar, A.R.W., Veillette, P., Meerts, W.L.: J. Mol. Spectrosc. **88** (1981) 372.

81 NS

Molecular constants of NS

Isotopic species	State		Parameter	Value	Ref.	Meth.	Remarks
	electronic	vibrational					
$^{14}N^{32}S$	X $^2\Pi_{1/2}$	$v=0$	$B_{0\,\mathrm{eff}}$	23072.229 (20) MHz	82LOV	MW	
			$D_{0\,\mathrm{eff}}$	36.3 (9) kHz	82LOV	MW	
			$p_{0\,\mathrm{eff}}$	397.409 (76) MHz	82LOV	MW	
	X $^2\Pi_{3/2}$	$v=0$	$B_{0\,\mathrm{eff}}$	23238.697 (20) MHz	82LOV	MW	
	X $^2\Pi$	$v=0$	$a-1/2(b+c)(^{14}N)$	67.60 (12) MHz	82LOV	MW	
			$a+1/2(b+c)(^{14}N)$	57.15 (20) MHz	82LOV	MW	
			$b(^{14}N)$	36.3 (74) MHz	82LOV	MW	
			$d(^{14}N)$	86.972 (90) MHz	82LOV	MW	
			$eq^{(a)}Q(^{14}N)$	−2.70 (20) MHz	82LOV	MW	
	X $^2\Pi_{1/2}$	$v=1$	$B_{1\,\mathrm{eff}}$	22884.69 (51) MHz	80MAT	IR	
			$D_{1\,\mathrm{eff}}$	35.7 (63) kHz	80MAT	IR	
			p_{eff}	397.5 (39) MHz	80MAT	IR	
	X $^2\Pi_{3/2}$	$v=1$	$B_{1\,\mathrm{eff}}$	23048.52 (78) MHz	80MAT	IR	
			$D_{1\,\mathrm{eff}}$	42.0 (90) kHz	80MAT	IR	
	X $^2\Pi$		$A_{v=1}-A_{v=0}$	−0.1863 (22) cm^{-1}	80MAT	IR	
			$Y_{10}+2\,Y_{20}$	1204.1824 (11) cm^{-1}	80MAT	IR	systematic calibration error $\pm 0.0007\ cm^{-1}$
			Y_{01}^{ev}	23250.37 (30) MHz	80MAT	IR	using microwave data on $v=0$ from [A1] of Vol. II/6, part 5.1.5
			Y_{11}^{ev}	−188.75 (48) MHz	80MAT	IR	
			r_e	1.49403 (4) Å	80MAT	IR	
$^{14}N^{34}S$	X $^2\Pi_{1/2}$	$v=0$	$B_{0\,\mathrm{eff}}$	22661.682 (42) MHz	82LOV	MW	
			$D_{0\,\mathrm{eff}}$	36.8 (35) kHz	82LOV	MW	
			$p_{0\,\mathrm{eff}}$	390.384 (102) MHz	82LOV	MW	
	X $^2\Pi_{3/2}$	$v=0$	$B_{0\,\mathrm{eff}}$	22822.292 (54) MHz	82LOV	MW	
	X $^2\Pi$	$v=0$	$a-1/2(b+c)(^{14}N)$	67.49 (18) MHz	82LOV	MW	
			$a+1/2(b+c)(^{14}N)$	57.23 (52) MHz	82LOV	MW	
			$b(^{14}N)$	33 (11) MHz	82LOV	MW	
			$d(^{14}N)$	87.08 (15) MHz	82LOV	MW	
			$eq^{(a)}Q(^{14}N)$	−2.80 (22) MHz	82LOV	MW	

Line frequencies of NS

Isotopic species	State: electronic	State: vibrational	Transition: parity	Transition: rotational J′	J″	Transition: hyperfine F_1'	F_1''	F'	F''	ν MHz	Ref.	Remarks
$^{14}N^{32}S$	X $^2\Pi_{1/2}$	$v=0$	− +	3/2	1/2			5/2	3/2	69002.99 (10)	82LOV	
			− +	3/2	1/2			3/2	1/2	69018.16 (10)		
			− +	3/2	1/2			3/2	3/2	69037.36 (10)		
			− +	3/2	1/2			1/2	1/2	69040.28 (10)		
			− +	3/2	1/2			1/2	3/2	69059.87 (8)		
			+ −	3/2	1/2			1/2	3/2	69283.14 (20)		
			+ −	3/2	1/2			3/2	3/2	69330.58 (8)		
			+ −	3/2	1/2			5/2	3/2	69411.89 (8)		
			+ −	3/2	1/2			1/2	1/2	69437.84 (10)		
			+ −	3/2	1/2			3/2	1/2	69485.14 (10)		
			+ −	5/2	3/2			7/2	5/2	115153.835 (80)		
			+ −	5/2	3/2			5/2	3/2	115156.799 (80)		
			+ −	5/2	3/2			3/2	1/2	115162.955 (80)		
			+ −	5/2	3/2			3/2	3/2	115185.394 (60)		
			+ −	5/2	3/2			5/2	5/2	115191.265 (60)		
			− +	5/2	3/2			5/2	5/2	115489.527 (80)		
			− +	5/2	3/2			3/2	3/2	115524.772 (80)		
			− +	5/2	3/2			7/2	5/2	115556.312 (60)		
			− +	5/2	3/2			5/2	3/2	115570.762 (50)		
			− +	5/2	3/2			3/2	1/2	115571.93 (6)		
			− +	7/2	5/2			9/2	7/2	161297.22 (10)		
			− +	7/2	5/2			7/2	5/2	161298.33 (10)		
			− +	7/2	5/2			5/2	3/2	161301.74 (5)		
			− +	7/2	5/2			5/2	5/2	161330.33 (10)		
			+ −	7/2	5/2			9/2	7/2	161697.33 (5)		
			+ −	7/2	5/2			7/2	5/2	161703.50 (10)		
			+ −	7/2	5/2			5/2	3/2	161704.00 (10)		
	X $^2\Pi_{3/2}$	$v=0$		5/2	3/2			5/2	5/2	116147.27 (10)	82LOV	
				5/2	3/2			3/2	3/2	116179.74 (15)		
				5/2	3/2			7/2	5/2	116181.24 (10)		
				5/2	3/2			5/2	3/2	116203.49 (10)		
				5/2	3/2			3/2	1/2	116214.95 (10)		

Line frequencies of NS (continued)

Isotopic species	State		Transition							ν MHz	Ref.	Remarks
	electronic	vibrational	parity	rotational J'	J''	hyperfine F_1'	F_1''	F'	F''			
$^{14}N^{34}S$	X $^2\Pi_{1/2}$	$v=0$	− +	3/2	1/2			5/2	3/2	67774.75 (3)	82LOV	
			− +	3/2	1/2			3/2	1/2	67789.69 (15)		
			− +	3/2	1/2			3/2	3/2	67809.42 (20)		
			− +	3/2	1/2			1/2	1/2	67812.44 (20)		
			+ −	3/2	1/2			3/2	3/2	68095.20 (20)		
			+ −	3/2	1/2			5/2	3/2	68176.88 (5)		
			+ −	3/2	1/2			1/2	1/2	68202.50 (20)		
			+ −	3/2	1/2			3/2	1/2	68249.97 (20)		
			+ −	5/2	3/2			7/2	5/2	113104.65 (10)		
			+ −	5/2	3/2			5/2	3/2	113107.63 (11)		
			+ −	5/2	3/2			3/2	1/2	113113.80 (19)		
			+ −	5/2	3/2			3/2	3/2	113136.13 (19)		
			+ −	5/2	3/2			5/2	5/2	113142.17 (15)		
			− +	5/2	3/2			5/2	5/2	113433.23 (15)		
			− +	5/2	3/2			3/2	3/2	113468.55 (19)		
			− +	5/2	3/2			7/2	5/2	113499.98 (10)		
			− +	5/2	3/2			5/2	3/2	113514.55 (10)		
			− +	5/2	3/2			3/2	1/2	113515.67 (10)		
	X $^2\Pi_{3/2}$	$v=0$		5/2	3/2			7/2	5/2	114099.34 (10)	82LOV	
				5/2	3/2			5/2	3/2	114121.39 (13)		
				5/2	3/2			3/2	1/2	114132.97 (15)		

References:
80MAT Matsumura,, K., Kawaguchi, K., Nagai, K., Yamada, C., Hirota, E.: J. Mol. Spectrosc. **84** (1980) 68.
82LOV Lovas, F.J., Suenram, R.D.: J. Mol. Spectrosc. **93** (1982) 416.

81a AsO

Molecular constants of AsO

Isotopic species	State		Parameter	Value	Ref.	Meth.	Remarks
	electronic	vibrational					
$^{75}As^{16}O$	X $^2\Pi_{3/2}$	$v=0, 1$	$\|h\|=\|a+1/2(b+c)\|(^{75}As)$	370 MHz	81UEH	LMR	
			$eq^{(a)}Q(^{75}As)$	−170 MHz	81UEH	LMR	

Reference to No. 81a (AsO): 81UEH Uehara, H.: Chem. Phys. Lett. **84** (1981) 539.

91 FO

Molecular constants of FO

Isotopic species	State electronic	State vibrational	Parameter	Value	Ref.	Meth.	Remarks
$^{19}F^{16}O$	$X\,^2\Pi_{3/2}$	$v=0$	$B_{0\,eff}$	31375.229 MHz	79McK	LMR	
		$v=1$	$B_{1\,eff}$	30973.687 MHz	79McK	LMR	
	$X\,^2\Pi$	$v=0$	A_0	−177.3 (57) cm^{-1}	79McK	LMR	
			B_0	31562.7 (57) MHz	79McK	LMR	
			D_0	0.12 (10) MHz	79McK	LMR	
			$a+1/2(b+c)(^{19}F)$	696 (18) MHz	79McK	LMR	
		equ.	α_e	404.0 (10) MHz	79McK	LMR	
			r_e	1.35789 (25) Å	79McK	LMR	

Reference: 79McK McKellar, A.R.W.: Can. J. Phys. **57** (1979) 2106.

92 ClO

Molecular constants of ClO

Isotopic species	State electronic	State vibrational	Parameter	Value	Ref.	Meth.	Remarks
$^{35}Cl^{16}O$	$X\,^2\Pi_{1/2}$	$v=0$	$B_{0\,eff}$	18623.639 (11) MHz	78KAK	MW	a combined fit of optical, infrared and microwave data is presented in [79COX]
			$D_{0\,eff}$	40.01 (31) kHz	78KAK	MW	
	$X\,^2\Pi_{3/2}$	$v=0$	$B_{0\,eff}$	18579.846 (15) MHz	78KAK	MW	
			$D_{0\,eff}$	39.42 (41) kHz	78KAK	MW	
	$X\,^2\Pi$	$v=0$	α_p^{eff}	−337.071 (26) MHz	78KAK	MW	
			β_p^{eff}	0.79 (13) MHz	78KAK	MW	
			$a+1/2(b+c)(^{35}Cl)$	112.528 (77) MHz	78KAK	MW	
			$a-1/2(b+c)(^{35}Cl)$	159.977 (68) MHz	78KAK	MW	
			$b(^{35}Cl)$	45.0 (67) MHz	78KAK	MW	
			$d(^{35}Cl)$	173.030 (59) MHz	78KAK	MW	
			$eq^{(a)}Q(^{35}Cl)$	−87.95 (19) MHz	78KAK	MW	
			$eq^{(b)}Q(^{35}Cl)$	−126.4 (65) MHz	78KAK	MW	sign was reversed according to [82END]

Molecular constants of ClO (continued)

Isotopic species	State electronic	State vibrational	Parameter	Value	Ref.	Meth.	Remarks
$^{35}Cl^{16}O$	$X\,^2\Pi_{1/2}$	$v=1$	$B_{1\,eff}$	18444.983 (48) MHz	78KAK	MW	
			$D_{1\,eff}$	39.9 (41) kHz	78KAK	MW	
	$X\,^2\Pi_{3/2}$	$v=1$	$B_{1\,eff}$	18402.046 (25) MHz	78KAK	MW	
			$D_{1\,eff}$	39.04 (72) kHz	78KAK	MW	
	$X\,^2\Pi$	$v=1$	α_p^{eff}	−334.555 (60) MHz	78KAK	MW	
			$a+1/2(b+c)(^{35}Cl)$	110.26 (16) MHz	78KAK	MW	
			$a-1/2(b+c)(^{35}Cl)$	157.10 (14) MHz	78KAK	MW	
			$b(^{35}Cl)$	49 (19) MHz	78KAK	MW	
			$d(^{35}Cl)$	169.69 (15) MHz	78KAK	MW	
			$eq^{(a)}Q(^{35}Cl)$	−88.48 (46) MHz	78KAK	MW	
	$X\,^2\Pi_{3/2}$	$v=2$	$B_{2\,eff}$	18227.4 (66) MHz	80LOW	LMR	
	An extension including new IR-measurements can be found in [82MAK].						
$^{37}Cl^{16}O$	$X\,^2\Pi_{1/2}$	$v=0$	$B_{0\,eff}$	18308.067 (19) MHz	78KAK	MW	a combined fit of optical, infrared, and microwave data is presented in [79COX]
			$D_{0\,eff}$	39.44 (49) kHz	78KAK	MW	
	$X\,^2\Pi_{3/2}$	$v=0$	$B_{0\,eff}$	18266.202 (16) MHz	78KAK	MW	
			$D_{0\,eff}$	38.32 (42) kHz	78KAK	MW	
	$X\,^2\Pi$	$v=0$	α_p^{eff}	−331.399 (41) MHz	78KAK	MW	
			$a+1/2(b+c)(^{37}Cl)$	93.684 (83) MHz	78KAK	MW	
			$a-1/2(b+c)(^{37}Cl)$	133.04 (48) MHz	78KAK	MW	
			$b(^{37}Cl)$	44.6 (94) MHz	78KAK	MW	
			$d(^{37}Cl)$	144.22 (33) MHz	78KAK	MW	
			$eq^{(a)}Q(^{37}Cl)$	−69.50 (34) MHz	78KAK	MW	
	$X\,^2\Pi_{3/2}$	$v=2$	$B_{2\,eff}$	17920.1 (96) MHz	80LOW	LMR	
ClO	$X\,^2\Pi$		r_e	1.56959 (1) Å	78KAK	MW	
	Additional data from IR-diode laser experiments in [77MEN, 78MAR, 78ROG]. Pressure broadening parameters with N_2 for different temperatures are given in [81PIC].						

Line frequencies of ClO

Isotopic species	State: electronic	State: vibrational	Transition: parity	Transition: rotational J'	J''	hyperfine F_1'	F_1''	F'	F''	ν MHz	Ref.	Remarks
$^{35}Cl^{16}O$	X $^2\Pi_{3/2}$	$v=0$		5/2	3/2			1	0	92990.35 (10)	78KAK	Λ-doubling splitting:
				5/2	3/2			2	1	92961.56 (10)	78KAK	
				5/2	3/2			3	2	92909.15 (10)	78KAK	
				5/2	3/2			4	3	92873.95 (10)	78KAK	
				5/2	3/2			1	1	92927.87 (10)	78KAK	
				5/2	3/2			2	2	92853.90 (10)	78KAK	
				5/2	3/2			3	3	92791.41 (10)	78KAK	0.404 MHz
				5/2	3/2			1	2	92820.15 (10)	78KAK	0.329 MHz
				5/2	3/2			2	3	92736.13 (10)	78KAK	
				9/2	7/2			3	2	167221.57 (10)	78KAK	
				9/2	7/2			4	3	167213.42 (10)	78KAK	
				9/2	7/2			5	4	167204.80 (10)	78KAK	
				9/2	7/2			6	5	167195.52 (10)	78KAK	
				9/2	7/2			3	3	167201.74 (10)	78KAK	
				9/2	7/2			4	4	167175.22 (10)	78KAK	
				9/2	7/2			5	5	167138.19 (10)	78KAK	0.900 MHz
				11/2	9/2			7	6	204346.000 (23)	81PIC	unresolved
			− +	11/2	9/2			6	5	204351.630 (20)	81PIC	
			+ −	11/2	9/2			6	5	204352.220 (20)	81PIC	
			− +	11/2	9/2			5	4	204356.860 (20)	81PIC	
			+ −	11/2	9/2			5	4	204357.480 (30)	81PIC	
				11/2	9/2			4	3	204362.557 (20)	81PIC	unresolved
			− +	15/2	13/2			9	8	278627.115 (27)	81PIC	
			+ −	15/2	13/2			9	8	278628.137 (14)	81PIC	
			− +	15/2	13/2			8	7	278630.099 (14)	81PIC	
			+ −	15/2	13/2			8	7	278631.219 (31)	81PIC	
			− +	15/2	13/2			7	6	278632.647 (22)	81PIC	
			+ −	15/2	13/2			7	6	278633.783 (23)	81PIC	
			− +	15/2	13/2			6	5	278635.570 (40)	81PIC	
			+ −	15/2	13/2			6	5	278636.580 (160)	81PIC	

Line frequencies of ClO (continued)

Isotopic species	State		Transition						ν MHz	Ref.	Remarks
	electronic	vibrational	parity	rotational J' – J''		hyperfine F_1' – F_1''	F' – F''				
$^{35}Cl^{16}O$	$X\,^2\Pi_{3/2}$	$v=1$		5/2	3/2		1	0	92100.36 (10)	78KAK	
				5/2	3/2		2	1	92071.28 (10)	78KAK	
				5/2	3/2		3	2	92019.43 (10)	78KAK	
				5/2	3/2		4	3	91985.59 (10)	78KAK	
				5/2	3/2		1	1	92038.47 (10)	78KAK	
				5/2	3/2		2	2	91965.36 (10)	78KAK	
				5/2	3/2		3	3	91904.46 (10)	78KAK	
				9/2	7/2		3	2	165621.14 (10)	78KAK	
				9/2	7/2		4	3	165613.06 (10)	78KAK	
				9/2	7/2		5	4	165604.60 (10)	78KAK	
				9/2	7/2		6	5	165595.78 (10)	78KAK	
	$X\,^2\Pi_{1/2}$	$v=0$	– +	3/2	1/2		2	1	55528.17 (10)	78KAK	
			+ –	3/2	1/2		2	1	56422.09 (10)	78KAK	
			– +	3/2	1/2		3	2	55488.67 (10)	78KAK	
			+ –	3/2	1/2		3	2	56197.00 (10)	78KAK	
			– +	3/2	1/2		1	1	55595.80 (10)	78KAK	
			+ –	3/2	1/2		1	1	56304.74 (10)	78KAK	
			– +	3/2	1/2		2	2	55545.83 (10)	78KAK	
			+ –	3/2	1/2		2	2	55978.03 (10)	78KAK	
			– +	3/2	1/2		0	1	55638.39 (10)	78KAK	
			+ –	3/2	1/2		0	1	56254.34 (10)	78KAK	
			– +	3/2	1/2		1	2	55613.46 (10)	78KAK	
			+ –	3/2	1/2		1	2	55860.74 (10)	78KAK	
			– +	5/2	3/2		1	0	93497.93 (10)	78KAK	
			+ –	5/2	3/2		2	1	92770.20 (10)	78KAK	
			– +	5/2	3/2		2	1	93504.99 (10)	78KAK	
			+ –	5/2	3/2		3	2	92764.64 (10)	78KAK	
			– +	5/2	3/2		3	2	93492.44 (10)	78KAK	
			+ –	5/2	3/2		4	3	92759.42 (10)	78KAK	
			– +	5/2	3/2		4	3	93448.12 (10)	78KAK	
			+ –	5/2	3/2		1	1	92831.60 (10)	78KAK	
			+ –	5/2	3/2		2	2	92837.83 (10)	78KAK	
			– +	5/2	3/2		2	2	93387.73 (10)	78KAK	

Line frequencies of ClO (continued)

Isotopic species	State: electronic	State: vibrational	Transition: parity	Transition: rotational J'	J''	hyperfine F_1' – F_1''	F'	F''	ν MHz	Ref.	Remarks
$^{35}Cl^{16}O$	$X\,^2\Pi_{1/2}$	$v=0$	+ −	5/2	3/2		3	3	92821.77 (10)	78KAK	
			− +	5/2	3/2		3	3	93273.44 (10)	78KAK	
			+ −	9/2	7/2		3	2	167263.63 (10)	78KAK	
			− +	9/2	7/2		3	2	167945.80 (10)	78KAK	
			+ −	9/2	7/2		4	3	167256.90 (10)	78KAK	
			− +	9/2	7/2		4	3	167946.82 (10)	78KAK	
			+ −	9/2	7/2		5	4	167254.99 (10)	78KAK	
			− +	9/2	7/2		5	4	167943.79 (10)	78KAK	
			+ −	9/2	7/2		6	5	167254.23 (10)	78KAK	
			− +	9/2	7/2		6	5	167933.28 (10)	78KAK	
		$v=1$	− +	3/2	1/2		2	1	54994.69	77AMA	
			+ −	3/2	1/2		2	1	55879.35	77AMA	
			− +	3/2	1/2		3	2	54956.01 (20)	78KAK	
			+ −	3/2	1/2		3	2	55658.84 (20)	78KAK	
			+ −	3/2	1/2		1	1	55764.54 (20)	78KAK	
			− +	3/2	1/2		2	2	55011.70 (20)	78KAK	
			+ −	3/2	1/2		2	2	55443.55	77AMA	
			+ −	5/2	3/2		2	1	91879.51 (20)	78KAK	
			− +	5/2	3/2		2	1	92608.02 (20)	78KAK	
			+ −	5/2	3/2		3	2	91874.14 (20)	78KAK	
			− +	5/2	3/2		3	2	92595.88 (20)	78KAK	
			− +	5/2	3/2		4	3	92552.43 (20)	78KAK	
$^{37}Cl^{16}O$	$X\,^2\Pi_{3/2}$	$v=0$		5/2	3/2		1	0	91405.42 (10)	78KAK	
				5/2	3/2		2	1	91382.32 (10)	78KAK	
				5/2	3/2		3	2	91339.57 (10)	78KAK	
				5/2	3/2		4	3	91309.31 (10)	78KAK	
				5/2	3/2		1	1	91354.08 (10)	78KAK	
				5/2	3/2		2	2	91293.45 (10)	78KAK	
				5/2	3/2		3	3	91240.74 (10)	78KAK	
				5/2	3/2		1	2	91265.19 (10)	78KAK	
				5/2	3/2		2	3	91194.64 (10)	78KAK	
				9/2	7/2		3	2	164396.16 (10)	78KAK	
				9/2	7/2		4	3	164389.50 (10)	78KAK	

Line frequencies of ClO (continued)

Isotopic species	State electronic	State vibrational	Transition parity	rotational J'	– J''	hyperfine F_1' – F_1''	F'	– F''	ν MHz	Ref.	Remarks
$^{37}Cl^{16}O$	$X\,^2\Pi_{3/2}$	$v=0$		9/2	7/2		5	4	164382.31 (10)	78KAK	
				9/2	7/2		6	5	164374.57 (10)	78KAK	
				9/2	7/2		3	3	164379.05 (10)	78KAK	
				9/2	7/2		4	4	164357.46 (10)	78KAK	
				9/2	7/2		5	5	164327.43 (10)	78KAK	
		$v=1$		5/2	3/2		4	3	90472.48 (20)	78KAK	
				5/2	3/2		3	2	90443.38 (20)	78KAK	
	$X\,^2\Pi_{1/2}$	$v=0$	– +	5/2	3/2		1	0	91906.95 (20)	78KAK	
			+ –	5/2	3/2		2	1	91199.68 (20)	78KAK	
			– +	5/2	3/2		2	1	91913.14 (20)	78KAK	
			+ –	5/2	3/2		3	2	91194.76 (20)	78KAK	
			– +	5/2	3/2		3	2	91902.43 (20)	78KAK	
			+ –	5/2	3/2		4	3	91190.25 (20)	78KAK	
			– +	5/2	3/2		4	3	91865.29 (20)	78KAK	
			– +	5/2	3/2		2	2	91814.68 (20)	78KAK	
			– +	5/2	3/2		3	3	91720.66 (20)	78KAK	
			+ –	9/2	7/2		3	2	164429.08 (20)	78KAK	
			– +	9/2	7/2		3	2	165098.30 (20)	78KAK	
			+ –	9/2	7/2		4	3	164423.47 (20)	78KAK	
			– +	9/2	7/2		4	3	165099.29 (20)	78KAK	
			+ –	9/2	7/2		5	4	164421.70 (20)	78KAK	
			– +	9/2	7/2		5	4	165096.67 (20)	78KAK	
			+ –	9/2	7/2		6	5	164421.09 (20)	78KAK	
			– +	9/2	7/2		6	5	165087.86 (20)	78KAK	

References

77AMA Amano, T., Hirota, E.: J. Mol. Spectrosc. **66** (1977) 185.

77MEN Menzies, R.T., Margolis, J.S., Hinkley, E.D., Toth, R.A.: Appl. Opt. **16** (1977) 523.

78KAK Kakar, R.K., Cohen, E.A., Geller, M.: J. Mol. Spectrosc. **70** (1978) 243.

78MAR Margolis, J.S., Menzies, R.T., Hinkley, E.D.: Appl. Opt. **17** (1978) 1680.

78ROG Rogowski, R.S., Bair, C.H., Wade, W.R., Hoell, J.M., Copeland, G.E.: Appl. Opt. **17** (1978) 1301.

79COX Coxon, J.A.: Can. J. Phys. **57** (1979) 1538.

80LOW Lowe, R.S., McKellar, A.R.W.: J. Mol. Spectrosc. **79** (1980) 424.

81PIC Pickett, H.M., Brinza, D.E., Cohen, E.A.: J. Geophys. Res. **86** (1981) 7279.

82END Endo, Y., Saito, S., Hirota, E.: J. Mol. Spectrosc. **94** (1982) 199.

82MAK Maki, A.G., Lovas, F.J., Olson, W.B.: J. Mol. Spectrosc. **52** (1982) 410.

93 BrO

Molecular constants of BrO

Isotopic species	State: electronic	State: vibrational	Parameter	Value	Ref.	Meth.	Remarks
$^{79}Br^{16}O$	$X\,^2\Pi_{3/2}$	$v=0$	$B_{0\,eff}$	12824.7381 (3) MHz	81COH	MW	
			$D_{0\,eff}$	17.7536 (17) kHz	81COH	MW	
			$\lambda \approx \frac{A_0}{B_0}$	−2270.8	81McK	LMR	
			$\frac{q_{eff}}{2-\lambda}$	−0.491 (21) kHz	81COH	MW	$q_{eff} = -1.12$ (5) MHz
			$a+1/2(b+c)(^{79}Br)$	504.419 (14) MHz	81COH	MW	this parameter contains a contribution of C_I(Br)
			$\frac{b}{2-\lambda}+C_I(^{79}Br)$	139.0 (8) kHz	81COH	MW	
			$eq^{(a)}Q(^{79}Br)$	650.045 (32) MHz	81COH	MW	
			$\frac{eq^{(b)}Q}{2-\lambda}(^{79}Br)$	−390.7 (41) kHz	81COH	MW	$eq^{(b)}Q = -888$ (9) MHz
		$v=1$	$B_{1\,eff}$	12715.651 (8) MHz	81COH	MW	
			$D_{1\,eff}$	17.18 (23) kHz	81COH	MW	
			$a+1/2(b+c)(^{79}Br)$	495.77 (9) MHz	81COH	MW	this parameter contains a contribution of C_I(Br)
			$eq^{(a)}Q(^{79}Br)$	654.02 (24) MHz	81COH	MW	
$^{81}Br^{16}O$	$X\,^2\Pi_{3/2}$	$v=0$	$B_{0\,eff}$	12771.5131 (4) MHz	81COH	MW	
			$D_{0\,eff}$	17.6074 (20) kHz	81COH	MW	
			$\lambda \approx \frac{A_0}{B_0}$	−2280.2	81McK	LMR	
			$\frac{q_{eff}}{2-\lambda}$	−0.433 (29) kHz	81COH	MW	$q_{eff} = -0.99$ (7) MHz
			$a+1/2(b+c)(^{81}Br)$	543.764 (16) MHz	81COH	MW	this parameter contains a contribution of C_I(Br)
			$\frac{b}{2-\lambda}+C_I(^{81}Br)$	148.6 (7) kHz	81COH	MW	
			$eq^{(a)}Q(^{81}Br)$	543.080 (40) MHz	81COH	MW	
			$\frac{eq^{(b)}Q}{2-\lambda}(^{81}Br)$	−331.7 (47) kHz	81COH	MW	$eq^{(b)}Q = -757$ (11) MHz

Molecular constants of BrO (continued)

Isotopic species	State: electronic	State: vibrational	Parameter	Value	Ref.	Meth.	Remarks
$^{81}Br^{16}O$	X $^2\Pi_{3/2}$	$v=1$	$B_{1\,\mathrm{eff}}$	12663.094 (8) MHz	81COH	MW	
			$D_{1\,\mathrm{eff}}$	16.95 (27) kHz	81COH	MW	
			$a+1/2(b+c)(^{81}Br)$	534.73 (8) MHz	81COH	MW	this parameter contains a contribution of C_I(Br)
			$eq^{(a)}Q(^{81}Br)$	546.30 (26) MHz	81COH	MW	

Line frequencies of BrO

Isotopic species	State: electronic	State: vibrational	Transition: parity	rotational J'	J''	hyperfine F_1'	F_1''	F'	F''	ν MHz	Ref.	Remarks
$^{79}Br^{16}O$	X $^2\Pi_{3/2}$	$v=0$		5/2	3/2			4	3	63949.920 (30)	81COH	
				5/2	3/2			3	2	63379.022 (30)	81COH	
				5/2	3/2			2	1	64369.301 (30)	81COH	
				5/2	3/2			1	0	64230.064 (30)	81COH	
				5/2	3/2			2	2	64096.564 (30)	81COH	
				5/2	3/2			1	1	64158.274 (30)	81COH	
				7/2	5/2			5	4	89678.110 (20)	81COH	
				7/2	5/2			4	3	89837.643 (20)	81COH	
				7/2	5/2			3	2	89889.839 (20)	81COH	
				7/2	5/2			2	1	89862.745 (25)	81COH	
				7/2	5/2			4	4	89534.133 (20)	81COH	Λ-doubling splitting is 0.650 (20) MHz
				7/2	5/2			3	3	89607.439 (50)	81COH	
				7/2	5/2			2	2	89651.815 (50)	81COH	
				9/2	7/2			6	5	115359.155 (20)	81COH	
				9/2	7/2			5	4	115439.887 (20)	81COH	
				9/2	7/2			4	3	115479.892 (20)	81COH	
				9/2	7/2			3	2	115476.438 (20)	81COH	
				9/2	7/2			5	5	115296.083 (50)	81COH	
				9/2	7/2			4	4	115249.762 (50)	81COH	
				9/2	7/2			3	3	115238.530 (50)	81COH	

Line frequencies of BrO (continued)

Isotopic species	State electronic	State vibrational	Transition parity	rotational J′	rotational J″	hyperfine F_1'	hyperfine F_1''	F′	F″	ν MHz	Ref.	Remarks
$^{79}Br^{16}O$	X $^2\Pi_{3/2}$	$v=0$		19/2	17/2			11	10	243595.785 (6)	81COH	
				19/2	17/2			10	9	243608.327 (6)	81COH	
				19/2	17/2			9	8	243618.819 (6)	81COH	
				19/2	17/2			8	7	243622.811 (6)	81COH	
				19/2	17/2			10	10	243663.346 (20)	81COH	Λ-doubling splitting is 1.600 (20) MHz
				19/2	17/2			9	9	243508.596 (20)	81COH	
				19/2	17/2			8	8	243397.833 (20)	81COH	Λ-doubling splitting is 2.075 (20) MHz
				23/2	21/2			13	12	294851.763 (25)	81COH	
				23/2	21/2			12	11	294859.722 (25)	81COH	
				23/2	21/2			11	10	294866.913 (25)	81COH	
				23/2	21/2			10	9	294870.136 (25)	81COH	
				31/2	29/2			17	16	397297.262 (40)	81COH	Λ-doubling splitting is 0.812 (40) MHz
				31/2	29/2			16	15	397301.224 (100)	81COH	Λ-doubling splitting is 0.752 (100) MHz
				31/2	29/2			15	14	397305.240 (100)	81COH	Λ-doubling splittings is 0.615 (100) MHz
				31/2	29/2			14	13	397307.214 (40)	81COH	Λ-doubling splitting is 0.860 (40) MHz
		$v=1$		7/2	5/2			5	4	88915.760 (100)	81COH	
				7/2	5/2			4	3	89073.753 (100)	81COH	
				7/2	5/2			3	2	89124.700 (40)	81COH	
				7/2	5/2			2	1	89096.730 (100)	81COH	

Line frequencies of BrO (continued)

Isotopic species	State: electronic	State: vibrational	Transition: parity	rotational J'	J''	hyperfine F_1'	F_1''	F'	F''	ν MHz	Ref.	Remarks
$^{79}Br^{16}O$	$X\ ^2\Pi_{3/2}$	$v=1$		9/2	7/2			6	5	114378.346 (70)	81COH	
				9/2	7/2			5	4	114458.233 (100)	81COH	
				9/2	7/2			4	3	114497.500 (150)	81COH	
				9/2	7/2			3	2	114493.546 (200)	81COH	
$^{81}Br^{16}O$	$X\ ^2\Pi_{3/2}$	$v=0$		5/2	3/2			4	3	63680.960 (70)	81COH	
				5/2	3/2			3	2	64103.032 (70)	81COH	
				5/2	3/2			2	1	64129.492 (50)	81COH	
				5/2	3/2			1	0	64019.792 (50)	81COH	
				5/2	3/2			2	2	63804.107 (50)	81COH	
				5/2	3/2			1	1	63911.069 (50)	81COH	
				7/2	5/2			5	4	89303.412 (20)	81COH	
				7/2	5/2			4	3	89462.074 (20)	81COH	
				7/2	5/2			3	2	89522.036 (20)	81COH	
				7/2	5/2			2	1	89506.470 (20)	81COH	
				7/2	5/2			4	4	89125.526 (20)	81COH	Λ-doubling splitting is 0.558 (20) MHz
				7/2	5/2			3	3	89223.194 (50)	81COH	
				7/2	5/2			2	2	89288.110 (50)	81COH	
				9/2	7/2			6	5	114878.566 (20)	81COH	
				9/2	7/2			5	4	114959.710 (20)	81COH	
				9/2	7/2			4	3	115002.520 (20)	81COH	
				9/2	7/2			3	2	115005.028 (20)	81COH	
				9/2	7/2			5	5	114781.707 (50)	81COH	
				9/2	7/2			4	4	114763.653 (50)	81COH	
				9/2	7/2			3	3	114771.031 (50)	81COH	
				19/2	17/2			11	10	242584.494 (6)	81COH	
				19/2	17/2			10	9	242597.394 (6)	81COH	
				19/2	17/2			9	8	242608.177 (6)	81COH	
				19/2	17/2			8	7	242613.109 (6)	81COH	
				19/2	17/2			10	10	242620.612 (40)	81COH	Λ-doubling splitting is 1.349 (40) MHz

Line frequencies of BrO (continued)

Isotopic species	State: electronic	State: vibrational	Transition: parity	rotational J'	J''	hyperfine F_1'	F_1''	F'	F''	ν MHz	Ref.	Remarks
$^{81}Br^{16}O$	$X\,^2\Pi_{3/2}$	$v=0$		19/2	17/2			9	9	242495.481 (10)	81COH	
				19/2	17/2			8	8	242406.786 (15)	81COH	Λ-doubling splitting is 1.789 (15) MHz
				23/2	21/2			13	12	293628.126 (25)	81COH	
				23/2	21/2			12	11	293636.372 (25)	81COH	
				23/2	21/2			11	10	293643.754 (25)	81COH	
				23/2	21/2			10	9	293647.538 (25)	81COH	
				31/2	29/2			17	16	395649.266 (40)	81COH	Λ-doubling splitting is 0.764 (60) MHz
				31/2	29/2			16	15	395653.410 (60)	81COH	Λ-doubling splitting is 0.480 (100) MHz
				31/2	29/2			15	14	395657.438 (60)	81COH	Λ-doubling splitting is 0.544 (100) MHz
				31/2	29/2			14	13	395659.787 (40)	81COH	Λ-doubling splitting is 0.766 (60) MHz
		$v=1$		7/2	5/2			5	4	88545.718 (50)	81COH	
				7/2	5/2			4	3	88702.600 (100)	81COH	
				7/2	5/2			3	2	88761.643 (100)	81COH	
				7/2	5/2			2	1	88745.108 (100)	81COH	
				9/2	7/2			6	5	113903.860 (70)	81COH	
				9/2	7/2			5	4	113984.045 (100)	81COH	
				9/2	7/2			4	3	114026.138 (150)	81COH	
				9/2	7/2			3	2	114028.000 (200)	81COH	

References
81COH Cohen, E.A., Pickett, H.M., Geller, M.: J. Mol. Spectrosc. **87** (1981) 459.
81McK McKellar, A.R.W.: J. Mol. Spectrosc. **86** (1981) 43.

94 IO

Molecular constants of IO

Isotopic species	State electronic	State vibrational	Parameter	Value	Ref.	Meth.	Remarks
$^{127}I^{16}O$	$X\,^2\Pi_{3/2}$	$v=0$	$B_{0\,\mathrm{eff}}$	10157.08 (18) MHz	73SAI	MW	
			$D_{0\,\mathrm{eff}}$	9 (15) kHz	73SAI	MW	
			$a+1/2(b+c)(^{127}I)$	583.92 (49) MHz	73SAI	MW	
			$eq^{(a)}Q(^{127}I)$	−1891.1 (38) MHz	73SAI	MW	
			r_e^{ev}	1.8677 (28) Å	73SAI	MW	

Line frequencies of IO

Isotopic species	State electronic	State vibrational	Transition parity	rotational J'	J''	hyperfine F_1'	F_1''	F'	F''	ν MHz	Ref.	Remarks
$^{127}I^{16}O$	$X\,^2\Pi_{3/2}$	$v=0$		5/2	3/2			4	4	50078.08 (10)	73SAI	
				5/2	3/2			3	3	50333.85 (30)	73SAI	
				5/2	3/2			5	4	50663.24 (10)	73SAI	
				5/2	3/2			4	3	50719.45 (10)	73SAI	
				5/2	3/2			2	2	50961.82 (20)	73SAI	
				5/2	3/2			3	2	51208.23 (20)	73SAI	
				5/2	3/2			1	1	51575.26 (30)	73SAI	
				5/2	3/2			2	1	51724.02 (30)	73SAI	
				7/2	5/2			5	5	70491.60 (10)	73SAI	
				7/2	5/2			4	4	70788.26 (20)	73SAI	
				7/2	5/2			6	5	71020.32 (10)	73SAI	
				7/2	5/2			5	4	71075.19 (10)	73SAI	
				7/2	5/2			4	3	71173.08 (20)	73SAI	

Reference

73SAI Saito, S.: J. Mol. Spectrosc. **48** (1973) 530.

95 SF

Molecular constants of SF

Isotopic species	State		Parameter	Value	Ref.	Meth.	Remarks
	electronic	vibrational					
$^{32}S^{19}F$	X $^2\Pi_{3/2}$	$v=0$	$B_{0\,\mathrm{eff}}$	16553.760 (52) MHz	73AMA	MW	
			μ	0.794 (12) D	73AMA	MW	
	X $^2\Pi$	$v=0$	B_0	16576.9140 (46) MHz	82END	MW	
			D_0	29.24 (10) kHz	82END	MW	
			p	3.409 (44) MHz	82END	MW	
			$a+1/2(b+c)(^{19}F)$	428.51 (12) MHz	82END	MW	
			$a-1/2(b+c)(^{19}F)$	535.67 (44) MHz	82END	MW	
			$b(^{19}F)$ [1]	210 (15) MHz	82END	MW	[1] large uncertainty is due to the assumed value $A=387\,(25)\ \mathrm{cm}^{-1}$ of the spin-orbit coupling constant
			$d(^{19}F)$	589.76 (92) MHz	82END	MW	
			$a^{ev}(^{19}F)$	482.09 (24) MHz	82END		
			$c^{ev}(^{19}F)$ [1]	−317 (15) MHz	82END		

Line frequencies of SF

Isotopic species	State		Transition							ν MHz	Ref.	Remarks
	electronic	vibrational	parity		rotational J'	J''	hyperfine F_1' − F_1''	F'	F''			
$^{32}S^{19}F$	X $^2\Pi_{1/2}$	$v=0$	+	−	3/2	1/2		2	1	49784.486	82END	
			+	−	3/2	1/2		1	0	50080.847		
			−	+	3/2	1/2		2	1	49742.764		
			−	+	5/2	3/2		3	2	82991.666		
			+	−	5/2	3/2		3	2	82971.962		
			+	−	5/2	3/2		2	1	83011.439		
			+	−	7/2	5/2		3	2	116218.374		
			−	+	7/2	5/2		4	3	116178.536		
			−	+	7/2	5/2		3	2	116198.545		
			−	+	9/2	7/2		5	4	149386.243		
			−	+	9/2	7/2		4	3	149401.895		
			+	−	9/2	7/2		5	4	149377.090		
			+	−	9/2	7/2		4	3	149389.390		
			+	−	11/2	9/2		6	5	182577.633		

Line frequencies of SF (continued)

Isotopic species	State		Transition								ν MHz	Ref.	Remarks
	electronic	vibrational	parity		rotational		hyperfine						
					J' –	J''	F_1' –	F_1''	F' –	F''			
$^{32}S^{19}F$	X $^2\Pi_{1/2}$	$v=0$	+	–	11/2	9/2			5	4	182587.883	82END	
			–	+	11/2	9/2			6	5	182570.296		
			–	+	11/2	9/2			5	4	182578.656		
	X $^2\Pi_{3/2}$	$v=0$			5/2	3/2			3	2	82732.093	82END	
					5/2	3/2			2	1	82511.597		
					5/2	3/2			2	2	82545.533		
					7/2	5/2			4	3	115853.472		
					7/2	5/2			3	2	115910.215		
					7/2	5/2			3	3	115689.709		
					9/2	7/2			5	4	148963.452		
					9/2	7/2			4	3	148996.466		
					11/2	9/2			6	5	182066.927		
					11/2	9/2			5	4	182088.511		

References

73AMA Amano, T., Hirota, E.: J. Mol. Spectrosc. **45** (1973) 417.

82END Endo, Y., Saito, S., Hirota, E.: J. Mol. Spectrosc. **92** (1982) 443.

4.1.6 The $^3\Pi$ electronic state

4.1.6.1 Preliminary remarks

1. Introduction

Molecules in $^3\Pi$ electronic states so far observed with microwave spectroscopy or related techniques belong mainly to the Hund's coupling case (*b*). The molecular levels in a vibrational state v are defined by the projection of the electronic angular momentum $|\Lambda|=1$, the electronic spin quantum number $S=1$, quantum number N which is the sum of the orbital angular momentum $\boldsymbol{L}$ and the rotational angular momentum $\boldsymbol{R}$, and the quantum number J of the total angular momentum which is constructed by the spin $\boldsymbol{S}$ and the total orbital angular momentum $\boldsymbol{N}$. Each J-state has two different parity states + or − called Λ-doublet. When including nuclear spins, $\boldsymbol{J}$ couples with $\boldsymbol{I}_1$ and $\boldsymbol{I}_2$ to give the total angular momentum $\boldsymbol{F}$ with quantum number F and the approximate intermediate quantum number F_1 resulting from $\boldsymbol{F}_1=\boldsymbol{J}+\boldsymbol{I}_1$, or I_T resulting from $\boldsymbol{I}_T=\boldsymbol{I}_1+\boldsymbol{I}_2$. In the case of homonuclear molecules the electronic state is further characterized by the inversion symmetry g or u. For some molecular states reported here the coupling case (*d*) is more appropriate. In this instance $|\Lambda|$ goes over into L. The Λ-doublet structure in the energy level scheme then disappears.

For CO, Br_2, I_2, and ICl we need Hund's coupling case (*a*) or (*c*), where the projection $|\Omega|$ of the total angular momentum $\boldsymbol{J}$ onto the molecular axis is a good quantum number and N looses its significance. Ω-doubling structure is present, therefore each J-state is twofold degenerate with different parity + or −.

The Hamiltonian for describing the rotational energy levels is principally given in 5.1.1 of Vol. II/6 Landolt-Börnstein, New Series and will not be repeated here. The effective Hamiltonian widely used for the interpretation of the fine structure spectra is reviewed by Miller and Freund [1] and simple formulas can also be found in Jette [2]. The discussion of the Born-Oppenheimer approximation of this electronic state can be found in Miller [3] and includes reinterpretations of the molecular parameters and their isotopic shifts. The special form of the effective Hamiltonian in a strong magnetic field is given in the paper by Miller and Freund [4] for the coupling case (*b*) or (*d*).

For Br_2, I_2, and ICl most experiments are performed on the $\Omega=0$ component of a $^3\Pi$-state for which the rotational energy is described by the same formula as for $^1\Sigma$-states (compare part 2.2).

A short description only of the parameters listed below is presented in Section 2. The precise definition can be found in the literature cited above. The reader should compare the original literature especially in cases of intermediate coupling.

2. List of parameters

A'	effective spin-orbit interaction constant which is measured as linear combination of the spin-orbit interaction constant A and the two spin-rotation interaction constants $\bar{\gamma}'$ and γ [2].
B_0, B_2	spin-spin coupling constant.
SS	spin-spin coupling constant defined in Hund's coupling case (*b*) and (*d*) for strong magnetic fields.
B	effective rotational constant, describing the energy separation between level N' and N''.
Y_{11}	Dunham-parameters to describe the rotational constant for the vibrational state v. (compare part 2.2)
Y_{12}	Dunham-parameters to describe the centrifugal distortion. (compare part 2.2).
$E(\mathrm{a})-E(\mathrm{b})$	energy difference of electronic state (a) and (b) which is measured by a microwave transition.
$a, b_F=b+c/3, c, d$	magnetic hyperfine coupling parameters.
eq_vQ	nuclear quadrupole coupling constant for the vibrational state v.
C_I	nuclear spin-rotation constant.
g_s'	g-factor of the electronic spin moment including relativistic effects.
g_L', g_N	g-factor of the orbital motion, which results in the effective value $g_N=\dfrac{g_L'\Lambda}{N(N+1)}$ for the specific state N. Normally g_N also includes the rotational g-factor g_{rot}.
χ_{eff}	parameter of the quadratic dependence of the spectral line positions by the magnetic field. For several experiments this constant is a linear combination of the anisotropy of the magnetic susceptibility and of the electric polarizability.
μ	electric dipole moment.
$\alpha_\perp, \alpha_\parallel$	electric polarizability of the molecule perpendicular and parallel to the molecular axis respectively.

3. Arrangement of substances

H_2, CO
Br_2, ICl, I_2

4. References

1 Miller, T.A., Freund, R.S.: Adv. Mag. Res. **9** (1977) 50.
2 Jette, A.N.: Chem. Phys. Lett. **25** (1974) 590; with some corrections in: Jette, A.N., Miller, T.A.: Chem. Phys. Lett. **29** (1974) 547.
3 Miller, T.A.: J. Chem. Phys. **59** (1973) 4078.
4 Miller, T.A., Freund, R.S.: J. Chem. Phys. **58** (1973) 2345.

4.1.6.2 Data and references

1 H_2

Molecular constants of H_2

Isotopic species	State		Parameter	Value	Ref.	Meth.	Remarks
	electronic	vibrational					
$^1H_2$(ortho)	c(2p) $^3\Pi_u$ $N=1$	$v=0$	$A+\bar{\gamma}'+2\gamma$	−3717.115 MHz	78LIC, 77MIL	MB	
			$B_0-\sqrt{6}B_2$	9562.480 MHz	78LIC, 77MIL	MB	
			$a(^1H)$	26.60 MHz	78LIC, 77MIL	MB	
			$b_F(^1H)$	450.479 MHz	78LIC, 77MIL	MB	
			$c-3d(^1H)$	104.177 MHz	78LIC, 77MIL	MB	$c=56.888$ MHz, $d=-15.763$ MHz obtained from theor. ratio of c and d
		$v=1$	$A+\bar{\gamma}'+2\gamma$	−3503.286 MHz	78LIC, 77MIL	MB	
			$B_0-\sqrt{6}B_2$	9169.165 MHz	78LIC, 77MIL	MB	
			$a(^1H)$	25.65 MHz	78LIC, 77MIL	MB	
			$b_F(^1H)$	439.504 MHz	78LIC, 77MIL	MB	
			$c-3d(^1H)$	99.042 MHz	78LIC, 77MIL	MB	$c=54.084$ MHz, $d=-14.986$ MHz obtained from theor. ratio of c and d
$^1H_2$(para)	c(2p) $^3\Pi_u$ $N=2$	$v=0$	$A+\bar{\gamma}'+6\gamma$	−3621.987 MHz	78LIC, 77MIL	MB	
			$B_0+\sqrt{6}B_2$	−12715.452 MHz	78LIC, 77MIL	MB	

Molecular constants of H_2 (continued)

Isotopic species	State		Parameter	Value	Ref.	Meth.	Remarks
	electronic	vibrational					
$^1H_2$(para)	c(2 p) $^3\Pi_u$ $N=2$	$v=1$	$A+\bar{\gamma}'+6\gamma$	−3412.786 MHz	78LIC, 77MIL	MB	
			$B_0+\sqrt{6}\,B_2$	−12131.197 MHz	78LIC, 77MIL	MB	
		$v=2$	$A+\bar{\gamma}'+6\gamma$	−3200.36 (2) MHz	78LIC	MB	
			$B_0+\sqrt{6}\,B_2$	−11556.01 (2) MHz	78LIC	MB	
$^1H_2$	c(2 p) $^3\Pi_u$	$v=0$	A	−3740.897 MHz	78LIC	MB	using approximation $\bar{\gamma}'\approx-\gamma$ compare [74JET1] and [74JET2]
			γ	23.782 MHz	78LIC	MB	
			B_0	−1576.486 MHz	78LIC	MB	
			B_2	−4547.464 MHz	78LIC	MB	
		$v=1$	A	−3525.911 MHz	78LIC	MB	using approximation $\bar{\gamma}'\approx-\gamma$ compare [74JET1] and [74JET2]
			γ	22.625 MHz	78LIC	MB	
			B_0	−1481.016 MHz	78LIC	MB	
			B_2	−4347.918 MHz	78LIC	MB	
$^1H_2$(ortho)	c(2 p) $^3\Pi_u$ $N=1$	$v=0$	$\alpha_\perp$	$5.00(36)\cdot10^3$ Å^3	76KAG	MB	calculated from Stark-shift coefficients given in [76KAG]
		$v=1$	$\alpha_\perp$	$2.29(16)\cdot10^3$ Å^3	76KAG	MB	
$^1H_2$(para)	c(2 p) $^3\Pi_u$ $N=2$	$v=0$	$\alpha_\perp$	4348 (43) Å^3	74ENG, 75ENG, 76KAG	MB	calculated from Stark-shift coefficients given in [76KAG]
		$v=1$	$\alpha_\perp$	2175 (24) Å^3	74ENG, 75ENG, 76KAG	MB	
		$v=2$	$\alpha_\perp$	1434 (26) Å^3	74ENG, 75ENG, 76KAG	MB	
$^1H_2$(para)	d(3 p) $^3\Pi_u$ $N=1$	$v=0$	$A+\bar{\gamma}'+2\gamma$	−706.8 (16) MHz	77MIL	MOMRIE	microwave optical magnetic resonance induced by electrons
			$B_0+\sqrt{6}\,B_2$	−2901.4 (32) MHz	77MIL	MOMRIE	
			g_L'	0.98728 (7)	77MIL	MOMRIE	field calibration error not included
			g_s'	2.0025 (4)	77MIL	MOMRIE	
			χ_{eff}	$17(3)\cdot10^3$ Hz/kG^2	77MIL	MOMRIE	

Molecular constants of H_2 (continued)

Isotopic species	State		Parameter	Value	Ref.	Meth.	Remarks
	electronic	vibrational					
$^1H_2$(para)	d(3 p) $^3\Pi_u$ $N=1$	$v=1$	$A+\bar{\gamma}'+2\gamma$	−668.7 (10) MHz	77MIL	MOMRIE	
			$B_0+\sqrt{6}\,B_2$	−2778.4 (19) MHz	77MIL	MOMRIE	
			g_L'	0.99130 (4)	77MIL	MOMRIE	field calibration error not included
			g_s'	2.0026 (2)	77MIL	MOMRIE	
			χ_{eff}	9 (2) · 10^3 Hz/kG^2	77MIL	MOMRIE	
		$v=2$	$A+\bar{\gamma}'+2\gamma$	−596.0 (7) MHz	77MIL	MOMRIE	
			$B_0+\sqrt{6}\,B_2$	−2606.0 (13) MHz	77MIL	MOMRIE	
			g_L'	0.98218 (3)	77MIL	MOMRIE	field calibration error not included
			g_s'	2.0028 (2)	77MIL	MOMRIE	
			χ_{eff}	10 (2) · 10^3 Hz/kG^2	77MIL	MOMRIE	
		$v=3$	$A+\bar{\gamma}'+2\gamma$	−519.1 (12) MHz	77MIL	MOMRIE	
			$B_0+\sqrt{6}\,B_2$	−2432.1 (22) MHz	77MIL	MOMRIE	
			g_L'	0.96226 (5)	77MIL	MOMRIE	field calibration error not included
			g_s'	2.0028 (3)	77MIL	MOMRIE	
			χ_{eff}	17 (3) · 10^3 Hz/kG^2	77MIL	MOMRIE	
$^1H_2$(ortho)	d(3 p) $^3\Pi_u$ $N=1$	$v=0$	$A+\bar{\gamma}'+2\gamma$	−788.7 (10) MHz	77MIL	MOMRIE	
			$B_0-\sqrt{6}\,B_2$	2241.0 (19) MHz	77MIL	MOMRIE	
			g_L'	0.99973 (4)	77MIL	MOMRIE	field calibration error not included
			g_s'	2.0023 (1)	77MIL	MOMRIE	
			χ_{eff}	8 (1) · 10^3 Hz/kG^2	77MIL	MOMRIE	
			$a(^1H)$	6.7 (9) MHz	77MIL	MOMRIE	
			$b_F(^1H)$	459.4 (4) MHz	77MIL	MOMRIE	
			$c-3d(^1H)$	73.0 (28) MHz	77MIL	MOMRIE	
		$v=1$	$A+\bar{\gamma}'+2\gamma$	−736.0 (8) MHz	77MIL	MOMRIE	
			$B_0-\sqrt{6}\,B_2$	2145.0 (15) MHz	77MIL	MOMRIE	
			g_L'	0.99973 (3)	77MIL	MOMRIE	field calibration error not included
			g_s'	2.0027 (1)	77MIL	MOMRIE	
			χ_{eff}	−4 (2) · 10^3 Hz/kG^2	77MIL	MOMRIE	
			$a(^1H)$	6.3 (7) MHz	77MIL	MOMRIE	
			$b_F(^1H)$	448.0 (4) MHz	77MIL	MOMRIE	
			$c-3d(^1H)$	68.3 (28) MHz	77MIL	MOMRIE	
		$v=2$	$A+\bar{\gamma}'+2\gamma$	−682.2 (8) MHz	77MIL	MOMRIE	
			$B_0-\sqrt{6}\,B_2$	2047.0 (17) MHz	77MIL	MOMRIE	
			g_L'	0.99971 (4)	77MIL	MOMRIE	field calibration error not included

Molecular constants of H_2 (continued)

Isotopic species	State: electronic	State: vibrational	Parameter	Value	Ref.	Meth.	Remarks
$^1H_2$(ortho)	d(3 p) $^3\Pi_u$ $N=1$	$v=2$	g_s'	2.0029 (1)	77MIL	MOMRIE	
			χ_{eff}	$-6(2)\cdot10^3$ Hz/kG^2	77MIL	MOMRIE	
			$a(^1H)$	6.6 (9) MHz	77MIL	MOMRIE	
			$b_F(^1H)$	437.9 (3) MHz	77MIL	MOMRIE	
			$c-3d(^1H)$	66.4 (28) MHz	77MIL	MOMRIE	
		$v=3$	$A+\bar{\gamma}'+2\gamma$	-625.8 (8) MHz	77MIL	MOMRIE	
			$B_0-\sqrt{6}B_2$	1952.0 (14) MHz	77MIL	MOMRIE	
			g_L'	0.99970 (3)	77MIL	MOMRIE	field calibration error not included
			g_s'	2.0025 (1)	77MIL	MOMRIE	
			χ_{eff}	$-4(1)\cdot10^3$ Hz/kG^2	77MIL	MOMRIE	
			$a(^1H)$	5.6 (7) MHz	77MIL	MOMRIE	
			$b_F(^1H)$	428.3 (3) MHz	77MIL	MOMRIE	
			$c-3d(^1H)$	64.5 (19) MHz	77MIL	MOMRIE	
$^1H_2$	d(3 p) $^3\Pi_u$	Redefinitions of the molecular parameters including breakdown of the Born-Oppenheimer approximation [73MIL1] can explain the differences in the values of ortho- and para-hydrogen molecule [77MIL] and [73MIL2].					
$^2D_2$(para)	d(3 p) $^3\Pi_u$ $N=1$	$v=0$	$A+\bar{\gamma}'+2\gamma$	-809.1 (5) MHz	77MIL	MOMRIE	
			$B_0-\sqrt{6}B_2$	2259.8 (10) MHz	77MIL	MOMRIE	
			g_L'	0.99996 (2)	77MIL	MOMRIE	field calibration error not included
			χ_{eff}	$-13(1)\cdot10^3$ Hz/kG^2	77MIL	MOMRIE	
		$v=1$	$A+\bar{\gamma}'+2\gamma$	-772.0 (21) MHz	77MIL	MOMRIE	
			$B_0-\sqrt{6}B_2$	2188.6 (25) MHz	77MIL	MOMRIE	
			g_L'	1.00001 (5)	77MIL	MOMRIE	field calibration error not included
			χ_{eff}	$-3(7)\cdot10^3$ Hz/kG^2	77MIL	MOMRIE	
		$v=2$	$A+\bar{\gamma}'+2\gamma$	-732.8 (13) MHz	77MIL	MOMRIE	
			$B_0-\sqrt{6}B_2$	2122.2 (22) MHz	77MIL	MOMRIE	
			g_L'	1.00011 (5)	77MIL	MOMRIE	field calibration error not included
			χ_{eff}	$-8(2)\cdot10^3$ Hz/kG^2	77MIL	MOMRIE	
		$v=3$	$A+\bar{\gamma}'+2\gamma$	-692.7 (21) MHz	77MIL	MOMRIE	
			$B_0-\sqrt{6}B_2$	2054.5 (41) MHz	77MIL	MOMRIE	
			g_L'	0.99967 (9)	77MIL	MOMRIE	field calibration error not included
			χ_{eff}	$-42(5)\cdot10^3$ Hz/kG^2	77MIL	MOMRIE	
		$v=4$	$A+\bar{\gamma}'+2\gamma$	-655.1 (15) MHz	77MIL	MOMRIE	
			$B_0-\sqrt{6}B_2$	1987.8 (28) MHz	77MIL	MOMRIE	

Molecular constants of H_2 (continued)

Isotopic species	State: electronic	State: vibrational	Parameter	Value	Ref.	Meth.	Remarks
$^2D_2$(para)	d(3 p) $^3\Pi_u$ $N=1$	$v=4$	g_L'	1.00011 (6)	77MIL	MOMRIE	field calibration error not included
			χ_{eff}	$-7(4)\cdot 10^3$ Hz/kG^2	77MIL	MOMRIE	
		$v=5$	$A+\bar{\gamma}'+2\gamma$	-614.4 (19) MHz	77MIL	MOMRIE	
			$B_0-\sqrt{6}B_2$	1919.5 (35) MHz	77MIL	MOMRIE	
			g_L'	0.99999 (8)	77MIL	MOMRIE	field calibration error not included
			χ_{eff}	$-3(3)\cdot 10^3$ Hz/kG^2	77MIL	MOMRIE	
$^2D_2$(ortho)	d(3 p) $^3\Pi_u$ $N=1$	$v=0$	$A+\bar{\gamma}'+2\gamma$	-767.6 (14) MHz	77MIL	MOMRIE	
			$B_0+\sqrt{6}B_2$	-2960.8 (25) MHz	77MIL	MOMRIE	
			g_L'	0.99427 (6)	77MIL	MOMRIE	field calibration error not included
			χ_{eff}	$16(3)\cdot 10^3$ Hz/kG^2	77MIL	MOMRIE	
		$v=1$	$A+\bar{\gamma}'+2\gamma$	-733.4 (8) MHz	77MIL	MOMRIE	
			$B_0+\sqrt{6}B_2$	-2861.2 (16) MHz	77MIL	MOMRIE	
			g_L'	0.99477 (4)	77MIL	MOMRIE	field calibration error not included
			χ_{eff}	$10(2)\cdot 10^3$ Hz/kG^2	77MIL	MOMRIE	
		$v=2$	$A+\bar{\gamma}'+2\gamma$	-689.9 (6) MHz	77MIL	MOMRIE	
			$B_0+\sqrt{6}B_2$	-2752.8 (13) MHz	77MIL	MOMRIE	
			g_L'	0.99218 (3)	77MIL	MOMRIE	field calibration error not included
			χ_{eff}	$11(2)\cdot 10^3$ Hz/kG^2	77MIL	MOMRIE	
		$v=3$	$A+\bar{\gamma}'+2\gamma$	-655.1 (8) MHz	77MIL	MOMRIE	
			$B_0+\sqrt{6}B_2$	-2656.9 (16) MHz	77MIL	MOMRIE	
			g_L'	0.99398 (4)	77MIL	MOMRIE	field calibration error not included
			χ_{eff}	$14(2)\cdot 10^3$ Hz/kG^2	77MIL	MOMRIE	
		$v=4$	$A+\bar{\gamma}'+2\gamma$	-612.7 (7) MHz	77MIL	MOMRIE	
			$B_0+\sqrt{6}B_2$	-2553.2 (13) MHz	77MIL	MOMRIE	
			g_L'	0.99203 (3)	77MIL	MOMRIE	field calibration error not included
			χ_{eff}	$11(1)\cdot 10^3$ Hz/kG^2	77MIL	MOMRIE	
$^2D_2$	d(3 p) $^3\Pi_u$	Redefinitions of the molecular parameters including breakdown of the Born-Oppenheimer approximation [73MIL1] can explain the differences in the values of ortho- and para-states; isotopic shifts between D_2 and H_2 parameters are discussed [77MIL, 73MIL2]					
$^1H_2$(para)	k(4 p) $^3\Pi_u$ $N=1$	$v=0$	$A+\bar{\gamma}'+2\gamma$	-213 (7) MHz	77MIL	MOMRIE	
			$B_0+\sqrt{6}B_2$	-1107 (13) MHz	77MIL	MOMRIE	
			g_L'	0.97645 (30)	77MIL	MOMRIE	field calibration error not included
			χ_{eff}	$90(20)\cdot 10^3$ Hz/kG^2	77MIL	MOMRIE	

Molecular constants of H_2 (continued)

Isotopic species	State: electronic	State: vibrational	Parameter	Value	Ref.	Meth.	Remarks
$^1H_2$(ortho)	k(4p) $^3\Pi_u$	$v=1$	$A+\bar{\gamma}'+2\gamma$	−265.6 (10) MHz	74MIL	MOMRIE	
	$N=1$		$B_0-\sqrt{6}\,B_2$	838.3 (19) MHz	74MIL	MOMRIE	
			g_L'	0.99987 (4)	74MIL	MOMRIE	field calibration error not included
			χ_{eff}	$-10(2)\cdot10^3$ Hz/kG^2	74MIL	MOMRIE	
			$a(^1H)$	1.9 (9) MHz	74MIL	MOMRIE	
			$c-3d(^1H)$	60.7 (28) MHz	74MIL	MOMRIE	
		$v=3$	$A+\bar{\gamma}'+2\gamma$	−235.0 (5) MHz	74MIL	MOMRIE	
			$B_0-\sqrt{6}\,B_2$	756.1 (9) MHz	74MIL	MOMRIE	
			g_L'	0.99964 (2)	74MIL	MOMRIE	field calibration error not included
			χ_{eff}	$-30(1)\cdot10^3$ Hz/kG^2	74MIL	MOMRIE	
			$a(^1H)$	1.1 (2) MHz	74MIL	MOMRIE	
			$c-3d(^1H)$	52.2 (19) MHz	74MIL	MOMRIE	
	$N=3$	$v=1$	$A+\bar{\gamma}'+12\gamma$	−6.0 (23) MHz	75MIL	level anticrossing spectroscopy and microwave transitions k(4p) $^3\Pi_u$ − p(4d) $^3\Sigma_g^+$	
			$B_0+2/3\sqrt{6}\,B_2$	−765.6 (72) MHz	75MIL		
			$E(k\,^3\Pi_u, v=1, N=3) - E(p\,^3\Sigma_g^+, v=1, N=5)$	8356 (2) MHz	75MIL		
$^1H_2$(ortho)	p(4d) $^3\Sigma_g^+$	$v=1$	$2A+3\gamma$	89.0 (90)	75MIL	level anticrossing spectroscopy and microwave transitions between k $^3\Pi_u$ $v=1$, $N=3$ and p $^3\Sigma_g^+$ $v=1$, $N=5$	
	$N=5$		SS	1.19 (5) MHz	75MIL		
			g_N	0.38464 (6)	75MIL	field calibration error not included	
			χ_{eff}	$6.7(12)\cdot10^3$ Hz/kG^2	75MIL		
						the p $^3\Sigma_g^+$ is close to Hund's coupling case (*d*)	
$^1H_2$(para)	i(3d) $^3\Pi_g$	$v=3$	$A+\bar{\gamma}'+6\gamma$	−376 (132) MHz	74FRE	level anticrossing spectroscopy and microwave transitions between d $^3\Pi_u$ and i $^3\Pi_g$	
	$N=2$		$B_0+\sqrt{6}\,B_2$	−85 (219) MHz	74FRE		
			g_N	−0.2202 (12)	74FRE	the state i $^3\Pi_g$ is intermediate between Hund's coupling case (*b*) and (*d*) but the parameters are given as those of pure coupling case (*b*), therefore they should be redefined	
			$E(i\,^3\Pi_g, v=3, N=2) - E(d\,^3\Pi_u, v=3, N=1)$	57691 (21) MHz	74FRE		
	$N=1$	equilibrium	$A+\bar{\gamma}'+2\gamma$	−145 (6) MHz	78WIK	molecular beam laser spectroscopy	
			$B_0-\sqrt{6}\,B_2$	958 (28) MHz	78WIK		

Molecular constants of H_2 (continued)

Isotopic species	State: electronic	State: vibrational	Parameter	Value	Ref.	Meth.	Remarks
$^2D_2$(para)	i(3 d) $^3\Pi_g$ $N=1$	$v=3$	$A+\bar{\gamma}'+2\gamma$	−37 (61) MHz	76FRE		level anticrossing spectroscopy and microwave transitions between d $^3\Pi_u$ and i $^3\Pi_g$
			SS	20 (4) MHz	76FRE		
			g_N	0.0930 (17)	76FRE		
			χ_{eff}	$17(12)\cdot 10^3$ Hz/kG^2	76FRE		
			$E(i\ ^3\Pi_g, v=3, N=1) - E(d\ ^3\Pi_u, v=3, N=1)$	26861 (8) MHz	76FRE		
$^2D_2$(ortho)	r(4 d) $^3\Pi_g$ $N=2$	$v=0$	g_N	0.1467 (3)	76MIL		level anticrossing spectroscopy with the G(3 d) $^1\Sigma_g^+$ state
			$b_F(^2D)$	73 (3) MHz	76MIL		
	Such experiments are used to determine the energy separation of the singlet and triplet manifold. Further details are given in [77MIL].						
$^2D_2$(ortho)	G(3 d) $^1\Sigma_g^+$ $N=2$	$v=4$	g_N	0.1963 (3)	76MIL		level anticrossing spectroscopy further data on this state in 2.9.1 [74MEL]

Line frequencies of H_2

Isotopic species	State: electronic	State: vibrational	Transition rotational N'	− N''	rotational J'	− J''	hyperfine $I'_T = I''_T$	F'	− F''	ν MHz	Ref.	Remarks
$^1H_2$(para)	c(2 p) $^3\Pi_u$	$v=0$	2	2	2	1	$I_T=0$			5150.398 (3)	78LIC	
					2	3				5898.105 (12)	78LIC	
		$v=1$	2	2	2	1	$I_T=0$			4928.003 (10)	78LIC	
					2	3				5605.709 (4)	78LIC	
		$v=2$	2	2	2	1	$I_T=0$			4711.218 (10)	78LIC	
					2	3				5314.611 (10)	78LIC	
$^1H_2$(ortho)	c(2 p) $^3\Pi_u$	$v=1$	1	1	1	1	$I_T=1$	1	0	272.741 (30)	78LIC	
					1	1	$I_T=1$	2	1	375.334 (170)	78LIC	
					2	2	$I_T=1$	2	1	450.625 (300)	78LIC	
					2	2	$I_T=1$	3	2	707.550 (90)	78LIC	
					1	0	$I_T=1$	0	1	4758.186 (40)	78LIC	
					1	0	$I_T=1$	1	1	5030.948 (6)	78LIC	

Line frequencies of H_2 (continued)

Isotopic species	State: electronic	State: vibrational	Transition: rotational N' – N''		rotational J' – J''		hyperfine I'_T = I''_T	F' – F''		ν MHz	Ref.	Remarks
$^1H_2$(ortho)	c(2 p) $^3\Pi_u$	$v=1$	1	1	1	0	$I_T=1$	2	1	5406.276 (4)	78LIC	
					1	2	$I_T=1$	2	3	6023.864 (25)	78LIC	
					1	2	$I_T=1$	1	2	6356.024 (3)	78LIC	
					1	2	$I_T=1$	0	1	6533.938 (7)	78LIC	
					1	2	$I_T=1$	2	2	6731.350 (5)	78LIC	
					1	2	$I_T=1$	1	1	6806.684 (7)	78LIC	
					1	2	$I_T=1$	2	1	7182.012 (13)	78LIC	

References

73MIL1 Miller, T.A.: J. Chem. Phys. **59** (1973) 4078.
73MIL2 Miller, T.A., Freund, R.S.: J. Chem. Phys. **59** (1973) 4093.
74ENG English, T.C., Kagann, R.H.: Phys. Rev. Lett. **33** (1974) 995.
74FRE Freund, R.S., Miller, T.A.: J. Chem. Phys. **60** (1974) 4900.
74JET1 Jette, A.N.: Chem. Phys. Lett. **25** (1974) 590.
74JET2 Jette, A.N., Miller, T.A.: Chem. Phys. Lett. **29** (1974) 547.
74MEL Melieres-Marechal, M.A., Lombardi, M.: J. Chem. Phys. **61** (1974) 2600.
74MIL Miller, T.A., Freund, R.S., Zegarski, B.R.: J. Chem. Phys. **60** (1974) 3195.
75ENG English, T.C., Albritton, D.L.: J. Phys. **B 8** (1975) 2123.
75MIL Miller, T.A., Freund, R.S.: J. Chem. Phys. **62** (1975) 2240.
76FRE Freund, R.S., Miller, T.A., Zegarski, B.R.: J. Chem. Phys. **64** (1976) 4069.
76KAG Kagann, R.H., English, T.C.: Phys. Rev. **A 13** (1976) 1451.
76MIL Miller, T.A., Freund, R.S., Zegarski, B.R.: J. Chem. Phys. **64** (1976) 1842.
77MIL Miller, T.A., Freund, R.S.: Adv. Magn. Reson. **9** (1977) 49, (and literature cited therein).
78LIC Lichten, W., Wik, T.: J. Chem. Phys. **69** (1978) 5428.
78WIK Wik, T., Bennett, W.R., Lichten, W.: Phys. Rev. Lett. **40** (1978) 1080.

The energy separation of the singlet and triplet manifold is measured in several experiments of level crossing spectroscopy which are reviewed in [77MIL].

59 CO

Molecular constants of CO

Isotopic species	State: electronic	State: vibrational	Parameter	Value	Ref.	Meth.	Remarks
$^{12}C^{16}O$	a $^3\Pi_2$	$v=6$	$\langle\mu\Omega\rangle$	2.73974 (50) D	75WIC	MB	effective stark-coefficient $v=6$ is probably undisturbed by a′ $^3\Sigma$.
		$v=7$	$\langle\mu\Omega\rangle$	2.74578 (50) D	75WIC	MB	$v=7$ is disturbed by a′ $^3\Sigma$, $v=4$

Line frequencies of CO

Isotopic species	State		Transition							ν MHz	Ref.	Remarks
	electronic	vibrational	parity		rotational J' – J''		hyperfine F_1' – F_1''		F' – F''			
$^{12}C^{16}O$	a $^3\Pi_2$	$v=6$	+	–	2	2				6.123 (2)	75WIC	
			–	+	3	3				28.524 (2)	75WIC	
			+	–	4	4				78.286 (2)	75WIC	
			–	+	5	5				164.627 (2)	75WIC	
		$v=7$	+	–	2	2				6.185 (2)	75WIC	

Reference

75WIC Wicke, B.G., Klemperer, Field, R.: J. Chem. Phys. **62** (1975) 3544.

100 ICl

Molecular constants of ICl

Isotopic species	State		Parameter	Value	Ref.	Meth.	Remarks
	electronic	vibrational					
$^{127}I^{35}Cl$	B $^3\Pi_{o^+}$	equ.	B_e^{cor}	2592.919 (25) MHz	80KNÖ	MODR	B_e^{cor} is corrected for the mass shift of the chlorine nucleus: $d_{Cl}=7.95$ (37) (for definition compare part 2.2)
			Y_{11}	–36.68 (10) MHz	80KNÖ	MODR	
			Y_{21}	1.876 (81) MHz	80KNÖ	MODR	
			Y_{31}	–1.668 (18) MHz	80KNÖ	MODR	
			Y_{02}	–1.860 (8) kHz	80KNÖ	MODR	
			Y_{12}	0.00 (3) kHz	80KNÖ	MODR	
			Y_{22}	–175.5 (16) Hz	80KNÖ	MODR	
$^{127}I^{35}Cl$	B $^3\Pi_{o^+}$	$v=0$	$eq_0Q(^{127}I)$	–664 (9) MHz	82HAN	MODR	$eq_vQ = -657.14 - 5.96(v+1/2) - 15.53\,(v+1/2)^2$ MHz [82KNÖ]
		$v=1$	$eq_1Q(^{127}I)$	–701 (1) MHz	80KNÖ	MODR	
		$v=2$	$eq_2Q(^{127}I)$	–769.0 (4) MHz	80KNÖ	MODR	
			$eq_2Q(^{35}Cl)$	–27 (6) MHz	80KNÖ	MODR	
			$C_I(^{127}I)$	0.16 (7) MHz	80KNÖ	MODR	
$^{127}I^{37}Cl$	B $^3\Pi_{o^+}$	$v=2$	$eq_2Q(^{127}I)$	–764.8 (4) MHz	80KNÖ	MODR	
ICl	B $^3\Pi_{o^+}$	equ.	r_e^{cor}	2.6663 (1) Å	82KNÖ	MODR	

Molecular constants of ICl (continued)

Isotopic species	State electronic	State vibrational	Parameter	Value	Ref.	Meth.	Remarks
ICl	A $^3\Pi_1$		Determination of the electric dipole moment in several vibrational states by the observation of Stark-effect in the optical spectrum. The results and the references are presented in part 2.6.1.				

References

79KNÖ Knöckel, H., Tiemann, E.: Chem. Phys. Lett. **64** (1979) 593.
80KNÖ Knöckel, H.: Dissertation University Hannover, **1980**.
82HAN Hansen, S.G., Howard, B.J.: Chem. Phys. Lett. **85** (1982) 249.
82KNÖ Knöckel, H., Tiemann, E.: Chem. Phys. **70** (1982) 345.

101 Br_2

Molecular constants of Br_2

Isotopic species	State electronic	State vibrational	Parameter	Value	Ref.	Meth.	Remarks
Br_2	B $^3\Pi_{0_u^+}$		Determination of the quadrupole hyperfine structure by high resolution laser spectroscopy. The results and references are given in part 2.7.1.				

103 I_2

Molecular constants of I_2

Isotopic species	State electronic	State vibrational	Parameter	Value	Ref.	Meth.	Remarks
I_2	B $^3\Pi_{0_u^+}$		Determination of the quadrupole hyperfine and magnetic hyperfine structure and the rotational g-factor by high resolution laser spectroscopy. The results and references are given in part 2.7.1, and 2.9.1.				

4.1.7 The $^1\Delta$ electronic state

4.1.7.1 Preliminary remarks

1. Introduction

Molecules in $^1\Delta$ electronic states so far observed with microwave spectroscopy and related techniques are fairly well described by Hund's coupling case (*a*) which leads to the electronic quantum numbers for the projection of the orbital angular momentum and total angular momentum onto the molecular axis which are then $|\Lambda| = |\Omega| = 2$. The molecular levels in a vibrational state v are defined by the total angular momentum quantum number J where $\boldsymbol{J}$ is the sum of the electronic angular momentum and the nuclear rotation $\boldsymbol{R}$. Each level should show-Λ-doubling structure but this is not resolved by the present experiments. Therefore the parity assignment to the levels is suppressed. In case of non-zero nuclear spin this couples with $\boldsymbol{J}$ and gives the quantum number F for the total angular momentum and possibly F_1 as intermediate quantum number if both nuclei have non-zero spins.

The Hamiltonian for describing the rotational energy levels is principally given in 5.1.1 of Vol. II/6 Landolt-Börnstein, New Series and will not be repeated here. Some details of contributions from other excited electronic states to the rotational energy are discussed by Yamada et al. [1].

2. List of parameters

$B_{v,\mathrm{e}}$	rotational constant for the vibrational state v or for the equilibrium configuration, respectively
D_v	centrifugal distortion constant of the rotational energy
Y_{lk}	Dunham energy coefficients, compare 2.2.1
a_i	magnetic hyperfine structure constant for the coupling of the electronic orbital angular motion of nucleus (i)
eqQ_{i}	nuclear quadrupole coupling constant of nucleus (i)
g_L	g-factor associated with the electronic orbital angular momentum
g_{rot}	rotational g-factor; the rotational magnetic moment is defined in nuclear magnetons (compare part 2.9)
μ_0	electric dipole moment for the vibrational state $v=0$
r_{e}	equilibrium internuclear distance derived from B_{e}
$r_{\mathrm{e}}^{\mathrm{cor}}$	equilibrium internuclear distance corrected for contributions of other electronic states to the rotational energy

3. Arrangement of substances

He_2, PH, NF, O_2, SO

4. Reference

1 Yamada, C., Kawaguchi, K., Hirota, E.: J. Chem. Phys. **69** (1978) 1942.

4.1.7.2 Data and references

7 He_2

Molecular constants of He_2

Isotopic species	State electronic	State vibrational	Parameter	Value	Ref.	Meth.	Remarks
$^4He_2$	F (3d) $^1\Delta_u$		measurement of the energy shift between the singlet and triplet levels		75MIL		observation of level anticrossing spectra between F (3 d) $^1\Delta_u$ and f (3 d) $^3\Delta_u$

Reference: 75MIL Miller, T.A., Freund, R.S., Zegarski, B.R., Jost, R., Lombardi, M., Derouard, J.: J. Chem. Phys. **63** (1975) 4042.

10 PH

Molecular constants of PH

Isotopic species	State electronic	State vibrational	Parameter	Value	Ref.	Meth.	Remarks
$^{31}P^1H$	a $^1\Delta_2$	$v=0$	B_0	253003.62 (14) MHz	75DAV, 79DAV	LMR	
			$a(^{31}P)$	768 (37) MHz	75DAV, 79DAV	LMR	
			$a(^1H)$	28 (15) MHz	75DAV, 79DAV	LMR	

References

75DAV Davies, P.B., Russel, D.K., Thrush, B.A.: Chem. Phys. Lett. **36** (1975) 280.

79DAV Davies, P.B., Russel, D.K., Smith, D.R., Thrush, B.A.: Can. J. Phys. **57** (1979) 522.

82 NF

Molecular constants of NF

Isotopic species	State electronic	State vibrational	Parameter	Value	Ref.	Meth.	Remarks
$^{14}N^{19}F$	a $^1\Delta_2$	$v=0$	$B_0(1-3\,g_{rot})$	36662 (15) MHz	73CUR	EPR	
			D_0	155.9 (30) kHz	81DAV2	IR	
			$g_L-1/2\,g_{rot}$	1.00005 (30)	73CUR	EPR	
			$a(^{19}F)$	758.06 (23) MHz	73CUR	EPR	
			$a(^{14}N)$	109.92 (14) MHz	73CUR	EPR	
			$eqQ(^{14}N)$	4.10 (23) MHz	73CUR	EPR	
			μ	0.37 (6) D	73CUR	EPR	

Molecular constants of NF (continued)

Isotopic species	State		Parameter	Value	Ref.	Meth.	Remarks
	electronic	vibrational					
$^{14}N^{19}F$	a $^1\Delta_2$	equ.	Y_{01} Y_{11}	36871.8 (21) MHz −433.80 (30) MHz	81DAV2 81DAV2	IR IR	

Line frequencies of NF

Isotopic species	State		Transition				ν MHz	Ref.	Remarks
	electronic	vibrational	parity	rotational J' − J''	hyperfine F_1' − F_1''	F' − F''			
$^{14}N^{19}F$	a $^1\Delta_2$	$v=0$		8 7			586145.8 (30)	81DAV1	extrapolation to zero magnetic field from LMR-measurements with hyperfine structure removed

References

73CUR Curran, A.H., Macdonald, R.G., Stone, A.J., Thrush, B.A.: Proc. Roy. Soc. London **A 332** (1973) 355.

81DAV1 Davies, P.B., Temps, F.: J. Chem. Phys. **74** (1981) 6556.

81DAV2 Davies, P.B., Hamilton, P.A., Okumura, M.: J. Chem. Phys. **75** (1981) 4294.

87 O_2

Molecular constants of O_2

Isotopic species	State		Parameter	Value	Ref.	Meth.	Remarks
	electronic	vibrational					
$^{16}O_2$	a $^1\Delta_{2g}$	$v=0$	B_0	42505.780 (25) MHz	81AMI	FIR	
			D_0	155.20 (17) kHz	81AMI	FIR	
			H	4.76 (29) Hz	81AMI	FIR	higher-order centrifuged distortion constant, compare 2.2.1
			q	50 (90) Hz	81SCA	LMR	estimation of Λ-doubling constant

Line frequencies of O_2

Isotopic species	State		Transition							ν MHz	Ref.	Remarks
	electronic	vibrational	parity		rotational J'	J''	hyperfine F_1' – F_1''	F' – F''				
$^{16}O^{16}O$	a $^1\Delta_{2g}$	$v=0$	+	+	4	3				340008.5 (6)	81SCA	extrapolation to zero magnetic field from LMR-measurements
			+	+	5	4				424981.0 (9)		
			+	+	7	6				594870 (1)		
			+	+	8	7				679780 (1)		
			+	+	9	8				764658 (1)		

References

81AMI Amiot, C., Verges, J.: Can. J. Phys. **59** (1981) 1391.
81SAC Scalabrin, A., Saykally, R.J., Evenson, K.M., Radford, H.E., Mizushima, M.: J. Mol. Spectrosc. **89** (1981) 344.

88 SO

Molecular constants of SO

Isotopic species	State electronic	State vibrational	Parameter	Value	Ref.	Meth.	Remarks
$^{32}S^{16}O$	a $^1\Delta_2$	$v=0$	B_0	21295.405 (24) MHz	76CLA	MW	
			D_0	35.0 (4) kHz	76CLA	MW	
			g_L	0.999821 (12)	78YAM	LMR	
			g_{rot}	−0.235 (9)	78YAM	LMR	
		$v=1$	B_1	21113.39 (78) MHz	78YAM	LMR	
			g_L	0.99947 (48)	78YAM	LMR	
			g_{rot}	−1.1 (5)	78YAM	LMR	
			band origin $\nu(v=0-v=1)$	1101.3942 (1) cm^{-1}	78YAM	LMR	
$^{34}S^{16}O$	a $^1\Delta_2$	$v=0$	LMR-measurements are consistent with the molecular constants which are calculated by isotopic relations from the results of $^{32}S^{16}O$ [78YAM].				
SO	a $^1\Delta_2$	equ.	r_e^{cor}	1.48886 (26) Å	78YAM		

Line frequencies of SO; Isotopic species $^{32}S^{16}O$

State		Transition			ν MHz	Ref.	Remarks
electronic	vibrational	parity	rotational J'	J''			
a $^1\Delta_2$	$v=0$		3	2	127770.416 (100)	76CLA	no error given in the original literature
			4	3	170356.467 (100)	76CLA	
			5	4	212939.337 (100)	76CLA	
			6	5	255518.037 (100)	76CLA	

References

76CLA Clark, W.W., De Lucia, F.C.: J. Mol. Spectrosc. **60** (1976) 332.
78YAM Yamada, C., Kawaguchi, K., Hirota, E.: J. Chem. Phys. **69** (1978) 1942.

4.1.8 The electronic states $^4\Pi$, $^2\Delta$, and $^3\Delta$

No microwave spectroscopy or related methods are applied for observations of molecules in electronic states $^4\Pi$, $^2\Delta$ or $^3\Delta$. But electric dipole moments are measured in a few cases:

NO in a $^4\Pi$
CH in A $^2\Delta$
CO in d $^3\Delta$

by different techniques. The results are included in part 2.6 in order to get a complete listing of this molecular parameter.

4.2 Polyatomic free radicals

4.2.0 Introduction

A free radical is defined to be a molecule in an open shell electronic state. It is often, although not necessarily, very reactive and short-lived in a laboratory environment. Several new species have been studied since the publication of the previous supplement, although the number for which microwave transition frequencies have been measured is still quite small. Experiments employing double resonance techniques (simultaneous irradiation with microwaves and either infrared or visible radiation) have made a notable contribution to the development of the field. The information about linear molecules, in $^2\Sigma$ or $^2\Pi$ states, is contained in Section 4.2.1. The non-linear radicals, almost all of which are triatomic, are presented in 4.2.2 (Non-linear triatomic) and 4.2.3 (Non-linear larger molecules).

Data that relate to molecular rotational energy levels have been fitted to the parameters of an appropriate effective Hamiltonian. Such a Hamiltonian must take account of all the many interactions that can arise for a molecule in a multiplet electronic state. The foundations for many of the features of this model have been securely laid by Van Vleck [51Va] with an important contribution for linear triatomic molecules in Π states (the Renner-Teller effect) having been made by Renner [34Re]. There have been many subsequent papers dealing with different aspects of the effective Hamiltonian, some of which are listed in the previous supplement. For a good overall description, the reader is referred to the third volume of Herzberg's book "Molecular Spectra and Molecular Structure" [66He]. The large number of significant experimental results on free radicals that have been obtained recently has prompted renewed activity on the definition of the correct form for the effective Hamiltonian. The effects of centrifugal distortion on the rotational energy levels cause indeterminacies among the parameters in the Hamiltonian. Watson [67Wa, 77Wa] has shown how these indeterminacies can be systematically removed. Watson's ideas about the expression of the effective Hamiltonian in terms of determinable parameters have been extended to the spin-rotation Hamiltonian and its centrifugal distortion corrections by Brown and Sears [79Br].

Values for the intertial defect and the electric dipole moment quoted in this section have been recalculated where necessary using the currently accepted values for Planck's constant, Avogadro's number and the experimental value for the dipole moment of $^{16}O^{12}C^{32}S$. The first two of these are $h = 6.626176\,(36) \cdot 10^{-34}$ J sec and $N_A = 6.022045\,(31) \cdot 10^{23}$ mol^{-1} [73Co]. The best determination of μ_{OCS} available to date is that by De Leeuw and Dymanus [70De] whose value is 0.71512(3) D. The accepted value for the speed of light is now $c = 2.997924580\,(12) \cdot 10^8$ m sec^{-1} [73Co]. The rotational constant · moment of inertia product on the ^{12}C basis is thus

$$B \cdot I = 5.05379\,(4) \cdot 10^5 \text{ MHz amu Å}^2 \equiv 16.85763\,(13) \text{ amu Å}^2 \text{ cm}^{-1}.$$

References

34Re Renner, R.: Z. Physik **92** (1934) 172.

51Va Van Vleck, J.H.: Rev. Mod. Physics **23** (1951) 213.

66He Herzberg, G.: "Molecular Spectra and Molecular Structure III. Electronic Spectra and Electronic Structure of Polyatomic Molecules", Van Nostrand Reinhold Company, New York, **1966**.

67Wa Watson, J.K.G.: J. Chem. Phys. **46** (1967) 1935.

70De De Leeuw, F.H., Dymanus, A.: Chem. Phys. Lett. **7** (1970) 288.

73Co Cohen, E.R., Taylor, B.N.: J. Phys. Chem. Ref. Data **2** (1973) 663.

77Wa Watson, J.K.G.: "Aspects of quartic and sextic centrifugal effects on rotational energy levels", in "Vibrational Spectra and Structure" (J.R. Durig, ed.), Vol. 6, Elsevier, Amsterdam, **1977**.

79Br Brown, J.M., Sears, T.J.: J. Mol. Spectrosc. **75** (1979) 111.

4.2.1 Linear polyatomic radicals

4.2.1.1 Preliminary remarks

1. Introduction

The rotational spectra of linear polyatomic molecules are very similar to those of diatomic molecules in the same electronic state. The reader is therefore referred to Sections 4.1.1, 4.1.2 and 4.1.5 for additional information. Data are included here for molecules in $^2\Sigma$ and $^2\Pi$ states only. In the former case, the description of the energy levels is identical to that for the corresponding diatomic molecule (except that there are more vibrational modes). In the latter case, the situation is made more complicated by the interaction between electron orbital and vibrational angular momenta, known as the Renner-Teller effect [34Re, 66He]. For a triatomic molecule, the effect involves the bending vibration ν_2 and the coupling leads to a pattern of vibronic energy levels which have been described in many other places e.g. [66He] and are labelled by the vibronic quantum number $K(=\Lambda + l)$ rather than the individual orbital and vibrational quantum numbers. There have been significant developments in the theoretical description of the vibronic energy levels over the past few years [77Br, 80Ju] but the details of this work falls outside the scope of the present compilation.

The data reported for the molecules in both the $^2\Sigma$ and $^2\Pi$ states have been analysed in terms of an effective Hamiltonian that refers to the rotational, spin and hyperfine levels of a particular vibronic state. The Hamiltonian is formulated in terms of the various angular momenta involved, namely N, L, S, G, J, I and F which are respectively the rotational, orbital, electron spin, vibrational, nuclear plus electronic, nuclear spin and total angular momenta (strictly speaking, $\boldsymbol{N}=\boldsymbol{R}+\boldsymbol{L}$ where $\boldsymbol{R}$ is the angular momentum of the nuclear framework). The effective Hamiltonian can be written

$$\mathcal{H}_{\text{eff}}=\mathcal{H}_{\text{rot}}+\mathcal{H}_{\text{so}}+\mathcal{H}_{\text{sr}}+\mathcal{H}_{\text{cd}}+\mathcal{H}_{\text{LD}}+\mathcal{H}_{\text{hfs}}+\mathcal{H}_{\text{Q}}.$$

The terms on the right hand side refer to the rotational kinetic energy, the spin-orbit interaction, the spin-rotation interaction, centrifugal distortion effects, lambda-type doubling, the magnetic nuclear hyperfine interactions and the electric quadrupole coupling term. They have been described in detail elsewhere (see Section 4.1.5 and refs. [62Ho], [78Br], [79Br1]). For molecules in Π electronic states, subject to the Renner-Teller effect, special considerations may be required in treating the various terms in $\mathcal{H}_{\text{eff}}$. Hougen has discussed the rotational Hamiltonian, $\mathcal{H}_{\text{rot}}$, from this point of view [62Ho] and Russell and Beaudet have considered the magnetic hyperfine interactions [74Ru]. A general formulation of the lambda doubling terms has been given by Brown and Merer [79Br2]:

$$\mathcal{H}_{\text{LD}}=\tfrac{1}{2}o_v(S_+^2+S_-^2)-\tfrac{1}{2}p_v(N_+S_++N_-S_-)+\tfrac{1}{2}q_v(N_+^2+N_-^2)$$

where o_v, p_v and q_v are the lambda-doubling parameters for the vibrational level v, $S_\pm=S_x\pm\mathrm{i}S_y$ and $N_\pm=N_x\pm\mathrm{i}N_y$ and the operators are defined on the implicit understanding that they link the $\Lambda=1$ and $\Lambda=-1$ components of the Π state only. The parameters can be related in turn to the electronic properties of the molecule.

The nuclear spin magnetic hyperfine interactions are represented by

$$\mathcal{H}_{\text{hfs}}=aI_zL_z+b_F\boldsymbol{I}\cdot\boldsymbol{S}+\tfrac{1}{3}c(3I_zS_z-\boldsymbol{I}\cdot\boldsymbol{S})-\tfrac{1}{2}d(S_+I_++S_-I_-)$$

where a, b_F, c and d are the four hyperfine parameters [52Fr, 78Br] and the angular momentum operators are as defined above. The four terms describe the nuclear spin orbit, the Fermi contact, the dipolar and the lambda doubling (dipolar) interactions respectively. Frosch and Foley [52Fr] originally formulated the Hamiltonian in terms of a slightly different parameter b, dependent on both the Fermi contact and dipolar interactions:

$$b=b_F-\tfrac{1}{3}c.$$

In many cases, the parameter pair b and c are better determined by the data than b_F and c even though the latter might be considered to have more physical significance. For molecules in $^2\Sigma$ states, the magnetic nuclear hyperfine effects can be described in terms of two parameters only, b_F and c (or b and c, if preferred).

2. List of tabulated parameters

B	rotational constant for the molecule in a particular vibrational level. The subscript v has been omitted in the tables because the vibrational or vibronic states are indicated separately
D	centrifugal distortion constant for a given vibrational level
A	spin orbit coupling constant for a given vibrational level
γ	spin rotation coupling constant for a given vibrational level
o, p, Q	lambda-type doubling parameters
a, b, c, d	nuclear spin magnetic hyperfine parameters
eq_0Q, eq_2Q	nuclear electric quadrupole coupling constants ($\Delta\Lambda=0$ and ±2 terms respectively), identical to $eq^{(a)}Q$ and $eq^{(b)}Q$ in 4.1.5.1
μ	electric dipole moment
ω_2	harmonic frequency for the bending vibration (ν_2)
ε	Renner-Teller coupling parameter

3. List of symbols used

$\tilde{X}\,^2\Sigma$, $\tilde{X}\,^2\Pi$	designation of $^2\Sigma$ or $^2\Pi$ electronic ground states
$^2\Pi$, $^2\Delta$	designation of vibronic states with $\lvert K\rvert=1$ or 2
v_1, v_2, v_3	vibrational quantum numbers for the three normal modes of a linear triatomic molecule. Mode number 2 is always associated with the bending vibration
Λ	quantum number associated with component of orbital angular momentum $\boldsymbol{L}$ along internuclear axis
l, Σ	corresponding quantum numbers for the components of the vibrational ($\boldsymbol{G}$) and spin ($\boldsymbol{S}$) angular momenta along nuclear axis
K	vibronic quantum number, valid in the presence of a strong Renner-Teller effect. Defined $K=\Lambda+l$
e, f	designation of parity. States with parity $(-1)^{J-\frac{1}{2}}$ are labelled e, those with parity $-(-1)^{J-\frac{1}{2}}$ are labelled f

4. Arrangement of substances

1. C_2H
2. C_3N
3. C_4H
4. NCO
5. BO_2

5. References

34Re Renner, R.: Z. Physik **92** (1934) 172.
52Fr Frosch, R.A., Foley, M.M.: Phys. Rev. **88** (1952) 1337.
62Ho Hougen, J.T.: J. Chem. Phys. **36** (1962) 519.
66He Herzberg, G.: "Molecular Spectra and Molecular Structure III. Electronic Spectra and Electronic Structure of Polyatomic Molecules", Van Nostrand Reinhold Company, New York, **1966**.
74Ru Russell, D.K., Beaudet, R.A.: Mol. Phys. **27** (1974) 1645.
75Br Brown, J.M., Hougen, J.T., Huber, K.-P., Johns, J.W.C., Kopp, I., LeFebvre-Brion, H., Merer, A.J., Ramsay, D.A., Rostas, J., Zare, R.N.: J. Mol. Spectrosc. **55** (1975) 500.
77Br Brown, J.M.: J. Mol. Spectrosc. **68** (1977) 412.
78Br Brown, J.M., Kopp, I., Malmberg, C., Rydh, B.: Phys. Scri. **17** (1978) 55.
79Br1 Brown, J.M., Colbourn, E.A., Watson, J.K.G., Wayne, F.D.: J. Mol. Spectrosc. **74** (1979) 294.
79Br2 Brown, J.M., Merer, A.J.: J. Mol. Spectrosc. **74** (1979) 488.
80Ju Jungen, Ch., Merer, A.J.: Mol. Phys., **40** (1980) 1 (and following papers).

4.2.1.2 Data and references

1 C_2H

Microwave data for $^{12}C_2{}^1H$

Transition			ν MHz	Ref.
rotational $N'-N''$	fine $J'-J''$	hyperfine $F'-F''$		
State: electronic X $^2\Sigma^+$; vibrational (0, 0, 0)				
1−0	1½− ½	2−1	87317.23 (4) [a]	80Wo
1−0	1½− ½	1−0	87328.70 (7)	74Tu
1−0	½− ½	1−1	87402.10 (7)	74Tu
1−0	½− ½	0−1	87407.23 (15)	74Tu
3−2	3½−2½	4−3	262004.52 (6)	81Zi
3−2	3½−2½	3−2	262006.76 (5)	81Zi
3−2	2½−1½	3−2	262065.12 (4)	81Zi
3−2	2½−1½	2−1	262067.65 (5)	81Zi
3−2	2½−1½	2−2	262080 (1)	81Zi
7−6			611334 [b]	78Sa

[a]) Author's estimate of experimental uncertainty, in units of the last quoted decimal place.
[b]) Observation by LMR spectroscopy (490 μm line of CD_3I).

Molecular parameters for $^{12}C_2{}^1H$

Parameter		Value	Method	Ref.
State: electronic X $^2\Sigma^+$; vibrational (0, 0, 0)				
B	[MHz]	43674.70 (10)	Radio-Astronomy	81Zi
D	[MHz]	0.1137 (6)		81Zi
γ	[MHz]	−62.62 (2)		81Zi
b	[MHz]	40.39 (14)		81Zi
c	[MHz]	12.32 (9)		81Zi

References for C_2H

74Tu Tucker, K.D., Kutner, M.L., Thaddeus, P.: Astrophys. J. **193** (1974) L115.

78Sa Saykally, R.J., Evenson, K.M.: 34th Symposium on Molecular Spectroscopy, Columbus, Ohio, Paper FB-8.

80Wo Wootten, A., Bozyan, E.P., Garrett, D.B., Loren, R.B., Snell, R.L.: Astrophys. J. **239** (1980) 844.

81Zi Ziurys, L.M., Saykally, R.J., Plambeck, R.L., Erickson, N.R.: Astrophys. J. **254** (1982) 94.

2 C_3N

Microwave data for $^{12}C_3{}^{14}N$

Transition			ν	Ref.
rotational $N'-N''$	fine $J'-J''$	hyperfine $F'-F''$	MHz	
State: electronic $\tilde{X}$ $^2\Sigma^+$; vibrational (0, 0, 0)				
1− 0	1½− ½	2½−1½	9885.89 (1)	82Gu
2− 1	2½− 1½	2½−1½	19780.800 (2.5)	82Gu
	2½− 1½	1½− ½	19780.826 (4)	82Gu
	2½− 1½	3½−2½	19781.094 (2.5)	82Gu
	1½− ½	1½− ½	19799.951 (5)	82Gu
	1½− ½	2½−1½	19800.121 (2.5)	82Gu
3− 2	3½− 2½	2½−1½	29676.14 (2)	82Gu
	3½− 2½	3½−2½	29676.14 (2)	82Gu
	3½− 2½	4½−3½	29676.28 (2)	82Gu
	2½− 1½	1½− ½	29694.99 (2)	82Gu
	2½− 1½	2½−1½	29695.13 (2)	82Gu
	2½− 1½	3½−2½	29695.13 (2)	82Gu
9− 8	8½− 7½ [a])	[b])	89045.7 (3)	77Gu
	9½− 8½	[b])	89064.4 (3)	77Gu
10− 9	9½− 8½	[b])	98939.9 (5)	77Gu
	10½− 9½	[b])	98958.6 (5)	77Gu
11−10	10½− 9½	[b])	108833.8	78Gu
	11½−10½	[b])	108852.5	78Gu

[a]) This assignment is based on the assumption that the spin-rotation parameter is positive.

[b]) Hyperfine structure not resolved.

Molecular parameters for $^{12}C_3{}^{14}N$

Parameter		Value	Method	Ref.
State: electronic $\tilde{X}\,^2\Sigma^+$; vibrational (0, 0, 0)				
B	[MHz]	4947.6190 (18)[a])	Radio-Astronomy	82Gu
D	[MHz]	0.68 (12) · 10^{-2}		
γ	[MHz]	−18.743 (8)		
b	[MHz]	−2.16 (10)		
c	[MHz]	2.85 (20)		
eQq_0	[MHz]	−4.34 (25)		

[a]) The figure in parentheses is the authors' estimate of experimental uncertainty.

References for C_3N

77Gu Guélin, M., Thaddeus, P.: Astrophys. J. **212** (1977) L81.
78Gu Guélin, M., Green, S., Thaddeus, P.: Astrophys. J. **224** (1978) L 27.
82Gu Guélin, M., Friberg, P., Mezaoui, A.: Astron. Astrophys. **109** (1982) 23.

3 C_4H

Microwave data for $^{12}C_4{}^1H$

Transition			ν MHz	Ref.
rotational $N'-N''$	fine $J'-J''$	hyperfine $F'-F''$		
State: electronic $\tilde{X}\,^2\Sigma^+$; vibrational (0, 0, 0)				
1− 0	½− ½	1−1	9563.0678 (150)	82Be
	½− ½	1−0	9547.1264 (150)	82Be
	1½− ½	1−1	9508.0113 (150)	82Be
	1½− ½	2−1	9497.6106 (150)	82Be
	1½− ½	1−0	9493.0643 (150)	82Be
2− 1	2½− 1½	2−1	19014.7204 (15)	82Gu
	2½− 1½	3−2	19015.1435 (15)	82Gu
	2½− 1½	2−2	19025.107 (4)	82Gu
	1½− ½	1−1	19044.760 (5)	82Gu
	1½− ½	2−1	19054.4762 (15)	82Gu
	1½− ½	1−0	19055.9468 (15)	82Gu
	1½− 1½	1−1	19099.656 (6)	82Gu
3− 2	3½− 2½	3−2	28532.31 (1)	82Gu
	3½− 2½	4−3	28532.46 (1)	82Gu
	2½− 1½	3−2	28571.37 (1)	82Gu
	2½− 1½	2−1	28571.53 (2)	82Gu
9− 8	8½− 7½	[a])	85672.4 (10)	78Gu
	9½− 8½	[a])	85633.9 (10)	78Gu
10− 9	9½− 8½	[a])	95189.0 (10)	78Gu
	10½− 9½	[a])	95149.5 (10)	78Gu
11−10	10½− 9½	[a])	104706.0	78Li
	11½−10½	[a])	104667.3	78Li
12−11	11½−10½	[a])	114221	78Sc
	12½−11½	[a])	114182	78Sc

[a]) Hyperfine components not resolved.

Molecular parameters for $^{12}C_4H$

Parameter		Value	Method	Ref.
State: electronic $\tilde{X}\ ^2\Sigma^+$; vibrational (0, 0, 0)				
B	[MHz]	4758.6567 (14)[a])	Radio-Astronomy	78Gu, 82Be, 82Gu
D	[MHz]	0.90 (5)		
γ	[MHz]	−38.554 (7)		
b_F	[MHz]	−19.09 (3)		
c	[MHz]	12.43 (4)		

[a]) The figures in parentheses are 1 standard deviation of the least-squares fit, in units of the last quoted decimal place.

References for C_4H

78Gu Guélin, M., Green, S., Thaddeus, P.: Astrophys. J. **224** (1978) L 27.
78Li Liszt, H.S.: Astrophys. J. **219** (1978) 454.
78Sc Scoville, N.Z., Solomon, P.M.: Astrophys. J. **220** (1978) L 103.
82Be Bell, M.B., Sears, T.J., Mathews, H.E.: Astrophys. J., **255** (1982) L 75.
82Gu Guélin, M., Friberg, P., Mezaoui, A.: Astron. Astrophys. **109** (1982) 23.

4 NCO

Microwave data for $^{14}N^{12}C^{16}O$

Transition			ν	Ref.
rotational $J'-J''$	parity	hyperfine $F'-F''$	MHz	
State: electronic $\tilde{X}\ ^2\Pi$; vibrational (0, 0, 0); vibronic $^2\Pi_{3/2}$				
2½−1½	[a])	3½−2½	58140.97 (4) [b])	81Sa
2½−1½	[a])	2½−1½	58161.49 (6)	81Sa
2½−1½	[a])	1½− ½	58171.71 (5)	81Sa
2½−1½	[a])	2½−2½	58108.44 (4)	81Sa
2½−1½	[a])	1½−1½	58138.64 (4)	81Sa
3½−2½	[a])	4½−3½	81404.09 (7)	70Sa
3½−2½	[a])	3½−2½	81413.15 (28)	70Sa
3½−2½	[a])	2½−1½	81418.46 (23)	70Sa
3½−2½	[a])	3½−3½	81380.88	72Am
3½−2½	[a])	2½−2½	81394.96 (21)	81Sa
4½−3½	[a])	5½−4½	104665.36 (6)	81Sa
4½−3½	[a])	4½−3½	104670.27 (6)	81Sa
4½−3½	[a])	3½−2½	104673.61 (10)	81Sa
4½−3½	[a])	4½−4½	104646.44 (6)	81Sa
4½−3½	[a])	3½−3½	104655.96 (7)	81Sa
5½−4½	−←+	6½−5½	127925.05 (2)	81Sa
5½−4½	−←+	5½−4½	127928.08 (4)	81Sa
5½−4½	−←+	4½−3½	127930.23 (3)	81Sa
5½−4½	+←−	6½−5½	127926.16 (2)	81Sa
5½−4½	+←−	5½−4½	127929.25 (3)	81Sa
5½−4½	+←−	4½−3½	127931.39 (2)	81Sa

[a]) Lambda-type doubling not resolved in these transitions.
[b]) The figures in parentheses are estimates of the experimental uncertainty, in units of the last quoted decimal place.

Microwave data for $^{14}N^{12}C^{16}O$ (continued)

Transition			ν MHz	Ref.
rotational $J'-J''$	parity	hyperfine $F'-F''$		
State: electronic $\tilde{X}\ ^2\Pi$; vibrational (0, 0, 0); vibronic $^2\Pi_{1/2}$				
1½ − ½	− ← +	2½ − 1½	35120.68 (2) [b]	81Sa
1½ − ½	− ← +	1½ − ½	35137.76 (5)	81Sa
1½ − ½	− ← +	1½ − 1½	35155.98 (8)	81Sa
1½ − ½	− ← +	½ − ½	35160.32 (4)	81Sa
1½ − ½	+ ← −	2½ − 1½	35208.33 (5)	81Sa
1½ − ½	+ ← −	1½ − ½	35284.76 (6)	81Sa
1½ − ½	+ ← −	1½ − 1½	35125.76 (3)	81Sa
1½ − ½	+ ← −	½ − ½	35236.20 (4)	81Sa
2½ − 1½	+ ← −	3½ − 2½	58578.53 (3)	81Sa
2½ − 1½	+ ← −	2½ − 1½	58581.98 (2)	81Sa
2½ − 1½	+ ← −	1½ − ½	58588.60 (4)	81Sa
2½ − 1½	+ ← −	2½ − 2½	58617.30 (10)	72Am
2½ − 1½	+ ← −	1½ − 1½	58611.15 (10)	72Am
2½ − 1½	− ← +	3½ − 2½	58659.18 (2)	81Sa
2½ − 1½	− ← +	2½ − 1½	58674.52 (2)	81Sa
2½ − 1½	− ← +	1½ − ½	58676.13 (10)	81Sa
2½ − 1½	− ← +	2½ − 2½	58591.98 (5)	81Sa
2½ − 1½	− ← +	1½ − 1½	58627.54 (10)	72Am
3½ − 2½	− ← +	4½ − 3½	82030.39 (2)	81Sa
3½ − 2½	− ← +	3½ − 2½	82031.92 (6)	81Sa
3½ − 2½	− ← +	2½ − 1½	82035.61 (6)	81Sa
3½ − 2½	− ← +	3½ − 3½	82070.51 (8)	81Sa
3½ − 2½	− ← +	2½ − 2½	82064.56 (5)	81Sa
3½ − 2½	+ ← −	4½ − 3½	82108.39 (8)	81Sa
3½ − 2½	+ ← −	3½ − 2½	82115.10 (6)	81Sa
3½ − 2½	+ ← −	2½ − 1½	82115.94 (6)	81Sa
3½ − 2½	+ ← −	3½ − 3½	82048.06 (10)	81Sa
3½ − 2½	+ ← −	2½ − 2½	82069.30 (6)	81Sa
4½ − 3½	+ ← −	5½ − 4½	105480.07 (7)	81Sa
4½ − 3½	+ ← −	4½ − 3½	105480.97 (8)	81Sa
4½ − 3½	+ ← −	3½ − 2½	105483.36 (10)	81Sa
4½ − 3½	+ ← −	4½ − 4½	105552.13 (9)	81Sa
4½ − 3½	+ ← −	3½ − 3½	105516.22 (8)	81Sa
4½ − 3½	− ← +	5½ − 4½	105556.95 (10)	81Sa
4½ − 3½	− ← +	4½ − 3½	105561.09 (20) [c]	81Sa
4½ − 3½	− ← +	3½ − 2½	105561.09 (20) [c]	81Sa
4½ − 3½	− ← +	4½ − 4½	105500.14 (3)	81Sa
4½ − 3½	− ← +	3½ − 3½	105515.44 (5)	81Sa

[b]) The figures in parentheses are estimates of the experimental uncertainty, in units of the last quoted decimal place.

[c]) Hyperfine components not fully resolved.

Molecular parameters for $^{14}N^{12}C^{16}O$

Parameter		Value	Method	Ref.
State: electronic $\tilde{X}\ ^2\Pi$; vibrational (0, 0, 0); vibronic $^2\Pi$				
B	[MHz]	11677.317 (8) [a]	MW, Opt	81Sa
D	[MHz]	$0.386 (19) \cdot 10^{-2}$		81Sa
A	[MHz]	−2865570 [b]	Opt	75Bo
$\tilde{\gamma}$	[MHz] [c]	−34.0 (92)		81Sa
p	[MHz]	75.966 (39)		81Sa
q	[MHz]	−1.728 (72)		81Sa
a	[MHz]	61.847 (80)		81Sa
$(b+c)$	[MHz]	−17.20 (16)		81Sa
b	[MHz]	33.3 (27)		81Sa
d	[MHz]	88.462 (84)		81Sa
eq_0Q	[MHz]	−2.23 (11)		81Sa
eq_2Q	[MHz]	16.9 (79)		81Sa
State: electronic $\tilde{X}\ ^2\Pi$; vibrational (0, 0, 0); vibronic $^2\Delta$				
B	[MHz]	11708.305 (10)	MW	75Bo
D	[MHz]	$0.579 (60) \cdot 10^{-2}$	Opt	75Bo
A	[MHz]	−2823775 (150)	Opt	75Bo
$\tilde{\gamma}$	[MHz] [c]	−79 (3)	MW	75Bo

[a]) The figures in parentheses are 1 standard deviation of the least-squares fit, in units of the last quoted decimal place.
[b]) Parameter constrained to the value from the optical spectrum in the fit.
[c]) Effective parameter, the parameter A_D being constrained to zero in the fit.

References for NCO

70Sa Saito, S., Amano, T.: J. Mol. Spectrosc. **34** (1970) 383.
72Am Amano, T., Hirota, E.: J. Chem. Phys. **57** (1972) 5608.
75Bo Bolman, P.S.H., Brown, J.M., Carrington, A., Kopp, I., Ramsay, D.A.: Proc. Roy. Soc. London A **343** (1975) 17.
81Sa Saito, S., Brown, J.M.: unpublished results **1981**.

5 BO_2

Molecular parameters for $^{10}BO_2$

Parameter		Value	Method	Ref.
State: electronic $\tilde{X}\ ^2\Pi_g$; vibrational (0, 0, 0)				
B	[MHz]	9874.35 (84) [a]	Diode Laser	81Ka
D	[MHz]	$0.426 (36) \cdot 10^{-2}$	IR	81Ka
A	[MHz]	−4454300	Opt	61Jo, 77Ru
A_D	[MHz]	−0.161 (30)	IR	81Ka
p	[MHz]	195 (39)	IR	81Ka
$\varepsilon\omega_2$	[GHz]	−2569 (12)	Opt	77Di

[a]) The figures in parentheses are 3 standard deviations of the least squares fit, in units of the last quoted decimal place.

Molecular parameters for $^{11}BO_2$

Parameter		Value	Method	Ref.
State: electronic $\tilde{X}\ ^2\Pi_g$; vibrational (0, 0, 0)				
B	[MHz]	9874.65 (45) [a])	Diode Laser IR	81Ka
D	[MHz]	$0.406\,(18)\cdot 10^{-2}$	IR	81Ka
A	[MHz]	−4454300	Opt	61Jo, 77Ru
A_D	[MHz]	−0.411 (190)	IR	81Ka
p	[MHz]	156 (12)	IR	81Ka
$\varepsilon\omega_2$	[GHz]	−2569 (12)	Opt	81Ka
a	[MHz]	0.5	Opt	79Lo
b	[MHz]	−71.8	Opt	79Lo
c	[MHz]	0.1	Opt	79Lo
d	[MHz]	−0.2	Opt	79Lo

[a]) The figures in parentheses are 3 standard deviations of the least squares fit, in units of the last quoted decimal place.

References for BO_2

61Jo Johns, J.W.C.: Can. J. Phys. **39** (1961) 1738.
77Di Dixon, R.N., Field, D., Noble, M.: Chem. Phys. Lett. **50** (1977) 1.
77Ru Russell, D.K., Kroll, M., Beaudet, R.A.: J. Chem. Phys. **66** (1977) 1999.
79Lo Lowe, R.S., Gerhardt, M., Dillenschneider, W., Curl, R.F., Tittel, F.K.: J. Chem. Phys. **70** (1979) 42.
81Ka Kawaguchi, K., Hirota, E., Yamada, C.: Mol. Phys., **44** (1981) 509.

4.2.2 Non-linear triatomic radicals

4.2.2.1 Preliminary remarks

1. Introduction

The Hamiltonian used to describe the microwave spectra of non-linear molecules in open-shell electronic states can be written as the sum of several separate terms:

$$\mathscr{H} = \mathscr{H}_{\rm rot} + \mathscr{H}_{\rm cd} + \mathscr{H}_{\rm ss} + \mathscr{H}_{\rm sr} + \mathscr{H}_{\rm srcd} + \mathscr{H}_{\rm hfs} + \mathscr{H}_{\rm Q}.$$

The individual terms in this effective Hamiltonian are taken to have the following form:

$\mathscr{H}_{\rm rot}$ *Rigid asymmetric rotor Hamiltonian*

The rigid rotor Hamiltonian for a molecule in vibrational level v has the form

$$\mathscr{H}_{\rm rot} = A_v N_a^2 + B_v N_b^2 + C_v N_c^2$$

where A_v, B_v and C_v are the rotational "constants" and N_a, N_b and N_c are the components of the rotational angular momentum $\boldsymbol{N}$, referred to the principal inertial axes [51Va]. The rotational constants are closely related to the principal moments of inertia

$$A \simeq \hbar^2/2hI_a, \quad B \simeq \hbar^2/2hI_b, \quad C \simeq \hbar^2/2hI_c.$$

The exact relationship depends on the form adopted for the corrections to the rotational energy levels for the effects of centrifugal distortion [67Wa, 77Wa2].

$\mathscr{H}_{\rm cd}$ *Centrifugal distortion effects*

Watson [67Wa] has shown that the number of centrifugal distortion parameters that can be determined from experimental data is less than the number of independent parameters in the fundamental Hamiltonian first formulated by Kivelson and Wilson [52Ki, 53Ki]. It is therefore necessary to reduce the Hamiltonian so that it contains only determinable combinations of parameters (otherwise indeterminacies manifest themselves in the least-squares fitting of experimental data). Separate reductions have to be performed at the levels of the quartic, sextic, etc. terms. Two different reductions have become generally accepted, called the symmetric (S) and asymmetric (A) reductions by Watson. The former is particularly appropriate to molecules with inertial properties near either the oblate or prolate symmetric top limits whereas the latter is better suited to describe the energy levels of strongly asymmetric rotors. The A-reduced Hamiltonian has the advantage that its matrix representation is simpler but its parameters do not correlate smoothly with those in the symmetric top limit.

The A-reduced form for the quartic and sextic centrifugal distortion terms is

$$\begin{aligned}\mathcal{H}_{cd}^{A} = & -\Delta_N (N^2)^2 - \Delta_{NK} N^2 N_z^2 - \Delta_K N_z^4 - \tfrac{1}{2}[\delta_N N^2 + \delta_K N_z^2, (N_+^2 + N_-^2)]_+ \\ & + \Phi_N (N^2)^3 + \Phi_{NK} (N^2)^2 N_z^2 + \Phi_{KN} N^2 N_z^4 + \Phi_K N_z^6 \\ & + \tfrac{1}{2}[\phi_N (N^2)^2 + \phi_{NK} N^2 N_z^2 + \phi_K N_z^4, (N_+^2 + N_-^2)]_+\end{aligned}$$

in a $\mathrm{I^r}$ representation, $a \leftrightarrow z$, $b \leftrightarrow x$, $c \leftrightarrow y$ [63Al]. The symbol $[\mathrm{A}, \mathrm{B}]_+$ stands for the anti-commutator (AB + BA) and the operator $N_\pm$ is defined by

$$N_\pm = (N_x \pm \mathrm{i}\, N_y).$$

Following Watson [77Wa2], the sextic coefficients have been denoted $\Phi_N, \ldots, \phi_K$ rather than the previous $H_N, \ldots, h_K$ [68Wa] because it seems preferable to reserve the latter for the corresponding terms of the symmetric top reduced Hamiltonian.

The symmetric top or S-reduced Hamiltonian for quartic and sextic centrifugal distortion terms has the form

$$\begin{aligned}\mathcal{H}_{cd} = & -D_N (N^2)^2 - D_{NK} N^2 N_z^2 - D_K N_z^4 + d_1 N^2 (N_+^2 + N_-^2) + d_2 (N_+^4 + N_-^4) \\ & + H_N (N^2)^3 + H_{NK} (N^2)^2 N_z^2 + H_{KN} N^2 N_z^4 + H_K N_z^6 \\ & + h_1 (N^2)^2 (N_+^2 + N_-^2) + h_2 N^2 (N_+^4 + N_-^4) + h_3 (N_+^6 + N_-^6),\end{aligned}$$

again in a $\mathrm{I^r}$ representation. Watson [77Wa1, 77Wa2] has given the relationships between the parameters in the A and S reduced Hamiltonians. It is thus possible to estimate the parameters for the alternative reduction from those determined in a particular fit to experimental data. Note that the values determined for the rotational constants also depend on the reduction employed.

In the fit of some data sets for light molecules which display large centrifugal distortion effects, it has been necessary to include the leading octic terms also. In the present compilation, these effects are described by

$$\mathcal{H}_{cd}^{(8)} = -L_K N_z^8 - \cdots.$$

$\mathcal{H}_{ss}$ *Electronic spin-spin (dipolar) interaction*

The effective spin dipolar Hamiltonian for asymmetric top molecules has been derived by Van Vleck [51Va]:

$$\begin{aligned}\mathcal{H}_{ss} = & \alpha (3 S_z^2 - S^2) + \beta (S_x^2 - S_y^2) + \gamma (S_x S_y + S_y S_x) \\ & + \delta (S_x S_z + S_z S_x) + \varepsilon (S_y S_z + S_z S_y).\end{aligned}$$

Symmetry arguments can be used to show that only the parameters α and β are non-zero for the symmetric triatomic molecule AB_2; only one of the three other parameters (δ for a $\mathrm{I^r}$ representation) is non-zero for the non-symmetric molecule ABC.

$\mathcal{H}_{sr}$ *Electronic spin-rotation interaction*

The effective spin-rotation Hamiltonian for asymmetric top molecules has been discussed by several authors [51Va, 61Cu, 79Br]. In general it takes the form

$$\mathcal{H}_{sr} = \tfrac{1}{2} \sum_{\alpha,\beta} \varepsilon_{\alpha\beta} (N_\alpha S_\beta + S_\beta N_\alpha)$$

where the indices α and β run separately over the molecule-fixed coordinates x, y, z and $\varepsilon_{\alpha\beta}$ are the spin-rotation parameters. There are in principle nine independent components of the spin rotation tensor but not all of them are determinable in practice. In the same way that the rotational and centrifugal distortion Hamiltonians can be reduced so that they are expressed in terms of determinable parameters only [67Wa], the spin-rotation Hamiltonian must be reduced by application of appropriate transformations in spin space. A solution to this problem has been given by Brown and Sears [79Br], who show that there are at the most six determinable parameters. If we restrict ourselves to triatomic molecules, the situation is even simpler. For an AB_2 molecule, with C_{2v} symmetry, the Hamiltonian is

$$\mathcal{H}_{sr} = \varepsilon_{aa} N_a S_a + \varepsilon_{bb} N_b S_b + \varepsilon_{cc} N_c S_c$$

where a, b and c are the principal axes of the inertial tensor (defined by symmetry). The ABC molecule has only one element of symmetry, the plane of reflection (C_s); the spin-rotation Hamiltonian is a little more complicated in this case. The preferred form [79Br] is:

$$\mathcal{H}_{sr} = \tilde{\varepsilon}_{aa} N_a S_a + \tilde{\varepsilon}_{bb} N_b S_b + \tilde{\varepsilon}_{cc} N_c S_c + \tfrac{1}{2} \tilde{\varepsilon}_{ab} ([N_a, S_b]_+ + [N_b, S_a]_+)$$

where $\tilde{\varepsilon}_{aa}$, $\tilde{\varepsilon}_{bb}$, etc. are reduced parameters, related to $\varepsilon_{\alpha\beta}$ as shown by Brown and Sears.

$\mathscr{H}_{\text{srcd}}$ *The centrifugal distortion corrections to the spin-rotation interaction.*

The spin-rotation interaction is modified as a molecule rotates faster by centrifugal distortion effects in much the same way as the rotational kinetic energy is altered. Such effects have been discussed by Dixon and Duxbury [67Di] and by Brown and Sears [77Br, 79Br]. In particular, the latter authors have considered the reduction of the quartic terms in the Hamiltonian so that only determinable combinations of parameters are retained. There are 6 quartic parameters for a C_{2v} molecule (AB_2) and 8 for a C_s molecule (ABC). Brown and Sears [79Br] give two possible reduced forms, constructed along the same lines as Watson's A- and S-reduced Hamiltonians for rotational motion.

The A-reduced form for C_{2v} molecules is

$$\mathscr{H}_{\text{srcd}} = \Delta_N^s N^2(\boldsymbol{N}\cdot\boldsymbol{S}) + \tfrac{1}{2}\Delta_{NK}^s[N^2, N_z S_z]_+ + \Delta_{KN}^s N_z^2(\boldsymbol{N}\cdot\boldsymbol{S}) + \Delta_K^s N_z^3 S_z + \delta_N^s(\boldsymbol{N}\cdot\boldsymbol{S})(N_+^2+N_-^2) + \tfrac{1}{2}\delta_K^s[N_z S_z, (N_+^2+N_-^2)]_+$$

where the coefficients have been defined by analogy with the corresponding coefficients in $\mathscr{H}_{\text{cd}}$ (note however the change of sign in this definition). The S-reduced form for C_{2v} molecules is

$$\mathscr{H}_{\text{srcd}} = D_N^s N^2(\boldsymbol{N}\cdot\boldsymbol{S}) + \tfrac{1}{2}D_{NK}^s[N^2, N_z S_z]_+ + D_{KN}^s N_z^2(\boldsymbol{N}\cdot\boldsymbol{S}) + D_K^s N_z^3 S_z + d_1^s(N_+^2+N_-^2)(\boldsymbol{N}\cdot\boldsymbol{S}) + d_2^s(N_+^3 S_+ + N_-^3 S_-).$$

Brown and Sears [79Br] have given the relationships between these parameters and the parameters $\eta_{\alpha\beta\gamma\delta}$ involved in the basic definition of centrifugal distortion corrections to the spin-rotation interaction.

$\mathscr{H}_{\text{hfs}}$ *Magnetic nuclear hyperfine interactions*

There are two contributions to the magnetic hyperfine interaction for each nucleus,

$$\mathscr{H}_{\text{hfs}} = \mathscr{H}_{\text{FC}} + \mathscr{H}_{\text{dip}}.$$

These are the Fermi contact and dipolar interactions respectively. The Fermi contact term describes the isotropic interaction

$$\mathscr{H}_{\text{FC}} = (O)_I\, \boldsymbol{I}\cdot\boldsymbol{S}$$

where $(O)_I = 2\mu_B g_N \mu_N/(3\mu_0^{-1})\sum_i g_i\langle\delta(\boldsymbol{r}_i)\rangle$,

g_N and g_i are the nuclear and electron spin g-factors, μ_N and μ_B are the nuclear and Bohr magnetons, μ_0 is the pearmeability of free space and $\langle\delta(\boldsymbol{r}_i)\rangle$ gives the electron density at the nucleus for electron i. The summation is performed over unpaired electrons only.

The dipolar interaction is anisotropic and can be represented in terms of the components of a cartesian tensor $(\alpha\beta)_I$, often written as $T_{\alpha\beta}$.

$$\mathscr{H}_{\text{dip}} = \sum_{\alpha,\beta}(\alpha\beta)_I S_\alpha I_\beta,$$

where α, β run separately over the cartesian coordinates a, b and c. The tensor components can, to a good approximation, be interpreted as the expectation values

$$(\alpha\beta)_I = -\frac{\mu_B g_N \mu_N}{4\pi\mu_0^{-1}}\sum_i g_i\langle(\delta_{\alpha\beta} - 3e_\alpha e_\beta)/r_i^3\rangle$$

where the summation again runs over open shell electrons only, $\delta_{\alpha\beta}$ is the Kronecker delta symbol, e_α, e_β are unit vectors along the principal axes α, β and r_i is the distance from the given nucleus to electron i. The operator inside the expectation value is often expressed in terms of spherical polar coordinates (r_i, θ_i, ϕ_i). For example, it becomes $\langle(1-3\cos^2\theta_i)/r_i^3\rangle$ for the component $(zz)_I$.

There is a pair of terms of the above form for each nucleus with non-zero spin present in the molecule. As before, symmetry can be used to reduce the number of non-zero components of the traceless tensor $(\alpha\beta)_I$. For a C_{2v} molecule (AB_2) with a single non-zero nucleus, there are 3 non-zero components $(aa)_I$, $(bb)_I$, $(cc)_I$ of which 2 are independent since $(aa)_I+(bb)_I+(cc)_I$ equals zero. For the planar ABC molecule, the additional components $(ab)_I$ and $(ba)_I$ are non-zero. In all analyses performed so far, it has been assumed that these components are equal. It is not clear that this is so and if not, what combinations of parameters are determinable. The $(ab)_I$ component is also non-zero for an AB_2 molecule with 2 equivalent, non-zero nuclei. If the nuclear spins are I_1 and I_2, the dipolar Hamiltonian is

$$\mathscr{H}_{\text{dip}} = (aa)_I I_a S_a + (bb)_I I_b S_b + (cc)_I I_c S_c + (ab)_I[(I_1-I_2)_a S_b + (I_1-I_2)_b S_a]$$

where I is the total nuclear spin, $\boldsymbol{I} = \boldsymbol{I}_1 + \boldsymbol{I}_2$. The additional term mixes ortho and para spin states. The nomenclature used here is due to Curl and Kinsey [61Cu].

$\mathscr{H}_Q$ *The nuclear electric quadrupole interaction*

The interaction of the nuclear electric quadrupole with the molecular electric field gradient can be described by the product of the nuclear quadrupole moment tensor $Q_{\alpha\beta}$ and the electric field gradient tensor $(\boldsymbol{\nabla E})_{\alpha\beta}$:

$$\mathscr{H}_Q = \mathrm{e}\,\boldsymbol{Q}:\boldsymbol{\nabla E}$$

where $-\mathrm{e}$ is the charge on the electron. The operator is often re-expressed in terms of the nuclear spin angular momentum operators by use of the Wigner-Eckart theorem:

$$\mathscr{H}_Q = \sum_{\alpha\beta} (\alpha\,\beta)_Q\, I_\alpha\, I_\beta$$

where α, β run over the principal coordinates of inertia and $(\alpha\,\beta)_Q$ are the quadrupole parameters [61Cu]. This form is in some ways more convenient for the evaluation of matrix elements. For C_{2v} molecules, it reduces to

$$\mathscr{H}_Q = (a\,a)_Q\, I_a^2 + (b\,b)_Q\, I_b^2 + (c\,c)_Q\, I_c^2$$

with $(a\,a)_Q + (b\,b)_Q + (c\,c)_Q = 0$. The parameters can be interpreted in terms of

$$(\alpha\,\beta)_Q = [\mathrm{e}\,Q/2\,I(2\,I-1)]\,\langle \partial^2\, V/\partial\alpha\,\partial\beta\rangle$$

where Q is the nuclear quadrupole moment.

Experimental data are fitted to the appropriate molecular Hamiltonian by numerical procedures which involve the determination of the eigenvalues and eigenfunctions of the Hamiltonian. The matrix elements required have been given in many places in the literature [61Cu, 71Go, 73Bo, 77Wa2, 79Br]. There are some mathematical and computational advantages in using spherical tensor techniques. A simple procedure for dealing with several nuclear spins has been described by Godfrey [71Go]. Three different types of angular momentum are involved ($\boldsymbol{N}, \boldsymbol{S}$ and $\boldsymbol{I}$) and they can interact through a variety of coupling schemes. However, in almost every case studied so far, the spin- rotation interaction is larger than the hyperfine effects so that natural scheme to use in the basis sets is

$$\boldsymbol{N}+\boldsymbol{S}=\boldsymbol{J}, \quad \boldsymbol{J}+\boldsymbol{I}=\boldsymbol{F}.$$

The quantum number J remains good enough that it can be unambiguously defined; it is used to label the fine structure transitions in the following tables. The situation becomes more complicated when two or more nuclear spins are involved. For the coupling of two nuclear spins to $\boldsymbol{J}$ it is usual to define an intermediate angular momentum $\boldsymbol{F}_1$:

$$\boldsymbol{J}+\boldsymbol{I}_1=\boldsymbol{F}_1, \quad \boldsymbol{F}_1+\boldsymbol{I}_2=\boldsymbol{F}.$$

Again, if the first spin is much more strongly coupled than the second, the quantum number F_1 remains good and can be used to label the levels involved in the transitions. However, it often happens that the two hyperfine interactions are of similar magnitude and F_1 is not well-defined. In this case, the fine and hyperfine substates of a given rotational level ($N_{K_aK_c}$) with the same value of F are identified by an energy ordering index i; the index increases with energy so that the lowest state has i=1. Such a definition is consistent with that adopted for the labelling electron spin fine structure states in molecules [66He].

In quoting the results of fits of experimental data to a Hamiltonian, it is important to specify the representation employed (usually $\mathrm{I^r}$ for a molecule nearer the prolate symmetric top limit and $\mathrm{III^r}$ or $\mathrm{III^l}$ for a molecule near the oblate limit) and the reduction(s) used in the centrifugal distortion and spin-rotation Hamiltonians (A or S in the present compilation).

2. List of tabulated parameters

A, B, C.	rotational constants for an asymmetric rotor in a given vibrational state; the subscripts v have been omitted in the tables since the vibrational state is shown separately
Δ_c	inertial defect with respect to the c axis, $I_{cc}-I_{aa}-I_{bb}$.
$\Delta_N, \Delta_{NK}, \Delta_K, \delta_N, \delta_K$	quartic centrifugal distortion parameters in an A-reduced Hamiltonian
$\Phi_N, \Phi_{NK}, \Phi_{KN}, \Phi_K, \phi_N, \phi_{NK}, \phi_K$	sextic centrifugal distortion parameters in an A-reduced Hamiltonian
$D_N, D_{NK}, D_K, d_1, d_2$	quartic centrifugal distortion parameters in an S-reduced Hamiltonian
$H_N, H_{NK}, H_{KN}, H_K, h_1, h_2, h_3$	sextic centrifugal distortion parameters in an S-reduced Hamiltonian
L_K	octic centrifugal distortion parameter
α, β	electronic spin-spin dipolar parameters

List of tabulated parameters (continued)

$\varepsilon_{aa}, \varepsilon_{bb}, \varepsilon_{cc}$	electron spin-rotation parameters for a C_{2v} molecule
$\tilde{\varepsilon}_{aa}, \tilde{\varepsilon}_{bb}, \tilde{\varepsilon}_{cc}, \tilde{\varepsilon}_{ab}$	reduced electron spin-rotation parameters for a non-symmetric triatomic molecule (C_s)
$\Delta^s_N, \Delta^s_{NK}, \Delta^s_{KN}, \Delta^s_K, \delta^s_N, \delta^s_K$	parameters for the quartic centrifugal distortion of the spin-rotation Hamiltonian using an A-reduction
$D^s_N, D^s_{NK}, D^s_{KN}, D^s_K, d^s_1, d^s_2$	parameters for the quartic centrifugal distortion of the spin-rotation Hamiltonian, using an S-reduction
$(O)_I$	isotropic nuclear magnetic hyperfine parameter
$(aa)_I, (bb)_I, (cc)_I, (ab)_I$	magnetic hyperfine dipolar interaction parameters, expressed in terms of the principal intertial axes
$(aa)_Q, (bb)_Q, (cc)_Q$	nuclear electric quadrupole coupling parameters
$g^{aa}_S, g^{bb}_S, g^{cc}_S$	principal components of the electron spin g-tensor
$g^{aa}_r, g^{bb}_r, g^{cc}_r$	principal components of the rotation g-tensor
$r_s(X-Y), r_o(X-Y), r_e(X-Y)$	substitution, effective ground state and equilibrium values of the X−Y internuclear distance
$\measuredangle_s(XYX), \measuredangle_o(XYX), \measuredangle_e(XYX)$	substitution, effective ground state and equilibrium values of the XYX bond angle
$\langle\delta(\boldsymbol{r})\rangle_I$	density of the unpaired electron at nucleus I
$\langle 1/r^3\rangle_I, \langle(1-3\cos^2\theta)/r^3\rangle_I$	expectation values of the odd electron position (r, θ) with respect to nucleus I

3. List of symbols used

$\tilde{X}\ ^2A_1, \tilde{X}\ ^2B_1$ etc.	designation of the electronic state involved. $\tilde{X}$ signifies the ground electronic state, $\tilde{A}$ the first excited state and so on
v_1, v_2, v_3	vibrational quantum numbers for the three normal modes of vibration. In most cases, v_2 refers to the bonding vibration
$'$, $''$	the upper state quantum numbers are indicated by a single prime ($'$), the lower state by a double prime ($''$)
N	quantum number associated with rotational angular momentum, $\boldsymbol{N}$
K_a	quantum number for rotational angular momentum about the a axis in the limit of a prolate symmetric top
K_c	quantum number for rotational angular momentum about the c axis in the limit of an oblate symmetric top
J	quantum number associated with the resultant angular momentum obtained by coupling the rotational and electron spin angular momenta, $\boldsymbol{J}=\boldsymbol{N}+\boldsymbol{S}$
F	quantum number associated with the total angular momentum F, inclusive of all nuclear spins
F_1	quantum number associated with an intermediate angular momentum, obtained by coupling the first of two nuclear spins to $\boldsymbol{J}(\boldsymbol{F}_1=\boldsymbol{J}+\boldsymbol{I}_1)$. This quantum number is well-defined if the interaction involving $\boldsymbol{I}_1$ is much larger than that involving $\boldsymbol{I}_2$
i	index of energy levels with the same value of F in a particular rotational level $N_{K_aK_c}$. The index labels the states in order of increasing energy (i = 1 is the lowest, then i = 2 and so on). This index is used when the intermediate quantum number J and/or F_1 are not well-defined, as is the case in NF_2 for example

4. Arrangement of molecules in the tables

1. NO_2
2. ClO_2
3. NF_2
4. NH_2, NHD
5. HCO, DCO
6. HO_2, DO_2
7. HSO, DSO
8. FSO
9. PH_2.
10. CH_2

For References, see next page.

5. References

51Va Van Vleck, J.H.: Rev. Mod. Phys. **23** (1951) 213.
52Ki Kivelson, D., Wilson, E.B., Jr.: J. Chem. Phys. **20** (1952) 1575.
53Ki Kivelson, D., Wilson, E.B., Jr.: J. Chem. Phys. **21** (1953) 1229.
61Cu Curl, R.F., Jr., Kinsey, J.L.: J. Chem. Phys. **35** (1961) 1758.
63Al Allen, H.C. and Cross, P.C.: "Molecular Vib. Rotors", Wiley, New York, **1963**.
66He Herzberg, G.: "Molecular Spectra and Molecular Structure III. Electronic Spectra and Electronic Structure of Polyatomic Molecules", Van Nostrand. Reinhold Company, New York, **1966**.
67Di Dixon, R.N., Duxbury, G.: Chem. Phys. Lett. **1** (1967) 330.
67Wa Watson, J.K.G.: J. Chem. Phys. **46** (1967) 1935.
68Wa Watson, J.K.G.: J. Chem. Phys. **48** (1968) 4517.
71Go Godfrey, P.D.: Ph.D. Thesis, Monash University, **1971**.
73Bo Bowater, I.C., Brown, J.M., Carrington, A.: Proc. Roy. Soc. London A **333** (1973) 265.
77Br Brown, J.M., Sears, T.J.: Mol. Phys. **34** (1977) 1595.
77Wa1 Watson, J.K.G.: J. Mol. Spectrosc. **65** (1977) 123.
77Wa2 Watson, J.K.G.: "Aspects of quartic and sextic centrifugal effects on rotational energy levels", in "Vibrational Spectra and Structure" (J.R. Durig, ed.), Vol. 6, Elsevier, Amsterdam, **1977**.
79Br Brown, J.M., Sears, T.J.: J. Mol. Spectrosc. **75** (1979) 111.

4.2.2.2 Data and references

1 NO_2

Microwave data for $^{14}N^{16}O_2$

Transition			ν	Ref.
rotational $N'_{K_aK_c}-N''_{K_aK_c}$	fine $J'-J''$	hyperfine $F'-F''$	MHz	
State: electronic $\tilde{X}\,^2A_1$; vibrational (0, 0, 0)				
$5_{15}-6_{06}$	4½–5½	3½–4½	69786.60 (5) [a])	74Ba
	4½–5½	4½–5½	69750.37 (5)	74Ba
	5½–6½	4½–5½	70646.19 (5)	74Ba
	4½–5½	5½–6½	69745.49 (5)	74Ba
	5½–6½	5½–6½	70654.05 (5)	74Ba
	5½–6½	6½–7½	70589.69 (5)	74Ba
	4½–6½	4½–5½	70186.46 (5)	74Ba
	5½–5½	4½–5½	70209.97 (5)	74Ba
	4½–6½	5½–6½	70057.28 (5)	74Ba
	5½–5½	5½–6½	70342.18 (5)	74Ba
	5½–4½	4½–4½	69704.8	74Ba
	5½–6½	5½–5½	70714.90 (5)	74Ba
	4½–5½	5½–5½	69682.08 (5)	74Ba
	5½–5½	5½–5½	70278.84 (5)	74Ba
	5½–6½	6½–6½	70718.67 (5)	74Ba
	5½–5½	6½–6½	70406.86 (5)	74Ba
$24_{1,23}-23_{2,22}$	24½–23½	25½–24½	26569.224 (20)	81Br
	24½–23½	24½–23½	26577.074 (20)	81Br
	24½–23½	23½–22½	26563.074 (20)	81Br
	23½–22½	24½–23½	26619.291 (20)	81Br
	23½–22½	23½–22½	26633.890 (20)	81Br
	23½–22½	22½–21½	26647.248 (20)	81Br
$21_{2,20}-22_{1,21}$	21½–22½	22½–23½	39246.961 (20)	81Br
	21½–22½	21½–22½	39192.090 (20)	81Br
	21½–22½	20½–21½	39235.755 (20)	81Br
	20½–21½	21½–22½	39141.533 (20)	81Br
	20½–21½	20½–21½	39097.379 (20)	81Br
	20½–21½	19½–20½	39066.485 (20)	81Br

[a]) Estimated experimental uncertainty, in units of the last quoted decimal place.

Microwave data for $^{14}N^{16}O_2$ (continued)

Transition rotational $N'_{K_aK_c}-N''_{K_aK_c}$	fine $J'-J''$	hyperfine $F'-F''$	ν MHz	Ref.
State: electronic $\tilde{X}\ ^2A_1$; vibrational (0, 0, 0) (continued)				
4_{22} [b])	4½ – 3½	5½ – 4½	4397.9 (5) [a])	81Br
	4½ – 3½	4½ – 4½	4307.9 (5)	81Br
	4½ – 3½	4½ – 3½	4230.3 (5)	81Br
	4½ – 3½	3½ – 3½	4154.4 (5)	81Br
	4½ – 3½	3½ – 2½	4091.6 (5)	81Br
4_{31} [b])	4½ – 3½	5½ – 4½	10196.5 (5)	81Br
	4½ – 3½	4½ – 3½	10037.3 (5)	81Br
	4½ – 3½	3½ – 2½	9908.1 (5)	81Br
8_{44} [b])	8½ – 7½	9½ – 8½	9043.2 (5)	81Br
	8½ – 7½	8½ – 7½	8883.7 (5)	81Br
	8½ – 7½	7½ – 6½	8739.7 (5)	81Br
10_{46} [b])	10½ – 9½	11½ – 10½	7127.6 (5)	81Br
	10½ – 9½	10½ – 9½	6968.2 (5)	81Br
	10½ – 9½	9½ – 8½	6819.5 (5)	81Br
$13_{4,10}$ [b])	13½ – 12½	14½ – 13½	5311.7 (5)	81Br
	13½ – 12½	13½ – 12½	5151.6 (5)	81Br
	13½ – 12½	12½ – 11½	5000.7 (5)	81Br
$15_{4,12}$ [b])	15½ – 14½	16½ – 15½	4474.5 (5)	81Br
	15½ – 14½	15½ – 14½	4315.1 (5)	81Br
	15½ – 14½	14½ – 13½	4161.3 (5)	81Br
$11_{5,7}$ [b])	11½ – 10½	12½ – 11½	10220.1 (5)	81Br
	11½ – 10½	11½ – 10½	10062.4 (5)	81Br
	11½ – 10½	10½ – 9½	9915.6 (5)	81Br
$20_{6,14}$ [b])	20½ – 19½	21½ – 20½	7631.0 (5)	81Br
	20½ – 19½	20½ – 19½	7468.5 (5)	81Br
	20½ – 19½	19½ – 18½	7320.3 (5)	81Br
$23_{6,18}$ [b])	23½ – 22½	24½ – 23½	6418.5 (5)	81Br
	23½ – 22½	23½ – 22½	6262.0 (5)	81Br
	23½ – 22½	22½ – 21½	6111.1 (5)	81Br
$18_{3,15}$ [b])	18½ – 17½	19½ – 18½	1724.6 (10)	80Al
	18½ – 17½	18½ – 17½	1568.7 (10)	80Al
	18½ – 17½	17½ – 16½	1410.0 (10)	80Al
$19_{2,18}$ [b])	19½ – 18½	20½ – 19½	319.6 (10)	80Al
	19½ – 18½	19½ – 18½	240.9 (10)	80Al
	19½ – 18½	18½ – 17½	281.1 (10)	80Al
	19½ – 18½	18½ – 18½	201.7 (10)	80Al
	19½ – 18½	17½ – 18½	157.3 (10)	80Al
$20_{4,16}$ [b])	20½ – 19½	20½ – 19½	2877.3 (10)	80Al
	20½ – 19½	19½ – 18½	2720.4 (10)	80Al
$33_{1,33}$ [b])	32½ – 33½	33½ – 34½	2196.5 (10)	80Al
	32½ – 33½	32½ – 33½	2344.8 (10)	80Al
	32½ – 33½	31½ – 32½	2479.0 (10)	80Al
$39_{4,36}$ [b])	39½ – 38½	40½ – 39½	479.4 (10)	80Al
	39½ – 38½	39½ – 38½	357.3 (10)	80Al
	39½ – 38½	38½ – 37½	174.7 (10)	80Al
	39½ – 38½	39½ – 39½	423.7 (10)	80Al
	39½ – 38½	38½ – 38½	291.2 (10)	80Al
	39½ – 39½	40½ – 39½	55.6 (10)	80Al

[a]) Estimated experimental uncertainty, in units of the last quoted decimal place.
[b]) Magnetic dipole transition between the two spin components of a given rotational level $N_{K_aK_c}$.

Microwave data for $^{14}N^{16}O_2$ (continued)

Transition rotational $N'_{K_aK_c}-N''_{K_aK_c}$	fine $J'-J''$	hyperfine $F'-F''$	ν MHz	Ref.
State: electronic $\tilde{X}\ ^2A_1$; vibrational (0, 0, 0) (continued)				
$39_{4,36}$ [b])	39½–39½	39½–38½	66.1 (10) [a]) [c])	80Al
	38½–38½	38½–39½	66.1 (10) [c])	80Al
$40_{6,34}$ [b])	40½–39½	41½–40½	2643.1 (10)	80Al
	40½–39½	40½–39½	2486.7 (10)	80Al
	40½–39½	39½–38½	2326.7 (10)	80Al
$44_{6,38}$ [b])	44½–43½	45½–44½	2100.4 (10) [d])	80Al
	44½–43½	44½–43½	1945.0 (10) [d])	80Al
	44½–43½	43½–42½	1783.2 (10) [d])	80Al
$1_{11}-0_{00}$	1½– ½	2½–1½	253338.12	82Bo
		1½– ½	253445.93	82Bo
		1½–1½	253225.03	82Bo
		½–1½	253155.07	82Bo
		½– ½	253375.94	82Bo
	½– ½	½–1½	249424.37	82Bo
		1½–1½	249388.60	82Bo
		½– ½	249645.23	82Bo
		1½– ½	249609.48	82Bo
$2_{11}-2_{02}$	2½–2½	3½–3½	229067.49	82Bo
		1½–1½	229089.17	82Bo
	1½–1½	2½–2½	226854.64	82Bo
		½– ½	226943.36	82Bo
	2½–2½	2½–3½	228961.71	82Bo
		1½–2½	229056.39	82Bo
		2½–1½	229172.14	82Bo
	1½–1½	1½–2½	226950.25	82Bo
		2½–1½	226814.28	82Bo
		1½– ½	226967.80	82Bo
	2½–1½	3½–2½	229014.91	82Bo
		2½–2½	228909.18	82Bo
		2½–1½	228868.86	82Bo
	1½–2½	2½–2½	227084.98	82Bo
		1½–1½	227213.18	82Bo
$4_{13}-4_{04}$	4½–4½	5½–5½	231229.96	82Bo
		4½–4½	231287.14	82Bo
		3½–3½	231252.98	82Bo
	3½–3½	4½–4½	229939.52	82Bo
		3½–3½	229977.32	82Bo
		2½–2½	230017.52	82Bo
$6_{15}-6_{06}$	6½–6½	7½–7½	235030.76	82Bo
		6½–6½	235078.04	82Bo
		5½–5½	235055.49	82Bo
	5½–5½	6½–6½	234039.71	82Bo
		5½–5½	234064.90	82Bo
		4½–4½	234108.49	82Bo

a) Estimated experimental uncertainty, un units of the last quoted decimal place.
b) Magnetic dipole transition between the two spin components of a given rotational level $N'_{K_aK_c}$.
c) Blended line, with multiple assignment.
d) Lines for NO_2 in (0, 0, 0) level blended with lines for NO_2 in (0, 0, 1) level, observation by microwave-IR double resonance [80Al].

Microwave data for $^{14}N^{16}O_2$ (continued)

Transition rotational $N'_{K_aK_c}-N''_{K_aK_c}$	fine $J'-J''$	hyperfine $F'-F''$	ν MHz	Ref.
State: electronic $\tilde{X}\ ^2A_1$; vibrational (0, 0, 0) (continued)				
$8_{17}-8_{08}$	8½−8½	9½−9½	240393.93 [a])	82Bo
		8½−8½	240437.04	82Bo
		7½−7½	240424.40	82Bo
	7½−7½	8½−8½	239529.51	82Bo
		7½−7½	239544.17	82Bo
		6½−6½	239594.52	82Bo
$10_{19}-10_{0,10}$	10½−10½	11½−11½	247355.37	82Bo
	9½−9½	10½−10½	246541.12	82Bo
		9½−9½	246543.06	82Bo
		8½−8½	246607.02	82Bo
$10_{28}-11_{1,11}$	10½−11½	11½−12½	427195.40	82Bo
		10½−11½	427227.99	82Bo
		9½−10½	427209.73	82Bo
	9½−10½	10½−11½	425343.38	82Bo
		9½−10½	425363.30	82Bo
		8½−9½	425397.92	82Bo
$11_{2,10}-12_{1,11}$	11½−12½	12½−13½	350699.56	82Bo
		11½−12½	350678.18	82Bo
		10½−11½	350652.30	82Bo
	10½−11½	11½−12½	349709.36	82Bo
		10½−11½	349733.68	82Bo
		9½−10½	349666.03	82Bo
$12_{1,11}-12_{0,12}$	12½−12½	13½−13½	255989.05	82Bo
		12½−12½	256037.73	82Bo
		11½−11½	256057.99	82Bo
	11½−11½	12½−12½	255181.35	82Bo
		11½−11½	255162.63	82Bo
		10½−10½	255252.98	82Bo
$14_{0,14}-13_{1,13}$	14½−13½	15½−14½	156172.35	82Bo
		14½−13½	156171.77	82Bo
		13½−12½	156163.73	82Bo
	13½−12½	14½−13½	156255.53	82Bo
		13½−12½	156264.65	82Bo
		12½−11½	156273.68	82Bo
$13_{2,12}-14_{1,13}$	13½−14½	14½−15½	290868.03	82Bo
		13½−14½	290834.91	82Bo
		12½−13½	290779.95	82Bo
	12½−13½	13½−14½	290137.16	82Bo
		12½−13½	290191.05	82Bo
		11½−12½	290082.44	82Bo
$14_{1,13}-14_{0,14}$	14½−14½	15½−15½	266391.99	82Bo
		14½−14½	266451.98	82Bo
		13½−13½	266312.77	82Bo
	13½−13½	14½−14½	265560.20	82Bo
		13½−13½	265700.70	82Bo
		12½−12½	265643.79	82Bo
$14_{2,12}-15_{1,15}$	14½−15½	15½−16½	347287.52	82Bo
		14½−15½	347308.17	82Bo
		13½−14½	347297.91	82Bo

[a]) Estimated experimental uncertainty, in units of the last quoted decimal place.

Microwave data for $^{14}N^{16}O_2$ (continued)

Transition			ν MHz	Ref.
rotational $N'_{K_aK_c}-N''_{K_aK_c}$	fine $J'-J''$	hyperfine $F'-F''$		
State: electronic $\tilde{X}\ ^2A_1$; vibrational (0, 0, 0) (continued)				
$14_{2,12}-15_{1,15}$	13½–14½	14½–15½	345710.01 [a])	82Bo
		13½–14½	345721.55	82Bo
		12½–13½	345754.84	82Bo
$16_{0,16}-15_{1,15}$	16½–15½	17½–16½	215246.55	82Bo
		16½–15½	215243.25	82Bo
		15½–14½	215236.47	82Bo
	15½–14½	16½–15½	215255.18	82Bo
		15½–14½	215262.83	82Bo
		14½–13½	215269.66	82Bo
$15_{2,14}-16_{1,15}$	15½–16½	16½–17½	229782.05	82Bo
		15½–16½	229730.49	82Bo
		14½–15½	229866.63	82Bo
	14½–15½	15½–16½	229263.98	82Bo
		14½–15½	229126.80	82Bo
		13½–14½	229191.10	82Bo
$16_{1,15}-16_{0,16}$	16½–16½	17½–17½	278677.88	82Bo
		16½–16½	278557.51	82Bo
		15½–15½	278621.91	82Bo
	15½–15½	16½–16½	277996.44	82Bo
		15½–15½	277933.26	82Bo
		14½–14½	277900.13	82Bo
$16_{2,14}-17_{1,17}$	16½–17½	17½–18½	310493.43	82Bo
		16½–17½	310515.09	82Bo
		15½–16½	310507.86	82Bo
	15½–16½	16½–17½	308960.86	82Bo
		15½–16½	308969.31	82Bo
		14½–15½	309007.80	82Bo
$18_{0,18}-17_{1,17}$	18½–17½	19½–18½	275010.50	82Bo
		18½–17½	275006.21	82Bo
		17½–16½	275000.03	82Bo
	17½–16½	18½–17½	274958.84	82Bo
		17½–16½	274965.72	82Bo
		16½–15½	274971.38	82Bo
$17_{2,16}-18_{1,17}$	17½–18½	18½–19½	167462.66	82Bo
		17½–18½	167384.62	82Bo
		16½–17½	167549.39	82Bo
	16½–17½	17½–18½	167129.84	82Bo
		16½–17½	166964.28	82Bo
		15½–16½	167031.00	82Bo
$18_{2,16}-19_{1,19}$	18½–19½	19½–20½	276052.51	82Bo
		18½–19½	276078.62	82Bo
		17½–18½	276077.64	82Bo
	17½–18½	18½–19½	274526.21	82Bo
		17½–18½	274528.17	82Bo
		16½–17½	274578.71	82Bo
$20_{0,20}-19_{1,19}$	20½–19½	21½–20½	335304.47	82Bo
		20½–19½	335299.90	82Bo
		19½–18½	335294.11	82Bo

[a]) Estimated experimental uncertainty, in units of the last quoted decimal place.

Microwave data for $^{14}N^{16}O_2$ (continued)

Transition rotational $N'_{K_aK_c}-N''_{K_aK_c}$	fine $J'-J''$	hyperfine $F'-F''$	ν MHz	Ref.
State: electronic $\tilde{X}\ ^2A_1$; vibrational (0, 0, 0) (continued)				
$20_{0,20}-19_{1,19}$	19½−18½	20½−19½	335204.83 [a]	82Bo
		19½−18½	335211.19	82Bo
		18½−17½	335216.08	82Bo
$20_{1,19}-20_{0,20}$	20½−20½	21½−21½	309411.88	82Bo
		20½−20½	309354.60	82Bo
		19½−19½	309373.79	82Bo
	19½−19½	20½−20½	308568.22	82Bo
		19½−19½	308550.15	82Bo
		18½−18½	308536.79	82Bo
$20_{2,18}-21_{1,21}$	20½−21½	21½−22½	244173.36	82Bo
		20½−21½	244209.09	82Bo
		19½−20½	244225.06	82Bo
	19½−20½	20½−21½	242620.40	82Bo
		19½−20½	242605.46	82Bo
		18½−19½	242683.86	82Bo
$22_{0,22}-21_{1,21}$	22½−21½	23½−22½	395959.12	82Bo
		21½−20½	395949.13	82Bo
	21½−20½	22½−21½	395822.29	82Bo
		21½−20½	395828.22	82Bo
		20½−19½	395832.59	82Bo
$22_{1,21}-22_{0,22}$	22½−22½	23½−23½	328131.07	82Bo
		22½−22½	328088.18	82Bo
		21½−21½	328096.68	82Bo
	21½−21½	22½−22½	327208.63	82Bo
		21½−21½	327201.10	82Bo
		20½−20½	327192.37	82Bo
$22_{2,20}-23_{1,23}$	22½−23½	23½−24½	215075.85	82Bo
		22½−23½	215131.38	82Bo
		21½−22½	214984.65	82Bo
	21½−22½	22½−23½	213463.76	82Bo
		21½−22½	213611.63	82Bo
		20½−21½	213548.24	82Bo
$24_{0,24}-23_{1,23}$	24½−23½	25½−24½	456800.84	82Bo
		24½−23½	456796.44	82Bo
		23½−22½	456791.39	82Bo
	23½−22½	24½−23½	456636.60	82Bo
		23½−22½	456642.04	82Bo
		22½−21½	456645.96	82Bo
$23_{3,21}-24_{2,22}$	23½−24½	23½−24½	509199.79	82Bo
		22½−23½	509146.58	82Bo
	22½−23½	21½−22½	508163.96	82Bo
$24_{1,23}-24_{0,24}$	24½−24½	25½−25½	349264.58	82Bo
		24½−24½	349230.34	82Bo
		23½−23½	349232.52	82Bo
	23½−23½	24½−24½	348260.83	82Bo
		23½−23½	348259.55	82Bo
		22½−22½	348254.11	82Bo

[a]) Estimated experimental uncertainty, in units of the last quoted decimal place.

Microwave data for $^{14}N^{16}O_2$ (continued)

Transition			ν MHz	Ref.
rotational $N'_{K_aK_c}-N''_{K_aK_c}$	fine $J'-J''$	hyperfine $F'-F''$		
State: electronic $\tilde{X}\ ^2A_1$; vibrational (0, 0, 0) (continued)				
$24_{3,21}-25_{2,24}$	24½ – 25½	25½ – 26½	509068.64 [a])	82Bo
		24½ – 25½	509108.98	82Bo
		23½ – 24½	509100.77	82Bo
$26_{0,26}-25_{1,25}$	26½ – 25½	27½ – 26½	517656.74	82Bo
		26½ – 25½	517652.66	82Bo
		25½ – 24½	517647.97	82Bo
	25½ – 24½	26½ – 25½	517473.81	82Bo
		25½ – 24½	517478.84	82Bo
		24½ – 23½	517482.38	82Bo
$25_{3,23}-26_{2,24}$	25½ – 26½	26½ – 27½	452774.59	82Bo
		25½ – 26½	452856.47	82Bo
		24½ – 25½	452830.74	82Bo
	24½ – 25½	25½ – 26½	451913.47	82Bo
		24½ – 25½	451938.92	82Bo
		23½ – 24½	451987.33	82Bo
$26_{1,25}-26_{0,26}$	26½ – 26½	27½ – 27½	372936.11	82Bo
		26½ – 26½	372907.54	82Bo
		25½ – 25½	372905.59	82Bo
	25½ – 25½	26½ – 26½	371847.06	82Bo
		25½ – 25½	371849.98	82Bo
		24½ – 24½	371847.06	82Bo
$26_{2,24}-27_{1,27}$	26½ – 27½	27½ – 28½	166124.18	82Bo
		26½ – 27½	166055.88	82Bo
		25½ – 26½	166075.03	82Bo
	25½ – 26½	26½ – 27½	164502.49	82Bo
		25½ – 26½	164484.30	82Bo
		24½ – 25½	164466.18	82Bo
$26_{3,23}-27_{2,26}$	26½ – 27½	26½ – 27½	461817.45	82Bo
		25½ – 26½	461815.10	82Bo
	25½ – 26½	26½ – 27½	460651.34	82Bo
		25½ – 26½	460653.72	82Bo
		24½ – 25½	460690.45	82Bo
$28_{0,28}-27_{1,27}$	28½ – 27½	29½ – 28½	578361.30	82Bo
		28½ – 27½	578357.48	82Bo
		27½ – 26½	578353.12	82Bo
	27½ – 26½	28½ – 27½	578167.26	82Bo
		27½ – 26½	578171.77	82Bo
		26½ – 25½	578175.25	82Bo
$28_{1,27}-27_{2,26}$	28½ – 27½	29½ – 28½	161361.53	82Bo
		28½ – 27½	161353.08	82Bo
		27½ – 26½	161347.57	82Bo
	27½ – 26½	28½ – 27½	161255.91	82Bo
		27½ – 26½	161261.87	82Bo
		26½ – 25½	161267.01	82Bo
$27_{3,25}-28_{2,26}$	27½ – 28½	28½ – 29½	395146.79	82Bo
		27½ – 28½	395215.07	82Bo
		26½ – 27½	395206.06	82Bo
	26½ – 27½	26½ – 27½	394432.26	82Bo
		25½ – 26½	394482.88	82Bo

[a]) Estimated experimental uncertainty, in units of the last quoted decimal place.

Microwave data for $^{14}N^{16}O_2$ (continued)

Transition rotational $N'_{K_aK_c}-N''_{K_aK_c}$	fine $J'-J''$	hyperfine $F'-F''$	ν MHz	Ref.
State: electronic $\tilde{X}\ ^2A_1$; vibrational (0, 0, 0) (continued)				
$28_{1,27}-28_{0,28}$	28½–28½	29½–29½	399252.43 a)	82Bo
		28½–28½	399227.47	82Bo
		27½–27½	399222.85	82Bo
	27½–27½	28½–28½	398074.59	82Bo
		27½–27½	398080.08	82Bo
		26½–26½	398078.88	82Bo
$28_{2,26}-29_{1,29}$	28½–29½	29½–30½	146706.57	82Bo
		28½–29½	146656.59	82Bo
		27½–28½	146662.97	82Bo
	27½–28½	28½–29½	144977.51	82Bo
		27½–28½	144972.04	82Bo
		26½–27½	144960.99	82Bo
$28_{3,25}-29_{2,28}$	28½–29½	29½–30½	415307.77	82Bo
		28½–29½	415345.22	82Bo
		27½–28½	415350.47	82Bo
	27½–28½	28½–29½	414215.40	82Bo
		27½–28½	414210.10	82Bo
		26½–27½	414256.93	82Bo
$30_{1,29}-29_{2,28}$	30½–29½	31½–30½	230210.02	82Bo
		30½–29½	230200.59	82Bo
		29½–28½	230195.25	82Bo
	29½–28½	30½–29½	230032.88	82Bo
		29½–28½	230038.62	82Bo
		28½–27½	230042.64	82Bo
$29_{3,27}-30_{2,28}$	29½–30½	30½–31½	336069.50	82Bo
		29½–30½	336138.26	82Bo
		28½–29½	336148.06	82Bo
	28½–29½	29½–30½	335464.44	82Bo
		28½–29½	335454.44	82Bo
		27½–28½	335523.34	82Bo
$30_{1,29}-30_{0,30}$	30½–30½	31½–31½	428295.68	82Bo
		30½–30½	428273.37	82Bo
		29½–29½	428266.84	82Bo
	29½–29½	30½–30½	427025.85	82Bo
$30_{2,28}-31_{1,31}$	30½–31½	31½–32½	130929.44	82Bo
	29½–30½	30½–31½	129089.49	82Bo
		29½–30½	129090.61	82Bo
		28½–29½	129083.83	82Bo
$30_{3,27}-31_{2,30}$	30½–31½	31½–32½	369776.43	82Bo
		30½–31½	369823.32	82Bo
		29½–30½	369846.89	82Bo
	29½–30½	30½–31½	368701.12	82Bo
		29½–30½	368677.45	82Bo
		28½–29½	368752.79	82Bo
$32_{1,31}-31_{2,30}$	32½–31½	33½–32½	299926.29	82Bo
		32½–31½	299916.75	82Bo
		31½–30½	299911.24	82Bo

a) Estimated experimental uncertainty, in units of the last quoted decimal place.

Microwave data for $^{14}N^{16}O_2$ (continued)

Transition			ν MHz	Ref.
rotational $N'_{K_aK_c}-N''_{K_aK_c}$	fine $J'-J''$	hyperfine $F'-F''$		
State: electronic $\tilde{X}\ ^2A_1$; vibrational (0, 0, 0) (continued)				
$32_{1,31}-31_{2,30}$	31½–30½	32½–31½	299684.74 [a])	82Bo
		31½–30½	299690.56	82Bo
		30½–29½	299694.11	82Bo
$31_{3,29}-32_{2,30}$	31½–32½	32½–33½	275434.06	82Bo
		31½–32½	275520.85	82Bo
		30½–31½	275373.65	82Bo
	30½–31½	31½–32½	274925.23	82Bo
		30½–31½	275072.21	82Bo
		29½–30½	275001.25	82Bo
$32_{1,31}-32_{0,32}$	32½–32½	33½–33½	460117.85	82Bo
		32½–32½	460097.44	82Bo
		31½–31½	460089.43	82Bo
	31½–31½	32½–32½	458754.18	82Bo
		31½–31½	458763.02	82Bo
		30½–30½	458765.09	82Bo
$32_{2,30}-33_{1,33}$	32½–33½	33½–34½	118966.52	82Bo
		32½–33½	118931.10	82Bo
		31½–32½	118926.98	82Bo
	31½–32½	32½–33½	117009.29	82Bo
		31½–32½	117014.25	82Bo
		30½–31½	117010.24	82Bo
$32_{3,29}-33_{2,32}$	32½–33½	33½–34½	325329.89	82Bo
		32½–33½	325402.29	82Bo
		31½–32½	325263.24	82Bo
	31½–32½	32½–33½	324243.89	82Bo
		31½–32½	324382.80	82Bo
		30½–31½	324321.94	82Bo
$34_{1,33}-33_{2,32}$	34½–33½	35½–34½	370422.33	82Bo
		34½–33½	370413.02	82Bo
		33½–32½	370407.44	82Bo
	33½–32½	34½–33½	370124.04	82Bo
		32½–31½	370133.36	82Bo
$33_{3,31}-34_{2,32}$	33½–34½	34½–35½	213155.54	82Bo
		33½–34½	213096.75	82Bo
		32½–31½	213138.17	82Bo
	32½–33½	33½–34½	212902.85	82Bo
		32½–33½	212861.19	82Bo
		31½–32½	212832.34	82Bo
$34_{1,33}-34_{0,34}$	34½–34½	35½–35½	494734.78	82Bo
		34½–34½	494715.90	82Bo
		33½–33½	494706.89	82Bo
	33½–33½	34½–34½	493277.42	82Bo
		33½–33½	493287.39	82Bo
		32½–32½	493290.47	82Bo
$34_{2,32}-35_{1,35}$	34½–35½	35½–36½	110963.06	82Bo
		34½–35½	110930.79	82Bo
		33½–34½	110924.10	82Bo

[a]) Estimated experimental uncertainty, in units of the last quoted decimal place.

Microwave data for $^{14}N^{16}O_2$ (continued)

Transition rotational $N'_{K_aK_c}-N''_{K_aK_c}$	fine $J'-J''$	hyperfine $F'-F''$	ν MHz	Ref.
State: electronic $\tilde{X}\ ^2A_1$; vibrational (0, 0, 0) (continued)				
$34_{2,32}-35_{1,35}$	33½ – 34½	34½ – 35½	108881.72 [a])	82Bo
		33½ – 34½	108889.29	82Bo
		32½ – 33½	108887.30	82Bo
$34_{3,31}-35_{2,34}$	34½ – 35½	35½ – 36½	282122.13	82Bo
		34½ – 35½	282049.38	82Bo
		33½ – 34½	282085.82	82Bo
	33½ – 34½	34½ – 35½	281187.03	82Bo
		33½ – 34½	281150.33	82Bo
		32½ – 33½	281120.66	82Bo
$36_{1,35}-35_{2,34}$	36½ – 35½	37½ – 36½	441598.94	82Bo
		36½ – 35½	441590.03	82Bo
		35½ – 34½	441584.08	82Bo
	35½ – 34½	36½ – 35½	441253.11	82Bo
		35½ – 34½	441259.13	82Bo
		34½ – 33½	441262.40	82Bo
$35_{3,33}-36_{2,34}$	35½ – 36½	36½ – 37½	149176.24	82Bo
		35½ – 36½	149157.73	82Bo
		34½ – 35½	149173.91	82Bo
	34½ – 35½	35½ – 36½	148990.24	82Bo
		34½ – 35½	148973.79	82Bo
		33½ – 34½	148959.17	82Bo
$36_{1,35}-36_{0,36}$	36½ – 36½	37½ – 37½	532123.51	82Bo
		36½ – 36½	532105.71	82Bo
		35½ – 35½	532095.74	82Bo
	35½ – 35½	36½ – 36½	530575.00	82Bo
		35½ – 35½	530585.99	82Bo
		34½ – 34½	530589.99	82Bo
$36_{2,34}-37_{1,37}$	36½ – 37½	37½ – 38½	107031.75	82Bo
		36½ – 37½	107001.60	82Bo
		35½ – 36½	106993.04	82Bo
	35½ – 36½	36½ – 37½	104820.34	82Bo
		35½ – 36½	104829.71	82Bo
		34½ – 35½	104829.29	82Bo
$36_{3,33}-37_{2,36}$	36½ – 37½	37½ – 38½	240317.51	82Bo
		36½ – 37½	240278.01	82Bo
		35½ – 36½	240292.29	82Bo
$38_{2,36}-39_{1,39}$	38½ – 39½	39½ – 40½	107250.88	82Bo
		38½ – 39½	107222.17	82Bo
		37½ – 38½	107212.15	82Bo
	37½ – 38½	38½ – 39½	104904.45	82Bo
		37½ – 38½	104915.26	82Bo
		36½ – 37½	104916.09	82Bo
$38_{3,35}-39_{2,38}$	38½ – 39½	39½ – 40½	200077.27	82Bo
		38½ – 39½	200050.60	82Bo
		37½ – 38½	200055.62	82Bo
	37½ – 38½	38½ – 39½	199111.30	82Bo
		37½ – 38½	199105.12	82Bo
		36½ – 37½	199093.97	82Bo

[a]) Estimated experimental uncertainty, in units of the last quoted decimal place.

Microwave data for $^{14}N^{16}O_2$ (continued)

Transition rotational $N'_{K_aK_c}-N''_{K_aK_c}$	fine $J'-J''$	hyperfine $F'-F''$	ν MHz	Ref.
State: electronic $\tilde{X}\ ^2A_1$; vibrational (0, 0, 0) (continued)				
$40_{2,38}-41_{1,41}$	40½–41½	41½–42½	111663.71 [a])	82Bo
		40½–41½	111636.04	82Bo
		39½–40½	111624.83	82Bo
	39½–40½	40½–41½	109178.56	82Bo
		39½–40½	109190.53	82Bo
		38½–39½	109192.40	82Bo
$41_{4,38}-42_{3,39}$	41½–42½	42½–43½	500442.38	82Bo
		41½–42½	500504.29	82Bo
$42_{4,38}-43_{3,41}$	42½–43½	41½–42½	493583.07	82Bo
	41½–42½	42½–43½	492739.47	82Bo
		41½–42½	492882.07	82Bo
		40½–41½	492807.65	82Bo
$44_{2,42}-43_{3,41}$	44½–43½	45½–44½	123941.93	82Bo
		44½–43½	123934.54	82Bo
		43½–42½	123930.07	82Bo
$43_{4,40}-44_{3,41}$	43½–44½	44½–45½	445303.56	82Bo
		42½–43½	445262.61	82Bo
	42½–43½	43½–44½	444582.81	82Bo
		42½–43½	444722.76	82Bo
		41½–42½	444677.63	82Bo
$44_{4,40}-45_{3,43}$	44½–45½	45½–46½	444220.37	82Bo
		43½–44½	444186.49	82Bo
	43½–44½	44½–45½	443275.85	82Bo
		43½–44½	443433.77	82Bo
		42½–43½	443399.99	82Bo
$45_{4,42}-46_{3,43}$	45½–46½	46½–47½	389201.33	82Bo
		45½–46½	389157.66	82Bo
		44½–45½	389183.46	82Bo
	44½–45½	45½–46½	388698.20	82Bo
		44½–45½	388672.26	82Bo
		43½–44½	388649.68	82Bo
$46_{4,42}-47_{3,45}$	46½–47½	47½–48½	395151.32	82Bo
		46½–47½	395114.33	82Bo
		45½–46½	395131.94	82Bo
	45½–46½	46½–47½	394391.58	82Bo
		45½–46½	394373.99	82Bo
		44½–45½	394354.95	82Bo
$47_{4,44}-48_{3,45}$	47½–48½	48½–49½	332008.34	82Bo
		47½–48½	331988.30	82Bo
		46½–47½	332000.13	82Bo
	46½–47½	47½–48½	331558.16	82Bo
		46½–47½	331546.30	82Bo
		45½–46½	331532.73	82Bo
$48_{4,44}-49_{3,47}$	48½–49½	49½–50½	346531.67	82Bo
		48½–49½	346510.73	82Bo
		47½–48½	346518.61	82Bo
	47½–48½	48½–49½	345774.66	82Bo
		47½–48½	345766.32	82Bo

[a]) Estimated experimental uncertainty, in units of the last quoted decimal place.

Microwave data for $^{14}N^{16}O_2$ (continued)

Transition			ν MHz	Ref.
rotational $N'_{K_aK_c}-N''_{K_aK_c}$	fine $J'-J''$	hyperfine $F'-F''$		
State: electronic $\tilde{X}\ ^2A_1$; vibrational (0, 0, 0) (continued)				
$49_{4,46}-50_{3,47}$	49½ – 50½	50½ – 51½	273595.30 [a])	82Bo
		49½ – 50½	273585.14	82Bo
		48½ – 49½	273592.10	82Bo
	48½ – 49½	49½ – 50½	273214.04	82Bo
		48½ – 49½	273206.97	82Bo
		47½ – 48½	273197.75	82Bo
State: electronic $\tilde{X}\ ^2A_1$; vibrational (0, 1, 0)				
$7_{17}-8_{09}$	7½ – 8½	8½ – 9½	26603.65 (50) [a])	76Hr
	7½ – 8½	7½ – 8½	26681.41 (50)	76Hr
	7½ – 8½	6½ – 7½	26702.85 (50)	76Hr
	6½ – 7½	7½ – 8½	25978.63 (50)	76Hr
	6½ – 7½	6½ – 7½	25955.17 (50)	76Hr
	6½ – 7½	5½ – 6½	26034.95 (50)	76Hr
	7½ – 7½	7½ – 8½	26288.52 (50)	76Hr
	7½ – 7½	6½ – 7½	26173.67 (50)	76Hr
	6½ – 8½	7½ – 8½	26371.83 (50)	76Hr
	6½ – 8½	6½ – 7½	26484.73 (50)	76Hr
$10_{0,10}-9_{19}$	10½ – 9½	11½ – 10½	29740.56 (50)	76Hr
	10½ – 9½	10½ – 9½	29791.65 (50)	76Hr
	10½ – 9½	9½ – 8½	29753.64 (50)	76Hr
	9½ – 8½	10½ – 9½	30042.52 (50)	76Hr
	9½ – 8½	9½ – 8½	30081.41 (50)	76Hr
	9½ – 8½	8½ – 7½	30113.90 (50)	76Hr
	9½ – 9½	10½ – 9½	30268.69 (50)	76Hr
	9½ – 9½	9½ – 8½	30374.98 (50)	76Hr
	10½ – 8½	10½ – 9½	29564.82 (50)	76Hr
	10½ – 8½	9½ – 8½	29460.85 (50)	76Hr
State: electronic $\tilde{X}\ ^2A_1$; vibrational (0, 0, 1)				
$19_{3,16}$ [b])	19½ – 18½	20½ – 19½	1479 (1) [a])	80Al
	19½ – 18½	19½ – 18½	1325 (1)	80Al
	19½ – 18½	18½ – 17½	1165.9 (10)	80Al
$20_{2,19}$ [b])	19½ – 20½	18½ – 19½	229.1 (10)	80Al
	19½ – 20½	19½ – 20½	217.8 (10)	80Al
	19½ – 20½	19½ – 19½	220.0 (10)	80Al
	19½ – 20½	20½ – 20½	212.5 (10)	80Al
	19½ – 20½	20½ – 19½	215.1 (10)	80Al
	19½ – 19½	18½ – 19½	11.6 (10)	80Al
	20½ – 19½	21½ – 20½	7.6 (10)	80Al
	20½ – 20½	20½ – 19½	3.0 (10)	80Al
$21_{4,17}$ [b])	21½ – 20½	22½ – 21½	2682.6 (10)	80Al
	21½ – 20½	21½ – 20½	2527.0 (10)	80Al
	21½ – 20½	20½ – 19½	2370.5 (10)	80Al
$32_{1,32}$ [b])	31½ – 32½	32½ – 33½	2108.7 (10)	80Al
	31½ – 32½	31½ – 32½	2256.4 (10)	80Al
	31½ – 32½	30½ – 31½	2389.7 (10)	80Al

[a]) Estimated experimental uncertainty, in units of the last quoted decimal place.
[b]) Magnetic dipole transition between the two spin components of a given rotational level $N_{K_aK_c}$.

Microwave data for $^{14}N^{16}O_2$ (continued)

Transition			ν	Ref.
rotational $N'_{K_aK_c}-N''_{K_aK_c}$	fine $J'-J''$	hyperfine $F'-F''$	MHz	
State: electronic $\tilde{X}\ ^2A_1$; vibrational (0, 0, 1) (continued)				
$40_{4,37}$ [b])	40½ − 39½	41½ − 40½	346.0 (10) [a])	80Al
	40½ − 39½	40½ − 39½	256.1 (10)	80Al
	40½ − 39½	39½ − 38½	62.0 (10)	80Al
	40½ − 39½	40½ − 40½	303.1 (10)	80Al
	40½ − 39½	39½ − 39½	209.2 (10)	80Al
	40½ − 40½	41½ − 40½	43.0 (10)	80Al
	40½ − 40½	40½ − 39½	46.9 (10)	80Al
	39½ − 39½	39½ − 40½	46.9 (10)	80Al
$41_{6,35}$ [b])	41½ − 40½	42½ − 41½	2357.7 (10)	80Al
	41½ − 40½	41½ − 40½	2203.2 (10)	80Al
	41½ − 40½	40½ − 39½	2043.8 (10)	80Al
$43_{6,37}$ [b])	43½ − 42½	44½ − 43½	2093.6 (10) [d])	80Al
	43½ − 42½	43½ − 42½	1941.5 (10) [d])	80Al
	43½ − 42½	42½ − 41½	1782.0 (10) [d])	80Al
State: electronic $^2B_2$; vibrational ( ) [e])				
$6_{16}-7_{07}$	6½ − 7½	7½ − 8½	42371.6 (20) [a])	76Ta
	6½ − 7½	6½ − 7½	42400.2 (20)	76Ta
	6½ − 7½	5½ − 6½	42425.8 (20)	76Ta
	6½ − 7½	7½ − 7½	42412.4 (20)	76Ta
	6½ − 7½	6½ − 6½	42436.0 (20)	76Ta
	5½ − 6½	6½ − 7½	25330.8 (20)	76Ta
	5½ − 6½	5½ − 6½	25328.4 (20)	76Ta
	5½ − 6½	4½ − 5½	25327.2 (20)	76Ta
	5½ − 6½	6½ − 6½	25291.4 (20)	76Ta
	5½ − 6½	5½ − 5½	25294.4 (20)	76Ta
	6½ − 6½	7½ − 7½	31604.6 (20)	76Ta
	6½ − 6½	6½ − 6½	31553.0 (20)	76Ta
	6½ − 6½	5½ − 5½	31508.7 (20)	76Ta
$9_{09}-8_{18}$	9½ − 8½	10½ − 9½	24399.0 (5)	74Ta
	9½ − 8½	9½ − 8½	24402.0 (5)	74Ta
	9½ − 8½	8½ − 7½	24403.5 (5)	74Ta
	9½ − 8½	9½ − 9½	24360.9 (5)	74Ta
	9½ − 8½	8½ − 8½	24368.1 (5)	74Ta
	8½ − 7½	9½ − 8½	33085.3 (5)	74Ta
	8½ − 7½	8½ − 7½	33090.0 (5)	74Ta
	8½ − 7½	7½ − 6½	33093.7 (5)	74Ta
	8½ − 7½	8½ − 8½	33125.0 (5) [c])	74Ta
	8½ − 7½	7½ − 7½	33125.0 (5) [c])	74Ta
	8½ − 8½	9½ − 9½	38119.6 (5)	74Ta
	8½ − 8½	8½ − 8½	38200.4 (5)	74Ta
	8½ − 8½	7½ − 7½	38270.8 (5)	74Ta

a) Estimated experimental uncertainty, in units of the last quoted decimal place.
b) Magnetic dipole transition between the two spin components of a given rotational level $N_{K_aK_c}$.
c) Blended line, with multiple assignment.
d) Lines for NO_2 in (0, 0, 0) level blended with lines for NO_2 in (0, 0, 1) level, observation by microwave-IR double resonance [80Al].
e) Vibrational state not assigned.
f) Identity of rotational level uncertain [74Ta].

Microwave data for $^{14}N^{16}O_2$ (continued)

Transition rotational $N'_{K_aK_c}-N''_{K_aK_c}$	fine $J'-J''$	hyperfine $F'-F''$	ν MHz	Ref.
State: electronic $^2B_2$; vibrational ( )[e]) (continued)				
x−9_{09} [f])	9½−9½	10½−10½	47054.5 (5) [a])	74Ta
	9½−9½	9½−9½	47051.3 (5)	74Ta
	9½−9½	8½−8½	47047.3 (5)	74Ta
	9½−9½	10½−9½	47092.5 (5)	74Ta
	9½−9½	9½−8½	47085.0 (5)	74Ta
	9½−9½	9½−10½	47013.5 (5) [c])	74Ta
	9½−9½	8½−9½	47013.5 (5) [c])	74Ta
x−9_{09} [f])	9½−8½	10½−9½	33334.2 (5)	74Ta
	9½−8½	9½−8½	33253.3 (5)	74Ta
	9½−8½	8½−7½	33180.0 (5)	74Ta
z−8_{18} [f])	8½−8½	9½−8½	24352.8 (5)	74Ta
	8½−8½	8½−7½	24379.5 (5)	74Ta
	8½−8½	8½−8½	24344.1 (5)	74Ta
State: electronic $(2B_2)$[g]); vibrational ( )[e])				
$5_{23}-5_{24}$	$J-J$ [h])	$J+1 \leftarrow J+1$	9267.18 (2)	74So
	$J-J$	$J \leftarrow J$	9267.80 (2)	74So
	$J-J$	$J-1 \leftarrow J-1$	9269.27 (2)	74So
	$J-J$	$(J+1) \leftarrow J$	9294.9 (1)	78So
	$J-J$	$(J-1) \leftarrow J$	9246.6 (1)	78So
	$J-J$	$J \leftarrow (J+1)$	9240.4 (1)	78So
	$J-J$	$J \leftarrow (J-1)$	9290.8 (1)	78So
[i])			7606.04 (2)	78So
[i])			7603.19 (2)	78So
[i])			7601.39 (2)	78So
[i])			10788.09 (4)	78So
[i])			10785.66 (4)	78So
[i])			10783.03 (5)	78So
[i])			10827.5 (5)	78So
[j])			10859.0 (5)	78So
[j])			10858.5 (5)	78So
[j])			10858.0 (5)	78So
[k])			8663.65 (2)	78So
[k])			8641.42 (2)	78So
[k])			10729.6 (2)	78So
[k])			10734.9 (2)	78So
[k])			10708.7 (2)	78So
[k])			10702.6 (2)	78So

[a]) Estimated experimental uncertainty, in units of the last quoted decimal place.
[c]) Blended line, with multiple assignment.
[e]) Vibrational state not assigned.
[f]) Identity of rotational level uncertain [74Ta].
[g]) Probable assignment of electronic state, pumped by 488 nm Ar^+ laser line.
[h]) Assignment of J-value uncertain [78So].
[i]) Not assigned, pumped by 496.5 nm Ar^+ laser line.
[j]) Not assigned, pumped by 472.7 nm Ar^+ laser line.
[k]) Not assigned, pumped by 457.9 nm Ar^+ laser line.

Molecular parameters for $^{14}N^{16}O_2$

Parameter [a])		Value	Method	Ref.
State: electronic $\tilde{X}\ ^2A_1$; vibrational (0, 0, 0)				
A	[MHz]	239904.433 (15) [b])	MW, IR, MODR	64Bi, 66Le, 74Ba, 77La, 81Br, 82Bo
B	[MHz]	13002.1917 (9)		
C	[MHz]	12304.7461 (9)		
Δ_K	[MHz]	80.5501 (42)		
Δ_{NK}	[MHz]	−0.589412 (48)		
Δ_N	[MHz]	$0.89727 (14) \cdot 10^{-2}$		
δ_K	[MHz]	0.120677 (91)		
δ_N	[MHz]	$0.95908 (10) \cdot 10^{-3}$		
Φ_K	[MHz]	$0.8925 (46) \cdot 10^{-1}$		
Φ_{KN}	[MHz]	$-0.8022 (26) \cdot 10^{-3}$		
Φ_{NK}	[MHz]	$0.594 (47) \cdot 10^{-6}$		
Φ_N	[MHz]	$0.1047 (64) \cdot 10^{-7}$		
ϕ_K	[MHz]	$0.5638 (96) \cdot 10^{-3}$		
ϕ_{NK}	[MHz]	0.0		
ϕ_N	[MHz]	$0.3301 (44) \cdot 10^{-8}$		
L_K	[MHz]	$0.44 (17) \cdot 10^{-4}$		
ε_{aa}	[MHz]	5406.539 (50)		
ε_{bb}	[MHz]	7.707 (30)		
ε_{cc}	[MHz]	−95.266 (25)		
Δ^s_K	[MHz]	−5.173 (19)		
$(\Delta^s_{NK}+\Delta^s_{KN})$	[MHz]	$0.228 (10) \cdot 10^{-1}$		
Δ^s_{NK}	[MHz]	$-0.455 (11) \cdot 10^{-1}$		
Δ^s_N	[MHz]	$0.293 (12) \cdot 10^{-4}$		
δ^s_K	[MHz]	$0.1887 (67) \cdot 10^{-1}$		
δ^s_N	[MHz]	$0.721 (53) \cdot 10^{-5}$		
$(O)_w$	[MHz]	147.263 (12)		
$(aa)_N$	[MHz]	−22.139 (42)		
$(bb)_N$	[MHz]	39.882 (25)		
$(aa)_Q$	[MHz]	0.412 (30)		
$(bb)_Q$	[MHz]	−1.622 (17)		
$(g_S^{aa}+g_S^{bb}+g_S^{cc})/3$		1.999840 (43)	EPR, LMR	77Br, 81Br
$(2\,g_S^{aa}-g_S^{bb}-g_S^{cc})$		−0.02702 (74)		
$(g_S^{bb}-g_S^{cc})$		−0.00328 (454)		
g_r^{aa}		−0.001254 [d])		
g_r^{bb}		−0.000002 [d])		
g_r^{cc}		−0.000022 [d])		
State: electronic $\tilde{X}\ ^2A_1$; vibrational (0, 1, 0)				
ε_{aa}	[MHz]	5920.4 (13)	MW	76Hr
ε_{bb}	[MHz]	6.4 (3)		
ε_{cc}	[MHz]	−97.1 (3)		
$(O)_N$	[MHz]	147.0 (2)		

[a]) The parameters have been determined in a fit to the A-reduced Hamiltonian.

[b]) The numbers in parentheses are 1 standard deviation of the least-squares fit, in units of the last quoted decimal place.

[d]) Theoretical value [81Br].

Molecular parameters for $^{14}N^{16}O_2$ (continued)

Parameter [a])		Value	Method	Ref.
State: electronic $\tilde{X}\ ^2A_1$; vibrational (0, 1, 0) (continued)				
$(aa)_N$	[MHz]	−22.2 (4)	MW	
$(cc)_N$	[MHz]	−18.1 (4)		
$(aa)_Q$	[MHz]	0.45 [e])		
$(cc)_Q$	[MHz]	1.26 [e])		
State: electronic $\tilde{X}\ ^2A_1$; vibrational (0, 0, 1)				
A	[MHz]	233087.20 [f])	IR	76Ca
B	[MHz]	12920.012 [f])		
C	[MHz]	12226.640 [f])		
Δ_K	[MHz]	76.75 [f])		
Δ_{NK}	[MHz]	−0.6473 [f])		
Δ_N	[MHz]	$0.9147 \cdot 10^{-2}$ [f])		
δ_K	[MHz]	$0.8090 \cdot 10^{-1}$ [f])		
δ_N	[MHz]	$0.9626 \cdot 10^{-3}$ [f])		
Φ_K	[MHz]	$0.938 \cdot 10^{-1}$ [f])		
Φ_{KN}	[MHz]	$0.108 \cdot 10^{-4}$ [f])		
Φ_N	[MHz]	$0.117 \cdot 10^{-7}$ [f])		
L_K	[MHz]	$0.165 \cdot 10^{-3}$ [f])		
ε_{aa}	[MHz]	5187.2 (22)	IRMWDR	80Al
ε_{bb}	[MHz]	6.885 (88)		
ε_{cc}	[MHz]	−94.666 (43)		
Δ_K^s	[MHz]	−6.337 (58)		
$(O)_N$	[MHz]	146.61 (28)		
$(aa)_N$	[MHz]	−20.54 (72)		
$(bb)_N$	[MHz]	37.84 (105)		
$(aa)_Q$	[MHz]	0.45 [e])	MW	
$(cc)_Q$	[MHz]	1.26 [e])		

a) The parameters have been determined in a fit to the A-reduced Hamiltonian.
e) Parameter constrained to the value for the (0, 0, 0) level in the fit.
f) Parameter constrained to the value obtained in the fit of IR data [76Ca].

References for NO_2

64Bi Bird, G.R., Baird, J.C., Jache, A.W., Hodgeson, J.A., Curl, R.F., Kunkle, A.C., Bransford, J.W., Rastrup-Andersen, J., Rosenthal, J.: J. Chem. Phys. **40** (1964) 3378.
66Le Lees, R.M., Curl, R.F., Baker, J.G.: J. Chem. Phys. **45** (1966) 2037.
74Ba Baron, P.A., Godfrey, P.D., Harris, D.O.: J. Chem. Phys. **60** (1974) 3723.
74So Solarz, R., Levy, D.H., Abe, K., Curl, R.F.: J. Chem. Phys. **60** (1974) 1158.
74Ta Tanaka, T., Field, R.W., Harris, D.O.: J. Chem. Phys. **61** (1974) 3401.
76Ca Cabana, A., Laurin, M., Pepin, C., Lafferty, W.J.: J. Mol. Spectrosc. **59** (1976) 13.
76Hr Hrubesh, L.W., Curl, R.F.: J. Mol. Spectrosc. **61** (1976) 144.
76Ta Tanaka, T., Harris, D.O.: J. Mol. Spectrosc. **59** (1976) 413.
77Br Brown, J.M., Sears, T.J.: Mol. Phys. **34** (1977) 1595.
77La Lafferty, W.J., Sams, R.L.: J. Mol. Spectrosc. **66** (1977) 478.
78So Solarz, R., Levy, D.H.: J. Mol. Spectrosc. **73** (1978) 374.
80Al Allegrini, M., McKellar, A.R.W., Pinson, P., Brown, J.M.: J. Chem. Phys. **73** (1980) 6086.
81Br Brown, J.M., Steimle, T.C., Coles, M.E., Curl, R.F.: J. Chem. Phys. **74** (1981) 3668.
82Bo Bowman, W.C., De Lucia, F.C.: J. Chem. Phys. **77** (1982) 92.

2 ClO_2
Microwave data for $^{35}ClO_2$

Transition rotational $N'_{K_a K_c} - N''_{K_a K_c}$	fine $J' - J''$	hyperfine $F' - F''$	ν MHz	Ref.
State: electronic $\tilde{X}\ ^2B_1$; vibrational (0, 0, 0)				
$1_{10} - 1_{01}$	½ − ½	1 − 1	44242.24 (5)[a]	82Ta
		1 − 2	44376.50 (5)	82Ta
		2 − 1	44429.98 (5)	82Ta
		2 − 2	44564.24 (5)	82Ta
	½ − 1½	1 − 1	45546.10 (5)	82Ta
		1 − 2	44515.80 (5)	82Ta
		2 − 2	44703.19 (5)	82Ta
		2 − 3	44627.99 (5)	82Ta
	1½ − ½	0 − 1	43112.89 (5)	82Ta
		1 − 1	43125.10 (5)	82Ta
		2 − 1	43140.65 (5)	82Ta
		2 − 2	43274.88 (5)	82Ta
		3 − 2	43285.45 (10)	82Ta
	1½ − 1½	0 − 1	43416.76 (5)	82Ta
		1 − 1	43428.78 (5)	82Ta
		1 − 2	43398.44 (5)	82Ta
		2 − 1	43444.45 (5)	82Ta
		2 − 2	43413.91 (5)	82Ta
		2 − 3	43338.41 (5)	82Ta
		3 − 2	43424.15 (5)	82Ta
		3 − 3	43348.77 (2)	78Hu
$2_{21} - 3_{12}$	1½ − 2½	0 − 1	69961.56 (10)	82Ta
		1 − 2	69978.50 (5)	82Ta
		2 − 3	70000.40 (5)	82Ta
		3 − 4	70002.65 (5)	82Ta
	2½ − 3½	1 − 2	68517.50 (5)	82Ta
		2 − 3	68531.97 (5)	82Ta
		3 − 4	68538.24 (5)	82Ta
		4 − 5	68520.26 (5)	82Ta
$3_{03} - 2_{12}$	2½ − 1½	1 − 0	13852.03 (5)	82Ta
		2 − 1	13858.46 (5)	82Ta
		3 − 2	13894.58 (5)	82Ta
		4 − 3	13953.75 (5)	82Ta
	3½ − 2½	2 − 1	14262.63 (5)	82Ta
		3 − 2	14256.50 (5)	82Ta
		4 − 3	14246.74 (5)	82Ta
		5 − 4	14231.55 (2)	78Hu
		3 − 3	14208.95 (10)	81Jo
		2 − 2	14237.43 (10)	81Jo
		4 − 2	14294.78 (10)	81Jo
		5 − 2	14362.93 (10)	81Jo
	2½ − 2½	3 − 3	14606.68 (10)	81Jo
		3 − 2	14654.60 (10)	81Jo
		2 − 2	14711.38 (10)	81Jo
		2 − 1	14736.44 (10)	81Jo
		1 − 2	14756.22 (10)	81Jo
		1 − 1	14781.52 (10)	81Jo

[a]) The figures in parentheses represent the authors' estimates of experimental uncertainty, in units of the last quoted decimal place.

Microwave data for $^{35}ClO_2$ (continued)

Transition rotational $N'_{K_aK_c} - N''_{K_aK_c}$	fine $J' - J''$	hyperfine $F' - F''$	ν MHz	Ref.
State: electronic $\tilde{X}\ ^2B_1$; vibrational (0, 0, 0) (continued)				
$3_{12} - 3_{03}$	2½ – 2½	1 – 1	48231.41 (5) [a]	82Ta
		2 – 2	48276.48 (5)	82Ta
		3 – 3	48335.23 (5)	82Ta
		4 – 4	48397.41 (5)	82Ta
	3½ – 3½	2 – 2	47778.01 (5)	82Ta
		3 – 3	47764.64 (5)	82Ta
		4 – 4	47739.69 (5)	82Ta
		5 – 5	47697.32 (2)	78Hu
$3_{21} - 4_{14}$	2½ – 3½	1 – 2	64757.71 (5)	82Ta
		2 – 3	64826.05 (5)	82Ta
		3 – 4	64900.27 (5)	82Ta
		4 – 5	64965.96 (5)	82Ta
	3½ – 4½	2 – 3	63539.31 (5)	82Ta
		3 – 4	63522.53 (5)	82Ta
		4 – 5	63487.77 (5)	82Ta
		5 – 6	63424.74 (5)	82Ta
$4_{23} - 5_{14}$	3½ – 4½	2 – 3	25869.99 (5)	82Ta
		3 – 4	25852.52 (5)	82Ta
		4 – 5	25828.48 (5)	82Ta
		5 – 6	25798.54 (5)	82Ta
	4½ – 5½	3 – 4	25260.96 (5)	82Ta
		4 – 5	25274.73 (5)	82Ta
		5 – 6	25290.10 (5)	82Ta
		6 – 7	25306.71 (2)	78Hu
		6 – 6	25334.08 (10)	81Jo
$5_{05} - 4_{14}$	4½ – 3½	3 – 2	55054.16 (5)	82Ta
		4 – 3	55074.45 (5)	82Ta
		5 – 4	55104.26 (5)	82Ta
		6 – 5	55139.00 (5)	82Ta
	5½ – 4½	4 – 3	55136.06 (5)	82Ta
		5 – 4	55127.04 (5)	82Ta
		6 – 5	55112.82 (5)	82Ta
		7 – 6	55090.47 (2)	78Hu
$5_{14} - 5_{05}$	4½ – 4½	3 – 3	56576.88 (5)	82Ta
		4 – 4	56624.10 (5)	82Ta
		5 – 5	56674.83 (5)	82Ta
		6 – 6	56723.64 (5)	82Ta
	5½ – 5½	4 – 4	56103.05 (5)	82Ta
		5 – 5	56083.70 (5)	82Ta
		6 – 6	56053.77 (5)	82Ta
		7 – 7	56010.46 (2)	78Hu
		5 – 4	56109.76 (10)	81Jo
		6 – 5	56098.66 (10)	81Jo
		7 – 6	56081.27 (10)	81Jo
		5 – 6	56039.46 (10)	81Jo
		6 – 7	55983.59 (10)	81Jo
$5_{23} - 5_{14}$	5½ – 5½	4 – 4	118191.45 (20)	80Jo1
		5 – 5	118202.45 (20)	80Jo1

[a] The figures in parentheses represent the authors' estimates of experimental uncertainty, in units of the last quoted decimal place.

Microwave data for $^{35}ClO_2$ (continued)

Transition			ν MHz	Ref.
rotational $N'_{K_aK_c}-N''_{K_aK_c}$	fine $J'-J''$	hyperfine $F'-F''$		
State: electronic $\tilde{X}\ ^2B_1$; vibrational (0, 0, 0) (continued)				
$5_{23}-5_{14}$	5½–5½	6–6	118215.08 (20) [a])	80Jo1
		7–7	118228.89 (20)	80Jo1
$5_{23}-6_{16}$	4½–5½	3–4	38679.84 (5)	82Ta
		4–5	38739.04 (5)	82Ta
		5–6	38797.72 (5)	82Ta
		6–7	38849.84 (5)	82Ta
		6–6	38772.97 (10)	81Jo
	5½–6½	4–5	37727.00 (5)	82Ta
		5–6	37703.76 (5)	82Ta
		6–7	37666.72 (5)	82Ta
		7–8	37611.40 (5)	82Ta
$7_{16}-6_{25}$	6½–5½	5–4	20688.01 (5)	82Ta
		6–5	20713.23 (5)	82Ta
		7–6	20741.69 (5)	82Ta
		8–7	20771.01 (5)	82Ta
	7½–6½	6–5	20915.71 (5)	82Ta
		7–6	20900.61 (5)	82Ta
		8–7	20881.45 (5)	82Ta
		9–8	20857.39 (5)	82Ta
$7_{16}-7_{07}$	6½–6½	5–5	70119.85 (5)	82Ta
		6–6	70171.21 (5)	82Ta
		7–7	70222.82 (5)	82Ta
		8–8	70270.02 (5)	82Ta
	7½–7½	6–6	69531.17 (5)	82Ta
		7–7	69506.48 (5)	82Ta
		8–8	69471.63 (5)	82Ta
		9–9	69423.54 (5)	82Ta
$7_{34}-8_{27}$	6½–7½	5–6	71028.82 (5)	82Ta
		6–7	71046.30 (5)	82Ta
		7–8	71060.42 (5)	82Ta
		8–9	71069.87 (5)	82Ta
	7½–8½	6–7	70137.33 (5)	82Ta
		7–8	70132.84 (5)	82Ta
		8–9	70123.42 (5)	82Ta
		9–10	70107.60 (5)	82Ta
$8_{36}-9_{27}$	7½–8½	6–7	39624.68 (5)	82Ta
		7–8	39613.60 (5)	82Ta
		8–9	39601.04 (5)	82Ta
		9–10	39586.72 (5)	82Ta
	8½–9½	7–8	39103.64 (5)	82Ta
		8–9	39112.60 (5)	82Ta
		9–10	39122.24 (5)	82Ta
		10–11	39132.48 (5)	82Ta
$9_{18}-8_{27}$	8½–7½	7–6	69495.73 (5)	82Ta
		8–7	69522.93 (5)	82Ta
		9–8	69550.28 (5)	82Ta
		10–9	69576.95 (5)	82Ta

[a]) The figures in parentheses represent the authors' estimates of experimental uncertainty, in units of the last quoted decimal place.

Microwave data for $^{35}ClO_2$ (continued)

Transition: rotational $N'_{K_aK_c}-N''_{K_aK_c}$	fine $J'-J''$	hyperfine $F'-F''$	ν MHz	Ref.
State: electronic $\tilde{X}\ ^2B_1$; vibrational (0, 0, 0) (continued)				
$9_{18}-8_{27}$	9½−8½	8−7	69491.25 (5) [a]	82Ta
		9−8	69476.06 (5)	82Ta
		10−9	69456.15 (5)	82Ta
		11−10	69430.51 (5)	82Ta
$9_{36}-10_{29}$	8½−9½	7−8	37974.85 (5)	82Ta
		8−9	37992.57 (5)	82Ta
		9−10	38007.61 (5)	82Ta
		10−11	38018.79 (5)	82Ta
	9½−10½	8−9	37229.43 (5)	82Ta
		9−10	37221.55 (5)	82Ta
		10−11	37209.42 (5)	82Ta
		11−12	37192.33 (5)	82Ta
$13_{2,11}-12_{3,10}$	12½−11½	11−10	58674.42 (5)	82Ta
		12−11	58698.00 (5)	82Ta
		13−12	58720.85 (5)	82Ta
		14−13	58742.00 (5)	82Ta
	13½−12½	12−11	58673.97 (5)	82Ta
		13−12	58659.33 (5)	82Ta
		14−13	58641.08 (5)	82Ta
		15−14	58618.50 (5)	82Ta
$14_{2,13}-13_{3,10}$	13½−12½	12−11	16593.38 (5)	82Ta
		13−12	16566.51 (5)	82Ta
		14−13	16542.72 (5)	82Ta
		15−14	16522.88 (5)	82Ta
	14½−13½	13−12	17379.56 (5)	82Ta
		14−13	17393.70 (5)	82Ta
		15−14	17413.19 (5)	82Ta
		16−15	17438.75 (5)	82Ta
$15_{2,13}-16_{1,16}$	14½−15½	13−14	33003.61 (5)	82Ta
		14−15	33088.42 (5)	82Ta
		15−16	33165.12 (5)	82Ta
		16−17	33231.03 (5)	82Ta
	15½−16½	14−15	31109.14 (5)	82Ta
		15−16	31060.55 (5)	82Ta
		16−17	30996.24 (5)	82Ta
		17−18	30913.40 (5)	82Ta
$16_{2,15}-15_{3,12}$	15½−14½	14−13	34462.73 (5)	82Ta
		15−14	34428.47 (5)	82Ta
		16−15	34397.99 (5)	82Ta
		17−16	34372.29 (5)	82Ta
	16½−15½	15−14	35399.50 (5)	82Ta
		16−15	35418.61 (5)	82Ta
		17−16	35444.09 (5)	82Ta
		18−17	35476.93 (2)	78Hu
$17_{2,15}-18_{1,18}$	16½−17½	15−16	60362.04 (5)	82Ta
		16−17	60447.63 (5)	82Ta
		17−18	60524.39 (5)	82Ta
		18−19	60590.03 (5)	82Ta

[a]) The figures in parentheses represent the authors' estimates of experimental uncertainty, in units of the last quoted decimal place.

Microwave data for $^{35}ClO_2$ (continued)

Transition rotational $N'_{K_aK_c} - N''_{K_aK_c}$	fine $J' - J''$	hyperfine $F' - F''$	ν MHz	Ref.
State: electronic $\tilde{X}\ ^2B_1$; vibrational (0, 0, 0) (continued)				
$17_{2,15} - 18_{1,18}$	17½ − 18½	16 − 17	58210.65 (5) [a]	82Ta
		17 − 18	58160.57 (5)	82Ta
		18 − 19	58094.94 (5)	82Ta
		19 − 20	58010.93 (5)	82Ta
$18_{2,17} - 17_{3,14}$	17½ − 16½	16 − 15	43397.48 (5)	82Ta
		17 − 16	43355.05 (5)	82Ta
		18 − 17	43317.30 (5)	82Ta
		19 − 18	43285.45 (5)	82Ta
	18½ − 17½	17 − 16	44558.30 (5)	82Ta
		18 − 17	44582.70 (5)	82Ta
		19 − 18	44614.95 (5)	82Ta
		20 − 19	44655.70 (5)	82Ta
$20_{2,19} - 19_{3,16}$	19½ − 18½	18 − 17	41894.72 (5)	82Ta
		19 − 18	41844.40 (5)	82Ta
		20 − 19	41799.76 (5)	82Ta
		21 − 20	41762.28 (5)	82Ta
	20½ − 19½	19 − 18	43330.10 (5)	82Ta
		20 − 19	43359.90 (5)	82Ta
		21 − 20	43398.25 (5)	82Ta
		22 − 21	43447.10 (5)	82Ta
$12_{49} - 13_{3,10}$	11½ − 12½	10 − 11	54778.20 (5)	82Ta
		11 − 12	54774.90 (5)	82Ta
		12 − 13	54770.35 (5)	82Ta
		13 − 14	54764.60 (5)	82Ta
	12½ − 13½	11 − 12	54225.75 (5)	82Ta
		12 − 13	54229.95 (5)	82Ta
		13 − 14	54233.75 (5)	82Ta
		14 − 15	54273.10 (5)	82Ta
$13_{2,11} - 14_{1,14}$	12½ − 13½	11 − 12	14549.15 (5)	82Ta
		12 − 13	14631.03 (5)	82Ta
		13 − 14	14705.82 (5)	82Ta
		14 − 15	14770.58 (5)	82Ta
	13½ − 14½	12 − 13	12929.31 (5)	82Ta
		13 − 14	12883.59 (5)	82Ta
		14 − 15	12822.45 (5)	82Ta
		15 − 16	12742.98 (5)	82Ta
$13_{49} - 14_{3,12}$	12½ − 13½	11 − 12	42846.48 (5)	82Ta
		12 − 13	42851.92 (5)	82Ta
		13 − 14	42855.76 (5)	82Ta
		14 − 15	42857.68 (5)	82Ta
	13½ − 14½	12 − 13	42210.08 (10) [b]	82Ta
		13 − 14	42210.08 (10) [b]	82Ta
		14 − 15	42207.36 (5)	82Ta
		15 − 16	42202.72 (5)	82Ta
$14_{4,11} - 15_{3,12}$	13½ − 14½	12 − 13	10896.36 (5)	82Ta
		13 − 14	10887.45 (5)	82Ta

[a]) The figures in parentheses represent the authors' estimates of experimental uncertainty, in units of the last quoted decimal place.

[b]) Unresolved hyperfine compounds.

Microwave data for $^{35}ClO_2$ (continued)

Transition			ν MHz	Ref.
rotational $N'_{K_aK_c}-N''_{K_aK_c}$	fine $J'-J''$	hyperfine $F'-F''$		
State: electronic $\tilde{X}\ ^2B_1$; vibrational (0, 0, 0) (continued)				
$14_{4,11}-15_{3,12}$	13½−14½	14−15	10878.25 (5)[a]	82Ta
		15−16	10868.92 (5)	82Ta
	14½−15½	13−14	10558.63 (5)	82Ta
		14−15	10565.45 (5)	82Ta
		15−16	10573.10 (5)	82Ta
		16−17	10581.66 (5)	82Ta
$16_{5,12}-17_{4,13}$	15½−16½	14−15	67998.96 (5)	82Ta
		15−16	67998.54 (5)	82Ta
		16−17	67997.03 (5)	82Ta
		17−18	67994.44 (5)	82Ta
	16½−17½	15−16	67420.70 (5)	82Ta
		16−17	67422.65 (5)	82Ta
		17−18	67424.01 (5)	82Ta
		18−19	67424.52 (5)	82Ta
$17_{3,14}-16_{4,13}$	16½−15½	15−14	37167.09 (5)	82Ta
		16−15	37181.53 (5)	82Ta
		17−16	37195.31 (5)	82Ta
		18−17	37208.06 (5)	82Ta
	17½−16½	16−15	37288.55 (5)	82Ta
		17−16	37279.04 (5)	82Ta
		18−17	37267.38 (5)	82Ta
		19−18	37253.51 (5)	82Ta
$17_{5,12}-18_{4,15}$	16½−17½	15−16	51232.48 (5)	82Ta
		16−17	51234.31 (10)[b]	82Ta
		17−18	51234.84 (5)	82Ta
		18−19	51234.31 (10)[b]	82Ta
	17½−18½	16−17	50651.24 (10)[b]	82Ta
		17−18	50651.24 (10)[b]	82Ta
		18−19	50651.24 (10)[b]	82Ta
		19−20	50649.76 (5)	82Ta
$18_{3,16}-17_{4,13}$	17½−16½	16−15	25767.45 (5)	82Ta
		17−16	25758.51 (5)	82Ta
		18−17	25751.28 (5)	82Ta
		19−18	25745.81 (5)	82Ta
	18½−17½	17−16	26337.59 (5)	82Ta
		18−17	26341.93 (5)	82Ta
		19−18	26348.18 (5)	82Ta
		20−19	26356.74 (5)	82Ta
$18_{5,14}-19_{4,15}$	17½−18½	16−17	26668.56 (5)	82Ta
		17−18	26665.08 (5)	82Ta
		18−19	26661.23 (5)	82Ta
		19−20	26656.97 (5)	82Ta
	18½−19½	17−18	26239.38 (5)	82Ta
		18−19	26242.70 (5)	82Ta
		19−20	26245.99 (5)	82Ta
		20−21	26249.35 (5)	82Ta

[a]) The figures in parentheses represent the authors' estimates of experimental uncertainty, in units of the last quoted decimal place.

[b]) Unresolved hyperfine compounds.

Microwave data for $^{35}ClO_2$ (continued)

Transition			ν MHz	Ref.
rotational $N'_{K_aK_c} - N''_{K_aK_c}$	fine $J' - J''$	hyperfine $F' - F''$		
State: electronic $\tilde{X}\ ^2B_1$; vibrational (0, 0, 0) (continued)				
$20_{3,18} - 19_{4,15}$	19½ − 18½	18 − 17	55251.70 (5) [a])	82Ta
		19 − 18	55238.75 (5)	82Ta
		20 − 19	55227.85 (5)	82Ta
		21 − 20	55219.20 (5)	82Ta
	20½ − 19½	19 − 18	55880.25 (5)	82Ta
		20 − 19	55887.15 (5)	82Ta
		21 − 20	55896.60 (5)	82Ta
		22 − 21	55908.90 (5)	82Ta
$21_{4,17} - 20_{5,16}$	20½ − 19½	19 − 18	17324.86 (5)	82Ta
		20 − 19	17331.56 (5)	82Ta
		21 − 20	17338.20 (5)	82Ta
		22 − 21	17344.62 (5)	82Ta
	21½ − 20½	20 − 19	17594.85 (5)	82Ta
		21 − 20	17589.59 (5)	82Ta
		22 − 21	17583.87 (5)	82Ta
		23 − 22	17577.24 (5)	82Ta
$21_{6,15} - 22_{5,18}$	20½ − 21½	19 − 20	61250.15 (5)	82Ta
		20 − 21	61250.60 (10) [b])	82Ta
		21 − 22	61250.60 (10) [b])	82Ta
		22 − 23	61249.56 (5)	82Ta
	21½ − 22½	20 − 21	60689.76 (5)	82Ta
		21 − 22	60690.71 (10) [b])	82Ta
		22 − 23	60690.71 (10) [b])	82Ta
		23 − 24	60690.39 (5)	82Ta
$22_{2,21} - 21_{3,18}$	21½ − 20½	20 − 19	29368.51 (5)	82Ta
		21 − 20	29311.56 (5)	82Ta
		22 − 21	29261.46 (5)	82Ta
		23 − 22	29218.98 (5)	82Ta
	22½ − 21½	21 − 20	31100.64 (5)	82Ta
		22 − 21	31134.75 (5)	82Ta
		23 − 22	31178.82 (5)	82Ta
		24 − 23	31234.11 (5)	82Ta
$22_{4,19} - 21_{5,16}$	21½ − 20½	20 − 19	24340.46 (5)	82Ta
		21 − 20	24338.37 (5)	82Ta
		22 − 21	24337.08 (5)	82Ta
		23 − 22	24336.73 (5)	82Ta
	22½ − 21½	21 − 20	24801.17 (5)	82Ta
		22 − 21	24801.65 (5)	82Ta
		23 − 22	24802.94 (5)	82Ta
		24 − 23	24805.04 (5)	82Ta
$22_{6,17} - 23_{5,18}$	21½ − 22½	20 − 21	39802.04 (5)	82Ta
		21 − 22	39800.80 (5)	82Ta
		22 − 23	39798.96 (5)	82Ta
		23 − 24	39796.80 (5)	82Ta
	22½ − 23½	21 − 22	39324.32 (5)	82Ta
		22 − 23	39325.60 (5)	82Ta
		23 − 24	39327.44 (5)	82Ta
		24 − 25	39328.40 (5)	82Ta

[a]) The figures in parentheses represent the authors' estimates of experimental uncertainty, in units of the last quoted decimal place. [b]) Unresolved hyperfine compounds.

Microwave data for $^{35}ClO_2$ (continued)

Transition			ν MHz	Ref.
rotational $N'_{K_aK_c} - N''_{K_aK_c}$	fine $J' - J''$	hyperfine $F' - F''$		
State: electronic $\tilde{X}\ ^2B_1$; vibrational (0, 0, 0) (continued)				
$23_{4,19} - 22_{5,18}$	22½ – 21½	21 – 20	65099.57 (5) [a]	82Ta
		22 – 21	65110.09 (5)	82Ta
		23 – 22	65120.00 (5)	82Ta
		24 – 23	65129.03 (5)	82Ta
	23½ – 22½	22 – 21	65190.33 (5)	82Ta
		23 – 22	65183.08 (5)	82Ta
		24 – 23	65174.51 (5)	82Ta
		25 – 24	65164.45 (5)	82Ta
$23_{6,17} - 24_{5,20}$	22½ – 23½	21 – 22	22151.14 (10) [b]	82Ta
		22 – 23	22151.14 (10) [b]	82Ta
		23 – 24	22150.76 (5)	82Ta
		24 – 25	22149.56 (5)	82Ta
	23½ – 24½	22 – 23	21668.80 (5)	82Ta
		23 – 24	21669.84 (5)	82Ta
		24 – 25	21670.24 (5)	82Ta
		25 – 26	21670.40 (5)	82Ta
$24_{4,21} - 23_{5,18}$	23½ – 22½	22 – 21	60395.28 (5)	82Ta
		23 – 22	60391.83 (5)	82Ta
		24 – 23	60398.18 (5)	82Ta
		25 – 24	60387.50 (5)	82Ta
	24½ – 23½	23 – 22	60850.50 (5)	82Ta
		24 – 23	60851.96 (5)	82Ta
		25 – 24	60854.31 (5)	82Ta
		26 – 25	60857.69 (5)	82Ta
$25_{3,22} - 26_{2,25}$	24½ – 25½	23 – 24	26358.95 (5)	82Ta
		24 – 25	26423.28 (5)	82Ta
		25 – 26	26479.69 (5)	82Ta
		26 – 27	26527.09 (5)	82Ta
	25½ – 26½	24 – 25	24071.24 (5)	82Ta
		25 – 26	24031.81 (5)	82Ta
		26 – 27	23981.18 (5)	82Ta
		27 – 28	23918.41 (5)	82Ta
$25_{7,18} - 26_{6,21}$	24½ – 25½	23 – 24	71856.62 (10) [b]	82Ta
		24 – 25	71856.62 (10) [b]	82Ta
		25 – 26	71856.62 (10) [b]	82Ta
		26 – 27	71855.64 (5)	82Ta
	25½ – 26½	24 – 25	71303.34 (5)	82Ta
		25 – 26	71304.23 (10) [b]	82Ta
		26 – 27	71304.23 (10) [b]	82Ta
		27 – 28	71304.23 (10) [b]	82Ta
$26_{7,20} - 27_{6,21}$	25½ – 26½	24 – 25	51596.15 (10) [b]	82Ta
		25 – 26	51596.15 (10) [b]	82Ta
		26 – 27	51595.15 (5)	82Ta
		27 – 28	51593.75 (5)	82Ta
	26½ – 27½	25 – 26	51094.90 (5)	82Ta
		26 – 27	51096.20 (5)	82Ta

[a]) The figures in parentheses represent the authors' estimates of experimental uncertainty, in units of the last quoted decimal place.

[b]) Unresolved hyperfine compounds.

Microwave data for $^{35}ClO_2$ (continued)

Transition rotational $N'_{K_aK_c}-N''_{K_aK_c}$	fine $J'-J''$	hyperfine $F'-F''$	ν MHz	Ref.
State: electronic $\tilde{X}\ ^2B_1$; vibrational (0, 0, 0) (continued)				
$26_{7,20}-27_{6,21}$	26½ – 27½	27 – 28	51097.10 (10) [b])	82Ta
		28 – 29	51097.10 (10) [b])	82Ta
$27_{5,22}-26_{6,21}$	26½ – 25½	25 – 24	45549.60 (5)[a])	82Ta
		26 – 25	45554.64 (5)	82Ta
		27 – 26	45559.56 (5)	82Ta
		28 – 27	45564.36 (5)	82Ta
	27½ – 26½	26 – 25	45790.96 (5)	82Ta
		27 – 26	45786.96 (5)	82Ta
		28 – 27	45782.60 (5)	82Ta
		29 – 28	45777.68 (5)	82Ta
$27_{3,24}-28_{2,27}$	26½ – 27½	25 – 26	66638.66 (5)	82Ta
		26 – 27	66703.92 (5)	82Ta
		27 – 28	66761.01 (5)	82Ta
		28 – 29	66808.87 (5)	82Ta
	27½ – 28½	26 – 27	64129.88 (5)	82Ta
		27 – 28	64089.61 (5)	82Ta
		28 – 29	64038.06 (5)	82Ta
		29 – 30	63974.27 (5)	82Ta
$27_{7,20}-28_{6,23}$	26½ – 27½	25 – 26	32503.59 (10) [b])	82Ta
		26 – 27	32503.59 (10) [b])	82Ta
		27 – 28	32502.75 (5)	82Ta
		28 – 29	32501.40 (5)	82Ta
	27½ – 28½	26 – 27	32023.32 (5)	82Ta
		27 – 28	32024.40 (5)	82Ta
		28 – 29	32025.18 (10) [b])	82Ta
		29 – 30	32025.18 (10) [b])	82Ta
$28_{5,24}-27_{6,21}$	27½ – 26½	26 – 25	56211.65 (10) [b])	82Ta
		27 – 26	56211.65 (10) [b])	82Ta
		28 – 27	56211.65 (10) [b])	82Ta
		29 – 28	56211.65 (10) [b])	82Ta
	28½ – 27½	27 – 26	56590.90 (10) [b])	82Ta
		28 – 27	56590.90 (10) [b])	82Ta
		29 – 28	56590.90 (10) [b])	82Ta
		30 – 29	56590.90 (10) [b])	82Ta
$28_{7,22}-29_{6,23}$	27½ – 28½	26 – 27	11184.60 (5)	82Ta
		27 – 28	11183.21 (5)	82Ta
		28 – 29	11181.49 (5)	82Ta
		29 – 30	11179.51 (5)	82Ta
	28½ – 29½	27 – 28	10767.40 (5)	82Ta
		28 – 29	10769.11 (5)	82Ta
		29 – 30	10770.65 (5)	82Ta
		30 – 31	10772.06 (5)	82Ta
$30_{8,23}-31_{7,24}$	29½ – 30½	28 – 29	62789.37 (10) [b])	82Ta
		29 – 30	62789.37 (10) [b])	82Ta
		30 – 31	62788.73 (5)	82Ta
		31 – 32	62787.63 (5)	82Ta

[a]) The figures in parentheses represent the authors' estimates of experimental uncertainty, in units of the last quoted decimal place.

[b]) Unresolved hyperfine components.

Microwave data for $^{35}ClO_2$ (continued)

Transition			ν MHz	Ref.
rotational $N'_{K_aK_c}-N''_{K_aK_c}$	fine $J'-J''$	hyperfine $F'-F''$		
State: electronic $\tilde{X}\ ^2B_1$; vibrational (0, 0, 0) (continued)				
$30_{8,23}-31_{7,24}$	30½ − 31½	29 − 30	62276.12 (5) [a])	82Ta
		30 − 31	62277.00 (5)	82Ta
		31 − 32	62277.51 (10) [b])	82Ta
		32 − 33	62277.51 (10) [b])	82Ta
$31_{6,25}-30_{7,24}$	30½ − 29½	29 − 28	30494.10 (5)	82Ta
		30 − 29	30496.50 (5)	82Ta
		31 − 30	30499.20 (5)	82Ta
		32 − 31	30501.90 (5)	82Ta
	31½ − 30½	30 − 29	30822.30 (5)	82Ta
		31 − 30	30820.05 (5)	82Ta
		32 − 31	30817.59 (5)	82Ta
		33 − 32	30815.10 (5)	82Ta
$31_{8,23}-32_{7,26}$	30½ − 31½	29 − 30	43276.56 (10) [b])	82Ta
		30 − 31	43276.56 (10) [b])	82Ta
		31 − 32	43275.84 (5)	82Ta
		32 − 33	43274.88 (5)	82Ta
	31½ − 32½	30 − 31	42791.12 (5)	82Ta
		31 − 32	42792.20 (5)	82Ta
		32 − 33	42792.80 (10) [b])	82Ta
		33 − 34	42792.80 (10) [b])	82Ta
$32_{3,30}-31_{4,27}$	31½ − 30½	31 − 30	53942.81 (5)	82Ta
		32 − 31	53899.68 (5)	82Ta
		33 − 32	53863.71 (5)	82Ta
	32½ − 31½	31 − 30	56137.47 (5)	82Ta
		32 − 31	56168.40 (5)	82Ta
		33 − 32	56207.76 (5)	82Ta
		34 − 33	52256.22 (5)	82Ta
$32_{6,27}-31_{7,24}$	31½ − 30½	30 − 29	47458.04 (5)	82Ta
		31 − 30	47459.00 (5)	82Ta
		32 − 31	47460.24 (5)	82Ta
		33 − 32	47461.84 (5)	82Ta
	32½ − 31½	31 − 30	47821.24 (5)	82Ta
		32 − 31	47819.80 (5)	82Ta
		33 − 32	47818.52 (5)	82Ta
		34 − 33	47817.44 (5)	82Ta
$32_{8,25}-33_{7,26}$	31½ − 32½	30 − 31	22955.70 (5)	82Ta
		31 − 32	22955.32 (5)	82Ta
		32 − 33	22954.10 (5)	82Ta
		33 − 34	22952.80 (5)	82Ta
	32½ − 33½	31 − 32	22511.40 (5)	82Ta
		32 − 33	22512.60 (5)	82Ta
		33 − 34	22513.60 (5)	82Ta
		34 − 35	22514.40 (5)	82Ta
$33_{6,27}-32_{7,26}$	32½ − 31½	31 − 30	73962.10 (5)	82Ta
		32 − 31	73966.04 (5)	82Ta
		33 − 32	73969.82 (5)	82Ta
		34 − 33	73973.35 (5)	82Ta

[a]) The figures in parentheses represent the authors' estimates of experimental uncertainty, in units of the last quoted decimal place.

[b]) Unresolved hyperfine components.

Microwave data for $^{35}ClO_2$ (continued)

Transition			ν MHz	Ref.
rotational $N'_{K_aK_c} - N''_{K_aK_c}$	fine $J' - J''$	hyperfine $F' - F''$		
State: electronic $\tilde{X}\ ^2B_1$; vibrational (0, 0, 0) (continued)				
$33_{6,27} - 32_{7,26}$	33½ – 32½	32 – 31	74190.11 (5) [a])	82Ta
		33 – 32	74187.03 (5)	82Ta
		34 – 33	74183.60 (5)	82Ta
		35 – 34	74179.71 (5)	82Ta
$34_{9,26} - 35_{8,27}$	33½ – 34½	32 – 33	73675.59 (10) [b])	82Ta
		33 – 34	73675.59 (10) [b])	82Ta
		34 – 35	73675.59 (10) [b])	82Ta
		35 – 36	73674.10 (5)	82Ta
	34½ – 35½	33 – 34	73155.03 (5)	82Ta
		34 – 35	73155.86 (5)	82Ta
		35 – 36	73156.35 (10) [b])	82Ta
		36 – 37	73156.35 (10) [b])	82Ta
$35_{9,26} - 36_{8,29}$	34½ – 35½	33 – 34	54097.70 (10) [b])	82Ta
		34 – 35	54097.70 (10) [b])	82Ta
		35 – 36	54097.15 (5)	82Ta
		36 – 37	54096.10 (5)	82Ta
	35½ – 36½	34 – 35	53605.10 (5)	82Ta
		36 – 34	53606.00 (5)	82Ta
		37 – 36	53606.60 (10) [b])	82Ta
		38 – 37	53606.60 (10) [b])	82Ta
$36_{7,30} - 35_{8,27}$	35½ – 34½	34 – 33	36853.80 (5)	82Ta
		35 – 34	36854.88 (5)	82Ta
		36 – 35	36856.24 (5)	82Ta
		37 – 36	36857.68 (5)	82Ta
	36½ – 35½	35 – 34	37223.08 (5)	82Ta
		36 – 35	37221.55 (5)	82Ta
		37 – 36	37220.60 (5)	82Ta
		38 – 37	37219.40 (5)	82Ta
$36_{9,28} - 37_{8,29}$	35½ – 36½	34 – 35	34190.85 (5)	82Ta
		35 – 36	34190.55 (5)	82Ta
		36 – 37	34189.68 (5)	82Ta
		37 – 38	34188.60 (5)	82Ta
	36½ – 37½	35 – 36	33729.75 (5)	82Ta
		36 – 37	33730.80 (5)	82Ta
		37 – 38	33731.55 (5)	82Ta
		38 – 39	33731.94 (5)	82Ta
$37_{7,30} - 36_{8,29}$	36½ – 35½	35 – 34	59396.66 (5)	82Ta
		36 – 35	59398.86 (5)	82Ta
		37 – 36	59401.14 (5)	82Ta
		38 – 37	59403.35 (5)	82Ta
	37½ – 36½	36 – 35	59698.92 (5)	82Ta
		37 – 36	59696.91 (5)	82Ta
		38 – 37	59694.80 (5)	82Ta
		39 – 38	59692.62 (5)	82Ta

[a]) The figures in parentheses represent the authors' estimates of experimental uncertainty, in units of the last quoted decimal place.

[b]) Unresolved hyperfine components.

Microwave data for $^{35}ClO_2$ (continued)

Transition			ν MHz	Ref.
rotational $N'_{K_aK_c}-N''_{K_aK_c}$	fine $J'-J''$	hyperfine $F'-F''$		
State: electronic $\tilde{X}\ ^2B_1$; vibrational (0, 0, 0) (continued)				
$39_{10,29}-40_{9,32}$	38½−39½	37−38	64827.24 (10) [b])	82Ta
		38−39	64827.24 (10) [b])	82Ta
		39−40	64826.71 (5) [a])	82Ta
		40−41	64826.01 (5)	82Ta
	39½−40½	38−39	64327.72 (5)	82Ta
		39−40	64328.52 (5)	82Ta
		40−41	64329.05 (10) [b])	82Ta
		41−42	64329.05 (10) [b])	82Ta
$40_{8,33}-39_{9,30}$	39½−38½	38−37	25672.98 (5)	82Ta
		39−38	25673.93 (5)	82Ta
		40−39	25675.15 (5)	82Ta
		41−40	25676.40 (5)	82Ta
	40½−39½	39−38	26057.09 (5)	82Ta
		40−39	26055.89 (5)	82Ta
		41−40	26054.79 (5)	82Ta
		42−41	26053.82 (5)	82Ta
State: electronic $\tilde{X}\ ^2B_1$; vibrational (1, 0, 0)				
$1_{10}-1_{01}$	1½−1½	0−1	43413.17 (10) [a])	81Jo
$1_{10}-1_{01}$	1½−1½	1−1	43425.47 (10)	81Jo
$1_{10}-1_{01}$	1½−1½	1−0	43432.58 (10)	81Jo
$1_{10}-1_{01}$	1½−1½	2−1	43441.16 (10)	81Jo
$1_{10}-1_{01}$	½−1½	1−1	44553.69 (10)	81Jo
$1_{10}-1_{01}$	½−1½	1−0	44560.3 (2)	81Jo
$1_{10}-1_{01}$	½−1½	2−1	44739.84 (10)	81Jo
$4_{23}-5_{14}$	3½−4½	2−3	26490.79 (10)	81Jo
$4_{23}-5_{14}$	3½−4½	3−4	26473.68 (10)	81Jo
$4_{23}-5_{14}$	3½−4½	4−5	26449.56 (10)	81Jo
$5_{23}-4_{14}$	5½−4½	7−6	228718.6 (5)	75Dy
$5_{23}-5_{14}$	5½−5½	7−7	118270.2 (2)	80Jo1
$5_{23}-5_{14}$	5½−5½	6−6	118256.3 (2)	80Jo1
$5_{23}-5_{14}$	5½−5½	5−5	118243.7 (2) [a])	80Jo1
$5_{23}-5_{14}$	5½−5½	4−4	118232 (1)	80Jo1
$5_{23}-6_{16}$	5½−6½	7−8	38209.90 (10)	80Jo1
$5_{23}-6_{16}$	5½−6½	6−7	38264.72 (10)	80Jo1
$5_{23}-6_{16}$	5½−6½	5−6	38301.45 (10)	80Jo1
$5_{23}-6_{16}$	5½−6½	7−7	38305.32 (10)	80Jo1
$5_{23}-6_{16}$	5½−6½	4−5	38324.32 (10)	80Jo1
$5_{23}-6_{16}$	5½−6½	6−6	38328.65 (10)	80Jo1
$5_{23}-6_{16}$	5½−6½	5−5	38341.70 (10)	81Jo
$5_{23}-6_{16}$	5½−5½	7−7	37897.30 (10)	81Jo
$5_{23}-6_{16}$	5½−5½	6−6	37780.77 (10)	81Jo
$5_{23}-6_{16}$	5½−5½	5−5	37667.92 (10)	81Jo
$7_{25}-7_{16}$	6½−6½	8−8	112617.64 (20)	81Jo
$7_{25}-7_{16}$	6½−6½	7−7	112631.18 (20)	81Jo
$7_{25}-8_{18}$	6½−7½	8−9	19988.93 (10)	81Jo

[a]) The figures in parentheses represent the authors' estimates of experimental uncertainty, in units of the last quoted decimal place.

[b]) Unresolved hyperfine components.

Microwave data for $^{35}ClO_2$ (continued)

Transition			ν MHz	Ref.
rotational $N'_{K_aK_c} - N''_{K_aK_c}$	fine $J' - J''$	hyperfine $F' - F''$		
State: electronic $\tilde{X}\ ^2B_1$; vibrational (1, 0, 0) (continued)				
$7_{25}-8_{18}$	6½ – 7½	7 – 8	19936.25 (10) [a])	81Jo
$7_{25}-8_{18}$	6½ – 7½	8 – 8	19916.54 (10)	81Jo
$7_{25}-8_{18}$	6½ – 7½	6 – 7	19876.66 (10)	81Jo
$7_{25}-8_{18}$	6½ – 7½	7 – 7	19851.44 (10)	81Jo
$13_{2,11}-14_{1,14}$	13½ – 14½	15 – 16	13606.34 (10)	81Jo
$13_{2,11}-14_{1,14}$	13½ – 14½	14 – 15	13685.23 (10)	81Jo
$13_{2,11}-14_{1,14}$	13½ – 14½	13 – 14	13745.99 (10)	81Jo
$13_{2,11}-14_{1,14}$	13½ – 14½	12 – 13	13790.78 (10)	81Jo
$11_{38}-12_{2,11}$	11½ – 12½	13 – 14	8709.47 (10)	81Jo
$11_{38}-12_{2,11}$	11½ – 12½	12 – 13	8729.18 (10)	81Jo
$11_{38}-12_{2,11}$	11½ – 12½	11 – 12	8743.73 (10)	81Jo
$11_{38}-12_{2,11}$	11½ – 12½	10 – 11	8753.55 (10)	81Jo
$11_{38}-12_{2,11}$	11½ – 12½	11 – 11	8773.81 (10)	81Jo
State: electronic $\tilde{X}\ ^2B_1$; vibrational (0, 1, 0)				
$3_{12}-3_{03}$	3½ – 3½	5 – 5	48757.06 (10) [a])	81Jo
$3_{12}-3_{03}$	3½ – 3½	4 – 4	48798.98 (10)	81Jo
$3_{12}-3_{03}$	3½ – 3½	3 – 3	48824.54 (10)	81Jo
$3_{12}-3_{03}$	3½ – 3½	3 – 4	48786.37 (10)	81Jo
$3_{12}-3_{03}$	3½ – 3½	4 – 5	48731.46 (10)	81Jo
$3_{12}-3_{03}$	3½ – 3½	3 – 2	48843.62 (10)	81Jo
$3_{12}-3_{03}$	3½ – 2½	5 – 4	48520.38 (10)	81Jo
$4_{23}-5_{14}$	3½ – 4½	5 – 6	28886.84 (10)	81Jo
$4_{23}-5_{14}$	3½ – 4½	4 – 5	28916.46 (10)	81Jo
$4_{23}-5_{14}$	3½ – 4½	3 – 4	28940.52 (10)	81Jo
$4_{23}-5_{14}$	3½ – 4½	2 – 3	28957.98 (10)	81Jo
$4_{23}-5_{14}$	3½ – 4½	3 – 3	28931.19 (10)	81Jo
$4_{23}-5_{14}$	3½ – 4½	4 – 4	28909.01 (10)	81Jo
$9_{27}-10_{1,10}$	8½ – 9½	7 – 8	11424.56 (10)	81Jo
$9_{27}-10_{1,10}$	8½ – 9½	8 – 9	11493.94 (10)	81Jo
$9_{27}-10_{1,10}$	8½ – 9½	9 – 10	11558.87 (10)	81Jo
$9_{27}-10_{1,10}$	8½ – 9½	10 – 11	11616.34 (10)	81Jo
$9_{27}-10_{1,10}$	8½ – 9½	9 – 9	11473.14 (10)	81Jo
$14_{2,13}-13_{3,10}$	13½ – 12½	12 – 11	11226.78 (10)	81Jo
$14_{2,13}-13_{3,10}$	13½ – 12½	13 – 12	11200.36 (10)	81Jo
$14_{2,13}-13_{3,10}$	13½ – 12½	14 – 13	11177.07 (10)	81Jo
$14_{2,13}-13_{3,10}$	13½ – 12½	15 – 14	11157.53 (10)	81Jo
$14_{2,13}-13_{3,10}$	13½ – 12½	12 – 12	11261.81 (10)	81Jo
$13_{49}-14_{3,12}$	12½ – 13½	14 – 15	50339.6 (2)	81Jo
$13_{49}-14_{3,12}$	12½ – 13½	13 – 14	50338.2 (2)	81Jo
$13_{49}-14_{3,12}$	12½ – 13½	12 – 13	50334.48 (10)	81Jo
$13_{49}-14_{3,12}$	12½ – 13½	11 – 12	50328.84 (10)	81Jo
$15_{4,11}-16_{3,14}$	14½ – 15½	13 – 14	14899.45 (10)	81Jo
$15_{4,11}-16_{3,14}$	14½ – 15½	14 – 15	14905.72 (10)	81Jo
$15_{4,11}-16_{3,14}$	14½ – 15½	15 – 16	14910.79 (10)	81Jo
$15_{4,11}-16_{3,14}$	14½ – 15½	16 – 17	14913.83 (10)	81Jo
Additional frequencies given in 5.2 of LB NS, Vol. II/6				70Ma

[a]) The figures in parentheses represent the authors' estimates of experimental uncertainty, in units of the last quoted decimal place.

Microwave data for $^{35}ClO_2$ (continued)

Transition rotational $N'_{K_aK_c}-N''_{K_aK_c}$	fine $J'-J''$	hyperfine $F'-F''$	ν MHz	Ref.
State: electronic $\tilde{X}\ ^2B_1$; vibrational (0, 0, 1)				
$1_{11}-0_{00}$	1½ − ½	2−2	59478.27 (10)[a]	81Jo
$1_{11}-0_{00}$	1½ − ½	3−2	59537.60 (10)	81Jo
$1_{11}-0_{00}$	1½ − ½	2−1	59569.49 (10)	81Jo
$1_{11}-0_{00}$	1½ − ½	3−1	59628.97 (10)	81Jo
$1_{11}-2_{02}$	1½ − 1½	2−1	4758.36 (20)	81Jo
$1_{11}-2_{02}$	1½ − 1½	2−2	4818.36 (20)	81Jo
$1_{11}-2_{02}$	1½ − 1½	3−2	4877.94 (20)	81Jo
$1_{11}-2_{02}$	1½ − 1½	2−3	4884.64 (20)	81Jo
$1_{11}-2_{02}$	1½ − 1½	3−3	4943.94 (20)	81Jo
$1_{11}-2_{02}$	1½ − 2½	3−4	5100.23 (20)	81Jo
$1_{11}-2_{02}$	1½ − 2½	2−3	5109.73 (20)	81Jo
$1_{11}-2_{02}$	1½ − 2½	2−2	5144.64 (20)	81Jo
$1_{11}-2_{02}$	1½ − 2½	3−3	5169.19 (20)	81Jo
$1_{11}-2_{02}$	1½ − 2½	3−2	5203.36 (20)	81Jo
$1_{11}-2_{02}$	1½ − 2½	3−1	5218.40 (20)	81Jo
$2_{11}-2_{02}$	2½ − 1½	3−2	44394.12 (50)	81Jo
$2_{11}-2_{02}$	2½ − 1½	4−3	44482.62 (50)	81Jo
$2_{11}-2_{02}$	2½ − 2½	3−4	44615.3 (5)	81Jo
$2_{11}-2_{02}$	2½ − 2½	4−4	44638.7 (5)	81Jo
$2_{11}-2_{02}$	2½ − 2½	2−3	44672.7 (5)	81Jo
$2_{11}-2_{02}$	2½ − 2½	3−3	44685.4 (5)	81Jo
$2_{11}-2_{02}$	2½ − 2½	4−3	44708.0 (5)	81Jo
$2_{11}-2_{02}$	1½ − 1½	2−1	45306.0 (5)	81Jo
$2_{11}-2_{02}$	1½ − 1½	1−2	45348.5 (5)	81Jo
$2_{11}-2_{02}$	1½ − 1½	2−2	45366.0 (5)	81Jo
$2_{11}-2_{02}$	1½ − 1½	3−2	45391.7 (5)	81Jo
$2_{11}-2_{02}$	1½ − 1½	2−3	45432.1 (5)	81Jo
$2_{11}-2_{02}$	1½ − 1½	3−3	45457.7 (5)	81Jo
$2_{11}-2_{02}$	1½ − 2½	2−3	45657.3 (5)	81Jo
$2_{11}-2_{02}$	1½ − 2½	3−3	45683.2 (5)	81Jo
State: electronic $\tilde{X}\ ^2B_1$; vibrational (1, 1, 0)				
$3_{03}-2_{12}$	3½ − 2½	5−4	12769.38 (10)[a]	81Jo
$3_{03}-2_{12}$	3½ − 2½	4−3	12784.74 (10)	81Jo
$3_{03}-2_{12}$	3½ − 2½	4−4	12701.80 (10)	81Jo
$3_{12}-3_{03}$	2½ − 2½	1−1	49289.98 (10)	81Jo
$3_{12}-3_{03}$	2½ − 2½	2−2	49335.23 (10)	81Jo
$3_{12}-3_{03}$	2½ − 2½	3−3	49392.19 (10)	81Jo
$3_{12}-3_{03}$	2½ − 2½	4−4	49453.95 (10)	81Jo
$3_{12}-3_{03}$	2½ − 2½	3−4	49447.65 (10)	81Jo
$3_{12}-3_{03}$	3½ − 3½	5−5	48753.47 (10)	81Jo
$3_{12}-3_{03}$	3½ − 3½	4−4	48795.52 (10)	81Jo
$3_{12}-3_{03}$	3½ − 3½	3−3	48820.99 (10)	81Jo
$3_{12}-3_{03}$	3½ − 3½	3−4	48781.76 (10)	81Jo
$3_{12}-3_{03}$	3½ − 3½	4−5	48727.90 (10)	81Jo
$3_{12}-3_{03}$	3½ − 3½	3−2	48838.24 (10)	81Jo

[a]) The figures in parentheses represent the authors' estimates of experimental uncertainty, in units of the last quoted decimal place.

Microwave data for $^{35}ClO_2$ (continued)

Transition rotational $N'_{K_aK_c}-N''_{K_aK_c}$	fine $J'-J''$	hyperfine $F'-F''$	ν MHz	Ref.
State: electronic $\tilde{X}\ ^2B_1$; vibrational (1, 1, 0) (continued)				
$3_{12}-3_{03}$	2½−3½	3−4	49760.33 (10) [a]	81Jo
$3_{12}-3_{03}$	2½−3½	4−5	49698.50 (10)	81Jo
$16_{2,15}-15_{3,12}$	15½−14½	17−16	27542.25 (10)	81Jo
$16_{2,15}-15_{3,12}$	15½−14½	16−15	27566.85 (10)	81Jo
$16_{2,15}-15_{3,12}$	15½−14½	15−14	27596.78 (10)	81Jo
$16_{2,15}-15_{3,12}$	15½−14½	14−13	27630.16 (10)	81Jo
$13_{49}-14_{3,12}$	12½−13½	14−15	52056.31 (10)	81Jo
$13_{49}-14_{3,12}$	12½−13½	13−14	52054.99 (10)	81Jo
$13_{49}-14_{3,12}$	12½−13½	12−13	52051.42 (10)	81Jo
$13_{49}-14_{3,12}$	12½−13½	11−12	52046.01 (10)	81Jo
$14_{4,11}-15_{3,12}$	13½−14½	15−16	20579.15 (10)	81Jo
$14_{4,11}-15_{3,12}$	13½−14½	14−15	20587.91 (10)	81Jo
$14_{4,11}-15_{3,12}$	13½−14½	13−14	20596.92 (10)	81Jo
$14_{4,11}-15_{3,12}$	13½−14½	12−13	20605.39 (10)	81Jo
$14_{4,11}-15_{3,12}$	13½−14½	14−14	20562.74 (10)	81Jo
$14_{4,11}-15_{3,12}$	13½−14½	13−13	20565.76 (10)	81Jo
$20_{3,18}-19_{4,15}$	20½−19½	22−21	46385.91 (10)	81Jo
$20_{3,18}-19_{4,15}$	20½−19½	21−20	46374.15 (10)	81Jo
$20_{3,18}-19_{4,15}$	20½−19½	20−19	46365.06 (10)	81Jo
$20_{3,18}-19_{4,15}$	20½−19½	19−18	46358.55 (10)	81Jo

[a]) The figures in parentheses represent the authors' estimates of experimental uncertainty, in units of the last quoted decimal place.

Microwave data for $^{37}Cl^{16}O_2$

Transition rotational $N'_{K_aK_c}-N''_{K_aK_c}$	fine $J'-J''$	hyperfine $F'-F''$	ν MHz	Ref.
State: electronic $\tilde{X}\ ^2B_1$; vibrational (0, 0, 0)				
$1_{10}-1_{01}$	½− ½	1−1	42948.55 (5) [a]	82Ta
		1−2	43060.25 (5)	82Ta
		2−1	43104.21 (5)	82Ta
		2−2	43216.07 (5)	82Ta
	½−1½	1−1	43227.51 (5)	82Ta
		1−2	43201.41 (5)	82Ta
		2−3	43294.72 (5)	82Ta
		1−0	43234.00 (10)	81Jo
		2−2	43356.94 (10)	81Jo
		2−1	43382.95 (10)	81Jo
	1½− ½	0−1	41833.52 (5)	82Ta
		2−1	41856.48 (5)	82Ta
		2−2	41968.17 (5)	82Ta
		3−2	41976.49 (5)	82Ta

[a]) The figures in parentheses represent the authors' estimates of experimental uncertainty, in units of the last quoted decimal place.

Microwave data for $^{37}Cl^{16}O_2$ (continued)				
Transition			ν MHz	Ref.
rotational $N'_{K_aK_c} - N''_{K_aK_c}$	fine $J' - J''$	hyperfine $F' - F''$		
State: electronic $\tilde{X}\ ^2B_1$; vibrational (0, 0, 0) (continued)				
$1_{10} - 1_{01}$	1½ − 1½	0 − 1	42112.55 (5)[a]	82Ta
		1 − 1	42122.59 (5)	82Ta
		1 − 2	42096.47 (5)	82Ta
		1 − 0	42129.30 (10)	81Jo
		2 − 1	42135.42 (5)	82Ta
		2 − 2	42109.59 (5)	82Ta
		2 − 3	42047.35 (5)	82Ta
		3 − 2	42117.73 (5)	82Ta
		3 − 3	42055.94 (5)	82Ta
$2_{21} - 3_{12}$	1½ − 2½	0 − 1	65957.93 (5)	82Ta
		1 − 2	65972.07 (5)	82Ta
		2 − 3	65989.68 (5)	82Ta
		3 − 4	65992.16 (5)	82Ta
	2½ − 3½	1 − 2	64558.56 (5)	82Ta
		2 − 3	64570.09 (5)	82Ta
		3 − 4	64575.05 (5)	82Ta
		4 − 5	64560.96 (5)	82Ta
$3_{12} - 3_{03}$	2½ − 2½	1 − 1	47037.33 (5)	82Ta
		2 − 2	47074.58 (5)	82Ta
		3 − 3	47123.35 (5)	82Ta
		4 − 4	47175.64 (5)	82Ta
	3½ − 3½	2 − 2	46571.28 (5)	82Ta
		3 − 3	46559.96 (5)	82Ta
		4 − 4	46539.11 (5)	82Ta
		5 − 5	46503.74 (5)	82Ta
$3_{21} - 4_{14}$	2½ − 3½	1 − 2	61122.50 (5)	82Ta
		2 − 3	61178.76 (5)	82Ta
		3 − 4	61240.23 (5)	82Ta
		4 − 5	61296.05 (5)	82Ta
	3½ − 4½	2 − 3	59908.98 (5)	82Ta
		3 − 4	59894.00 (5)	82Ta
		4 − 5	59864.78 (5)	82Ta
		5 − 6	59812.85 (5)	82Ta
$4_{23} - 5_{14}$	3½ − 4½	2 − 3	21798.29 (5)	82Ta
		3 − 4	21784.00 (5)	82Ta
		4 − 5	21764.20 (5)	82Ta
		5 − 6	21739.30 (5)	82Ta
	4½ − 5½	3 − 4	21226.13 (5)	82Ta
		4 − 5	21237.62 (5)	82Ta
		5 − 6	21250.47 (5)	82Ta
		6 − 7	21264.31 (5)	82Ta
$5_{05} - 4_{14}$	4½ − 3½	3 − 2	56285.55 (5)	82Ta
		4 − 3	56302.19 (5)	82Ta
		5 − 4	56326.31 (5)	82Ta
		6 − 5	56354.95 (5)	82Ta
	5½ − 4½	4 − 3	56348.96 (5)	82Ta
		5 − 4	56341.43 (5)	82Ta
		6 − 5	56329.61 (5)	82Ta
		7 − 6	56311.29 (5)	82Ta

[a]) The figures in parentheses represent the authors' estimates of experimental uncertainty, in units of the last quoted decimal place.

Microwave data for $^{37}Cl^{16}O_2$ (continued)

Transition			ν	Ref.
rotational $N'_{K_aK_c}-N''_{K_aK_c}$	fine $J'-J''$	hyperfine $F'-F''$	MHz	
State: electronic $\tilde{X}\ ^2B_1$; vibrational (0, 0, 0) (continued)				
$5_{14}-5_{05}$	4½−4½	3−3	55623.34 (5) [a]	82Ta
		4−4	55662.48 (5)	82Ta
		5−5	55704.74 (5)	82Ta
		6−6	55745.99 (5)	82Ta
	5½−5½	4−4	55133.53 (5)	82Ta
		5−5	55117.16 (5)	82Ta
		6−6	55091.82 (5)	82Ta
		7−7	55055.60 (5)	82Ta
$5_{23}-6_{16}$	4½−5½	3−4	35446.78 (5)	82Ta
		4−5	35496.03 (5)	82Ta
		5−6	35545.11 (5)	82Ta
		6−7	35589.53 (5)	82Ta
	5½−6½	4−5	34487.69 (5)	82Ta
		5−6	34467.51 (5)	82Ta
		6−7	34436.48 (5)	82Ta
		7−8	34390.35 (5)	82Ta
$7_{16}-6_{25}$	6½−5½	5−4	24856.19 (5)	82Ta
		6−5	24876.81 (5)	82Ta
		7−6	24900.24 (5)	82Ta
		8−7	24924.48 (5)	82Ta
	7½−6½	6−5	25054.54 (5)	82Ta
		7−6	25042.04 (5)	82Ta
		8−7	25026.20 (5)	82Ta
		9−8	25006.16 (5)	82Ta
$7_{16}-7_{07}$	6½−6½	5−5	69609.41 (5)	82Ta
		6−6	69652.25 (5)	82Ta
		7−7	69695.42 (5)	82Ta
		8−8	69735.64 (5)	82Ta
	7½−7½	6−6	68998.82 (5)	82Ta
		7−7	68978.01 (5)	82Ta
		8−8	68948.45 (5)	82Ta
		9−9	68908.15 (5)	82Ta
$7_{25}-7_{16}$	7½−7½	6−6	108258.48 (20)	80Jo1
		7−7	108263.95 (20)	80Jo1
		8−8	108271.08 (20)	80Jo1
		9−9	108279.14 (20)	80Jo1
$7_{25}-8_{18}$	7½−8½	6−7	15537.06 (10)	81Jo
		7−8	15511.86 (10)	81Jo
		8−9	15475.65 (10)	81Jo
		9−10	15425.96 (10)	81Jo
		8−8	15533.76 (10)	81Jo
		9−9	15507.35 (10)	81Jo
$7_{34}-8_{27}$	6½−7½	5−6	64846.49 (5)	82Ta
		6−7	64860.96 (5)	82Ta
		7−8	64872.91 (5)	82Ta
		8−9	64881.42 (5)	82Ta
	7½−8½	6−7	63973.26 (5)	82Ta
		7−8	63969.27 (5)	82Ta

[a]) The figures in parentheses represent the authors' estimates of experimental uncertainty, in units of the last quoted decimal place.

Microwave data for $^{37}Cl^{16}O_2$ (continued)

Transition rotational $N'_{K_aK_c}-N''_{K_aK_c}$	fine $J'-J''$	hyperfine $F'-F''$	ν MHz	Ref.
State: electronic $\tilde{X}\ ^2B_1$; vibrational (0, 0, 0) (continued)				
$7_{34}-8_{27}$	7½ − 8½	8−9	63961.36 (5) [a]	82Ta
		9−10	63948.07 (5)	82Ta
$8_{36}-9_{27}$	7½ − 8½	6−7	32550.57 (5)	82Ta
		7−8	32541.09 (5)	82Ta
		8−9	32530.19 (5)	82Ta
		9−10	32517.61 (5)	82Ta
	8½ − 9½	7−8	32067.37 (5)	82Ta
		8−9	32074.94 (5)	82Ta
		9−10	32083.34 (5)	82Ta
		10−11	32092.63 (5)	82Ta
$9_{18}-8_{27}$	8½ − 7½	7−6	73751.03 (5)	82Ta
		8−7	73773.01 (5)	82Ta
		9−8	73795.67 (5)	82Ta
		10−9	73817.74 (5)	82Ta
	9½ − 8½	8−7	73723.78 (5)	82Ta
		9−8	73711.24 (5)	82Ta
		10−9	73694.86 (5)	82Ta
		11−10	73674.10 (5)	82Ta
$9_{36}-10_{29}$	8½ − 9½	7−8	32159.18 (5)	82Ta
		8−9	32174.12 (5)	82Ta
		9−10	32187.08 (5)	82Ta
		10−11	32197.14 (5)	82Ta
	9½ − 10½	8−9	31420.31 (5)	82Ta
		9−10	31413.63 (5)	82Ta
		10−11	31403.48 (5)	82Ta
		11−12	31389.19 (5)	82Ta
$13_{2,11}-12_{3,10}$	12½ − 11½	11−10	66770.70 (5)	82Ta
		12−11	66790.30 (5)	82Ta
		13−12	66809.64 (5)	82Ta
		14−13	66827.79 (5)	82Ta
	13½ − 12½	12−11	66737.50 (5)	82Ta
		13−12	66725.15 (5)	82Ta
		14−13	66709.59 (5)	82Ta
		15−14	66690.65 (5)	82Ta
$14_{2,13}-13_{3,10}$	13½ − 12½	12−11	20705.45 (5)	82Ta
		13−12	20682.20 (5)	82Ta
		14−13	20661.30 (5)	82Ta
		15−14	20643.66 (5)	82Ta
	14½ − 13½	13−12	21509.80 (5)	82Ta
		14−13	21522.62 (5)	82Ta
		15−14	21539.62 (5)	82Ta
		16−15	21561.96 (5)	82Ta
$15_{2,13}-16_{1,16}$	14½ − 15½	13−14	36278.18 (5)	82Ta
		14−15	36348.59 (5)	82Ta
		15−16	36412.73 (5)	82Ta
		16−17	36468.54 (5)	82Ta
	15½ − 16½	14−15	34345.38 (5)	82Ta
		15−16	34304.09 (5)	82Ta

[a]) The figures in parentheses represent the authors' estimates of experimental uncertainty, in units of the last quoted decimal place.

Microwave data for $^{37}Cl^{16}O_2$ (continued)

Transition rotational $N'_{K_aK_c}-N''_{K_aK_c}$	fine $J'-J''$	hyperfine $F'-F''$	ν MHz	Ref.
State: electronic $\tilde{X}\ ^2B_1$; vibrational (0, 0, 0) (continued)				
$15_{2,13}-16_{1,16}$	15½−16½	16−17	34250.12 (5) [a]	82Ta
		17−18	34181.30 (5)	82Ta
$16_{2,15}-15_{3,12}$	15½−14½	14−13	36967.41 (5)	82Ta
		15−14	36937.81 (5)	82Ta
		16−15	36911.21 (5)	82Ta
		17−16	36888.17 (5)	82Ta
	16½−15½	15−14	37938.60 (5)	82Ta
		16−15	37955.69 (5)	82Ta
		17−16	37978.09 (5)	82Ta
		18−17	38006.65 (5)	82Ta
$17_{2,15}-18_{1,18}$	16½−17½	15−16	65279.63 (5)	82Ta
		16−17	65350.24 (5)	82Ta
		17−18	65413.94 (5)	82Ta
		18−19	65469.02 (5)	82Ta
	17½−18½	16−17	63096.42 (5)	82Ta
		17−18	63054.06 (5)	82Ta
		18−19	62999.43 (5)	82Ta
		19−20	62930.04 (5)	82Ta
$18_{2,17}-17_{3,14}$	17½−16½	16−15	43707.84 (5)	82Ta
		17−16	43671.43 (5)	82Ta
		18−17	43638.36 (5)	82Ta
		19−18	43610.19 (5)	82Ta
	18½−17½	17−16	44917.45 (5)	82Ta
		18−17	44939.08 (5)	82Ta
		19−18	44967.33 (5)	82Ta
		20−19	45002.56 (5)	82Ta
$20_{2,19}-19_{3,16}$	19½−18½	18−17	39535.58 (5)	82Ta
		19−18	39492.52 (5)	82Ta
		20−19	39454.04 (5)	82Ta
		21−20	39421.06 (5)	82Ta
	20½−19½	19−18	41029.29 (5)	82Ta
		20−19	41055.22 (5)	82Ta
		21−20	41088.80 (5)	82Ta
		22−21	41130.55 (5)	82Ta
$11_{29}-10_{38}$	10½−9½	9−8	15002.69 (5)	82Ta
		10−9	15018.71 (5)	82Ta
		11−10	15035.15 (5)	82Ta
		12−11	15051.18 (5)	82Ta
	11½−10½	10−9	15200.50 (5)	82Ta
		11−10	15189.91 (5)	82Ta
		12−11	15177.11 (5)	82Ta
		13−12	15161.77 (5)	82Ta
$11_{47}-12_{3,10}$	10½−11½	9−10	70569.50 (5)	82Ta
		10−11	70574.54 (5)	82Ta
		11−12	70577.98 (5)	82Ta
		12−13	70579.38 (5)	82Ta
	11½−12½	10−11	69837.42 (5)	82Ta
		11−12	69836.96 (5)	82Ta

[a]) The figures in parentheses represent the authors' estimates of experimental uncertainty, in units of the last quoted decimal place.

Microwave data for $^{37}Cl^{16}O_2$ (continued)

Transition			ν MHz	Ref.
rotational $N'_{K_aK_c}-N''_{K_aK_c}$	fine $J'-J''$	hyperfine $F'-F''$		
State: electronic $\tilde{X}\ ^2B_1$; vibrational (0, 0, 0) (continued)				
$11_{47}-12_{3,10}$	11½–12½	12–13	69834.48 (5) [a])	82Ta
		13–14	69829.89 (5)	82Ta
$12_{49}-13_{3,10}$	11½–12½	10–11	45128.99 (5)	82Ta
		11–12	45125.40 (5)	82Ta
		12–13	45121.01 (5)	82Ta
		13–14	45115.45 (5)	82Ta
	12½–13½	11–12	44612.13 (5)	82Ta
		12–13	44615.91 (5)	82Ta
		13–14	44619.55 (5)	82Ta
		14–15	44622.96 (5)	82Ta
$13_{49}-14_{3,12}$	12½–13½	11–12	34133.18 (5)	82Ta
		12–13	34137.77 (5)	82Ta
		13–14	34141.24 (5)	82Ta
		14–15	34142.96 (5)	82Ta
	13½–14½	12–13	33500.74 (5)	82Ta
		13–14	33503.69 (5)	82Ta
		14–15	33504.40 (5)	82Ta
		15–16	33502.82 (5)	82Ta
$16_{5,12}-17_{4,13}$	15½–16½	14–15	55919.30 (5)	82Ta
		15–16	55918.49 (5)	82Ta
		16–17	55916.93 (5)	82Ta
		17–18	55914.59 (5)	82Ta
	16½–17½	15–16	55373.01 (5)	82Ta
		16–17	55374.81 (5)	82Ta
		17–18	55376.23 (5)	82Ta
		18–19	55376.99 (5)	82Ta
$17_{3,14}-16_{4,13}$	16½–15½	15–14	48409.01 (5)	82Ta
		16–15	48421.69 (5)	82Ta
		17–16	48433.93 (5)	82Ta
		18–17	48445.28 (5)	82Ta
	17½–16½	16–15	48485.80 (5)	82Ta
		17–16	48477.16 (5)	82Ta
		18–17	48466.91 (5)	82Ta
		19–18	48454.64 (5)	82Ta
$17_{5,12}-18_{4,15}$	16½–17½	15–16	39629.66 (5)	82Ta
		16–17	39631.30 (10) [b])	82Ta
		17–18	39631.30 (10) [b])	82Ta
		18–19	39631.30 (10) [b])	82Ta
	17½–18½	16–17	39068.71 (10) [b])	82Ta
		17–18	39068.71 (10) [b])	82Ta
		18–19	39068.71 (10) [b])	82Ta
		19–20	39067.48 (5)	82Ta
$18_{3,16}-17_{4,13}$	17½–16½	16–15	33354.37 (5)	82Ta
		17–16	33346.27 (5)	82Ta
		18–17	33339.46 (5)	82Ta
		19–18	33334.16 (5)	82Ta
	18½–17½	17–16	33930.28 (5)	82Ta
		18–17	33934.43 (5)	82Ta

[a]) The figures in parentheses represent the authors' estimates of experimental uncertainty, in units of the last quoted decimal place.

Microwave data for $^{37}Cl^{16}O_2$ (continued)

Transition			ν MHz	Ref.
rotational $N'_{K_aK_c} - N''_{K_aK_c}$	fine $J' - J''$	hyperfine $F' - F''$		
State: electronic $\tilde{X}\ ^2B_1$; vibrational (0, 0, 0) (continued)				
$18_{3,16} - 17_{4,13}$	18½ – 17½	19 – 18	33940.24 (5) [a]	82Ta
		20 – 19	33948.05 (5)	82Ta
$20_{3,18} - 19_{4,15}$	19½ – 18½	18 – 17	61610.60 (5)	82Ta
		19 – 18	61599.05 (5)	82Ta
		20 – 19	61588.94 (5)	82Ta
		21 – 20	61580.67 (5)	82Ta
	20½ – 19½	19 – 18	62260.92 (5)	82Ta
		20 – 19	62267.40 (5)	82Ta
		21 – 20	62276.12 (5)	82Ta
		22 – 21	62287.47 (5)	82Ta
$20_{6,15} - 21_{5,16}$	19½ – 20½	18 – 19	65314.44 (10) [b]	82Ta
		19 – 20	65314.44 (10) [b]	82Ta
		20 – 21	65313.90 (5)	82Ta
		21 – 22	65312.48 (5)	82Ta
	20½ – 21½	19 – 20	64756.73 (5)	82Ta
		20 – 21	64757.85 (5)	82Ta
		21 – 22	64758.50 (10) [b]	82Ta
		22 – 23	64758.50 (10) [b]	82Ta
$21_{4,17} - 20_{5,16}$	20½ – 19½	19 – 18	30756.29 (5)	82Ta
		20 – 19	30762.61 (5)	82Ta
		21 – 20	30768.78 (5)	82Ta
		22 – 21	30774.74 (5)	82Ta
	21½ – 20½	20 – 19	30981.84 (5)	82Ta
		21 – 20	30977.11 (5)	82Ta
		22 – 21	30971.66 (5)	82Ta
		23 – 22	30965.48 (5)	82Ta
$21_{6,15} - 22_{5,18}$	20½ – 21½	19 – 20	46868.21 (10) [b]	82Ta
		20 – 21	46868.21 (10) [b]	82Ta
		21 – 22	46868.21 (10) [b]	82Ta
		22 – 23	46867.56 (5)	82Ta
	21½ – 22½	20 – 21	46331.28 (5)	82Ta
		21 – 22	46332.03 (10) [b]	82Ta
		22 – 23	46332.03 (10) [b]	82Ta
		23 – 24	46332.03 (10) [b]	82Ta
$22_{2,21} - 21_{3,18}$	21½ – 20½	20 – 19	24101.59 (5)	82Ta
		21 – 20	24053.45 (5)	82Ta
		22 – 21	24010.47 (5)	82Ta
		23 – 22	24973.70 (5)	82Ta
	22½ – 21½	21 – 20	25893.56 (5)	82Ta
		22 – 21	25923.32 (5)	82Ta
		23 – 22	25961.19 (5)	82Ta
		24 – 23	26008.12 (5)	82Ta
$22_{4,19} - 21_{5,16}$	21½ – 20½	20 – 19	35459.97 (5)	82Ta
		21 – 20	35457.87 (5)	82Ta
		22 – 21	35456.51 (5)	82Ta
		23 – 22	35455.96 (5)	82Ta

[a]) The figures in parentheses represent the authors' estimates of experimental uncertainty, in units of the last quoted decimal place.

[b]) Unresolved hyperfine components.

Microwave data for $^{37}Cl^{16}O_2$ (continued)

Transition rotational $N'_{K_aK_c}-N''_{K_aK_c}$	fine $J'-J''$	hyperfine $F'-F''$	ν MHz	Ref.
State: electronic $\tilde{X}\ ^2B_1$; vibrational (0, 0, 0) (continued)				
$22_{4,19}-21_{5,16}$	22½ – 21½	21 – 20	35914.60 (5) [a]	82Ta
		22 – 21	35915.15 (5)	82Ta
		23 – 22	35916.41 (5)	82Ta
		24 – 23	35918.49 (5)	82Ta
$22_{6,17}-23_{5,18}$	21½ – 22½	20 – 21	24903.62 (5)	82Ta
		21 – 22	24902.38 (5)	82Ta
		22 – 23	24900.62 (5)	82Ta
		23 – 24	24898.54 (5)	82Ta
	22½ – 23½	21 – 22	24457.08 (5)	82Ta
		22 – 23	24458.82 (5)	82Ta
		23 – 24	24460.38 (5)	82Ta
		24 – 25	24461.75 (5)	82Ta
$24_{4,21}-23_{5,18}$	23½ – 22½	22 – 21	70829.37 (5)	82Ta
		23 – 22	70825.87 (5)	82Ta
		24 – 23	70823.09 (5)	82Ta
		25 – 24	70821.23 (5)	82Ta
	24½ – 23½	23 – 22	71290.09 (5)	82Ta
		24 – 23	71291.69 (5)	82Ta
		25 – 24	71294.14 (5)	82Ta
		26 – 25	71297.62 (5)	82Ta
$24_{7,18}-25_{6,19}$	23½ – 24½	22 – 23	74005.43 (10) [b]	82Ta
		23 – 24	74005.43 (10) [b]	82Ta
		24 – 25	74005.43 (10) [b]	82Ta
		25 – 26	74004.50 (5)	82Ta
	24½ – 25½	23 – 24	73445.81 (5)	82Ta
		24 – 25	73446.72 (10) [b]	82Ta
		25 – 26	73446.72 (10) [b]	82Ta
		26 – 27	73446.72 (10) [b]	82Ta
$25_{3,22}-26_{2,25}$	24½ – 25½	23 – 24	37084.68 (5)	82Ta
		24 – 25	37138.22 (5)	82Ta
		25 – 26	37185.56 (5)	82Ta
		26 – 27	37225.11 (5)	82Ta
	25½ – 26½	24 – 25	34754.87 (5)	82Ta
		25 – 26	34721.14 (5)	82Ta
		26 – 27	34678.69 (5)	82Ta
		27 – 28	34626.29 (5)	82Ta
$25_{7,18}-26_{6,21}$	24½ – 25½	23 – 24	54792.11 (10) [b]	82Ta
		24 – 25	54792.11 (10) [b]	82Ta
		25 – 26	54792.11 (10) [b]	82Ta
		26 – 27	54791.05 (5)	82Ta
	25½ – 26½	24 – 25	54263.79 (5)	82Ta
		25 – 26	54264.73 (10) [b]	82Ta
		26 – 27	54264.73 (10) [b]	82Ta
		27 – 28	54264.73 (10) [b]	82Ta
$26_{5,22}-25_{6,19}$	25½ – 24½	24 – 23	31492.19 (10) [b]	82Ta
		25 – 24	31492.19 (10) [b]	82Ta
		26 – 25	31492.71 (5)	82Ta
		27 – 26	31493.60 (5)	82Ta

[a]) The figures in parentheses represent the authors' estimates of experimental uncertainty, in units of the last quoted decimal place. [b]) Unresolved hyperfine components.

Microwave data for $^{37}Cl^{16}O_2$ (continued)

Transition			ν MHz	Ref.
rotational $N'_{K_aK_c}-N''_{K_aK_c}$	fine $J'-J''$	hyperfine $F'-F''$		
State: electronic $\tilde{X}\,^2B_1$; vibrational (0, 0, 0) (continued)				
$26_{5,22}-25_{6,19}$	26½ − 25½	25 − 24	31900.80 (5) [a]	82Ta
		26 − 25	31899.51 (5)	82Ta
		27 − 26	31898.55 (5)	82Ta
		28 − 27	31897.99 (5)	82Ta
$26_{7,20}-27_{6,21}$	25½ − 26½	24 − 25	34258.71 (5)	82Ta
		25 − 26	34258.12 (5)	82Ta
		26 − 27	34257.19 (5)	82Ta
		27 − 28	34255.85 (5)	82Ta
	26½ − 27½	25 − 26	33785.78 (5)	82Ta
		26 − 27	33786.91 (5)	82Ta
		27 − 28	33787.80 (5)	82Ta
		28 − 29	33788.39 (5)	82Ta
$27_{5,22}-26_{6,21}$	26½ − 25½	25 − 24	61974.05 (5)	82Ta
		26 − 25	61978.95 (5)	82Ta
		27 − 26	61983.59 (5)	82Ta
		28 − 27	61988.01 (5)	82Ta
	27½ − 26½	26 − 25	62169.48 (5)	82Ta
		27 − 26	62165.79 (5)	82Ta
		28 − 27	62161.56 (5)	82Ta
		29 − 28	62156.79 (5)	82Ta
$28_{5,24}-27_{6,21}$	27½ − 26½	26 − 25	70353.48 (10) [b]	82Ta
		27 − 26	70353.48 (10) [b]	82Ta
		28 − 27	70353.48 (10) [b]	82Ta
		29 − 28	70353.95 (5)	82Ta
	28½ − 27½	27 − 26	70725.47 (10) [b]	82Ta
		28 − 27	70725.47 (10) [b]	82Ta
		29 − 28	70725.47 (10) [b]	82Ta
		30 − 29	70725.47 (10) [b]	82Ta
$29_{8,21}-30_{7,24}$	28½ − 29½	27 − 28	62903.97 (10) [b]	82Ta
		28 − 29	62903.97 (10) [b]	82Ta
		29 − 30	62903.97 (10) [b]	82Ta
		30 − 31	62903.04 (5)	82Ta
	29½ − 30½	28 − 29	62377.98 (5)	82Ta
		29 − 30	62378.79 (10) [b]	82Ta
		30 − 31	62378.79 (10) [b]	82Ta
		31 − 32	62378.79 (10) [b]	82Ta
$30_{3,28}-29_{4,25}$	29½ − 28½	28 − 27	71284.37 (5)	82Ta
		29 − 28	71245.84 (5)	82Ta
		30 − 29	71211.76 (5)	82Ta
		31 − 30	71182.94 (5)	82Ta
	30½ − 29½	29 − 28	73210.12 (5)	82Ta
		30 − 29	73234.55 (5)	82Ta
		31 − 30	73265.56 (5)	82Ta
		32 − 31	73303.38 (5)	82Ta
$30_{6,25}-29_{7,22}$	29½ − 28½	28 − 27	24621.07 (5)	82Ta
		29 − 28	24621.73 (5)	82Ta
		30 − 29	24622.69 (5)	82Ta

[a]) The figures in parentheses represent the authors' estimates of experimental uncertainty, in units of the last quoted decimal place.
[b]) Unresolved hyperfine components.

Microwave data for $^{37}Cl^{16}O_2$ (continued)

Transition: rotational $N'_{K_aK_c}-N''_{K_aK_c}$	fine $J'-J''$	hyperfine $F'-F''$	ν MHz	Ref.
State: electronic $\tilde{X}\ ^2B_1$; vibrational (0, 0, 0) (continued)				
$30_{6,25}-29_{7,22}$	29½−28½	31−30	24623.85 (5) [a]	82Ta
	30½−29½	29−28	25016.24 (5)	82Ta
		30−29	25015.16 (5)	82Ta
		31−30	25014.26 (5)	82Ta
		32−31	25013.56 (5)	82Ta
$30_{8,23}-31_{7,24}$	29½−30½	28−29	42938.15 (10) [b]	82Ta
		29−30	42938.15 (10) [b]	82Ta
		30−31	42937.52 (5)	82Ta
		31−32	42936.49 (5)	82Ta
	30½−31½	29−30	42452.12 (5)	82Ta
		30−31	42453.02 (5)	82Ta
		31−32	42453.66 (10) [b]	82Ta
		32−33	42453.66 (10) [b]	82Ta
$31_{6,25}-30_{7,24}$	30½−29½	29−28	48841.06 (5)	82Ta
		30−29	48843.66 (5)	82Ta
		31−30	48846.21 (5)	82Ta
		32−31	48848.75 (5)	82Ta
	31½−30½	30−29	49132.78 (5)	82Ta
		31−30	49130.59 (5)	82Ta
		32−31	49128.38 (5)	82Ta
		33−32	49125.74 (5)	82Ta
$31_{8,23}-32_{7,26}$	30½−31½	29−30	23415.32 (10) [b]	82Ta
		30−31	23415.32 (10) [b]	82Ta
		31−32	23414.60 (5)	82Ta
		32−33	23413.60 (5)	82Ta
$32_{3,30}-31_{4,27}$	31½−30½	30−29	39626.51 (5)	82Ta
		32−31	39547.34 (5)	82Ta
		33−32	39515.94 (5)	82Ta
	32½−31½	31−30	41849.84 (5)	82Ta
		32−31	41877.04 (5)	82Ta
		33−32	41910.80 (5)	82Ta
		34−33	41952.08 (5)	82Ta
$32_{6,27}-31_{7,24}$	31½−30½	30−29	64837.35 (5)	82Ta
		31−30	64838.12 (5)	82Ta
		32−31	64839.16 (5)	82Ta
		33−32	64840.47 (5)	82Ta
	32½−31½	32−31	65182.12 (5)	82Ta
		33−32	65181.04 (5)	82Ta
		34−33	65180.00 (5)	82Ta
$33_{9,24}-34_{8,27}$	32½−33½	31−32	70997.37 (10) [b]	82Ta
		32−33	70997.37 (10) [b]	82Ta
		33−34	70997.37 (10) [b]	82Ta
		34−35	70996.64 (5)	82Ta
	33½−34½	32−33	70471.36 (5)	82Ta
		33−34	70472.16 (10) [b]	82Ta
		34−35	70472.16 (10) [b]	82Ta
		35−36	70472.16 (10) [b]	82Ta

[a]) The figures in parentheses represent the authors' estimates of experimental uncertainty, in units of the last quoted decimal place.

[b]) Unresolved hyperfine components.

Microwave data for $^{37}Cl^{16}O_2$ (continued)

Transition			ν MHz	Ref.
rotational $N'_{K_aK_c} - N''_{K_aK_c}$	fine $J' - J''$	hyperfine $F' - F''$		
State: electronic $\tilde{X}\ ^2B_1$; vibrational (0, 0, 0) (continued)				
$34_{9,26} - 35_{8,27}$	33½ – 34½	32 – 33	51282.81 (10) [b]	82Ta
		33 – 34	51282.81 (10) [b]	82Ta
		34 – 35	51282.81 (10) [b]	82Ta
		35 – 36	51281.76 (5) [a]	82Ta
	34½ – 35½	33 – 34	50789.20 (5)	82Ta
		34 – 35	50789.94 (5)	82Ta
		35 – 36	50790.39 (10) [b]	82Ta
		36 – 37	50790.39 (10) [b]	82Ta
$35_{7,28} - 34_{8,27}$	34½ – 33½	33 – 32	38259.74 (5)	82Ta
		34 – 33	38261.15 (5)	82Ta
		35 – 34	38262.74 (5)	82Ta
		36 – 35	38264.38 (5)	82Ta
	35½ – 34½	34 – 33	38604.12 (5)	82Ta
		35 – 34	38602.66 (5)	82Ta
		36 – 35	38601.18 (5)	82Ta
		37 – 36	38599.71 (5)	82Ta
$35_{4,31} - 36_{3,34}$	34½ – 35½	33 – 34	49678.56 (5)	82Ta
		34 – 35	49723.03 (5)	82Ta
		35 – 36	49762.15 (5)	82Ta
		36 – 37	49794.88 (5)	82Ta
	35½ – 36½	34 – 35	46998.19 (5)	82Ta
		35 – 36	46969.40 (5)	82Ta
		36 – 37	46933.18 (5)	82Ta
		37 – 38	46888.99 (5)	82Ta
$35_{9,26} - 36_{8,29}$	34½ – 35½	33 – 34	31635.61 (10) [b]	82Ta
		34 – 35	31635.61 (10) [b]	82Ta
		35 – 36	31635.01 (5)	82Ta
		36 – 37	31634.13 (5)	82Ta
	35½ – 36½	34 – 35	31168.75 (5)	82Ta
		35 – 36	31169.55 (5)	82Ta
		36 – 37	31170.11 (10) [b]	82Ta
		37 – 38	31170.11 (10) [b]	82Ta
$36_{7,30} - 35_{8,27}$	35½ – 34½	34 – 33	57119.45 (5)	82Ta
		35 – 34	57120.46 (5)	82Ta
		36 – 35	57121.59 (5)	82Ta
		37 – 36	57122.89 (5)	82Ta
	36½ – 35½	35 – 34	57466.09 (5)	82Ta
		36 – 35	57464.91 (5)	82Ta
		37 – 36	57463.70 (5)	82Ta
		38 – 37	57462.69 (5)	82Ta
$38_{10,29} - 39_{9,30}$	37½ – 38½	36 – 37	59417.90 (10) [b]	82Ta
		37 – 38	59417.90 (10) [b]	82Ta
		38 – 39	59417.90 (10) [b]	82Ta
		39 – 40	59417.00 (5)	82Ta
	38½ – 39½	37 – 38	58918.91 (5)	82Ta
		38 – 39	58919.69 (10) [b]	82Ta

[a]) The figures in parentheses represent the authors' estimates of experimental uncertainty, in units of the last quoted decimal place.

[b]) Unresolved hyperfine components.

Microwave data for $^{37}Cl^{16}O_2$ (continued)

Transition rotational $N'_{K_aK_c}-N''_{K_aK_c}$	fine $J'-J''$	hyperfine $F'-F''$	ν MHz	Ref.
State: electronic $\tilde{X}\,^2B_1$; vibrational (0, 0, 0) (continued)				
$38_{10,29}-39_{9,30}$	38½–39½	39–40	58919.69 (10) [b]	82Ta
		40–41	58919.69 (10) [b]	82Ta
$39_{8,31}-38_{9,30}$	38½–37½	37–36	28851.44 (5) [a]	82Ta
		38–37	28852.34 (5)	82Ta
		39–38	28853.42 (5)	82Ta
		40–39	28854.67 (5)	82Ta
	39½–38½	38–37	29226.35 (5)	82Ta
		39–38	29225.25 (5)	82Ta
		40–39	29224.25 (5)	82Ta
		41–40	29223.29 (5)	82Ta
$40_{8,33}-39_{9,30}$	39½–38½	38–37	48623.59 (5)	82Ta
		39–38	48624.46 (5)	82Ta
		40–39	48625.50 (5)	82Ta
		41–40	48626.73 (5)	82Ta
	40½–39½	39–38	48982.74 (5)	82Ta
		40–39	48981.66 (5)	82Ta
		41–40	48980.68 (5)	82Ta
		42–41	48979.78 (5)	82Ta
Additional frequencies given in 5.2 of LB NS, Vol. II/6				61Cu, 62Cu, 62Pi
State: electronic $\tilde{X}\,^2B_1$; vibrational (1, 0, 0)				
$7_{16}-6_{25}$	7½–6½	8–8	24133.02 (10) [a]	81Jo
$7_{16}-6_{25}$	7½–6½	7–7	24162.08 (10)	81Jo
$7_{16}-6_{25}$	7½–6½	6–6	24184.96 (10)	81Jo
$7_{16}-6_{25}$	7½–6½	9–8	24156.03 (10)	80Jo1
$7_{16}-6_{25}$	7½–6½	8–7	24175.95 (10)	80Jo1
$7_{16}-6_{25}$	7½–6½	7–6	24191.66 (10)	80Jo1
$7_{16}-6_{25}$	7½–6½	6–5	24203.90 (10)	80Jo1
$22_{2,21}-21_{3,18}$	21½–20½	23–22	22836.10 (10)	81Jo
$22_{2,21}-21_{3,18}$	21½–20½	22–21	22872.45 (10)	81Jo
$22_{2,21}-21_{3,18}$	21½–20½	21–20	22914.79 (10)	81Jo
$22_{2,21}-21_{3,18}$	21½–20½	20–19	22962.31 (10)	81Jo
State: electronic $\tilde{X}\,^2B_1$; vibrational (0, 1, 0)				
$3_{12}-3_{03}$	3½–3½	5–5	47536.52 (10) [a]	81Jo
$3_{12}-3_{03}$	3½–3½	4–4	47571.67 (10)	81Jo
$3_{12}-3_{03}$	3½–3½	3–3	47592.69 (10)	81Jo
$3_{12}-3_{03}$	3½–3½	2–2	47604.28 (10)	81Jo
$3_{12}-3_{03}$	2½–2½	1–1	48066.94 (10)	81Jo
$3_{12}-3_{03}$	2½–2½	2–2	48103.84 (10)	81Jo
$3_{12}-3_{03}$	2½–2½	3–3	48152.38 (10)	81Jo
$3_{12}-3_{03}$	2½–2½	4–4	48204.54 (10)	81Jo
$5_{05}-4_{14}$	4½–3½	5–4	55257.2 (20)	81Jo
$5_{05}-4_{14}$	4½–3½	6–5	55286.65 (10)	81Jo
$5_{05}-4_{14}$	5½–4½	4–3	55276.00 (10)	81Jo

[a]) The figures in parentheses represent the authors' estimates of experimental uncertainty, in units of the last quoted decimal place.

[b]) Unresolved hyperfine components.

Microwave data for $^{37}Cl^{16}O_2$ (continued)

Transition rotational $N'_{K_aK_c}-N''_{K_aK_c}$	fine $J'-J''$	hyperfine $F'-F''$	ν MHz	Ref.
State: electronic $\tilde{X}\ ^2B_1$; vibrational (0, 1, 0) (continued)				
$5_{05}-4_{14}$	5½−4½	5−4	55268.71 (10) [a]	81Jo
$5_{05}-4_{14}$	5½−4½	6−5	55257.24 (10)	81Jo
$5_{05}-4_{14}$	5½−4½	7−6	55238.67 (10)	81Jo
$5_{14}-5_{05}$	4½−4½	3−3	56715.19 (10)	81Jo
$5_{14}-5_{05}$	4½−4½	4−4	56753.89 (10)	81Jo
$5_{14}-5_{05}$	4½−4½	5−5	56796.03 (10)	81Jo
$5_{14}-5_{05}$	4½−4½	6−6	56837.07 (10)	81Jo
$5_{14}-5_{05}$	5½−5½	4−4	56227.23 (10)	81Jo
$5_{14}-5_{05}$	5½−5½	6−6	56185.50 (10)	81Jo
$5_{14}-5_{05}$	5½−5½	7−7	56149.05 (10)	81Jo
$4_{23}-5_{14}$	4½−5½	3−4	24243.28 (10)	81Jo
$4_{23}-5_{14}$	4½−5½	4−5	24254.80 (10)	81Jo
$4_{23}-5_{14}$	4½−5½	5−6	24267.58 (10)	81Jo
$4_{23}-5_{14}$	4½−5½	6−7	24281.53 (10)	81Jo
$7_{16}-6_{25}$	6½−5½	5−4	21883.65 (10)	81Jo
$7_{16}-6_{25}$	6½−5½	6−5	21904.65 (10)	81Jo
$7_{16}-6_{25}$	6½−5½	7−6	21927.99 (10)	81Jo
$7_{16}-6_{25}$	6½−5½	8−7	21952.23 (10)	81Jo
$7_{16}-6_{25}$	7½−6½	9−8	22025.88 (10)	81Jo
$7_{16}-6_{25}$	7½−6½	8−7	22045.81 (10)	81Jo
$7_{16}-6_{25}$	7½−6½	7−6	22061.82 (10)	81Jo
$7_{16}-6_{25}$	7½−6½	6−5	22074.19 (10)	81Jo
$9_{27}-10_{1,10}$	9½−10½	8−9	8707.15 (10)	81Jo
$9_{27}-10_{1,10}$	9½−10½	9−10	8676.90 (10)	81Jo
$9_{27}-10_{1,10}$	9½−10½	10−11	8635.30 (10)	81Jo
$18_{5,14}-19_{4,15}$	18½−19½	17−18	23321.60 (10)	81Jo
$18_{5,14}-19_{4,15}$	18½−19½	18−19	23324.54 (10)	81Jo
$18_{5,14}-19_{4,15}$	18½−19½	19−20	23327.56 (10)	81Jo
$18_{5,14}-19_{4,15}$	18½−19½	20−21	23330.54 (10)	81Jo
State: electronic $\tilde{X}\ ^2B_1$; vibrational (0, 0, 1)				
$1_{11}-0_{00}$	1½− ½	2−2	58140.52 (10)	80Jo2
$1_{11}-0_{00}$	1½− ½	2−1	58216.56 (10)	80Jo2
$1_{11}-0_{00}$	1½− ½	1−1	58183.00 (10)	80Jo2
$1_{11}-0_{00}$	1½− ½	1−2	58106.96 (10)	80Jo2
$1_{11}-0_{00}$	1½− ½	0−1	58166.08 (10)	80Jo2
$1_{11}-2_{02}$	1½−1½	1−0	3469.02 (10)	80Jo2
$1_{11}-2_{02}$	1½−1½	0−1	3480.57 (10)	80Jo2
$1_{11}-2_{02}$	1½−1½	1−1	3497.73 (10)	80Jo2
$1_{11}-2_{02}$	1½−1½	2−1	3531.24 (10)	80Jo2
$1_{11}-2_{02}$	1½−1½	2−2	3581.55 (10)	80Jo2
$1_{11}-2_{02}$	1½−1½	3−2	3630.60 (10)	80Jo2
$1_{11}-2_{02}$	½−1½	2−1	4538.68 (10)	80Jo2
$1_{11}-2_{02}$	½−1½	1−0	4626.56 (10)	80Jo2
$1_{11}-2_{02}$	½−1½	1−1	4655.36 (10)	80Jo2
$2_{11}-2_{02}$	1½−1½	1−0	44018.31 (10)	80Jo2

[a]) The figures in parentheses represent the authors' estimates of experimental uncertainty, in units of the last quoted decimal place.

Microwave data for $^{37}Cl^{16}O_2$ (continued)

Transition rotational $N'_{K_aK_c}-N''_{K_aK_c}$	fine $J'-J''$	hyperfine $F'-F''$	ν MHz	Ref.
State: electronic $\tilde{X}\ ^2B_1$; vibrational (0, 1, 0) (continued)				
$2_{11}-2_{02}$	1½−1½	0−1	44039.40 (10) [a]	80Jo2
$2_{11}-2_{02}$	1½−1½	1−1	44046.65 (10)	80Jo2
$2_{11}-2_{02}$	1½−1½	2−1	44061.18 (10)	80Jo2
$2_{11}-2_{02}$	1½−1½	1−2	44096.6 (2)	80Jo2
$2_{11}-2_{02}$	1½−1½	2−2	44110.7 (2)	80Jo2
$2_{11}-2_{02}$	2½−1½	1−0	43062.04 (10)	80Jo2
$2_{11}-2_{02}$	2½−1½	1−1	43090.32 (10)	80Jo2
$2_{11}-2_{02}$	2½−1½	2−1	43096.45 (10)	80Jo2
$6_{24}-7_{17}$	6½−7½	5−6	23325.00 (10)	81Jo
$6_{24}-7_{17}$	6½−7½	6−7	23302.88 (10)	81Jo
$6_{24}-7_{17}$	6½−7½	7−8	23269.79 (10)	81Jo
$6_{24}-7_{17}$	6½−7½	8−9	23222.30 (10)	81Jo
State: electronic $\tilde{X}\ ^2B_1$; vibrational (1, 1, 0)				
$7_{25}-8_{18}$	7½−8½	6−7	19722.26 (10) [a]	81Jo
$7_{25}-8_{18}$	7½−8½	7−8	19697.57 (10)	81Jo
$7_{25}-8_{18}$	7½−8½	8−9	19662.21 (10)	81Jo

[a]) The figures in parentheses represent the authors' estimates of experimental uncertainty, in units of the last quoted decimal place.

Molecular parameters for $^{35}Cl^{16}O_2$

Parameter [a])		Value	Method	Ref.
State: electronic $\tilde{X}\ ^2B_1$; vibrational (0, 0, 0)				
A	[MHz]	52081.264 (16) [b])	IRMWDR, MW	81Jo, 82Ta
B	[MHz]	9952.6668 (31)		
C	[MHz]	8334.1654 (28)		
Δ_N	[MHz]	$0.8756\ (14)\cdot 10^{-2}$		
Δ_{NK}	[MHz]	−0.11500 (19)		
Δ_K	[MHz]	2.05461 (37)		
δ_N	[MHz]	$0.229825\ (66)\cdot 10^{-2}$		
δ_K	[MHz]	$0.29529\ (90)\cdot 10^{-1}$		
Φ_{NK}	[Hz]	−1.155 (130)		
Φ_{KN}	[Hz]	−15.3 (11)		
Φ_K	[Hz]	253.2 (72)		
ϕ_{NK}	[Hz]	0.403 (77)		
ϕ_K	[Hz]	−1.7 (45)		
ε_{aa}	[MHz]	−1388.186 (27)		
ε_{bb}	[MHz]	−216.904 (12)		
ε_{cc}	[MHz]	4.600 (12)		

[a]) The parameters have been determined in a fit to the A-reduced Hamiltonian.

[b]) The numbers in parentheses are 1 standard deviation of the least-squares fit, in units of the last quoted decimal place.

Molecular parameters for $^{35}Cl^{16}O_2$ (continued)

Parameter[a])		Value	Method	Ref.
State: electronic $\tilde{X}\ ^2B_1$; vibrational (0, 0, 0) (continued)				
Δ^s_{NK}	[kHz]	−38.8 (46)[b])	IRMWDR, MW	81Jo, 82Ta
Δ^s_K	[kHz]	−10.3 (22)		
δ^s_N	[kHz]	−0.0817 (65)		
δ^s_K	[kHz]	−7.87 (75)		
$(O)_{Cl}$	[MHz]	46.115 (28)		
$(aa)_{Cl}$	[MHz]	−77.742 (30)		
$(bb)_{Cl}$	[MHz]	−83.102 (24)		
$(aa)_Q$	[MHz]	−8.681 (11)		
$(bb)_Q$	[MHz]	0.468 (11)		
State: electronic $\tilde{X}\ ^2B_1$; vibrational (1, 0, 0)				
A	[MHz]	52027.28 (11)[b])	IRMWDR	81Jo
B	[MHz]	9892.617 (30)		
C	[MHz]	8280.669 (30)		
Δ_N	[MHz]	$0.850 (20) \cdot 10^{-2}$		
Δ_{NK}	[MHz]	−0.1095 (30)		
Δ_K	[MHz]	2.095 (35)		
δ_N	[MHz]	$0.2302 (30) \cdot 10^{-2}$		
δ_K	[MHz]	$0.2790 \cdot 10^{-1}$ [c])		
ε_{aa}	[MHz]	−1397.66 (24)		
ε_{bb}	[MHz]	−220.828 (77)		
ε_{cc}	[MHz]	4.731 (25)		
Δ^s_K	[MHz]	$-0.287 \cdot 10^{-1}$ [c])		
$(O)_{Cl}$	[MHz]	46.007 (83)[b])		
$(aa)_{Cl}$	[MHz]	−77.42 (24)		
$(bb)_{Cl}$	[MHz]	−82.25 (15)		
$(aa)_Q$	[MHz]	−8.613 [c])		
$(bb)_Q$	[MHz]	0.427 [c])		
State: electronic $\tilde{X}\ ^2B_1$; vibrational (0, 1, 0)				
A	[MHz]	53079.547 (64)[b])	IRMWDR	81Jo
B	[MHz]	9943.567 (14)		
C	[MHz]	8310.136 (14)		
Δ_N	[MHz]	$0.867 (29) \cdot 10^{-2}$		
Δ_{NK}	[MHz]	−0.1207 (44)		
Δ_K	[MHz]	2.2708 (87)		
δ_N	[MHz]	$0.2295 (26) \cdot 10^{-2}$		
δ_K	[MHz]	$0.3529 (60) \cdot 10^{-1}$		
ε_{aa}	[MHz]	−1376.00 (15)		
ε_{bb}	[MHz]	−217.370 (51)		
ε_{cc}	[MHz]	4.655 (49)		
Δ^s_K	[MHz]	$-0.287 \cdot 10^{-1}$ [c])		
$(O)_{Cl}$	[MHz]	46.365 (94)		
$(aa)_{Cl}$	[MHz]	−77.71 (16)		
$(bb)_{Cl}$	[MHz]	−82.91 (12)		
$(aa)_Q$	[MHz]	−8.620 (54)		
$(bb)_Q$	[MHz]	0.454 (37)		

[a]) The parameters have been determined in a fit to the A-reduced Hamiltonian.
[b]) The numbers in parentheses are 1 standard deviation of the least-squares fit, in units of the last quoted decimal place.
[c]) Parameter constrained to the value for the (0, 0, 0) level in the fit.

Molecular parameters for $^{35}Cl^{16}O_2$ (continued)

Parameter[a])		Value	Method	Ref.
State: electronic $\tilde{X}\ ^2B_1$; vibrational (0, 0, 1)				
A	[MHz]	51599.759 (81) [b])	IRMWDR	81Jo
B	[MHz]	9898.020 (42)		
C	[MHz]	8291.311 (42)		
Δ_N	[MHz]	$0.8372 \cdot 10^{-2}$ [c])		
Δ_{NK}	[MHz]	−0.1194 [c])		
Δ_K	[MHz]	2.0561 (32)		
δ_N	[MHz]	$0.22783 \cdot 10^{-2}$ [c])		
δ_K	[MHz]	$0.2790 \cdot 10^{-1}$ [c])		
ε_{aa}	[MHz]	−1414.61 (42) [b])		
ε_{bb}	[MHz]	−219.645 (60)		
ε_{cc}	[MHz]	4.604 [c])		
Δ_K^s	[MHz]	$-0.287 \cdot 10^{-1}$ [c])		
$(O)_{Cl}$	[MHz]	45.588 (35)		
$(aa)_{Cl}$	[MHz]	−76.92 (12)		
$(bb)_{Cl}$	[MHz]	−82.40 (21)		
$(aa)_Q$	[MHz]	−8.613 [c])		
$(bb)_Q$	[MHz]	0.427 [c])		
State: electronic $\tilde{X}\ ^2B_1$; vibrational (1, 1, 0)				
A	[MHz]	53043.23 (13) [b])	IRMWDR	81Jo
B	[MHz]	9883.434 (32)		
C	[MHz]	8256.840 (26)		
Δ_N	[MHz]	$0.8071 (28) \cdot 10^{-2}$		
Δ_{NK}	[MHz]	−0.12072 [d])		
Δ_K	[MHz]	2.3136 (62)		
δ_N	[MHz]	$0.2295 \cdot 10^{-2}$ [d])		
δ_K	[MHz]	$0.3529 \cdot 10^{-1}$ [d])		
ε_{aa}	[MHz]	−1384.75 (72)		
ε_{bb}	[MHz]	−221.09 (20)		
ε_{cc}	[MHz]	4.655 [d])		
Δ_K^s	[MHz]	$-0.287 \cdot 10^{-1}$		
$(O)_{Cl}$	[MHz]	45.99 (47)		
$(aa)_{Cl}$	[MHz]	−77.0 (11)		
$(bb)_{Cl}$	[MHz]	−82.01 (65)		
$(aa)_Q$	[MHz]	−8.620 [d])		
$(bb)_Q$	[MHz]	0.454 [d])		
State: electronic $\tilde{X}\ ^2B_1$; vibrational: equilibrium				
Δ_e	[amu Å^2]	−0.0067		81Jo
r_e(Cl-O)	[Å]	1.47001 (13)		
$\measuredangle_e$(OClO)	[degrees]	117° 23.24′ (700)		

a) The parameters have been determined in a fit to the A-reduced Hamiltonian.

b) The numbers in parentheses are 1 standard deviation of the least-squares fit, in units of the last quoted decimal place.

c) Parameter constrained to the value for the (0, 0, 0) level in the fit.

d) Parameter constrained to the value for the (0, 1, 0) level in the fit.

Molecular Parameters for $^{37}Cl^{16}O_2$

Parameter [a]		Value	Method	Ref.
State: electronic $\tilde{X}\ ^2B_1$; vibrational (0, 0, 0)				
A	[MHz]	50736.837 (11) [b]	IRMWDR, MW	81Jo, 82Ta
B	[MHz]	9953.1733 (23)		
C	[MHz]	8299.0754 (20)		
Δ_N	[MHz]	$0.86535\,(60) \cdot 10^{-2}$		
Δ_{NK}	[MHz]	$-0.111103\,(80)$		
Δ_K	[MHz]	1.95060(19)		
δ_N	[MHz]	$0.232160\,(54) \cdot 10^{-2}$		
δ_K	[MHz]	0.28363 (56)		
Φ_{NK}	[MHz]	$-0.1261\,(67) \cdot 10^{-5}$		
Φ_{KN}	[MHz]	$-0.1804\,(75) \cdot 10^{-4}$		
Φ_K	[MHz]	$0.2644\,(51) \cdot 10^{-3}$		
ϕ_{NK}	[MHz]	$0.648\,(32) \cdot 10^{-6}$		
ϕ_K	[MHz]	$-0.155\,(23) \cdot 10^{-4}$		
ε_{aa}	[MHz]	$-1352.267\,(18)$		
ε_{bb}	[MHz]	$-212.8410\,(87)$		
ε_{cc}	[MHz]	4.5991 (84)		
Δ^s_{NK}	[MHz]	$-0.293\,(23) \cdot 10^{-1}$		
Δ^s_K	[MHz]	$-0.129\,(12) \cdot 10^{-1}$		
δ^s_N	[MHz]	$-0.797\,(34) \cdot 10^{-4}$		
δ^s_K	[MHz]	$-0.772\,(48) \cdot 10^{-2}$		
$(O)_{Cl}$	[MHz]	38.497 (29)		
$(aa)_{Cl}$	[MHz]	$-64.627\,(33)$		
$(bb)_{Cl}$	[MHz]	$-69.210\,(27)$		
$(cc)_{Cl}$	[MHz]	133.837 (25)		
$(aa)_Q$	[MHz]	$-6.846\,(18)$		
$(bb)_Q$	[MHz]	0.374 (12)		
$(cc)_Q$	[MHz]	6.472 (12)		
State: electronic $\tilde{X}\ ^2B_1$; vibrational (1, 0, 0)				
A	[MHz]	50670.261 (57) [b]	IRMWDR	81Jo
B	[MHz]	9893.18 [d]		
C	[MHz]	8246.249 (1)		
Δ_N	[MHz]	$0.8873 \cdot 10^{-2}$ [c]		
Δ_{NK}	[MHz]	-0.1140 [c]		
Δ_K	[MHz]	1.9696 [c]		
δ_N	[MHz]	$0.2135 \cdot 10^{-2}$ [c]		
δ_K	[MHz]	$0.443 \cdot 10^{-1}$ [c]		
ε_{aa}	[MHz]	-1352.157 [c]		
ε_{bb}	[MHz]	-216.764 [c]		
ε_{cc}	[MHz]	4.815 [c]		
Δ^s_K	[MHz]	$-0.272 \cdot 10^{-1}$ [d]		
$(O)_{Cl}$	[MHz]	38.576 [c]		
$(aa)_{Cl}$	[MHz]	-64.597 [c]		
$(bb)_{Cl}$	[MHz]	-69.229 [c]		
$(aa)_{Cl}$	[MHz]	-6.817 [c]		
$(bb)_Q$	[MHz]	0.156 [c]		

a) The parameters have been determined in a fit to the A-reduced Hamiltonian.

b) The numbers in parentheses are 1 standard deviation of the least-squares fit, in units of the last quoted decimal place.

c) Parameter constrained to the value for the (0, 0, 0) level in the fit.

d) Parameter constrained to a value estimated from $^{35}ClO_2$.

Molecular parameters for $^{37}Cl^{16}O_2$ (continued)

Parameter[a])		Value	Method	Ref.
State: electronic $\tilde{X}\ ^2B_1$; vibrational (0, 1, 0)				
A	[MHz]	51708.05 (11) [b])	IRMWDR	81Jo
B	[MHz]	9944.086 (22)		
C	[MHz]	8275.010 (30)		
Δ_N	[MHz]	$0.898\,(21) \cdot 10^{-2}$		
Δ_{NK}	[MHz]	−0.1140 [c])		
Δ_K	[MHz]	2.162 (16)		
δ_N	[MHz]	$0.2326\,(41) \cdot 10^{-2}$		
δ_K	[MHz]	$0.443 \cdot 10^{-1}$ [c])		
ε_{aa}	[MHz]	−1339.54 (97)		
ε_{bb}	[MHz]	−217.29 (36)		
ε_{cc}	[MHz]	4.70 (44)		
Δ_K^s	[MHz]	$-0.272 \cdot 10^{-1}$ [d])		
$(O)_{Cl}$		38.576 [c])		
$(aa)_{Cl}$		−64.597 [c])		
$(bb)_{Cl}$		−69.229 [c])		
$(aa)_Q$		−6.817 [c])		
$(bb)_Q$		0.156 [c])		
State: electronic $\tilde{X}\ ^2B_1$; vibrational (0, 0, 1)				
A	[MHz]	50278.19 (17) [b])	IRMWDR	81Jo
B	[MHz]	9899.154 (31)		
C	[MHz]	8256.842 (62)		
Δ_N	[MHz]	$0.8873 \cdot 10^{-2}$ [c])		
Δ_{NK}	[MHz]	−0.1140 [c])		
Δ_K	[MHz]	1.75 (15)		
δ_N	[MHz]	$0.2135 \cdot 10^{-2}$ [c])		
δ_K	[MHz]	$0.443 \cdot 10^{-1}$ [c])		
ε_{aa}	[MHz]	−1377.47 (27)		
ε_{bb}	[MHz]	−219.65 (18)		
ε_{cc}	[MHz]	4.816 [c])		
Δ_K^s	[MHz]	$-0.272 \cdot 10^{-1}$ [d])		
$(O)_{Cl}$	[MHz]	38.083 (85)		
$(aa)_{Cl}$	[MHz]	−63.67 (27)		
$(bb)_{Cl}$	[MHz]	−68.61 (23)		
$(aa)_Q$	[MHz]	−6.817 [c])		
$(bb)_Q$	[MHz]	0.156 [c])		
State: electronic $\tilde{X}\ ^2B_1$; vibrational: equilibrium				
Δ_e	[amu Å^2]	−0.0059		81Jo
r_e(Cl–O)	[Å]	1.46998 (13)		
$\measuredangle_e$(OClO)	[degrees]	117° 23.37′ (700)		

[a]) The parameters have been determined in a fit to the A-reduced Hamiltonian.
[b]) The numbers in parentheses are 1 standard deviation of the least-squares fit, in units of the last quoted decimal place.
[c]) Parameter constrained to the value for the (0, 0, 0) level in the fit.
[d]) Parameter constrained to a value estimated from $^{35}ClO_2$.

For References, see next page.

References for ClO_2

61Cu Curl, R.F., Kinsey, J.L., Baker, J.G., Baird, J.C., Bird, G.R., Heidelberg, R.F., Sugden, T.M., Jenkins, D.R., Kenney, C.N.: Phys. Rev. **121** (1961) 1119.
62Cu Curl, R.F., Heidelberg, R.F., Kinsey, J.L.: Phys. Rev. **125** (1962) 1993.
62Pi Pillai, M.G. Krishna, Curl, R.F.: J. Chem. Phys. **37** (1962) 2921.
70Ma Mariella, R.P., Curl, R.F.: J. Chem. Phys. **52** (1970) 757.
75Dy Dyubko, S.F., Svich, V.A., Fresenko, L.D.: Pis'ma Zh. Tek. Fiz. **1** (1975) 409.
78Hu Hüttner, W., Lienert, W., Peting, K.M.: Mol. Phys. **36** (1978) 915.
80Jo1 Jones, H.: Chem. Phys. Lett. **69** (1980) 483.
80Jo2 Jones, H., Lienert, W.: Chem. Phys. Lett. **73** (1980) 417.
81Jo Jones, H., Brown, J.M.: J. Mol. Spectrosc. **90** (1981) 222.
82Ta Tanoura, M., Chiba, K., Tanaka, K., Tanaka, T.: J. Mol. Spectrosc. **95** (1982) 157.

3 NF_2

Microwave data for $^{14}N^{19}F_2$

Transition		ν MHz	Ref.
rotational [a]) $N'_{K_aK_c}-N''_{K_aK_c}$	fine, hyperfine $F(i)'-F(i)''$		
State: electronic $\tilde{X}\,^2B_1$; vibrational (0, 0, 0)			
$1_{10}-1_{01}$	1½ (1) − 2½ (1)	60074.66	74Br
$1_{10}-1_{01}$	2½ (1) − 1½ (2)	60076.69	
$1_{10}-1_{01}$	2½ (1) − 2½ (1)	60101.68	
$1_{10}-1_{01}$	½ (1) − 1½ (1)	60120.44	
$1_{10}-1_{01}$	1½ (1) − 1½ (1)	60129.98	
$1_{10}-1_{01}$	½ (1) − ½ (1)	60142.30	
$1_{10}-1_{01}$	1½ (1) − 1½ (1)	60151.83	
$1_{10}-1_{01}$	½ (2) − 1½ (2)	60801.71	
$1_{10}-1_{01}$	1½ (2) − ½ (2)	60816.01	
$1_{10}-1_{01}$	1½ (2) − 1½ (2)	60878.32	
$1_{10}-1_{01}$	1½ (2) − 2½ (1)	60903.14	
$1_{10}-1_{01}$	1½ (2) − 1½ (1)	60958.5	
$2_{11}-2_{02}$	3½ (1) − 3½ (2)	61843.9	
$2_{11}-2_{02}$	4½ (1) − 4½ (1)	61871.49	
$2_{11}-2_{02}$	3½ (2) − 3½ (3)	61884.87	
$2_{11}-2_{02}$	2½ (3) − 2½ (4)	61889.57	
$2_{11}-2_{02}$	4½ (1) − 3½ (3)	61905.0	
$2_{11}-2_{02}$	3½ (2) − 2½ (4)	61916.0	
$2_{11}-2_{02}$	2½ (3) − 1½ (3)	61929.5	
$2_{11}-2_{02}$	2½ (3) − 2½ (3)	61942.0	
$2_{11}-2_{02}$	3½ (2) − 3½ (2)	61948.9	
$2_{11}-2_{02}$	1½ (3) − 1½ (4)	62268.76	
$2_{11}-2_{02}$	1½ (3) − 2½ (5)	62287.7	
$2_{11}-2_{02}$	½ (3) − 1½ (4)	62303.4	
$2_{11}-2_{02}$	1½ (4) − 1½ (4)	62311.48	
$2_{11}-2_{02}$	1½ (4) − 2½ (5)	62330.5	
$2_{11}-2_{02}$	1½ (5) − ½ (2)	62331.5	
$2_{11}-2_{02}$	2½ (4) − 2½ (5)	62341.82	
$2_{11}-2_{02}$	2½ (5) − 1½ (4)	62363.90	
$2_{11}-2_{02}$	3½ (3) − 2½ (5)	62394.2	
$2_{11}-2_{02}$	3½ (3) − 4½ (1)	62437.2	

[a]) Because the two F nuclei are equivalent in NF_2, the rotational states can be divided into ortho ($I=1$, $K_a K_c$ even even or odd odd) and para ($I=0$, $K_a K_c$ eo or oe) spin states. The para levels therefore show only ^{14}N hyperfine structure.

Microwave data for $^{14}N^{19}F_2$ (continued)

Transition		ν MHz	Ref.
rotational[a]) $N'_{K_aK_c}-N''_{K_aK_c}$	fine, hyperfine $F(i)'-F(i)''$		
State: electronic $\tilde{X}\ ^2B_1$; vibrational (0, 0, 0) (continued)			
$2_{11}-2_{02}$	2½ (5)−3½ (3)	62459.3	74Br
$2_{11}-2_{02}$	3½ (4)−3½ (3)	62468.32	
$2_{11}-2_{20}$	1½ (1)−2½ (2)	61951.5	
$2_{11}-2_{20}$	3½ (2)−2½ (3)	61958.5	
$2_{11}-2_{20}$	4½ (1)−3½ (2)	61969.3	
$2_{11}-2_{20}$	2½ (1)−3½ (1)	61975.0	
$2_{11}-2_{20}$	1½ (2)−1½ (2)	61996.75	
$2_{11}-2_{20}$	2½ (2)−2½ (2)	61998.42	
$2_{11}-2_{20}$	3½ (1)−2½ (2)	62001.6	
$2_{11}-2_{20}$	1½ (1)−2½ (1)	62002.9	
$2_{11}-2_{20}$	2½ (1)−2½ (1)	62013.43	
$2_{11}-2_{20}$	3½ (1)−3½ (1)	62014.3	
$2_{11}-2_{20}$	½ (1)−1½ (1)	62033.2	
$2_{11}-2_{20}$	1½ (1)−1½ (1)	62036.6	
$2_{11}-2_{20}$	2½ (1)−1½ (1)	62047.15	
$2_{11}-2_{20}$	2½ (2)−2½ (1)	62049.60	
$2_{11}-2_{20}$	3½ (1)−2½ (1)	62052.89	
$2_{11}-2_{20}$	½ (1)− ½ (1)	62055.7	
$2_{11}-2_{20}$	1½ (1)− ½ (1)	62058.95	
$2_{11}-2_{20}$	1½ (3)− ½ (3)	62127.0	
$2_{11}-2_{20}$	½ (2)−1½ (5)	62165.5	
$2_{11}-2_{20}$	2½ (4)−1½ (5)	62223.2	
$2_{11}-2_{20}$	1½ (4)−2½ (3)	62480.53	
$2_{11}-2_{20}$	2½ (4)−3½ (2)	62482.4	
$2_{11}-2_{20}$	2½ (4)−2½ (3)	62492.02	
$2_{11}-2_{20}$	1½ (5)−1½ (3)	62509.59	
$2_{11}-2_{20}$	2½ (5)−2½ (3)	62532.95	
$2_{11}-2_{20}$	3½ (3)−3½ (2)	62534.71	
$2_{11}-2_{20}$	2½ (5)−2½ (2)	62681.28	
$2_{11}-2_{20}$	3½ (3)−3½ (1)	62705.29	
$1_{11}-2_{02}$	2½ (2)−3½ (3)	14425.6	
$1_{11}-2_{02}$	1½ (3)−2½ (4)	14428.75	
$1_{11}-2_{02}$	3½ (1)−4½ (1)	14431	
$1_{11}-2_{02}$	1½ (1)−2½ (1)	14438.6	
$1_{11}-2_{02}$	½ (2)−1½ (2)	14455.6	
$1_{11}-2_{02}$	1½ (2)−2½ (2)	14469.08	
$3_{12}-3_{03}$	4½ (1)−4½ (1)	64695.19	
$3_{12}-3_{03}$	3½ (1)−3½ (1)	64714.84	
$3_{12}-3_{03}$	2½ (1)−2½ (1)	64729.65	
$3_{12}-3_{03}$	1½ (1)−1½ (1)	65033.38	
$3_{12}-3_{03}$	2½ (2)−2½ (2)	65058.46	
$3_{12}-3_{03}$	3½ (2)−3½ (2)	65094.62	
$4_{04}-3_{13}$	1½ (1)− ½ (1)	32978.86	
$4_{04}-3_{13}$	2½ (3)−1½ (3)	32988.15	
$4_{04}-3_{13}$	3½ (5)−2½ (5)	33011.36	
$4_{04}-3_{13}$	2½ (2)−1½ (2)	33057.56	
$4_{04}-3_{13}$	3½ (4)−2½ (4)	33072.19	

[a]) Because the two F nuclei are equivalent in NF_2, the rotational states can be divided into ortho ($I=1$, K_aK_c even even or odd odd) and para ($I=0$, K_aK_c eo or oe) spin states. The para levels therefore show only ^{14}N hyperfine structure.

Microwave data for $^{14}N^{19}F_2$ (continued)

Transition		ν MHz	Ref.
rotational[a]) $N'_{K_aK_c}-N''_{K_aK_c}$	fine, hyperfine $F(i)'-F(i)''$		
State: electronic $\tilde{X}\ ^2B_1$; vibrational (0, 0, 0) (continued)			
$4_{04}-3_{13}$	4½ (5) − 3½ (5)	33094.08	74Br
$4_{04}-3_{13}$	5½ (1) − 4½ (2)	33110.15	
$4_{04}-3_{13}$	3½ (3) − 2½ (3)	33180.95	
$4_{04}-3_{13}$	4½ (3) − 3½ (4)	33192.3	
$4_{04}-3_{13}$	5½ (1) − 4½ (1)	33214.1	
$4_{04}-3_{13}$	6½ (1) − 5½ (1)	33221.92	
$4_{04}-3_{13}$	4½ (3) − 3½ (3)	33223.9	
$4_{04}-3_{13}$	4½ (4) − 3½ (4)	33224.9	
$4_{04}-3_{13}$	5½ (3) − 4½ (3)	33235.07	
$4_{04}-3_{13}$	4½ (2) − 3½ (2)	33252.50	
$4_{04}-3_{13}$	4½ (4) − 3½ (3)	33255.45	
$4_{04}-3_{13}$	5½ (2) − 4½ (2)	33256.25	
$4_{04}-3_{13}$	3½ (2) − 2½ (2)	33272.29	
$4_{04}-3_{13}$	4½ (1) − 3½ (1)	33301.50	
$4_{04}-3_{13}$	3½ (1) − 2½ (1)	33316.98	
$4_{04}-3_{13}$	2½ (1) − 1½ (1)	33325.35	
$5_{24}-6_{15}$	3½ (1) − 4½ (1)	28461.8	
$5_{24}-6_{15}$	4½ (1) − 5½ (1)	28475.2	
$5_{24}-6_{15}$	5½ (1) − 6½ (1)	28489.7	
$5_{24}-6_{15}$	4½ (2) − 5½ (2)	28513.8	
$5_{24}-6_{15}$	5½ (2) − 6½ (2)	28526.6	
$5_{24}-6_{15}$	6½ (1) − 7½ (1)	28542.6	
$5_{24}-6_{15}$	5½ (3) − 6½ (3)	28552.9	
$5_{24}-6_{15}$	6½ (2) − 7½ (2)	28563.6	
$5_{24}-6_{15}$	7½ (1) − 8½ (1)	28575.6	
$5_{24}-6_{15}$	6½ (3) − 7½ (3)	28749.2	
$5_{24}-6_{15}$	5½ (4) − 6½ (4)	28772.1	
$5_{24}-6_{15}$	4½ (4) − 5½ (3)	28788.7	
$5_{24}-6_{15}$	5½ (5) − 6½ (5)	28836	
$5_{24}-6_{15}$	4½ (4) − 5½ (5)	28848	
$5_{24}-6_{15}$	4½ (4) − 5½ (4)	28864.6	
$5_{24}-6_{15}$	3½ (2) − 4½ (2)	28876.5	
$8_{17}-7_{26}$	5½ (1) − 4½ (1)	27263.08	
$8_{17}-7_{26}$	6½ (3) − 5½ (3)	27280.23	
$8_{17}-7_{26}$	7½ (5) − 6½ (5)	27300.56	
$8_{17}-7_{26}$	6½ (2) − 5½ (2)	27347.12	
$8_{17}-7_{26}$	7½ (4) − 6½ (4)	27363.57	
$8_{17}-7_{26}$	8½ (5) − 7½ (5)	27386.84	
$8_{17}-7_{26}$	7½ (3) − 6½ (3)	27435.45	
$8_{17}-7_{26}$	10½ (1) − 9½ (1)	27447.04	
$8_{17}-7_{26}$	8½ (4) − 7½ (4)	27452.51	
$8_{17}-7_{26}$	9½ (2) − 8½ (2)	27460.41	
$8_{17}-7_{26}$	8½ (3) − 7½ (3)	27473.08	
$8_{17}-7_{26}$	9½ (3) − 8½ (3)	27473.90	
$8_{17}-7_{26}$	9½ (1) − 8½ (1)	27493.13	
$8_{17}-7_{26}$	8½ (2) − 7½ (2)	27508.60	
$8_{17}-7_{26}$	7½ (2) − 6½ (2)	27522.77	

[a]) Because the two F nuclei are equivalent in NF_2, the rotational states can be divided into ortho ($I=1$, K_aK_c even even or odd odd) and para ($I=0$, K_aK_c eo or oe) spin states. The para levels therefore show only ^{14}N hyperfine structure.

Microwave data for $^{14}N^{19}F_2$ (continued)

Transition: rotational[a]) $N'_{K_aK_c}-N''_{K_aK_c}$	Transition: fine, hyperfine $F(i)'-F(i)''$	ν MHz	Ref.
State: electronic $\tilde{X}\ ^2B_1$; vibrational (0, 0, 0) (continued)			
$8_{17}-7_{26}$	8½ (1)−7½ (1)	27552.69	74Br
$8_{17}-7_{26}$	7½ (1)−6½ (1)	27567.27	
$8_{17}-7_{26}$	6½ (1)−5½ (1)	27581.79	
$8_{26}-9_{19}$	10½ (1)−11½ (1)	28404.8	
$8_{26}-9_{19}$	9½ (2)−10½ (2)	28430.1	
$8_{26}-9_{19}$	8½ (3)−9½ (3)	28457.3	
$8_{26}-9_{19}$	9½ (1)−10½ (1)	28520.7	
$8_{26}-9_{19}$	8½ (2)−9½ (2)	28550.6	
$8_{26}-9_{19}$	7½ (2)−8½ (2)	28579.2	
$8_{26}-9_{19}$	8½ (1)−9½ (1)	28647.8	
$8_{26}-9_{19}$	7½ (1)−8½ (1)	28675.5	
$8_{26}-9_{19}$	6½ (1)−7½ (1)	28704.2	
$10_{28}-11_{1,11}$	12½ (1)−13½ (1)	13845.52	
$10_{28}-11_{1,11}$	11½ (1)−12½ (1)	13978.85	
$10_{28}-11_{1,11}$	10½ (2)−11½ (2)	14010.95	
$10_{28}-11_{1,11}$	9½ (2)−10½ (2)	14043.65	
$10_{28}-11_{1,11}$	10½ (1)−11½ (1)	14122.11	
$10_{28}-11_{1,11}$	9½ (1)−10½ (1)	14153.25	
$10_{28}-11_{1,11}$	8½ (1)− 9½ (1)	14185.87	
$10_{28}-11_{1,11}$	7½ (1)− 8½ (1)	14469.08	
$10_{28}-11_{1,11}$	8½ (3)− 9½ (3)	14514.75	
$10_{28}-11_{1,11}$	9½ (5)−10½ (5)	14553.2	
$10_{28}-11_{1,11}$	8½ (2)− 9½ (2)	14654.83	
$10_{28}-11_{1,11}$	9½ (4)−10½ (4)	14690.61	
$10_{28}-11_{1,11}$	10½ (5)−11½ (5)	14732.73	
$10_{28}-11_{1,11}$	9½ (3)−10½ (3)	14840.03	
$10_{28}-11_{1,11}$	10½ (4)−11½ (4)	14877.1	
$10_{28}-11_{1,11}$	11½ (3)−12½ (3)	14915.13	
$14_{2,12}-15_{1,15}$	16½ (1)−17½ (1)	16036 [b])	
$14_{2,12}-15_{1,15}$	15½ (2)−16½ (2)	16078 [b])	
$14_{2,12}-15_{1,15}$	14½ (3)−15½ (3)	16117 [b])	
$14_{2,12}-15_{1,15}$	15½ (1)−16½ (1)	16210 [b])	
$14_{2,12}-15_{1,15}$	14½ (2)−15½ (2)	16250 [b])	
$14_{2,12}-15_{1,15}$	13½ (2)−14½ (2)	16290 [b])	
$14_{2,12}-15_{1,15}$	12½ (2)−13½ (2)	16929 [b])	
$14_{2,12}-15_{1,15}$	13½ (4)−14½ (4)	16963 [b])	
$14_{2,12}-15_{1,15}$	14½ (5)−15½ (5)	17016 [b])	
$14_{2,12}-15_{1,15}$	13½ (3)−14½ (3)	17137 [b])	
$14_{2,12}-15_{1,15}$	14½ (4)−15½ (4)	17179 [b])	
$14_{2,12}-15_{1,15}$	15½ (3)−16½ (3)	17229 [b])	
$12_{39}-13_{2,12}$	14½ (1)−15½ (1)	24495.3	
$12_{39}-13_{2,12}$	13½ (2)−14½ (2)	25503.4	
$12_{39}-13_{2,12}$	12½ (3)−13½ (3)	25511.8	
$12_{39}-13_{2,12}$	13½ (1)−14½ (1)	25534.5	
$12_{39}-13_{2,12}$	12½ (2)−13½ (2)	25542.3	
$12_{39}-13_{2,12}$	11½ (2)−12½ (2)	25550.9	

[a]) Because the two F nuclei are equivalent in NF_2, the rotational states can be divided into ortho ($I=1$, K_aK_c even even or odd odd) and para ($I=0$, K_aK_c eo or oe) spin states. The para levels therefore show only ^{14}N hyperfine structure.

[b]) These measurements are approximate only, made with a wavemeter.

Microwave data for $^{14}N^{19}F_2$ (continued)

Transition: rotational[a]) $N'_{K_aK_c}-N''_{K_aK_c}$	Transition: fine, hyperfine $F(i)'-F(i)''$	ν MHz	Ref.
State: electronic $\tilde{X}\ ^2B_1$; vibrational (0, 0, 0) (continued)			
$12_{39}-13_{3,12}$	12½ (1)−13½ (1)	25572.7	74Br
$12_{39}-13_{3,12}$	11½ (1)−12½ (1)	25580.9	
$12_{39}-13_{3,12}$	10½ (1)−11½ (1)	25589.9	
$12_{39}-13_{3,12}$	9½ (1)−10½ (1)	25942.4	
$12_{39}-13_{3,12}$	10½ (3)−11½ (3)	25951.4	
$12_{39}-13_{3,12}$	11½ (5)−12½ (5)	25961.9	
$12_{39}-13_{2,12}$	10½ (2)−11½ (2)	25988.9	
$12_{39}-13_{2,12}$	11½ (4)−12½ (4)	25997.8	
$12_{39}-13_{2,12}$	12½ (5)−13½ (5)	26008.5	
$12_{39}-13_{2,12}$	11½ (3)−12½ (3)	26036.5	
$12_{39}-13_{2,12}$	12½ (4)−13½ (4)	26045.1	
$12_{39}-13_{2,12}$	13½ (3)−14½ (3)	26054.2	
$10_{37}-11_{2,10}$	12½ (1)−13½ (1)	62182.0	
$10_{37}-11_{2,10}$	11½ (2)−12½ (2)	62188.2	
$10_{37}-11_{2,10}$	10½ (3)−11½ (3)	62194.2	
$10_{37}-11_{2,10}$	11½ (1)−12½ (1)	62214.4	
$10_{37}-11_{2,10}$	10½ (2)−11½ (2)	62219.1	
$10_{37}-11_{2,10}$	9½ (2)−10½ (2)	62229.7	
$10_{37}-11_{2,10}$	7½ (1)− 8½ (1)	62664.58	
$10_{37}-11_{2,10}$	8½ (3)− 9½ (3)	62671.52	
$10_{37}-11_{2,10}$	9½ (5)−10½ (5)	62680.03	
$10_{37}-11_{2,10}$	8½ (2)− 9½ (2)	62699.65	
$10_{37}-11_{2,10}$	9½ (4)−10½ (4)	62706.47	
$10_{37}-11_{2,10}$	10½ (5)−11½ (5)	62714.79	
$10_{37}-11_{2,10}$	9½ (3)−10½ (3)	62737.32	
$10_{37}-11_{2,10}$	10½ (4)−11½ (4)	62743.73	
$10_{37}-11_{2,10}$	11½ (3)−12½ (3)	62751.4	
$15_{4,11}-16_{3,14}$	16½ (1)−17½ (1)	61820.6	
$15_{4,11}-16_{3,14}$	15½ (1)−16½ (1)	61821.9	
$15_{4,11}-16_{3,14}$	14½ (1)−15½ (1)	61823.5	
$15_{4,11}-16_{3,14}$	13½ (1)−14½ (1)	62230.5	
$15_{4,11}-16_{3,14}$	14½ (2)−15½ (2)	62232.0	
$15_{4,11}-16_{3,14}$	15½ (2)−16½ (2)	62234.0	
$20_{5,15}-21_{4,18}$	22½ (1)−23½ (1)	64633.10 c)	
$20_{5,15}-21_{4,18}$	21½ (2)−22½ (2)	64633.10 c)	
$20_{5,15}-21_{4,18}$	20½ (3)−21½ (3)	64633.10 c)	
$20_{5,15}-21_{4,18}$	21½ (1)−22½ (1)	64634.2 c)	
$20_{5,15}-21_{4,18}$	20½ (2)−21½ (2)	64634.2 c)	
$20_{5,15}-21_{4,18}$	19½ (2)−20½ (2)	64634.2 c)	
$20_{5,15}-21_{4,18}$	20½ (1)−21½ (1)	64634.38 c)	
$20_{5,15}-21_{4,18}$	19½ (1)−20½ (1)	64634.38 c)	
$20_{5,15}-21_{4,18}$	18½ (1)−19½ (1)	64634.38 c)	
$20_{5,15}-21_{4,18}$	17½ (1)−18½ (1)	65002.02 c)	
$20_{5,15}-21_{4,18}$	18½ (3)−19½ (3)	65002.02 c)	
$20_{5,15}-21_{4,18}$	19½ (5)−20½ (5)	65002.02 c)	

a) Because the two F nuclei are equivalent in NF_2, the rotational states can be divided into ortho ($I=1$, K_aK_c even even or odd odd) and para ($I=0$, K_aK_c eo or oe) spin states. The para levels therefore show only ^{14}N hyperfine structure.

c) Unresolved hyperfine structure.

Note that the earlier assignments of Hrubesh, Rinehart and Anderson [70Hr] have been found to be incorrect.

Microwave data for $^{14}N^{19}F_2$ (continued)

Transition: rotational[a]) $N'_{K_aK_c}-N''_{K_aK_c}$	Transition: fine, hyperfine $F(i)'-F(i)''$	ν MHz	Ref.
State: electronic $\tilde{X}\ ^2B_1$; vibrational (0, 0, 0) (continued)			
$20_{5,15}-21_{4,18}$	18½ (2) – 19½ (2)	65002.02 [c])	74Br
$20_{5,15}-21_{4,18}$	19½ (4) – 20½ (4)	65002.02 [c])	
$20_{5,15}-21_{4,18}$	20½ (5) – 21½ (5)	65002.02 [c])	
$20_{5,15}-21_{4,18}$	19½ (3) – 20½ (3)	65002.02 [c])	
$20_{5,15}-21_{4,18}$	20½ (4) – 21½ (4)	65002.02 [c])	
$20_{5,15}-21_{4,18}$	21½ (3) – 22½ (3)	65002.02 [c])	
$21_{5,17}-22_{4,18}$	19½ (1) – 20½ (1)	33266.4	
$21_{5,17}-22_{4,18}$	20½ (1) – 21½ (1)	33269.0	
$21_{5,17}-22_{4,18}$	21½ (1) – 22½ (1)	33271.7	
$21_{5,17}-22_{4,18}$	20½ (2) – 21½ (2)	33277.9	
$21_{5,17}-22_{4,18}$	21½ (2) – 22½ (2)	33280.5	
$21_{5,17}-22_{4,18}$	22½ (1) – 23½ (1)	33283.2	
$21_{5,17}-22_{4,18}$	21½ (3) – 22½ (3)	33288.4	
$21_{5,17}-22_{4,18}$	22½ (2) – 23½ (2)	33290.9	
$21_{5,17}-22_{4,18}$	23½ (1) – 24½ (1)	33293.4	
$21_{5,17}-22_{4,18}$	22½ (3) – 23½ (3)	33545.3	
$21_{5,17}-22_{4,18}$	21½ (4) – 22½ (4)	33548.4	
$21_{5,17}-22_{4,18}$	20½ (3) – 21½ (3)	33551.2	
$21_{5,17}-22_{4,18}$	21½ (5) – 22½ (5)	33558.9	
$21_{5,17}-22_{4,18}$	20½ (4) – 21½ (4)	33562.0	
$21_{5,17}-22_{4,18}$	19½ (2) – 20½ (2)	33564.9	
$21_{5,17}-22_{4,18}$	20½ (5) – 21½ (5)	33572.7	
$21_{5,17}-22_{4,18}$	19½ (3) – 20½ (3)	33575.6	
$21_{5,17}-22_{4,18}$	18½ (1) – 19½ (1)	33578.3	
$25_{4,22}-24_{5,19}$	26½ (3) – 25½ (3)	25506.6	
$25_{4,22}-24_{5,19}$	25½ (4) – 24½ (4)	25507.2	
$25_{4,22}-24_{5,19}$	24½ (3) – 23½ (3)	25507.5	
$25_{4,22}-24_{5,19}$	25½ (5) – 24½ (5)	25509.2	
$25_{4,22}-24_{5,19}$	24½ (4) – 23½ (4)	25509.8	
$25_{4,22}-24_{5,19}$	23½ (2) – 22½ (2)	25510.2	
$25_{4,22}-24_{5,19}$	24½ (5) – 23½ (5)	25511.8	
$25_{4,22}-24_{5,19}$	23½ (3) – 22½ (3)	25512.6	
$25_{4,22}-24_{5,19}$	22½ (1) – 21½ (1)	25513.0	
$25_{4,22}-24_{5,19}$	23½ (1) – 22½ (1)	25809.8	
$25_{4,22}-24_{5,19}$	24½ (1) – 23½ (1)	25810.6	
$25_{4,22}-24_{5,19}$	25½ (1) – 24½ (1)	25811.2	
$25_{4,22}-24_{5,19}$	24½ (2) – 23½ (2)	25813.4	
$25_{4,22}-24_{5,19}$	25½ (2) – 24½ (2)	25814.2	
$25_{4,22}-24_{5,19}$	26½ (1) – 25½ (1)	25814.8	
$25_{4,22}-24_{5,19}$	25½ (3) – 24½ (3)	25817.4	
$25_{4,22}-24_{5,19}$	26½ (2) – 25½ (2)	25818.2	
$25_{4,22}-24_{5,19}$	27½ (1) – 26½ (1)	25818.8	
$27_{6,22}-28_{5,23}$	25½ (1) – 26½ (1)	17354 [b])	
$27_{6,22}-28_{5,23}$	26½ (1) – 27½ (1)	17356 [b])	

[a]) Because the two F nuclei are equivalent in NF_2, the rotational states can be divided into ortho ($I=1$, K_aK_c even even or odd odd) and para ($I=0$, K_aK_c eo or oe) spin states. The para levels therefore show only ^{14}N hyperfine structure.

[b]) These measurements are approximate only, made with a wavemeter.

[c]) Unresolved hyperfine structure.

Note that the earlier assignments of Hrubesh, Rinehart and Anderson [70Hr] have been found to be incorrect.

Microwave data for $^{14}N^{19}F_2$ (continued)

Transition rotational[a]) $N'_{K_aK_c}-N''_{K_aK_c}$	fine, hyperfine $F(i)'-F(i)''$	ν MHz	Ref.
State: electronic $\tilde{X}\ ^2B_1$; vibrational (0, 0, 0) (continued)			
$27_{6,22}-28_{5,23}$	27½ (1)–28½ (1)	17358 [b])	74Br
$27_{6,22}-28_{5,23}$	26½ (2)–27½ (2)	17362 [b])	
$27_{6,22}-28_{5,23}$	27½ (2)–28½ (2)	17364 [b])	
$27_{6,22}-28_{5,23}$	28½ (1)–29½ (1)	17367 [b])	
$27_{6,22}-28_{5,23}$	27½ (3)–28½ (3)	17371 [b])	
$27_{6,22}-28_{5,23}$	28½ (2)–29½ (2)	17373 [b])	
$27_{6,22}-28_{5,23}$	29½ (1)–30½ (1)	17375 [b])	
$27_{6,22}-28_{5,23}$	28½ (3)–29½ (3)	17632 [b])	
$27_{6,22}-28_{5,23}$	27½ (4)–28½ (4)	17633 [b])	
$27_{6,22}-28_{5,23}$	26½ (3)–27½ (3)	17634 [b])	
$27_{6,22}-28_{5,23}$	27½ (5)–28½ (5)	17640 [b])	
$27_{6,22}-28_{5,23}$	26½ (4)–27½ (4)	17641 [b])	
$27_{6,22}-28_{5,23}$	25½ (2)–26½ (2)	17642 [b])	
$27_{6,22}-28_{5,23}$	26½ (5)–27½ (5)	17648 [b])	
$27_{6,22}-28_{5,23}$	25½ (5)–26½ (5)	17649 [b])	
$27_{6,22}-28_{5,23}$	24½ (1)–25½ (1)	17650 [b])	
$30_{5,25}-29_{6,24}$	27½ (1)–26½ (1)	34069.5	
$30_{5,25}-29_{6,24}$	28½ (3)–27½ (3)	34072.3	
$30_{5,25}-29_{6,24}$	29½ (5)–28½ (5)	34075.1	
$30_{5,25}-29_{6,24}$	28½ (2)–27½ (2)	34082.65	
$30_{5,25}-29_{6,24}$	29½ (4)–28½ (4)	34085.39	
$30_{5,25}-29_{6,24}$	30½ (5)–29½ (5)	34088.28	
$30_{5,25}-29_{6,24}$	29½ (3)–28½ (3)	34095.56	
$30_{5,25}-29_{6,24}$	30½ (4)–29½ (4)	34098.31	
$30_{5,25}-29_{6,24}$	31½ (3)–30½ (3)	34101.22	
$30_{5,25}-29_{6,24}$	32½ (1)–31½ (1)	34286.52	
$30_{5,25}-29_{6,24}$	31½ (2)–30½ (2)	34289.02	
$30_{5,25}-29_{6,24}$	30½ (3)–29½ (3)	34291.57	
$30_{5,25}-29_{6,24}$	31½ (1)–30½ (1)	34297.42	
$30_{5,25}-29_{6,24}$	30½ (2)–29½ (2)	34299.95	
$30_{5,25}-29_{6,24}$	29½ (2)–28½ (2)	34302.56	
$30_{5,25}-29_{6,24}$	30½ (1)–29½ (1)	34309.00	
$30_{5,25}-29_{6,24}$	29½ (1)–28½ (1)	34311.54	
$30_{5,25}-29_{6,24}$	28½ (1)–27½ (1)	34314.12	
$32_{7,26}-33_{6,27}$	31½ (1)–32½ (1)	25604.6	
$32_{7,26}-33_{6,27}$	32½ (1)–33½ (1)	25606.0	
$32_{7,26}-33_{6,27}$	33½ (1)–34½ (1)	25607.1	
$32_{7,26}-33_{6,27}$	32½ (2)–33½ (2)	25893.6	
$32_{7,26}-33_{6,27}$	31½ (2)–32½ (2)	25894.8	
$32_{7,26}-33_{6,27}$	30½ (1)–31½ (1)	25896.0	
$32_{7,25}-33_{6,28}$	30½ (1)–31½ (1)	27600.49	
$32_{7,25}-33_{6,28}$	31½ (1)–32½ (1)	27601.12	
$32_{7,25}-33_{6,28}$	32½ (1)–33½ (1)	27601.88	

[a]) Because the two F nuclei are equivalent in NF_2, the rotational states can be divided into ortho ($I=1$, K_aK_c even even or odd odd) and para ($I=0$, K_aK_c eo or oe) spin states. The para levels therefore show only ^{14}N hyperfine structure.

[b]) These measurements are approximate only, made with a wavemeter.

Microwave data for $^{14}N^{19}F_2$ (continued)

Transition rotational[a]) $N'_{K_aK_c}-N''_{K_aK_c}$	fine, hyperfine $F(i)'-F(i)''$	ν MHz	Ref.
State: electronic $\tilde{X}\ ^2B_1$; vibrational (0, 0, 0) (continued)			
$32_{7,25}-33_{6,28}$	31½ (2)–32½ (2)	27603.56	74Br
$32_{7,25}-33_{6,28}$	32½ (2)–33½ (2)	27604.25	
$32_{7,25}-33_{6,28}$	33½ (1)–34½ (1)	27604.98	
$32_{7,25}-33_{6,28}$	32½ (3)–33½ (3)	27606.27	
$32_{7,25}-33_{6,28}$	33½ (2)–34½ (2)	27606.95	
$32_{7,25}-33_{6,28}$	34½ (1)–35½ (1)	27607.68	
$32_{7,25}-33_{6,28}$	33½ (3)–34½ (3)	27903.13	
$32_{7,25}-33_{6,28}$	32½ (4)–33½ (4)	27903.90	
$32_{7,25}-33_{6,28}$	31½ (3)–32½ (3)	27904.67	
$32_{7,25}-33_{6,28}$	32½ (5)–33½ (5)	27906.50	
$32_{7,25}-33_{6,28}$	31½ (4)–32½ (4)	27907.27	
$32_{7,25}-33_{6,28}$	30½ (2)–31½ (2)	27908.08	
$32_{7,25}-33_{6,28}$	31½ (5)–32½ (5)	27910.10	
$32_{7,25}-33_{6,28}$	30½ (3)–31½ (3)	27910.76	
$32_{7,25}-33_{6,28}$	29½ (1)–30½ (1)	27911.55	
$37_{8,30}-38_{7,31}$	35½ (1)–36½ (1)	32946.6	
$37_{8,30}-38_{7,31}$	36½ (1)–37½ (1)	32947.5	
$37_{8,30}-38_{7,31}$	37½ (1)–38½ (1)	32948.3	
$37_{8,30}-38_{7,31}$	36½ (2)–37½ (2)	32950.2	
$37_{8,30}-38_{7,31}$	37½ (2)–38½ (2)	32951.1	
$37_{8,30}-38_{7,31}$	38½ (1)–39½ (1)	32951.9	
$37_{8,30}-38_{7,31}$	37½ (3)–38½ (3)	32953.7	
$37_{8,30}-38_{7,31}$	38½ (2)–39½ (2)	32954.5	
$37_{8,30}-38_{7,31}$	39½ (1)–40½ (1)	32955.3	
$37_{8,30}-38_{7,31}$	38½ (3)–39½ (3)	33241.0	
$37_{8,30}-38_{7,31}$	37½ (4)–38½ (4)	33241.9	
$37_{8,30}-38_{7,31}$	36½ (3)–37½ (3)	33242.8	
$37_{8,30}-38_{7,31}$	37½ (5)–38½ (5)	33245.0	
$37_{8,30}-38_{7,31}$	36½ (4)–37½ (4)	33245.9	
$37_{8,30}-38_{7,31}$	35½ (2)–36½ (2)	33246.7	
$37_{8,30}-38_{7,31}$	36½ (5)–37½ (5)	33249.0	
$37_{8,30}-38_{7,31}$	35½ (3)–36½ (3)	33249.9	
$37_{8,30}-38_{7,31}$	34½ (1)–35½ (1)	33250.8	

[a]) Because the two F nuclei are equivalent in NF_2, the rotational states can be divided into ortho ($I=1$, K_aK_c even even or odd odd) and para ($I=0$, K_aK_c eo or oe) spin states. The para levels therefore show only ^{14}N hyperfine structure.

Molecular parameters for $^{14}N^{19}F_2$

Parameter [a])		Value	Method	Ref.
State: electronic $\tilde{X}\ ^2B_1$; vibrational (0, 0, 0)				
A	[MHz]	70496 (2) [b])	MW	74Br
B	[MHz]	11872.94 (20)		
C	[MHz]	10136.46 (20)		
Δ_c	[amu Å^2]	0.1204		
τ_{aaaa}	[MHz]	−7.75 (20)		
τ_{bbbb}	[MHz]	$-0.81\ (20) \cdot 10^{-1}$		
τ_{aabb}	[MHz]	0.297 (30)		
τ_{abab}	[MHz]	−0.126 (60)		
ε_{aa}	[MHz]	−951.79		
ε_{bb}	[MHz]	−92.86		
ε_{cc}	[MHz]	4.49		
$(O)_F$	[MHz]	164.39		
$(aa)_F$	[MHz]	−241.75		
$(bb)_F$	[MHz]	−226.48		
$(cc)_F$	[MHz]	468.22		
$(O)_N$	[MHz]	46.57		
$(aa)_N$	[MHz]	−47.72		
$(bb)_N$	[MHz]	−50.47		
$(cc)_N$	[MHz]	98.19		
$(aa)_Q$	[MHz]	5.59		
$(bb)_Q$	[MHz]	−0.73		
$(cc)_Q$	[MHz]	−4.86		
μ_b	[D]	0.136 (10)		
r_0(N–F)	[Å]	1.3494		
$\sphericalangle_0$(FNF)	[degrees]	103° 20′		

a) The parameters have been determined from a fit of part only of the microwave data set. The transitions involved are:

$1_{10}-1_{01}$	$1_{11}-2_{02}$	$2_{11}-2_{02}$	$3_{12}-3_{03}$
$4_{04}-3_{13}$	$8_{17}-7_{26}$	$10_{28}-11_{1,11}$	$5_{24}-6_{15}$
$12_{39}-13_{2,12}$	$10_{37}-11_{2,10}$		

The four τ parameters are quartic centrifugal distortion parameters. They can be related to the five parameters Δ_N, etc. by imposition of the planarity constraint [77Wa] which only holds strictly for equilibrium values of these parameters.

b) The numbers in parentheses represent one standard deviation of the least-squares fit, in units of the last quoted decimal place.

References for NF_2

70Hr Hrubesh, L.W., Rinehart, E.A., Anderson, R.E.: J. Mol. Spectrosc. **36** (1970) 954.

74Br Brown, R.D., Burden, F.R., Godfrey, P.D., Gillard, I.R.: J. Mol. Spectrosc. **25** (1974) 301.

4 NH_2

Microwave data for $^{14}N^{1}H_2$

Transition rotational $N'_{K_aK_c} - N''_{K_aK_c}$	fine $J' - J''$	hyperfine $F'_1 - F''_1$ [a])	$F' - F''$	ν MHz	Ref.
State: electronic $\tilde{X}\ ^2B_1$; vibrational (0, 0, 0)					
$5_{23} - 6_{16}$	5½ − 6½		6½ − 7½	8749.68 (50) [b])	76Co
	5½ − 6½		5½ − 6½	8790.30 (50)	76Co
	5½ − 6½		6½ − 6½	8797.71 (50)	76Co
	5½ − 6½		4½ − 5½	8819.81 (50)	76Co
	5½ − 6½		5½ − 5½	8828.89 (50)	76Co
	4½ − 5½		5½ − 6½	16043.84 (10)	76Hi
	4½ − 5½		4½ − 5½	15991.22 (30)	76Hi
	4½ − 5½		3½ − 4½	15942.63 (30)	76Hi
	5½ − 5½		6½ − 6½	5490.3 (10)	82Hi1
	5½ − 5½		5½ − 5½	5428.5 (10)	82Hi1
	5½ − 5½		4½ − 4½	5371.2 (10)	82Hi1
$1_{10} - 1_{01}$	1½ − ½		2½ − 1½	461465.03 (16)	81Ch
	1½ − 1½		1½ − 2½	462425.42 (6)	81Ch
	1½ − 1½		2½ − 2½	462433.51 (6)	81Ch
	1½ − 1½		½ − 1½	462449.04 (10)	81Ch
	1½ − 1½		1½ − 1½	462455.58 (6)	81Ch
	1½ − 1½		½ − ½	462467.02 (7)	81Ch
	½ − ½		1½ − 1½	469440.62 (6)	81Ch
	½ − ½		1½ − ½	469383.37 (6)	81Ch
	½ − ½		½ − 1½	469366.12 (6)	81Ch
$3_{13} - 2_{20}$	2½ − 1½	2½ − 1½	3½ − 2½	229405.86 (7)	81Ch
	2½ − 1½	3½ − 2½	4½ − 3½	229409.87 (3)	81Ch
	2½ − 1½	3½ − 2½	3½ − 2½	229474.94 (6)	81Ch
	2½ − 1½	3½ − 1½	3½ − 2½	229494.38 (3)	81Ch
	2½ − 1½	3½ − 2½	2½ − 1½	229528.04 (6)	81Ch
	2½ − 1½	2½ − ½	2½ − 1½	229481.40 (6)	81Ch
	3½ − 2½	4½ − 3½	3½ − 2½	241533.95 (10) [c])	81Ch
	3½ − 2½	3½ − 2½	2½ − 1½	241533.95 (10) [c])	81Ch
	3½ − 2½	2½ − 1½	1½ − ½	241538.80 (50)	81Ch
	3½ − 2½	4½ − 3½	4½ − 3½	241557.68 (6) [c])	81Ch
	3½ − 2½	3½ − 2½	3½ − 2½	241557.68 (6) [c])	81Ch
	3½ − 2½	2½ − 1½	2½ − 1½	241561.83 (7)	81Ch
	3½ − 2½	3½ − 2½	4½ − 3½	241591.33 (9) [c])	81Ch
	3½ − 2½	4½ − 3½	5½ − 4½	241591.33 (9) [c])	81Ch
	3½ − 2½	2½ − 1½	3½ − 2½	241596.00 (26)	81Ch
3_{03} [d])	2½ − 3½		3½ − 4½	2310.68 (50)	77Co
	2½ − 3½		2½ − 3½	2397.48 (50)	77Co
	2½ − 3½		1½ − 2½	2461.04 (50)	77Co

[a]) For NH_2 in the $\tilde{X}\ ^2B_1$ state, the hyperfine interactions involving the protons are larger than those involving the ^{14}N nucleus. In the $\tilde{A}\ ^2A_1$ state on the other hand, the ^{14}N angular momentum is more strongly coupled than that of the protons. Thus the intermediate angular momentum quantum number refers to the coupling of $\boldsymbol{I}_H$ to $\boldsymbol{J}$ in the $\tilde{X}$ state and to the coupling of $\boldsymbol{I}_N$ to $\boldsymbol{J}$ in the $\tilde{A}$ state. For NH_2 in the $\tilde{X}\ ^2B_1$ state, the ortho rotational levels are those with $K_a K_c$ even even or odd odd and the para states are eo or oe. The association is exactly the opposite for the $\tilde{A}\ ^2A_1$ state.

[b]) The figures in parentheses represent the experimental uncertainty, in units of the last quoted decimal place.

[c]) Line blended with at least one other hyperfine component.

[d]) Magnetic dipole transition between the two spin components of a given rotational level $N_{K_aK_c}$.

Microwave data for $^{14}N^1H_2$ (continued)

Transition rotational $N'_{K_aK_c}-N''_{K_aK_c}$	fine $J'-J''$	hyperfine $F_1'-F_1''$ a)	$F'-F''$	ν MHz	Ref.
State: electronic $\tilde{X}\ ^2B_1$; vibrational (0, 0, 0) (continued)					
4_{04} d)	3½−4½	2½−3½	3½−4½	2690.57 (50) b)	77Co
	3½−4½	2½−3½	2½−3½	2783.66 (50)	77Co
	3½−4½	2½−3½	1½−2½	2854.20 (50)	77Co
	3½−4½	3½−4½	4½−5½	2756.67 (50)	77Co
	3½−4½	3½−4½	3½−4½	2848.44 (50)	77Co
	3½−4½	3½−4½	2½−3½	2923.19 (50)	77Co
	3½−4½	4½−5½	5½−6½	2835.27 (50)	77Co
	3½−4½	4½−5½	4½−5½	2927.86 (50)	77Co
	3½−4½	4½−5½	3½−4½	3002.22 (50)	77Co
	3½−4½	2½−3½	3½−3½	2733.04 (50)	77Co
	3½−4½	4½−4½	4½−4½	2886.59 (50) c)	77Co
	3½−4½	3½−3½	2½−2½	2886.59 (50) c)	77Co
5_{05} d)	4½−5½		5½−6½	3033.34 (50)	77Co
	4½−5½		4½−5½	3131.49 (50)	77Co
	4½−5½		3½−4½	3213.11 (50)	77Co
6_{06} d)	5½−6½	4½−5½	5½−6½	3092.06 (50)	77Co
	5½−6½	4½−5½	4½−5½	3193.71 (50)	77Co
	5½−6½	4½−5½	3½−4½	3279.32 (50)	77Co
	5½−6½	5½−6½	6½−7½	3160.43 (50)	77Co
	5½−6½	5½−6½	5½−6½	3260.69 (50)	77Co
	5½−6½	5½−6½	4½−5½	3345.35 (50)	77Co
	5½−6½	6½−7½	7½−8½	3237.09 (50)	77Co
	5½−6½	6½−7½	6½−7½	3337.31 (50)	77Co
	5½−6½	6½−7½	5½−6½	3423.99 (50)	77Co
7_{07} d)	6½−7½		7½−8½	3239.30 (50)	77Co
	6½−7½		6½−7½	3342.35 (50)	77Co
	6½−7½		5½−6½	3433.60 (50)	77Co
8_{08} d)	7½−8½	6½−7½	7½−8½	3202.75 (50)	77Co
	7½−8½	7½−8½	8½−9½	3272.56 (50)	77Co
	7½−8½	7½−8½	7½−8½	3376.67 (50)	77Co
	7½−8½	7½−8½	6½−7½	3468.65 (50)	77Co
	7½−8½	8½−9½	9½−10½	3348.00 (50)	77Co
	7½−8½	8½−9½	8½−9½	3451.30 (50)	77Co
	7½−8½	8½−9½	7½−8½	3544.88 (50)	77Co
1_{11} d)	½−1½	1½−2½	2½−3½	6933.25 (50)	77Co
	½−1½	1½−2½	1½−2½	7012.66 (50)	77Co
1_{11} d)	½−1½	1½−2½	½−1½	7035.24 (50)	77Co
	½−1½	1½−1½	1½−1½	6961.59 (50) c)	77Co
	½−1½	1½−2½	2½−2½	6961.59 (50) c)	77Co
	½−1½	1½−2½	1½−1½	7029.51 (50)	77Co

a) For NH_2 in the $\tilde{X}\ ^2B_1$ state, the hyperfine interactions involving the protons are larger than those involving the ^{14}N nucleus. In the $\tilde{A}\ ^2A_1$ state on the other hand, the ^{14}N angular momentum is more strongly coupled than that of the protons. Thus the intermediate angular momentum quantum number refers to the coupling of $\boldsymbol{I}_H$ to $\boldsymbol{J}$ in the $\tilde{X}$ state and to the coupling of $\boldsymbol{I}_N$ to $\boldsymbol{J}$ in the $\tilde{A}$ state. For NH_2 in the $\tilde{X}\ ^2B_1$ state, the ortho rotational levels are those with K_aK_c even even or odd odd and the para states are eo or oe. The association is exactly the opposite for the $\tilde{A}\ ^2A_1$ state.

b) The figures in parentheses represent the experimental uncertainty, in units of the last quoted decimal place.

c) Line blended with at least one other hyperfine component.

d) Magnetic dipole transition between the two spin components of a given rotational level $N_{K_aK_c}$.

Microwave data for $^{14}N^{1}H_2$ (continued)

Transition rotational $N'_{K_aK_c} - N''_{K_aK_c}$	fine $J' - J''$	hyperfine $F_1' - F_1''$ [a]	$F' - F''$	ν MHz	Ref.
State: electronic $\tilde{X}\ ^2B_1$; vibrational (0, 0, 0) (continued)					
1_{11} [d]	½–1½	1½–1½	1½– ½	6974.08 (50) [b]	77Co
2_{12} [d]	1½–2½		2½–3½	4357.74 (50)	77Co
	1½–2½		2½–2½	4395.18 (50)	77Co
	1½–2½		1½–2½	4452.24 (50)	77Co
	1½–2½		1½–1½	4477.63 (50)	77Co
	1½–2½		½–1½	4515.18 (50)	77Co
3_{13} [d]	2½–3½	1½–2½	2½–3½	3531.52 (50)	77Co
	2½–3½	1½–2½	1½–2½	3268.63 (50)	77Co
	2½–3½	2½–3½	3½–4½	3594.27 (50)	77Co
	2½–3½	2½–3½	2½–3½	3685.76 (50)	77Co
	2½–3½	2½–3½	1½–2½	3753.51 (50)	77Co
	2½–3½	3½–4½	4½–5½	3670.69 (50)	77Co
	2½–3½	3½–4½	3½–4½	3765.29 (50)	77Co
	2½–3½	3½–4½	2½–3½	3835.03 (50)	77Co
	2½–3½	3½–3½	2½–2½	3791.28 (50)	77Co
	2½–3½	1½–2½	1½–1½	3654.54 (50) [c]	77Co
	2½–3½	1½–2½	2½–2½	3654.54 (50) [c]	77Co
	2½–3½	3½–3½	3½–4½	3678.38 (50) [c]	77Co
	2½–3½	2½–4½	3½–4½	3678.38 (50) [c]	77Co
	2½–3½	3½–4½	4½–4½	3714.10 (50) [c]	77Co
	2½–3½	2½–3½	2½–2½	3714.10 (50) [c]	77Co
	2½–3½	3½–3½	3½–3½	3720.30 (50) [c]	77Co
	2½–3½	2½–2½	1½–1½	3720.30 (50) [c]	77Co
	2½–3½	3½–4½	4½–4½	3720.30 (50) [c]	77Co
4_{14} [d]	3½–4½		4½–5½	3348.94 (50)	77Co
	3½–4½		3½–4½	3446.67 (50)	77Co
	3½–4½		2½–3½	3523.98 (50)	77Co
5_{15} [d]	4½–5½	3½–4½	4½–5½	3191.20 (50)	77Co
	4½–5½	3½–4½	3½–4½	3291.80 (50)	77Co
	4½–5½	3½–4½	2½–3½	3373.82 (50)	77Co
	4½–5½	4½–5½	5½–6½	3258.50 (50)	77Co
	4½–5½	4½–5½	4½–5½	3356.93 (50)	77Co
	4½–5½	4½–5½	3½–4½	3436.73 (50) [c]	77Co
	4½–5½	5½–6½	5½–6½	3436.73 (50) [c]	77Co
	4½–5½	5½–6½	6½–7½	3335.68 (50)	77Co
	4½–5½	5½–6½	4½–5½	3517.15 (50)	77Co
	4½–5½	3½–5½	4½–5½	3270.18 (50) [c]	77Co
	4½–5½	4½–4½	4½–5½	3270.18 (50) [c]	77Co
	4½–5½	4½–4½	5½–4½	3270.18 (50) [c]	77Co
	4½–5½	5½–5½	4½–4½	3474.74 (50) [c]	77Co

[a]) For NH_2 in the $\tilde{X}\,^2B_1$ state, the hyperfine interactions involving the protons are larger than those involving the ^{14}N nucleus. In the $\tilde{A}\,^2A_1$ state on the other hand, the ^{14}N angular momentum is more strongly coupled than that of the protons. Thus the intermediate angular momentum quantum number refers to the coupling of $\boldsymbol{I}_H$ to $\boldsymbol{J}$ in the $\tilde{X}$ state and to the coupling of $\boldsymbol{I}_N$ to $\boldsymbol{J}$ in the $\tilde{A}$ state. For NH_2 in the $\tilde{X}\,^2B_1$ state, the ortho rotational levels are those with K_aK_c even even or odd odd and the para states are eo or oe. The association is exactly the opposite for the $\tilde{A}\,^2A_1$ state.

[b]) The figures in parentheses represent the experimental uncertainty, in units of the last quoted decimal place.

[c]) Line blended with at least one other hyperfine component.

[d]) Magnetic dipole transition between the two spin components of a given rotational level $N_{K_aK_c}$.

Microwave data for $^{14}N^{1}H_2$ (continued)

Transition				ν MHz	Ref.
rotational	fine	hyperfine			
$N'_{K_aK_c} - N''_{K_aK_c}$	$J' - J''$	$F_1' - F_1''$ [a])	$F' - F''$		
State: electronic $\tilde{X}\ ^2B_1$; vibrational (0, 0, 0) (continued)					
5_{15} [d])	4½ − 5½	5½ − 6½	5½ − 5½	3474.74 (50) [c])	77Co
6_{16} [d])	5½ − 6½		6½ − 7½	3259.61 (50) [b])	77Co
	5½ − 6½		5½ − 6½	3361.93 (50)	77Co
	5½ − 6½		4½ − 5½	3448.67 (50)	77Co
7_{17} [d])	6½ − 7½	5½ − 6½	6½ − 7½	3196.52 (50)	77Co
	6½ − 7½	5½ − 6½	5½ − 6½	3299.96 (50)	77Co
	6½ − 7½	5½ − 6½	4½ − 5½	3389.13 (50)	77Co
	6½ − 7½	6½ − 7½	7½ − 8½	3265.07 (50)	77Co
	6½ − 7½	6½ − 7½	6½ − 7½	3367.80 (50)	77Co
	6½ − 7½	6½ − 7½	5½ − 6½	3457.27 (50)	77Co
	6½ − 7½	7½ − 8½	8½ − 9½	3341.28 (50)	77Co
	6½ − 7½	7½ − 8½	7½ − 8½	3443.96 (50)	77Co
	6½ − 7½	7½ − 8½	6½ − 7½	3534.38 (50)	77Co
1_{10} [d])	½ − 1½		½ − 1½	7910.53 (50)	77Co
	½ − 1½		½ − ½	7916.63 (50)	77Co
	½ − 1½		1½ − 2½	7975.56 (50)	77Co
	½ − 1½		1½ − 1½	7983.72 (50)	77Co
2_{11} [d])	1½ − 2½	1½ − 1½	1½ − 2½	6046.45 (50)	77Co
	1½ − 2½	1½ − 2½	½ − 1½	6096.09 (50)	77Co
	1½ − 2½	1½ − 2½	2½ − 3½	6087.46 (50)	77Co
	1½ − 2½	2½ − 3½	1½ − 2½	6181.50 (50) [c])	77Co
	1½ − 2½	2½ − 3½	2½ − 3½	6181.50 (50) [c])	77Co
	1½ − 2½	2½ − 3½	3½ − 4½	6181.50 (50) [c])	77Co
	1½ − 2½	2½ − 2½	1½ − 1½	6129.44 (50) [c])	77Co
	1½ − 2½	2½ − 2½	2½ − 2½	6129.44 (50) [c])	77Co
	1½ − 2½	2½ − 2½	3½ − 3½	6129.44 (50) [c])	77Co
	1½ − 2½	2½ − 2½	1½ − 2½	6129.44 (50) [c])	77Co
	1½ − 2½	½ − 2½	½ − 1½	6070.99 (50) [c])	77Co
	1½ − 2½	½ − 2½	1½ − 2½	6070.99 (50) [c])	77Co
	1½ − 2½	1½ − 1½	1½ − 1½	6062.03 (50)	77Co
2_{11} [d])	1½ − 2½	1½ − 1½	2½ − 2½	6052.40 (50)	77Co
	1½ − 2½	½ − 1½	1½ − 1½	6036.56 (50) [c])	77Co
	1½ − 2½	½ − 1½	½ − ½	6036.56 (50) [c])	77Co
	1½ − 2½	½ − 1½	½ − 1½	6027.23 (50) [c])	77Co
	1½ − 2½	½ − 1½	1½ − 2½	6027.23 (50) [c])	77Co
	1½ − 2½	1½ − 3½	1½ − 2½	6149.51 (50) [c])	77Co
	1½ − 2½	1½ − 3½	2½ − 3½	6149.51 (50) [c])	77Co
3_{12} [d])	2½ − 3½		3½ − 4½	6070.82 (50)	77Co
	2½ − 3½		2½ − 3½	6087.66 (50)	77Co
	2½ − 3½		1½ − 2½	6099.20 (50)	77Co

[a]) For NH_2 in the $\tilde{X}\ ^2B_1$ state, the hyperfine interactions involving the protons are larger than those involving the ^{14}N nucleus. In the $\tilde{A}\ ^2A_1$ state on the other hand, the ^{14}N angular momentum is more strongly coupled than that of the protons. Thus the intermediate angular momentum quantum number refers to the coupling of $\boldsymbol{I}_H$ to $\boldsymbol{J}$ in the $\tilde{X}$ state and to the coupling of $\boldsymbol{I}_N$ to $\boldsymbol{J}$ in the $\tilde{A}$ state. For NH_2 in the $\tilde{X}\ ^2B_1$ state, the ortho rotational levels are those with K_aK_c even even or odd odd and the para states are eo or oe. The association is exactly the opposite for the $\tilde{A}\ ^2A_1$ state.

[b]) The figures in parentheses represent the experimental uncertainty, in units of the last quoted decimal place.

[c]) Line blended with at least one other hyperfine component.

[d]) Magnetic dipole transition between the two spin components of a given rotational level $N_{K_aK_c}$.

Microwave data for $^{14}N^{1}H_2$ (continued)

Transition				ν MHz	Ref.
rotational	fine	hyperfine			
$N'_{K_aK_c}-N''_{K_aK_c}$	$J'-J''$	$F_1'-F_1''$ [a]	$F'-F''$		
State: electronic $\tilde{X}\,^2B_1$; vibrational (0, 0, 0) (continued)					
4_{13} [d]	3½−4½	2½−3½	3½−4½	6234.81 (50) [b]	77Co
	3½−4½	2½−3½	2½−3½	6259.57 (50)	77Co
	3½−4½	2½−3½	1½−2½	6274.76 (50)	77Co
	3½−4½	3½−4½	4½ 5½	6300.75 (50)	77Co
	3½−4½	3½−4½	3½−4½	6324.25 (50)	77Co
	3½−4½	3½−4½	2½−3½	6343.51 (50)	77Co
	3½−4½	4½−5½	5½−6½	6381.26 (50)	77Co
	3½−4½	4½−5½	4½−5½	6404.53 (50)	77Co
	3½−4½	4½−5½	3½−4½	6424.05 (50)	77Co
	3½−4½	3½−4½	4½−4½	6316.39 (50)	77Co
5_{14} [d]	4½−5½		5½−6½	7351.21 (50)	77Co
	4½−5½		4½−5½	7389.63 (50)	77Co
	4½−5½		3½−4½	7422.73 (50)	77Co
6_{15} [d]	5½−6½	4½−5½	5½−6½	8075.05 (50)	77Co
	5½−6½	4½−5½	4½−5½	8126.18 (50)	77Co
	5½−6½	4½−5½	3½−4½	8168.56 (50)	77Co
	5½−6½	5½−6½	6½−7½	8144.94 (50)	77Co
	5½−6½	5½−6½	5½−6½	8194.20 (50)	77Co
	5½−6½	5½−6½	4½−5½	8235.64 (50)	77Co
	5½−6½	6½−7½	7½−8½	8227.24 (50)	77Co
	5½−6½	6½−7½	6½−7½	8274.16 (50)	77Co
	5½−6½	6½−7½	5½−6½	8217.91 (50)	77Co
2_{21} [d]	1½−2½		2½−3½	15842.38 (50) [c]	77Co
	1½−2½		1½−2½	15842.38 (50) [c]	77Co
	1½−2½		½−1½	15842.38 (50) [c]	77Co
3_{22} [d]	2½−3½	1½−2½	2½−3½	12208.93 (50)	77Co
	2½−3½	1½−2½	1½−2½	12240.39 (50)	77Co
	2½−3½	1½−2½	½−1½	12265.07 (50) [c]	77Co
	2½−3½	2½−3½	3½−4½	12265.07 (50) [c]	77Co
	2½−3½	2½−3½	2½−3½	12297.53 (50)	77Co
	2½−3½	2½−3½	1½−2½	12318.51 (50)	77Co
	2½−3½	3½−4½	4½−5½	12341.25 (50)	77Co
	2½−3½	3½−4½	3½−4½	12371.78 (50)	77Co
	2½−3½	3½−4½	2½−3½	12395.64 (50)	77Co
	2½−3½	2½−2½	3½−3½	12233.74 (50)	77Co
4_{23} [d]	3½−4½		4½−5½	10698.10 (50)	77Co
	3½−4½		3½−4½	10741.97 (50)	77Co
	3½−4½		2½−3½	10777.55 (50)	77Co
5_{24} [d]	4½−5½	3½−4½	4½−5½	9881.00 (50)	77Co
	4½−5½	3½−4½	3½−4½	9933.39 (50)	77Co

[a]) For NH_2 in the $\tilde{X}\,^2B_1$ state, the hyperfine interactions involving the protons are larger than those involving the ^{14}N nucleus. In the $\tilde{A}\,^2A_1$ state on the other hand, the ^{14}N angular momentum is more strongly coupled than that of the protons. Thus the intermediate angular momentum quantum number refers to the coupling of $\boldsymbol{I}_H$ to $\boldsymbol{J}$ in the $\tilde{X}$ state and to the coupling of $\boldsymbol{I}_N$ to $\boldsymbol{J}$ in the $\tilde{A}$ state. For NH_2 in the $\tilde{X}\,^2B_1$ state, the ortho rotational levels are those with K_aK_c even even or odd odd and the para states are eo or oe. The association is exactly the opposite for the $\tilde{A}\,^2A_1$ state.

[b]) The figures in parentheses represent the experimental uncertainty, in units of the last quoted decimal place.

[c]) Line blended with at least one other hyperfine component.

[d]) Magnetic dipole transition between the two spin components of a given rotational level $N_{K_aK_c}$.

Microwave data for $^{14}N^{1}H_2$ (continued)

Transition				ν	Ref.
rotational	fine	hyperfine		MHz	
$N'_{K_aK_c} - N''_{K_aK_c}$	$J' - J''$	$F'_1 - F''_1$ [a]	$F' - F''$		
State: electronic $\tilde{X}\ ^2B_1$; vibrational (0, 0, 0) (continued)					
5_{24} [d]	4½ – 5½	3½ – 4½	2½ – 3½	9975.75 (50) [b]	77Co
	4½ – 5½	4½ – 5½	5½ – 6½	9946.27 (50)	77Co
	4½ – 5½	4½ – 5½	4½ – 5½	9997.29 (50)	77Co
	4½ – 5½	4½ – 5½	3½ – 4½	10040.56 (50)	77Co
	4½ – 5½	5½ – 6½	6½ – 7½	10023.96 (50)	77Co
	4½ – 5½	5½ – 6½	5½ – 6½	10074.88 (50)	77Co
	4½ – 5½	5½ – 6½	4½ – 5½	10118.30 (50)	77Co
6_{25} [d]	5½ – 6½		6½ – 7½	9645.66 (50)	77Co
	5½ – 6½		5½ – 6½	9704.72 (50)	77Co
	5½ – 6½		4½ – 5½	9754.59 (50)	77Co
7_{26} [d]	6½ – 7½	5½ – 6½	6½ – 7½	9480.72 (50)	77Co
	6½ – 7½	5½ – 6½	5½ – 6½	9547.07 (50) [c]	77Co
	6½ – 7½	5½ – 6½	4½ – 5½	9601.54 (50)	77Co
	6½ – 7½	6½ – 7½	7½ – 8½	9547.07 (50) [c]	77Co
	6½ – 7½	6½ – 7½	6½ – 7½	9613.37 (50)	77Co
	6½ – 7½	6½ – 7½	5½ – 6½	9668.92 (50)	77Co
	6½ – 7½	7½ – 8½	8½ – 9½	9625.40 (50)	77Co
	6½ – 7½	7½ – 8½	7½ – 8½	9689.33 (50)	77Co
	6½ – 7½	7½ – 8½	6½ – 7½	9746.85 (50)	77Co
2_{20} [d]	1½ – 2½	2½ – 3½	3½ – 4½	15850.66 (50)	77Co
	1½ – 2½	½ – 3½	2½ – 2½	15828.51 (50)	77Co
	1½ – 2½	2½ – 2½	3½ – 3½	15809.56 (50)	77Co
	1½ – 2½	2½ – 2½	2½ – 3½	15796.71 (50)	77Co
	1½ – 2½	1½ – 2½	2½ – 2½	15784.56 (50)	77Co
	1½ – 2½	1½ – 2½	2½ – 3½	15778.90 (50)	77Co
	1½ – 2½	½ – 2½	1½ – 1½	15763.98 (50)	77Co
	1½ – 2½	½ – 2½	½ – 1½	15755.41 (50)	77Co
	1½ – 2½	½ – 1½	1½ – 1½	15729.18 (50)	77Co
3_{21} [d]	2½ – 3½		3½ – 4½	12272.98 (50) [c]	77Co
	2½ – 3½		2½ – 3½	12272.98 (50) [c]	77Co
	2½ – 3½		1½ – 2½	12272.98 (50) [c]	77Co
4_{22} [d]	3½ – 4½	2½ – 3½	3½ – 4½	10797.40 (50)	77Co
	3½ – 4½	2½ – 3½	2½ – 3½	10804.92 (50)	77Co
	3½ – 4½	2½ – 3½	1½ – 2½	10808.91 (50)	77Co
	3½ – 4½	3½ – 4½	4½ – 5½	10863.49 (50)	77Co
	3½ – 4½	3½ – 4½	3½ – 4½	10870.52 (50)	77Co
	3½ – 4½	3½ – 4½	2½ – 3½	10876.23 (50)	77Co
	3½ – 4½	4½ – 5½	5½ – 6½	10945.58 (50)	77Co
	3½ – 4½	4½ – 5½	4½ – 5½	10951.64 (50)	77Co
	3½ – 4½	4½ – 5½	3½ – 4½	10956.81 (50)	77Co

[a]) For NH_2 in the $\tilde{X}\ ^2B_1$ state, the hyperfine interactions involving the protons are larger than those involving the ^{14}N nucleus. In the $\tilde{A}\ ^2A_1$ state on the other hand, the ^{14}N angular momentum is more strongly coupled than that of the protons. Thus the intermediate angular momentum quantum number refers to the coupling of $\boldsymbol{I}_H$ to $\boldsymbol{J}$ in the $\tilde{X}$ state and to the coupling of $\boldsymbol{I}_N$ to $\boldsymbol{J}$ in the $\tilde{A}$ state. For NH_2 in the $\tilde{X}\ ^2B_1$ state, the ortho rotational levels are those with K_aK_c even even or odd odd and the para states are eo or oe. The association is exactly the opposite for the $\tilde{A}\ ^2A_1$ state.

[b]) The figures in parentheses represent the experimental uncertainty, in units of the last quoted decimal place.

[c]) Line blended with at least one other hyperfine component.

[d]) Magnetic dipole transition between the two spin components of a given rotational level $N_{K_aK_c}$.

Microwave data for $^{14}N^1H_2$ (continued)

Transition				ν MHz	Ref.
rotational	fine	hyperfine			
$N'_{K_aK_c} - N''_{K_aK_c}$	$J' - J''$	$F_1' - F_1''$[a]	$F' - F''$		
State: electronic $\tilde{X}\ ^2B_1$; vibrational (0, 0, 0) (continued)					
4_{22}[d]	3½–4½	3½–3½	2½–3½	10828.80 (50)[b][c]	77Co
	3½–4½	3½–3½	4½–4½	10828.80 (50)[c]	77Co
	3½–4½	3½–3½	3½–4½	10828.80 (50)[c]	77Co
	3½–4½	3½–3½	2½–2½	10841.46 (50)[c]	77Co
	3½–4½	2½–4½	3½–4½	10841.46 (50)[c]	77Co
	3½–4½	2½–3½	2½–3½	10848.11 (50)[c]	77Co
	3½–4½	3½–3½	3½–3½	10848.11 (50)[c]	77Co
5_{23}[d]	4½–5½		5½–6½	10553.66 (50)	77Co
	4½–5½		4½–5½	10562.33 (50)	77Co
	4½–5½		3½–4½	10570.95 (50)	77Co
6_{24}[d]	5½–6½	4½–5½	5½–6½	10776.22 (50)	77Co
	5½–6½	4½–5½	4½–5½	10790.85 (50)	77Co
	5½–6½	4½–5½	3½–4½	10803.06 (50)	77Co
	5½–6½	5½–6½	6½–7½	10846.48 (50)	77Co
	5½–6½	5½–6½	5½–6½	10861.43 (50)	77Co
	5½–6½	5½–6½	4½–5½	10873.34 (50)	77Co
	5½–6½	6½–7½	7½–8½	10928.20 (50)	77Co
	5½–6½	6½–7½	6½–7½	10942.76 (50)	77Co
	5½–6½	6½–7½	5½–6½	10955.57 (50)	77Co
7_{25}[d]	6½–7½		7½–8½	11537.14 (50)	77Co
	6½–7½		6½–7½	11558.85 (50)	77Co
	6½–7½		5½–6½	11578.69 (50)	77Co
9_{09}[d]	8½–9½		9½–10½	3301.1 (10)	82Hi1
			8½–9½	3407.6 (10)	82Hi1
			7½–8½	3502.6 (10)	82Hi1
8_{18}[d]	7½–8½		8½–9½	3290.1 (5)	82Hi1
			7½–8½	3395.1 (5)	82Hi1
			6½–7½	3488.45 (50)	82Hi1
2_{12}[d]	1½–2½		½–1½	4514.35 (50)	82Hi1
7_{16}[d]	6½–7½		7½–8½	8795.2 (5)	82Hi1
			6½–7½	8855.7 (5)	82Hi1
			5½–6½	8909.05 (50)	82Hi1
8_{27}[d]	7½–8½		8½–9½	9577.7 (10)	82Hi1
			7½–8½	9647.7 (10)	82Hi1
			6½–7½	9710.1 (10)	82Hi1
8_{08}[d]	7½–8½	6½–7½	6½–7½	3307.7 (5)	82Hi1
			5½–6½	3399.7 (5)	82Hi1
3_{13}[d]	2½–3½	1½–2½	1½–2½	3628.6 (5)	82Hi1
			½–1½	3694.8 (5)	82Hi1

[a]) For NH_2 in the $\tilde{X}\ ^2B_1$ state, the hyperfine interactions involving the protons are larger than those involving the ^{14}N nucleus. In the $\tilde{A}\ ^2A_1$ state on the other hand, the ^{14}N angular momentum is more strongly coupled than that of the protons. Thus the intermediate angular momentum quantum number refers to the coupling of $\boldsymbol{I}_H$ to $\boldsymbol{J}$ in the $\tilde{X}$ state and to the coupling of $\boldsymbol{I}_N$ to $\boldsymbol{J}$ in the $\tilde{A}$ state. For NH_2 in the $\tilde{X}\ ^2B_1$ state, the ortho rotational levels are those with K_aK_c even even or odd odd and the para states are eo or oe. The association is exactly the opposite for the $\tilde{A}\ ^2A_1$ state.

[b]) The figures in parentheses represent the experimental uncertainty, in units of the last quoted decimal place.

[c]) Line blended with at least one other hyperfine component.

[d]) Magnetic dipole transition between the two spin components of a given rotational level $N_{K_aK_c}$.

Microwave data for $^{14}N^1H_2$ (continued)

Transition rotational $N'_{K_aK_c} - N''_{K_aK_c}$	fine $J' - J''$	hyperfine $F_1' - F_1''$ [a])	$F' - F''$	ν MHz	Ref.
State: electronic $\tilde{X}\ ^2B_1$; vibrational (0, 0, 0) (continued)					
9_{19} [d])	8½−9½	7½−8½	8½−9½	3230.6 (5) [b])	82Hi1
			7½−8½	3336.5 (5)	82Hi1
			6½−7½	3431.4 (5)	82Hi1
		8½−9½	9½−10½	3299.3 (5)	82Hi1
			8½−9½	3405.2 (5)	82Hi1
			7½−8½	3500.2 (5)	82Hi1
		9½−10½	10½−11½	3375.5 (5)	82Hi1
			9½−10½	3441.1 (5)	82Hi1
			8½−9½	3481.1 (5)	82Hi1
8_{17} [d])	7½−8½	6½−7½	7½−8½	9144.1 (5)	82Hi1
			6½−7½	9213.0 (10)	82Hi1
			5½−6½	9273.0 (5)	82Hi1
		7½−8½	8½−9½	9213.0 (10)	82Hi1
			7½−8½	9281.0 (5)	82Hi1
			6½−7½	9341.1 (5)	82Hi1
		8½−9½	9½−10½	9290.2 (5)	82Hi1
			8½−9½	9358.1 (5)	82Hi1
			7½−8½	9418.6 (5)	82Hi1
$3_{31}-4_{13}$ [d])	3½−3½	4½−4½	3½−3½	5285.1 (10)	78Hi
	3½−3½	4½−4½	4½−4½	5292.1 (10)	78Hi
	3½−3½	4½−4½	5½−5½	5303.0 (10)	78Hi
	3½−3½	3½−3½	2½−2½	5355.0 (10)	78Hi
	3½−3½	3½−3½	3½−3½	5363.5 (10)	78Hi
	3½−3½	3½−3½	4½−4½	5375.1 (10)	78Hi
	3½−3½	2½−2½	2½−2½	5419.1 (10)	78Hi
	3½−3½	2½−2½	3½−3½	5429.6 (10)	78Hi
	3½−4½	2½−3½	3½−4½	11668.1 (20)	78Hi
	3½−4½	2½−3½	2½−3½	11677.4 (20) [c])	78Hi
	3½−4½	3½−4½	4½−5½	11677.4 (20) [c])	78Hi
	3½−4½	2½−3½	1½−2½	11686.8 (20) [c])	78Hi
	3½−4½	3½−4½	3½−4½	11686.8 (20) [c])	78Hi
	3½−4½	4½−5½	5½−6½	11686.8 (20) [c])	78Hi
	3½−4½	3½−4½	2½−3½	11697.4 (20) [c])	78Hi
	3½−4½	4½−5½	4½−5½	11697.4 (20) [c])	78Hi
State: electronic $\tilde{A}\ ^2A_1$; vibrational (0, 10, 0)					
2_{20}(u) [e])-1_{10}	1½− ½	2½−1½	1½−2½	6348.31 (20)	79Lo
	1½− ½	2½−1½	2½−2½	6349.11 (20)	79Lo

a) For NH_2 in the $\tilde{X}\ ^2B_1$ state, the hyperfine interactions involving the protons are larger than those involving the ^{14}N nucleus. In the $\tilde{A}\ ^2A_1$ state on the other hand, the ^{14}N angular momentum is more strongly coupled than that of the protons. Thus the intermediate angular momentum quantum number refers to the coupling of $\boldsymbol{I}_H$ to $\boldsymbol{J}$ in the $\tilde{X}$ state and to the coupling of $\boldsymbol{I}_N$ to $\boldsymbol{J}$ in the $\tilde{A}$ state. For NH_2 in the $\tilde{X}\ ^2B_1$ state, the ortho rotational levels are those with K_aK_c even even or odd odd and the para states are eo or oe. The association is exactly the opposite for the $\tilde{A}\ ^2A_1$ state.

b) The figures in parentheses represent the experimental uncertainty, in units of the last quoted decimal place.

c) Line blended with at least one other hyperfine component.

d) Magnetic dipole transition between the two spin components of a given rotational level $N_{K_aK_c}$.

e) The precise identity of the u-state is uncertain [79Lo].

Microwave data for $^{14}N^{1}H_2$ (continued)

Transition: rotational $N'_{K_aK_c}-N''_{K_aK_c}$	fine $J'-J''$	hyperfine $F_1'-F_1''$ [a]	hyperfine $F'-F''$	ν MHz	Ref.
State: electronic $\tilde{A}\ ^2A_1$; vibrational (0, 10, 0) (continued)					
2_{20}(u)[e]) -1_{10}	1½ – ½	2½ – 1½	3½ – 2½	6349.71 (20) [b])	79Lo
	1½ – ½	2½ – 1½	1½ – 1½	6373.04 (20)	79Lo
	1½ – ½	2½ – 1½	2½ – 1½	6373.89 (20)	79Lo
	1½ – ½	2½ – 1½	1½ – ½	6395.67 (20)	79Lo
	1½ – ½	1½ – 1½	1½ – 2½	6378.58 (20)	79Lo
	1½ – ½	1½ – 1½	2½ – 2½	6379.32 (20)	79Lo
	1½ – ½	1½ – 1½	½ – 1½	6403.16 (20)	79Lo
	1½ – ½	1½ – 1½	1½ – 1½	6403.80 (20)	79Lo
	1½ – ½	1½ – 1½	2½ – 1½	6404.29 (20)	79Lo
	1½ – ½	1½ – 1½	1½ – ½	6426.25 (20) [c])	79Lo
	1½ – ½	1½ – 1½	½ – ½	6426.25 (20) [c])	79Lo
	1½ – ½	½ – 1½	1½ – 2½	6397.15 (20)	79Lo
	1½ – ½	½ – 1½	½ – 1½	6421.14 (20)	79Lo
	1½ – ½	½ – 1½	1½ – 1½	6422.23 (20)	79Lo
	1½ – ½	½ – 1½	1½ – ½	6444.81 (20) [c])	79Lo
	1½ – ½	½ – 1½	½ – ½	6444.81 (20) [c])	79Lo
	1½ – ½	2½ – ½	1½ – ½	6466.32 (20) [c])	79Lo
	1½ – ½	2½ – ½	1½ – 1½	6489.45 (20)	79Lo
	1½ – ½	2½ – ½	2½ – 1½	6490.05 (20)	79Lo
	1½ – ½	1½ – ½	1½ – ½	6496.30 (20) [c])	79Lo
	1½ – ½	1½ – ½	½ – ½	6496.30 (20) [c])	79Lo
	1½ – ½	1½ – ½	1½ – 1½	6519.84 (20) [c])	79Lo
	1½ – ½	1½ – ½	½ – 1½	6519.84 (20) [c])	79Lo
	1½ – ½	1½ – ½	2½ – 1½	6520.40 (20)	79Lo
	1½ – ½	½ – ½	½ – ½	6514.42 (20)	79Lo
	1½ – ½	½ – ½	1½ – ½	6515.70 (20)	79Lo
	1½ – ½	½ – ½	½ – 1½	6537.28 (20)	79Lo
	1½ – ½	½ – ½	1½ – 1½	6538.16 (20)	79Lo
	1½ – 1½	2½ – 2½	2½ – 3½	15625.92 (50) [c])	79Lo
	1½ – 1½	2½ – 2½	3½ – 3½	15625.92 (50) [c])	79Lo
	1½ – 1½	2½ – 2½	1½ – 2½	15649.90 (50) [c])	79Lo
	1½ – 1½	2½ – 2½	2½ – 2½	15649.90 (50) [c])	79Lo
	1½ – 1½	2½ – 2½	3½ – 2½	15649.90 (50) [c])	79Lo
	1½ – 1½	2½ – 2½	1½ – 1½	15667.96 (50) [c])	79Lo
	1½ – 1½	2½ – 2½	2½ – 1½	15667.96 (50) [c])	79Lo
	1½ – 1½	2½ – 1½	1½ – 2½	15749.87 (50) [c])	79Lo
	1½ – 1½	2½ – 1½	2½ – 2½	15749.87 (50) [c])	79Lo
	1½ – 1½	2½ – 1½	3½ – 2½	15749.87 (50) [c])	79Lo
	1½ – 1½	2½ – 1½	2½ – 1½	15768.10 (50) [c])	79Lo
	1½ – 1½	2½ – 1½	1½ – 1½	15768.10 (50) [c])	79Lo

[a]) For NH_2 in the $\tilde{X}\ ^2B_1$ state, the hyperfine interactions involving the protons are larger than those involving the ^{14}N nucleus. In the $\tilde{A}\ ^2A_1$ state on the other hand, the ^{14}N angular momentum is more strongly coupled than that of the protons. Thus the intermediate angular momentum quantum number refers to the coupling of $\boldsymbol{I}_H$ to $\boldsymbol{J}$ in the $\tilde{X}$ state and to the coupling of $\boldsymbol{I}_N$ to $\boldsymbol{J}$ in the $\tilde{A}$ state. For NH_2 in the $\tilde{X}\ ^2B_1$ state, the ortho rotational levels are those with K_aK_c even even or odd odd and the para states are eo or oe. The association is exactly the opposite for the $\tilde{A}\ ^2A_1$ state.

[b]) The figures in parentheses represent the experimental uncertainty, in units of the last quoted decimal place.

[c]) Line blended with at least one other hyperfine component.

[e]) The precise identity of the u-state is uncertain [79Lo].

Microwave data for $^{14}N^1H_2$ (continued)

Transition				ν MHz	Ref.
rotational	fine	hyperfine			
$N'_{K_aK_c} - N''_{K_aK_c}$	$J' - J''$	$F_1' - F_1''$ [a]	$F' - F''$		
State: electronic $\tilde{A}\,^2A_1$; vibrational (0, 10, 0) (continued)					
2_{20}(u)[e]) $- 1_{10}$	1½−1½	2½−1½	1½− ½	15782.32 (50) [b])	79Lo
	1½−1½	2½− ½	1½−1½	15821.47 (50) [c])	79Lo
	1½−1½	2½− ½	2½−1½	15821.47 (50) [c])	79Lo
	1½−1½	2½− ½	1½− ½	15848.31 (50)	79Lo
	1½−1½	1½−2½	2½−3½	15596.28 (50)	79Lo
	1½−1½	1½−2½	1½−2½	15619.44 (50) [c])	79Lo
	1½−1½	1½−2½	2½−2½	15619.44 (50) [c])	79Lo
	1½−1½	1½−2½	½−1½	15637.96 (50) [c])	79Lo
	1½−1½	1½−2½	1½−1½	15637.96 (50) [c])	79Lo
	1½−1½	1½−2½	2½−1½	15637.96 (50) [c])	79Lo
	1½−1½	1½−1½	1½−2½	15720.23 (50) [c])	79Lo
	1½−1½	1½−1½	2½−2½	15720.23 (50) [c])	79Lo
	1½−1½	1½−1½	½−1½	15736.75 (50) [c])	79Lo
	1½−1½	1½−1½	1½−1½	15736.75 (50) [c])	79Lo
	1½−1½	1½−1½	2½−1½	15736.75 (50) [c])	79Lo
	1½−1½	1½−1½	½−1½	15752.46 (50) [c])	79Lo
	1½−1½	1½−1½	1½−1½	15752.46 (50) [c])	79Lo
	1½−1½	1½− ½	½−1½	15792.13 (50) [c])	79Lo
	1½−1½	1½− ½	1½−1½	15792.13 (50) [c])	79Lo
	1½−1½	1½− ½	2½−1½	15792.13 (50) [c])	79Lo
	1½−1½	1½− ½	½− ½	15818.65 (50) [c])	79Lo
	1½−1½	1½− ½	1½− ½	15818.65 (50) [c])	79Lo
	1½−1½	½−2½	1½−2½	15601.44 (50)	79Lo
	1½−1½	½−2½	½−1½	15620.44 (50) [c])	79Lo
	1½−1½	½−2½	1½−1½	15620.44 (50) [c])	79Lo
	1½−1½	½−1½	1½−2½	15701.67 (50)	79Lo
	1½−1½	½−1½	½−1½	15720.23 (50) [c])	79Lo
	1½−1½	½−1½	1½−1½	15720.23 (50) [c])	79Lo
	1½−1½	½−1½	½− ½	15733.90 (50) [c])	79Lo
	1½−1½	½−1½	1½− ½	15733.90 (50) [c])	79Lo
	1½−1½	½− ½	½−1½	15774.58 (50) [c])	79Lo
	1½−1½	½− ½	1½−1½	15774.58 (50) [c])	79Lo
	1½−1½	½− ½	½− ½	15800.84 (50) [c])	79Lo
	1½−1½	½− ½	1½− ½	15800.84 (50) [c])	79Lo
State: electronic $\tilde{A}\,^2A_1$; vibrational (0, 9, 0)					
3_{03} [d])	3½−2½	3½−2½	3½−2½	2167.21 (50)	82Hi
	3½−2½	2½−1½	1½− ½	1987.63 (50)	82Hi

[a]) For NH_2 in the $\tilde{X}\,^2B_1$ state, the hyperfine interactions involving the protons are larger than those involving the ^{14}N nucleus. In the $\tilde{A}\,^2A_1$ state on the other hand, the ^{14}N angular momentum is more strongly coupled than that of the protons. Thus the intermediate angular momentum quantum number refers to the coupling of $\boldsymbol{I}_H$ to $\boldsymbol{J}$ in the $\tilde{X}$ state and to the coupling of $\boldsymbol{I}_N$ to $\boldsymbol{J}$ in the $\tilde{A}$ state. For NH_2 in the $\tilde{X}\,^2B_1$ state, the ortho rotational levels are those with K_aK_c even even or odd odd and the para states are eo or oe. The association is exactly the opposite for the $\tilde{A}\,^2A_1$ state.

[b]) The figures in parentheses represent the experimental uncertainty, in units of the last quoted decimal place.

[c]) Line blended with at least one other hyperfine component.

[d]) Magnetic dipole transition between the two spin components of a given rotational level $N_{K_aK_c}$.

[e]) The precise identity of the u-state is uncertain [79Lo].

Microwave data for $^{14}N^1H_2$ (continued)

Transition rotational $N'_{K_aK_c} - N''_{K_aK_c}$	fine $J' - J''$	hyperfine $F_1' - F_1''$ [a]	$F' - F''$	ν MHz	Ref.
State: electronic $\tilde{A}\ ^2\tilde{A}_1$; vibrational (0, 9, 0) (continued)					
3_{03} [d]	3½–2½	2½–1½	2½–1½	2006.75 (50) [b]	82Hi
	3½–2½	2½–1½	3½–2½	2033.04 (50)	82Hi
	3½–2½	3½–2½	2½–1½	2146.77 (50)	82Hi
	3½–2½	3½–2½	4½–3½	2191.50 (50)	82Hi
	3½–2½	4½–3½	3½–2½	2348.93 (50)	82Hi
	3½–2½	4½–3½	4½–3½	2369.74 (50)	82Hi
	3½–2½	4½–3½	5½–4½	2393.85 (50)	82Hi
	3½–2½	3½–3½	4½–4½	2289.87 (50)	82Hi
	3½–2½	2½–2½	3½–3½	2106.14 (50)	82Hi
	3½–2½	2½–1½	2½–2½	2012.08 (50)	82Hi
	3½–2½	2½–2½	1½–1½	2058.42 (50)	82Hi
4_{04} [d]	4½–3½		5½–4½	5708.26 (50)	82Hi
	4½–3½		4½–4½	5605.05 (50)	82Hi
	4½–3½		4½–3½	5509.68 (50)	82Hi
	4½–3½		3½–3½	5422.84 (50)	82Hi
	4½–3½		3½–2½	5347.85 (50)	82Hi
7_{07} [d]	6½–7½	5½–6½	4½–5½	6990.72 (50)	82Hi
	6½–7½	5½–6½	5½–6½	6965.70 (50)	82Hi
	6½–7½	5½–6½	6½–7½	6937.62 (50)	82Hi
	6½–7½	6½–7½	5½–6½	6821.04 (50)	82Hi
	6½–7½	6½–7½	6½–7½	6797.69 (50)	82Hi
	6½–7½	6½–7½	7½–8½	6770.68 (50)	82Hi
	6½–7½	7½–8½	6½–7½	6628.47 (50)	82Hi
	6½–7½	7½–8½	7½–8½	6603.13 (50)	82Hi
	6½–7½	7½–8½	8½–9½	6574.53 (50)	82Hi
8_{08} [d]	7½–8½		8½–9½	7627.64 (50)	82Hi
	7½–8½		7½–8½	7828.45 (50)	82Hi
	7½–8½		6½–7½	8005.47 (50)	82Hi
State: electronic $\tilde{A}\ ^2A_1$; vibrational (0, 10, 0)					
2_{11} [d]	1½–2½		2½–3½	10397.53 (50)	82Hi
	1½–2½		2½–2½	10518.31 (50)	82Hi
	1½–2½		1½–2½	10626.88 (50)	82Hi
	1½–2½		1½–1½	10708.98 (50)	82Hi
	1½–2½		½–1½	10777.06 (50)	82Hi
3_{12} [d]	2½–3½	1½–2½	½–1½	6228.03 (50)	82Hi
	2½–3½	1½–2½	1½–2½	6199.81 (50)	82Hi
	2½–3½	1½–2½	2½–3½	6155.62 (50)	82Hi
	2½–3½	2½–3½	1½–2½	6054.80 (50)	82Hi
	2½–3½	2½–3½	2½–3½	6024.97 (50)	82Hi

[a]) For NH_2 in the $\tilde{X}\ ^2B_1$ state, the hyperfine interactions involving the protons are larger than those involving the ^{14}N nucleus. In the $\tilde{A}\ ^2A_1$ state on the other hand, the ^{14}N angular momentum is more strongly coupled than that of the protons. Thus the intermediate angular momentum quantum number refers to the coupling of $\boldsymbol{I}_H$ to $\boldsymbol{J}$ in the $\tilde{X}$ state and to the coupling of $\boldsymbol{I}_N$ to $\boldsymbol{J}$ in the $\tilde{A}$ state. For NH_2 in the $\tilde{X}\ ^2B_1$ state, the ortho rotational levels are those with K_aK_c even even or odd odd and the para states are eo or oe. The association is exactly the opposite for the $\tilde{A}\ ^2A_1$ state.

[b]) The figures in parentheses represent the experimental uncertainty, in units of the last quoted decimal place.

[d]) Magnetic dipole transition between the two spin components of a given rotational level $N_{K_aK_c}$.

Microwave data for $^{14}N^1H_2$ (continued)

Transition				ν MHz	Ref.
rotational	fine	hyperfine			
$N'_{K_aK_c} - N''_{K_aK_c}$	$J' - J''$	$F_1' - F_1''$ [a]	$F' - F''$		
State: electronic $\tilde{A}\ ^2A_1$; vibrational (0, 10, 0) (continued)					
3_{12} [d]	2½ – 3½	2½ – 3½	3½ – 4½	5985.97 (50) [b]	82Hi
	2½ – 3½	3½ – 4½	2½ – 3½	5827.50 (50)	82Hi
	2½ – 3½	3½ – 4½	3½ – 4½	5794.23 (50)	82Hi
	2½ – 3½	3½ – 4½	4½ – 5½	5753.68 (50)	82Hi
	2½ – 3½	1½ – 2½	1½ – 1½	6218.51 (50)	82Hi
	2½ – 3½	1½ – 2½	2½ – 2½	6183.12 (50)	82Hi
	2½ – 3½	2½ – 2½	1½ – 1½	6145.63 (50)	82Hi
	2½ – 3½	2½ – 2½	2½ – 2½	6116.05 (50)	82Hi
	2½ – 3½	2½ – 2½	3½ – 3½	6074.57 (50)	82Hi
	2½ – 3½	2½ – 3½	2½ – 2½	6044.71 (50)	82Hi
	2½ – 3½	2½ – 3½	3½ – 3½	6010.80 (50)	82Hi
	2½ – 3½	3½ – 3½	2½ – 2½	5946.99 (50)	82Hi
	2½ – 3½	3½ – 3½	3½ – 3½	5915.72 (50)	82Hi
	2½ – 3½	3½ – 3½	4½ – 4½	5876.24 (50)	82Hi
	2½ – 3½	3½ – 4½	3½ – 3½	5815.74 (50)	82Hi
	2½ – 3½	3½ – 4½	4½ – 4½	5779.37 (50)	82Hi
	2½ – 3½	2½ – 2½	3½ – 2½	6102.28 (50)	82Hi
4_{13} [d]	3½ – 4½		2½ – 3½	3434.14 (50)	82Hi
	3½ – 4½		3½ – 3½	3348.24 (50)	82Hi
	3½ – 4½		3½ – 4½	3258.34 (50)	82Hi
	3½ – 4½		4½ – 4½	3151.70 (50)	82Hi
	3½ – 4½		4½ – 5½	3032.68 (50)	82Hi
5_{14} [d]	4½ – 5½	3½ – 4½	2½ – 3½	2569.53 (50)	82Hi
	4½ – 5½	3½ – 4½	3½ – 4½	2539.19 (50)	82Hi
	4½ – 5½	3½ – 4½	4½ – 5½	2500.23 (50)	82Hi
	4½ – 5½	4½ – 5½	3½ – 4½	2387.70 (50)	82Hi
	4½ – 5½	4½ – 5½	4½ – 5½	2357.51 (50)	82Hi
	4½ – 5½	4½ – 5½	5½ – 6½	2321.19 (50)	82Hi
	4½ – 5½	5½ – 6½	4½ – 5½	2165.44 (50)	82Hi
	4½ – 5½	5½ – 6½	5½ – 6½	2133.06 (50)	82Hi
	4½ – 5½	5½ – 6½	6½ – 7½	2095.43 (50)	82Hi
	4½ – 5½	5½ – 5½	6½ – 6½	2214.82 (50)	82Hi
7_{16} [d]	6½ – 7½	7½ – 8½	8½ – 9½	3550.96 (50)	82Hi
	6½ – 7½	7½ – 8½	7½ – 8½	3579.72 (50)	82Hi
	6½ – 7½	7½ – 8½	6½ – 7½	3605.28 (50)	82Hi
	6½ – 7½	6½ – 7½	7½ – 8½	3757.32 (50)	82Hi
	6½ – 7½	6½ – 7½	6½ – 7½	3786.53 (50)	82Hi
	6½ – 7½	6½ – 7½	5½ – 6½	3812.25 (50)	82Hi
	6½ – 7½	5½ – 6½	6½ – 7½	3934.93 (50)	82Hi
	6½ – 7½	5½ – 6½	5½ – 6½	3966.16 (50)	82Hi

[a]) For NH_2 in the $\tilde{X}\ ^2B_1$ state, the hyperfine interactions involving the protons are larger than those involving the ^{14}N nucleus. In the $\tilde{A}\ ^2A_1$ state on the other hand, the ^{14}N angular momentum is more strongly coupled than that of the protons. Thus the intermediate angular momentum quantum number refers to the coupling of $\boldsymbol{I}_H$ to $\boldsymbol{J}$ in the $\tilde{X}$ state and to the coupling of $\boldsymbol{I}_N$ to $\boldsymbol{J}$ in the $\tilde{A}$ state. For NH_2 in the $\tilde{X}\ ^2B_1$ state, the ortho rotational levels are those with K_aK_c even even or odd odd and the para states are eo or oe. The association is exactly the opposite for the $\tilde{A}\ ^2A_1$ state.

[b]) The figures in parentheses represent the experimental uncertainty, in units of the last quoted decimal place.

[d]) Magnetic dipole transition between the two spin components of a given rotational level $N_{K_aK_c}$.

Microwave data for $^{14}N^{1}H_2$ (continued)

Transition				ν MHz	Ref.
rotational	fine	hyperfine			
$N'_{K_aK_c} - N''_{K_aK_c}$	$J' - J''$	$F'_1 - F''_1$ [a]	$F' - F''$		
State: electronic $\tilde{A}\,^2A_1$; vibrational (0, 10, 0) (continued)					
7_{16} [d]	6½–7½	5½–6½	4½–5½	3992.39 (50) [b]	82Hi
9_{18} [d]	8½–9½	9½–10½	10½–11½	5218.35 (50)	82Hi
	8½–9½	9½–10½	9½–10½	5250.92 (50)	82Hi
	8½–9½	9½–10½	8½–9½	5279.98 (50)	82Hi
	8½–9½	8½–9½	9½–10½	5428.49 (50)	82Hi
	8½–9½	8½–9½	8½–9½	5460.48 (50)	82Hi
	8½–9½	8½–9½	7½–8½	5489.12 (50)	82Hi
	8½–9½	7½–8½	8½–9½	5614.55 (50)	82Hi
	8½–9½	7½–8½	7½–8½	5646.94 (50)	82Hi
	8½–9½	7½–8½	6½–7½	5676.10 (50)	82Hi
2_{12} [d]	1½–2½	½–1½	½–1½	17082.49 (50) [c]	82Hi
	1½–2½	½–1½	1½–1½	17082.49 (50) [c]	82Hi
	1½–2½	½–1½	1½–2½	17055.89 (50)	82Hi
	1½–2½	1½–2½	½–1½	16989.23 (50) [c]	82Hi
	1½–2½	1½–2½	1½–1½	16989.23 (50) [c]	82Hi
	1½–2½	1½–2½	1½–2½	16971.79 (50) [c]	82Hi
	1½–2½	1½–2½	2½–2½	16971.79 (50) [c]	82Hi
	1½–2½	1½–2½	2½–3½	16950.42 (50)	82Hi
	1½–2½	2½–3½	1½–2½	16833.48 (50) [c]	82Hi
	1½–2½	2½–3½	2½–2½	16833.48 (50) [c]	82Hi
	1½–2½	2½–3½	2½–3½	16814.05 (50) [c]	82Hi
	1½–2½	2½–3½	3½–3½	16814.05 (50) [c]	82Hi
	1½–2½	2½–3½	3½–4½	16790.59 (50)	82Hi
	1½–2½	½–1½	½– ½	17097.85 (50) [c]	82Hi
	1½–2½	2½–2½	2½–2½	16915.18 (50) [c]	82Hi
	1½–2½	2½–2½	3½–2½	16915.18 (50) [c]	82Hi
	1½–2½	2½–2½	1½–2½	16915.18 (50) [c]	82Hi
	1½–2½	2½–2½	3½–3½	16893.44 (50) [c]	82Hi
	1½–2½	2½–2½	2½–3½	16893.44 (50) [c]	82Hi
	1½–2½	½–1½	1½– ½	17097.85 (50) [c]	82Hi
4_{14} [d]	3½–4½	2½–3½	1½–2½	17315.62 (50) [c]	82Hi
	3½–4½	2½–3½	2½–2½	17315.62 (50) [c]	82Hi
	3½–4½	2½–3½	2½–3½	17302.86 (50) [c]	82Hi
	3½–4½	2½–3½	3½–3½	17302.86 (50) [c]	82Hi
	3½–4½	2½–3½	3½–4½	17285.21 (50)	82Hi
	3½–4½	3½–4½	2½–3½	17189.74 (50) [c]	82Hi
	3½–4½	3½–4½	3½–3½	17189.74 (50) [c]	82Hi
	3½–4½	3½–4½	3½–4½	17176.61 (50) [c]	82Hi
	3½–4½	3½–4½	4½–4½	17176.61 (50) [c]	82Hi

[a]) For NH_2 in the $\tilde{X}\,^2B_1$ state, the hyperfine interactions involving the protons are larger than those involving the ^{14}N nucleus. In the $\tilde{A}\,^2A_1$ state on the other hand, the ^{14}N angular momentum is more strongly coupled than that of the protons. Thus the intermediate angular momentum quantum number refers to the coupling of $\boldsymbol{I}_H$ to $\boldsymbol{J}$ in the $\tilde{X}$ state and to the coupling of $\boldsymbol{I}_N$ to $\boldsymbol{J}$ in the $\tilde{A}$ state. For NH_2 in the $\tilde{X}\,^2B_1$ state, the ortho rotational levels are those with K_aK_c even even or odd odd and the para states are eo or oe. The association is exactly the opposite for the $\tilde{A}\,^2A_1$ state.

[b]) The figures in parentheses represent the experimental uncertainty, in units of the last quoted decimal place.

[c]) Line blended with at least one other hyperfine component.

[d]) Magnetic dipole transition between the two spin components of a given rotational level $N_{K_aK_c}$.

Microwave data for $^{14}N^{1}H_2$ (continued)

Transition				ν MHz	Ref.
rotational	fine	hyperfine			
$N'_{K_aK_c} - N''_{K_aK_c}$	$J' - J''$	$F_1' - F_1''$ [a])	$F' - F''$		
State: electronic Ã $^2A_1$; vibrational (0, 10, 0) (continued)					
4_{14} [d])	3½ – 4½	3½ – 4½	4½ – 5½	17159.89 (50) [b])	82Hi
	3½ – 4½	4½ – 5½	3½ – 4½	17032.06 (50) [c])	82Hi
	3½ – 4½	4½ – 5½	4½ – 4½	17032.06 (50) [c])	82Hi
	3½ – 4½	4½ – 5½	4½ – 5½	17018.26 (50) [c])	82Hi
	3½ – 4½	4½ – 5½	5½ – 5½	17018.26 (50) [c])	82Hi
	3½ – 4½	4½ – 5½	5½ – 6½	17001.56 (50)	82Hi
	3½ – 4½	4½ – 4½	3½ – 3½	17120.44 (50)	82Hi
	3½ – 4½	4½ – 4½	4½ – 4½	17106.35 (50)	82Hi

a) For NH_2 in the X̃ $^2B_1$ state, the hyperfine interactions involving the protons are larger than those involving the ^{14}N nucleus. In the Ã $^2A_1$ state on the other hand, the ^{14}N angular momentum is more strongly coupled than that of the protons. Thus the intermediate angular momentum quantum number refers to the coupling of $\boldsymbol{I}_H$ to $\boldsymbol{J}$ in the X̃ state and to the coupling of $\boldsymbol{I}_N$ to $\boldsymbol{J}$ in the Ã state. For NH_2 in the X̃ $^2B_1$ state, the ortho rotational levels are those with K_aK_c even even or odd odd and the para states are eo or oe. The association is exactly the opposite for the Ã $^2A_1$ state.

b) The figures in parentheses represent the experimental uncertainty, in units of the last quoted decimal place.

c) Line blended with at least one other hyperfine component.

d) Magnetic dipole transition between the two spin components of a given rotational level $N_{K_aK_c}$.

Microwave data for $^{14}N^{1}H^{2}D$

Transition					ν MHz	Ref.
rotational	fine	hyperfine [a])				
$N'_{K_aK_c} - N''_{K_aK_c}$	$J' - J''$	$F_H' - F_H''$	$F_N' - F_N''$	$F_D' - F_D''$		
State: electronic X̃ $^2A''$; vibrational (0, 0, 0)						
$1_{10} - 1_{11}$	1½ – 1½	2 – 2	3 – 3	4 – 4	72760.4 (5) [b])	80St
	½ – ½				73596.9 (8)	80Br
					73629.1 (5)	80Br
$2_{20} - 2_{21}$	2½ – 2½	3 – 2	4 – 3	4 – 3	9926.8 (5)	80St
	2½ – 2½	3 – 3	4 – 4	4 – 4	9969.7 (5)	80St
	2½ – 2½	2 – 3	3 – 4	2 – 3	10014.5 (5)	80St
	1½ – 1½	1 – 2	1 – 2	2 – 3	10001.0 (5)	80St
	1½ – 1½	1 – 2	2 – 3	3 – 4	10008.4 (5)	80St
	1½ – 1½	2 – 2	3 – 3	4 – 4	10045.6 (5)	80St
	1½ – 1½	2 – 1	3 – 2	2 – 1	10083.0 (5)	80St
$4_{31} - 4_{32}$	4½ – 4½	5 – 5	6 – 6	7 – 7	5977.7 (5)	80St
	3½ – 3½	4 – 4	5 – 5	6 – 6	5968.4 (5)	80St
1_{10} [c])	½ – 1½	1 – 2	0 – 1	1 – 1	6119.3 (5)	80St
	½ – 1½	1 – 2	0 – 1	1 – 2	6126.3 (5)	80St
	½ – 1½	1 – 2	2 – 3	1 – 2	6174.2 (5)	80St
	½ – 1½	1 – 2	2 – 3	2 – 3	6179.2 (5)	80St
	½ – 1½	1 – 2	2 – 3	3 – 4	6186.3 (5)	80St

a) The strongest hyperfine component contributing to an observed feature is listed only.

b) The figures in parentheses represent the experimental uncertainty, in units of the last quoted decimal place.

c) Magnetic dipole transition between the two spin components of a given rotational level $N_{K_aK_c}$.

Microwave data for $^{14}N^1H^2D$ (continued)

Transition rotational $N'_{K_aK_c} - N''_{K_aK_c}$	fine $J' - J''$	hyperfine[a]) $F'_H - F''_H$	$F'_N - F''_N$	$F'_D - F''_D$	ν MHz	Ref.
State: electronic $\tilde{X}\ ^2A''$; vibrational (0, 0, 0) (continued)						
1_{11} [c])	½−1½	0−1	1−2	2−3	5186.1 (5) [b])	80St
	½−1½	0−1	1−1	2−2	5218.8 (5)	80St
	½−1½	1−2	2−3	2−3	5262.5 (5)	80St
	½−1½	1−2	2−3	3−4	5268.5 (5)	80St
	½−1½	1−2	1−2	2−3	5347.6 (5)	80St
	½−1½	1−2	0−1	1−1	5366.7 (5)	80St
3_{12} [c])	2½−3½	2−3	3−4	3−4	4829.3 (5)	80St
	2½−3½	2−3	3−4	4−5	4843.9 (5)	80St
	2½−3½	3−4	4−5	4−5	4880.2 (5)	80St
	2½−3½	3−4	4−5	5−6	4894.9 (5)	80St
	2½−3½	3−4	3−4	4−5	4907.1 (5)	80St
3_{22} [c])	2½−3½	2−3	3−4	2−3	9431.9 (5)	80St
	2½−3½	2−3	3−4	3−4	9439.8 (5)	80St
	2½−3½	2−3	3−4	4−5	9451.1 (5)	80St
	2½−3½	2−3	2−3	3−4	9481.3 (5)	80St
	2½−3½	3−4	4−5	3−4	9497.5 (5)	80St
	2½−3½	3−4	4−5	4−5	9506.3 (5)	80St
	2½−3½	3−4	4−5	5−6	9517.1 (5)	80St
	2½−3½	3−4	3−4	4−5	9547.2 (5)	80St
	2½−3½	3−4	2−3	3−4	9570.3 (5)	80St
4_{13} [c])	3½−4½	3−4	4−5	4−5	5173.2 (5)	80St
	3½−4½	3−4	4−5	5−6	5192.4 (5)	80St
	3½−4½	4−5	5−6	4−5	5208.5 (5)	80St
	3½−4½	4−5	5−6	5−6	5224.7 (5)	80St
	3½−4½	4−5	5−6	6−7	5240.4 (5)	80St
	3½−4½	4−5	4−5	5−6	5258.7 (5)	80St
4_{22} [c])	3½−4½	3−4	4−5	5−6	8773.3 (5)	80St
	3½−4½	3−4	3−4	4−5	8787.3 (5)	80St
	3½−4½	4−5	5−6	4−5	8802.0 (5)	80St
	3½−4½	4−5	5−6	5−6	8814.2 (5)	80St
	3½−4½	4−5	4−5	3−4	8819.5 (5)	80St
	3½−4½	4−5	5−6	6−7	8829.0 (5)	80St
	3½−4½	4−5	4−5	5−6	8843.9 (5)	80St
	3½−4½	4−5	3−4	4−5	8856.8 (5)	80St
4_{23} [c])	3½−4½	3−4	4−5	3−4	8266.3 (5)	80St
	3½−4½	3−4	4−5	4−5	8275.4 (5)	80St
	3½−4½	3−4	4−5	5−6	8287.3 (5)	80St
	3½−4½	4−5	5−6	4−5	8329.4 (5)	80St
	3½−4½	4−5	5−6	5−6	8337.4 (5)	80St
	3½−4½	4−5	5−6	6−7	8350.1 (5)	80St
	3½−4½	3−4	2−3	3−4	8364.4 (5)	80St
	3½−4½	4−5	4−5	3−4	8369.4 (5)	80St
	3½−4½	4−5	4−5	4−5	8379.3 (5)	80St
	3½−4½	4−5	4−5	5−6	8391.7 (5)	80St
	3½−4½	4−5	3−4	2−3	8403.6 (5)	80St
	3½−4½	4−5	3−4	3−4	8412.4 (5)	80St

[a]) The strongest hyperfine component contributing to an observed feature is listed only.

[b]) The figures in parentheses represent the experimental uncertainty, in units of the last quoted decimal place.

[c]) Magnetic dipole transition between the two spin components of a given rotational level $N_{K_aK_c}$.

Microwave data for $^{14}N^{1}H^{2}D$ (continued)

Transition					ν MHz	Ref.
rotational	fine	hyperfine a)				
$N'_{K_aK_c} - N''_{K_aK_c}$	$J' - J''$	$F'_H - F''_H$	$F'_N - F''_N$	$F'_D - F''_D$		
State: electronic $\tilde{X}\ ^2A''$; vibrational (0, 0, 0) (continued)						
4_{23} c)	3½ − 4½	4 − 5	3 − 4	4 − 5	8425.9 (5) b)	80St
5_{23} c)	4½ − 5½	4 − 5	5 − 6	4 − 5	8549.3 (5)	80St
	4½ − 5½	4 − 5	5 − 6	5 − 6	8565.2 (5)	80St
	4½ − 5½	4 − 5	4 − 5	4 − 5	8578.5 (5)	80St
	4½ − 5½	4 − 5	4 − 5	5 − 6	8593.9 (5)	80St
	4½ − 5½	5 − 6	6 − 7	5 − 6	8604.5 (5)	80St
	4½ − 5½	5 − 6	6 − 7	6 − 7	8616.7 (5)	80St
	4½ − 5½	5 − 6	5 − 6	5 − 6	8630.8 (5)	80St
	4½ − 5½	5 − 6	5 − 6	6 − 7	8645.0 (5)	80St
	4½ − 5½	5 − 6	4 − 5	5 − 6	8658.6 (5)	80St
6_{15} c)	5½ − 6½	5 − 6	6 − 7	5 − 6	6040.5 (50)	80St
	5½ − 6½	6 − 7	6 − 7	7 − 8	6153.9 (50)	80St

a) The strongest hyperfine component contributing to an observed feature is listed only.
b) The figures in parentheses represent the experimental uncertainty, in units of the last quoted decimal place.
c) Magnetic dipole transition between the two spin components of a given rotational level $N_{K_aK_c}$.

Molecular parameters for $^{14}N^{1}H_2$

Parameter a)		Value	Method	Ref.
State: electronic $\tilde{X}\ ^2B_1$; vibrational (0, 0, 0)				
A	[MHz]	710311.6 (70) b)	MW, FIR, Opt	77Da, 77Co, 79Bi
B	[MHz]	388289.5 (50)		
C	[MHz]	245015.7 (50)		
Δ_N	[MHz]	31.56 (19)		
Δ_{NK}	[MHz]	−124.6 (10)		
Δ_K	[MHz]	659.0 (13)		
δ_N	[MHz]	12.63 (9)		
δ_K	[MHz]	30.0 (11)		
Φ_N	[MHz]	$1.23 (26) \cdot 10^{-2}$		
Φ_{NK}	[MHz]	$-1.9 (34) \cdot 10^{-2}$		
Φ_{KN}	[MHz]	−0.34 (10)		
Φ_K	[MHz]	1.937 (80)		
ϕ_N	[MHz]	$0.56 (10) \cdot 10^{-2}$		
ϕ_{NK}	[MHz]	$-1.3 (17) \cdot 10^{-2}$		
ϕ_K	[MHz]	0.57 (10)		
L_K	[MHz]	$0.443 (30) \cdot 10^{-2}$		
ε_{aa}	[MHz]	−9267.41 (35)		82Hi1
ε_{bb}	[MHz]	−1353.89 (49)		
ε_{cc}	[MHz]	11.83 (20)		
Δ^s_N	[MHz]	0.307 (6)		
$(\Delta^s_{NK} + \Delta^s_{KN})$	[MHz]	−3.29 (80)		
Δ^s_{NK}	[MHz]	−1.28 (108)		
Δ^s_K	[MHz]	32.84 (12)		

a) The parameters have been determined in a fit to the A-reduced Hamiltonian.
b) The numbers in parentheses are 1 standard deviation of the least-squares fit, in the units of the last decimal place.

Molecular parameters for $^{14}N^1H_2$ (continued)

Parameter [a])		Value	Method	Ref.
State: electronic $\tilde{X}\ ^2B_1$; vibrational (0, 0, 0) (continued)				
δ_N^s	[MHz]	0.152 (3) [b])	MW, FIR, Opt	82Hi1
δ_K^s	[MHz]	0.503 (50)		
$(O)_H$	[MHz]	−67.2 (4)	MODR	77Co
$(aa)_H$	[MHz]	18.6 (13)		
$(bb)_H$	[MHz]	−13.4 (9)		
$(cc)_H$	[MHz]	−5.2 (6)		
$(ab)_H$	[MHz]	58.5 (25)		80St
$(O)_N$	[MHz]	28.0 (4)		77Co
$(aa)_N$	[MHz]	−42.8 (13)		
$(bb)_N$	[MHz]	−44.7 (10)		
$(cc)_N$	[MHz]	87.5 (5)		
$(aa)_Q$	[MHz]	0.1 (13)		
$(bb)_Q$	[MHz]	−1.6 (11)		
$(cc)_Q$	[MHz]	1.5 (6)		
μ	[D]	1.82 (5)		79Br

a) See at bottom of this page.
b) See at bottom of this page.

Molecular parameters for $^{14}N^1H^2D$

Parameter [a])		Value	Method	Ref.
State: electronic $\tilde{X}\ ^2A''$; vibrational (0, 0, 0)				
A	[MHz]	603033 (52) [b])	MODR, Opt	80St, 79Ra
B	[MHz]	243175 (17)		
C	[MHz]	169906 (9)		
Δ_N	[MHz]	9.61 (34)		
Δ_{NK}	[MHz]	29.2 (15)		
Δ_K	[MHz]	264.2 (58)		
δ_N	[MHz]	3.07 (7)		
δ_K	[MHz]	45.5 (12)		
$\tilde{\varepsilon}_{aa}$	[MHz]	−7039 (6)		
$\tilde{\varepsilon}_{bb}$	[MHz]	−1150 (3)		
$\tilde{\varepsilon}_{cc}$	[MHz]	9 (4)		
$\lvert\tilde{\varepsilon}_{ab}\rvert$	[MHz]	1070 (85)		
Δ_K^s	[MHz]	14.5 (37)		
δ_N^s	[MHz]	0.18 (7)		
$(O)_H$	[MHz]	−67.2		
$(aa)_H$	[MHz]	−25.5		
$(bb)_H$	[MHz]	30.5		
$(O)_N$	[MHz]	28.0		
$(aa)_N$	[MHz]	−43.0		
$(bb)_N$	[MHz]	−44.4		
$(O)_D$	[MHz]	−10.3		
$(aa)_D$	[MHz]	8.2		
$(bb)_D$	[MHz]	−7.4		
μ_a	[D]	0.665 (2)	Opt. Stark	79Br

a) The parameters have been determined in a fit to the A-reduced Hamiltonian.
b) The numbers in parentheses are 1 standard deviation of the least-squares fit, in units of the last quoted decimal place.

For References, see next page.

References for NH_2

76Co Cook, J.M., Hills, G.W., Curl, R.F.: Astrophys. J. Lett. **207** (1976) L 139.
76Hi Hills, G.W., Cook, J.M.: Astrophys. J. Lett. **209** (1976) L 157.
77Co Cook, J.M., Hills, G.W., Curl, R.F.: J. Chem. Phys. **67** (1977) 1450.
77Da Davies, P.B., Russell, D.K., Thrush, B.A., Radford, H.E.: Proc. R. Soc. London **A 353**, (1977) 299.
78Hi Hills, G.W., Lowe, R.S., Cook, J.M., Curl, R.F.: J. Chem. Phys. **68** (1978) 4073.
79Bi Birss, F.W., Ramsay, D.A., Ross, S.C., Zauli, C.: J. Mol. Spectrosc. **78** (1979) 344.
79Br Brown, J.M., Chalkley, S.W., Wayne, F.D.: Mol. Phys. **38** (1979) 1521.
79Lo Lowe, R.S., Kasper, J.V.V., Hills, G.W., Dillenschneider, W., Curl, R.F.: J. Chem. Phys. **70** (1979) 3356.
79Ra Ramsay, D.A., Wayne, F.D.: Can. J. Phys. **57** (1979) 761.
80Br Brown, J.M., Steimle, T.C.: Astrophys. J. **236** (1980) L 101.
80St Steimle, T.C., Brown, J.M., Curl, R.F.: J. Chem. Phys. **73** (1980) 2552.
81Ch Charo, A., Sastry, K.V.L.N., Herbst, E., De Lucia, F.C.: Astrophys. J. **244** (1981) L 111.
82Hi Hills, G.W., Cook, J.M., Brown, J.M., Brazier, C.R., Curl, R.F.: J. Chem. Phys. **76** (1982) 240.
82Hi1 Hills, G.W., Cook, J.M.: J. Mol. Spectrosc. **94** (1982) 456.

5 HCO

Microwave data for $^1H^{12}C^{16}O$

Transition			ν MHz	Ref.
rotational $N'_{K_aK_c} - N''_{K_aK_c}$	fine $J' - J''$	hyperfine $F' - F''$		
State: electronic $\tilde{X}\ ^2A'$; vibrational (0, 0, 0)				
$1_{01} - 0_{00}$	1½ − ½	2 − 1	86670.82 (4) [a]	78Pi
	1½ − ½	1 − 0	86708.35 (4)	78Pi
	½ − ½	1 − 1	86777.43 (4)	78Pi
	½ − ½	0 − 1	86805.75 (10)	78Pi
$2_{02} - 1_{01}$	2½ − 1½	3 − 2	173377.5 (10)	81Ho
	2½ − 1½	2 − 1	173406.3 (10)	81Ho
$1_{10} - 1_{11}$	½ − ½	1 − 1	2723.41 (5)	74Au
	1½ − 1½	1 − 1	2888.78 (5)	74Au
	1½ − 1½	2 − 2	2891.78 (5)	74Au
$2_{11} - 2_{12}$	1½ − 1½	1 − 1	8347.7 (3)	74Au
	1½ − 1½	2 − 2	8340.7 (5)	74Au
	2½ − 2½	2 − 2	8620.6 (3)	74Au
	2½ − 2½	3 − 3	8624.7 (3)	74Au
	2½ − 2½	3 − 2	8846.0 (5)	77Bo
$3_{12} - 3_{13}$	2½ − 2½	2 − 2	16804.2 (8)	74Au
	2½ − 2½	3 − 3	16798.9 (8)	74Au
	3½ − 3½	3 − 3	17190.4 (9)	74Au
	3½ − 3½	4 − 4	17196.4 (9)	74Au

[a]) The figures in parentheses represent the estimate of experimental uncertainty, in units of the last quoted decimal place.

Far infrared LMR data for $^1H^{12}C^{16}O$

Laser: Gas	λ μm	ν GHz	Rotational transition	Ref.
State: electronic $\tilde{X}\ ^2A'$; vibrational (0, 0, 0)				
D_2O [a])	108	278.0666	$8_{26} - 7_{17}$	76Co
N_2H_4	193	1554.0760	$5_{23} - 6_{16}$	82Br
CH_3OD	295	1016.8972	$4_{14} - 3_{03}$	
CH_3NH_2	315	952.1850	$11_{2,10} - 10_{2,9}$	
			$11_{38} - 10_{37}$	
			$11_{39} - 10_{38}$	
			$11_4 - 10_4$	
			$11_5 - 10_5$	
CH_2CHCl	386	776.8471	$1_{11} - 0_{00}$	
CH_3OH	392	764.6426	$1_{11} - 0_{00}$	
			$10_{19} - 10_{0,10}$	
HCOOH	394	761.6065	$1_{11} - 0_{00}$	
HCOOH	419	716.1564	$6_{15} - 6_{06}$	
HCOOH	433	692.9505	$1_{10} - 1_{01}$	
			$3_{12} - 3_{03}$	
			$8_{08} - 7_{07}$	
			$7_{16} - 7_{07}$	
			$8_{26} - 7_{25}$	
			$8_{36} - 7_{35}$	
			$8_{35} - 7_{34}$	
			$8_4 - 7_4$	
N_2H_4	436	687.9574	$1_{10} - 1_{01}$	82Br
			$2_{11} - 2_{02}$	

[a]) Discharge laser.

EPR data for $^1H^{12}C^{16}O$

Spectrometer frequency	Rotational transitions	Ref.
State: electronic $\tilde{X}\ ^2A'$; vibrational (0, 0, 0)		
9 GHz	$1_{10} - 1_{11}$, $J = 1\frac{1}{2} - 1\frac{1}{2}$	78Bo
	$1\frac{1}{2} - \frac{1}{2}$	
	$2_{20} - 2_{21}$, $J = 2\frac{1}{2} - 2\frac{1}{2}$	
	$2\frac{1}{2} - 1\frac{1}{2}$	
	$3_3 - 3_3$, $J = 3\frac{1}{2} - 3\frac{1}{2}$	
	$3\frac{1}{2} - 2\frac{1}{2}$	
	$4_3 - 4_3$, $J = 4\frac{1}{2} - 4\frac{1}{2}$	
	$4\frac{1}{2} - 3\frac{1}{2}$	
	$5_3 - 5_3$, $J = 5\frac{1}{2} - 4\frac{1}{2}$	
	$4_4 - 4_4$, $J = 4\frac{1}{2} - 4\frac{1}{2}$	

Microwave data for $^2D^{12}C^{16}O$

Transition: rotational $N'_{K_aK_c}-N''_{K_aK_c}$	fine $J'-J''$	hyperfine $F'-F''$	ν MHz	Ref.
State: electronic $\tilde{X}\,^2A'$; vibrational (0, 0, 0)				
$1_{10}-0_{00}$	1½ − ½	1½ − 1½	73429.57 (20)	72Sa
	1½ − ½	2½ − 1½	73491.66 (10)	72Sa
	1½ − ½	½ − ½	73498.24 (10)	72Sa
	1½ − ½	1½ − ½	73518.05 (10)	72Sa
	½ − ½	1½ − 1½	73592.71 (20)	72Sa
	½ − ½	½ − 1½	73610.92 (20)	72Sa
	½ − ½	1½ − ½	73680.43 (20)	72Sa

EPR data for $^2D^{12}C^{16}O$

Spectrometer frequency	Rotational transitions	Ref.
State: electronic $\tilde{X}\,^2A'$; vibrational (0, 0, 0)		
9 GHz	$1_{10}-1_{11}$, $J=1½-1½$	76Bo
	$J=1½-½$	
	$2_{20}-2_{21}$, $J=2½-2½$	
	$J=2½-1½$	
	3_3-3_3, $J=3½-3½$	
	$J=3½-2½$	
	4_4-4_4, $J=4½-3½$	

Molecular parameters for $^1H^{12}C^{16}O$

Parameter [a])		Value	Method	Ref.
State: electronic $\tilde{X}\,^2A'$; vibrational (0, 0, 0)				
A	[MHz]	729365.7 (12) [b])	MW, EPR, LMR	82Br
B	[MHz]	44778.727 (37)		
C	[MHz]	41939.822 (37)		
D_K	[MHz]	919.81 (33)		
D_{NK}	[MHz]	0.498 (27)		
D_N	[MHz]	0.1133 (14)		
d_1	[MHz]	$-0.1109\,(43)\cdot10^{-1}$		
d_2	[MHz]	$-0.278\,(65)\cdot10^{-2}$		
H_{KN}	[MHz]	$-0.146\,(17)\cdot10^{-1}$		
H_{NK}	[MHz]	$0.17\,(17)\cdot10^{-3}$		
$\tilde{\varepsilon}_{aa}$	[MHz]	11626.0 (13)		
$\tilde{\varepsilon}_{bb}$	[MHz]	18.91 (13)		
$\tilde{\varepsilon}_{cc}$	[MHz]	−205.74 (13)		
D^s_K	[MHz]	−49.02 (59)		
$(D^s_{NK}+D^s_{KN})$	[MHz]	−0.011 (104)		
d^s_1	[MHz]	$0.150\,(81)\cdot10^{-1}$		
H^s_K	[MHz]	$0.25\,(26)\cdot10^{-1}$		

a) The parameters have been determined in a fit to the S-reduced Hamiltonian.

b) The numbers in parentheses are 1 standard deviation of the least-squares fit, in units of the last quoted decimal place.

Molecular parameters for $^{1}H^{12}C^{16}O$ (continued)

Parameter[a])		Value	Method	Ref.
State: electronic $\tilde{X}\ ^2A'$; vibrational (0, 0, 0) (continued)				
$(O)_H$	[MHz]	390.76 (80) [b])	MW, EPR, LMR	82Br
$(a\,a)_H$	[MHz]	11.6 (38)		
$(b\,b)_H$	[MHz]	3.9 (20)		
g_S^{aa}		1.9943 [c])		
g_S^{bb}		2.0021 [c])		
g_S^{cc}		2.0048 [c])		
$-10/\sqrt{3}\,(g_r^{aa}+g_r^{bb}+g_r^{cc})$		0.300 (31)		
$10/\sqrt{6}\,(2\,g_r^{aa}-g_r^{bb}-g_r^{cc})$		−0.348 (33)		
μ_a	[D]	1.3626 (39)	Laser Stark	77La

[a]) The parameters have been determined in a fit to the S-reduced Hamiltonian.
[b]) The numbers in parentheses are 1 standard deviation of the least-squares fit, in units of the last quoted decimal place.
[c]) Parameters constrained to these theoretical values in the fit.

Molecular parameters for $^{2}D^{12}C^{16}O$

Parameter [a])		Value	Method	Ref.
State: electronic $\tilde{X}\ ^2A'$; vibrational (0, 0, 0)				
A	[MHz]	440455	Opt	75Br
B	[MHz]	38415.64	EPR	73Bo, 78Bo
C	[MHz]	35118.85	EPR	73Bo, 78Bo
ε_{aa}	[MHz]	7125.3	EPR	78Bo
ε_{bb}	[MHz]	5.67	EPR, MW	78Bo
ε_{cc}	[MHz]	−174.02	EPR, MW	78Bo
Δ_K^s	[MHz]	−18.33	EPR	78Bo
$(O)_D$	[MHz]	58.70 (20)	MW	72Sa
$(a\,a)_D$	[MHz]	1.40 (40)	MW	72Sa
$(b\,b)_D$	[MHz]	0.82	MW	72Sa

[a]) Parameters determined in a fit to the A-reduced Hamiltonian.

References for HCO

72Sa Saito, S.: Astrophys. J. **178** (1972) L 95.
73Bo Bolman, P.S.H., Brown, J.M., Carrington, A., Lycett, G.J.: Proc. Roy. Soc. London A **335** (1973) 113.
74Au Austin, J.A., Levy, D.H., Gottlieb, C.A., Radford, H.E.: J. Chem. Phys. **60** (1974) 207.
75Br Brown, J.M., Ramsay, D.A.: Can. J. Phys. **53** (1975) 2232.
76Co Cook, J.M., Evenson, K.M., Howard, C.J., Curl, R.F.: J. Chem. Phys. **64** (1976) 1381.
77Bo Boland, B.J., Brown, J.M., Carrington, A.: Mol. Phys. **34** (1977) 453.
77La Landsberg, B.M., Merer, A.J., Oka, T.: J. Mol. Spectrosc. **67** (1977) 459.
78Bo Boland, B.J., Brown, J.M., Carrington, A., Nelson, A.C.: Proc. Roy. Soc. London A **360** (1978) 507.
78Pi Pickett, H.M., Boyd, T.L.: Chem. Phys. Lett. **58** (1978) 446.
81Ho Hollis, J.M.: Astrophys. J., to be published.
82Br Brown, J.M., Sears, T.J., Radford, H.E.: J. Mol. Spectrosc., to be published.
82Ch Charo, A., De Lucia, F.C.: J. Mol. Spectrosc. **94** (1982) 426.

6 HO_2

Microwave data for HO_2

Transition rotational $N'_{K_aK_c}-N''_{K_aK_c}$	fine $J'-J''$	hyperfine $F'-F''$	ν MHz	Ref.
State: electronic $\tilde{X}\ ^2A''$; vibrational (0, 0, 0)				
$1_{01}-0_{00}$	1½ − ½	1−0	65070.85 (8) [a]	77Sa
	1½ − ½	2−1	65081.82 (4)	77Sa
	1½ − ½	1−1	65098.44 (9)	77Sa
	½ − ½	1−0	65373.01 (10)	77Sa
	½ − ½	0−1	65396.15 (10)	77Sa
	½ − ½	1−1	65400.63 (10)	77Sa
$2_{02}-1_{01}$	2½ − 1½	2−1	130258.13 (20)	77Sa
	2½ − 1½	3−2	130260.07 (20)	77Sa
	1½ − ½	1−0	130463.68 (20)	77Sa
	1½ − ½	2−1	130467.41 (20)	77Sa
$2_{12}-1_{11}$	2½ − 1½	2−1	132959.56 (8)	77Sa
	2½ − 1½	3−2	132961.99 (8)	77Sa
	1½ − ½	1−1	119137.04 (7)	77Sa
	1½ − ½	2−1	119153.74 (4)	77Sa
	1½ − ½	1−0	119159.19 (7)	77Sa
$2_{11}-1_{10}$	2½ − 1½	2−1	136492.09 (12)	77Sa
	2½ − 1½	3−2	136495.97 (9)	77Sa
	1½ − ½	1−1	122856.61 (9)	77Sa
	1½ − ½	1−0	122858.26 (7)	77Sa
	1½ − ½	2−1	122858.92 (7)	77Sa
$6_{16}-7_{07}$	6½ − 7½	7−8	98117.82 (5)	77Sa
	6½ − 7½	6−7	98121.15 (5)	77Sa
	5½ − 6½	6−7	107640.22 (10)	77Sa
	5½ − 6½	5−6	107635.51 (10)	77Sa
$7_{17}-8_{08}$	7½ − 8½	8−9	27474.24 (6)	77Sa
	7½ − 8½	7−8	27477.69 (4)	77Sa
	6½ − 7½	7−8	35535.41 (4)	77Sa
	6½ − 7½	6−7	35530.88 (4)	77Sa
$9_{09}-8_{18}$	9½ − 8½	10−9	44073.10 (10)	77Sa
	9½ − 8½	9−8	44069.56 (4)	77Sa
	8½ − 7½	9−8	37138.59 (7)	77Sa
	8½ − 7½	8−7	37142.60 (6)	77Sa
$10_{0,10}-9_{19}$	10½ − 9½	11−10	116451.73 (10)	77Sa
	10½ − 9½	10−9	116448.40 (10)	77Sa
	9½ − 8½	10−9	110472.20 (4)	77Sa
	9½ − 8½	9−8	110476.90 (7)	77Sa
$3_{03}-2_{02}$	2½ − 1½	3−2	195629.55	82Ch
		2−1	195628.34	82Ch
	3½ − 2½	4−3	195423.56	82Ch
		3−2	195422.64	82Ch
$3_{12}-2_{11}$	2½ − 1½	3−2	193940.83	82Ch
		2−1	193939.56	82Ch
		2−2	193937.06	82Ch
	3½ − 2½	4−3	200617.47	82Ch
		3−2	200615.72	82Ch
$3_{13}-2_{12}$	2½ − 1½	3−2	188350.71	82Ch
		2−1	188351.58	82Ch

[a]) The figures in parentheses are the estimates of experimental uncertainty, in units of the last quoted decimal place.

Microwave data for HO_2 (continued)

Transition: rotational $N'_{K_aK_c}-N''_{K_aK_c}$	fine $J'-J''$	hyperfine $F'-F''$	ν MHz	Ref.
State: electronic $\tilde{X}\ ^2A''$; vibrational (0, 0, 0) (continued)				
$3_{13}-2_{12}$	3½–2½	4–3	195220.22	82Ch
		3–2	195219.25	82Ch
$3_{13}-4_{04}$	2½–3½	3–4	321833.31	82Ch
		2–3	321826.88	82Ch
	3½–4½	4–5	303438.00	82Ch
		3–4	303441.59	82Ch
$3_{21}-2_{20}$	2½–1½	3–2	184212.64	82Ch
		2–1	184214.87	82Ch
	3½–2½	4–3	202888.18	82Ch
		3–2	202885.94	82Ch
$3_{22}-2_{21}$	2½–1½	3–2	184194.55	82Ch
		2–1	184196.76	82Ch
	3½–2½	4–3	202872.19	82Ch
		3–2	202869.98	82Ch
$4_{04}-3_{03}$	3½–2½	4–3, 3–2	260770.14	82Ch
	4½–3½	5–4, 4–3	260565.86	82Ch
$4_{13}-3_{12}$	3½–2½	4–3, 3–2	262004.09	82Ch
	4½–3½	5–4	265770.21	82Ch
		4–3	265769.20	82Ch
$4_{14}-3_{13}$	3½–2½	4–3, 3–2	254551.53	82Ch
	4½–3½	5–4, 4–3	258522.94	82Ch
$4_{14}-5_{05}$	3½–4½	4–5	250502.47	82Ch
		3–4	250496.86	82Ch
	4½–5½	5–6	236280.92	82Ch
		4–5	236284.42	82Ch
$4_{22}-3_{21}$	3½–2½	4–3, 3–2	253233.72	82Ch
	4½–3½	5–4, 4–3	265731.52	82Ch
$4_{23}-3_{22}$	3½–2½	4–3, 3–2	253189.10	82Ch
	4½–3½	5–4, 4–3	265690.53	82Ch
$4_{32}-3_{31}$, $4_{31}-3_{30}$	3½–2½	4–3	248431.39	82Ch
		3–2	248433.15	82Ch
	4½–3½	5–4	269500.65	82Ch
		4–3	269498.83	82Ch
$5_{05}-4_{04}$	4½–3½	5–4, 4–3	325882.22	82Ch
	5½–4½	6–5, 5–4	325680.17	82Ch
$5_{14}-4_{13}$	4½–3½	5–4, 4–3	328995.55	82Ch

Microwave data for HO_2 (continued)

Transition rotational $N'_{K_aK_c}-N''_{K_aK_c}$	fine $J'-J''$	hyperfine $F'-F''$	ν MHz	Ref.
State: electronic $\tilde{X}\ ^2A''$; vibrational (0, 0, 0) (continued)				
$5_{14}-4_{13}$	5½ − 4½	6 − 5, 5 − 4	331333.01	82Ch
$5_{15}-4_{14}$	4½ − 3½	5 − 4, 4 − 3	316694.62	82Ch
	5½ − 4½	6 − 5, 5 − 4	322242.67	82Ch
$5_{15}-6_{06}$	4½ − 5½	5 − 6	179238.50	82Ch
		4 − 5	179233.33	82Ch
	5½ − 6½	6 − 7	167765.03	82Ch
		5 − 6	167768.28	82Ch
$5_{23}-4_{22}$	4½ − 3½	5 − 4, 4 − 3	320720.20	82Ch
	5½ − 4½	6 − 5, 5 − 4	329456.21	82Ch
$5_{24}-4_{23}$	4½ − 3½	5 − 4, 4 − 3	320631.39	82Ch
	5½ − 4½	6 − 5, 5 − 4	329373.04	82Ch
$5_{32}-4_{31}$, $5_{33}-4_{32}$	4½ − 3½	5 − 4, 4 − 3	316548.27	82Ch
	5½ − 4½	6 − 5, 5 − 4	332466.93	82Ch
$5_{41}-4_{40}$, $5_{42}-4_{41}$	4½ − 3½	5 − 4	312884.54	82Ch
		4 − 3	312886.24	82Ch
	5½ − 4½	6 − 5	335323.63	82Ch
		5 − 4	335321.98	82Ch
$6_{06}-5_{05}$	5½ − 4½	6 − 5, 5 − 4	390958.21	82Ch
	6½ − 5½	7 − 6, 6 − 5	390758.87	82Ch
$6_{15}-5_{14}$	5½ − 4½	6 − 5, 5 − 4	395535.68	82Ch
	6½ − 5½	7 − 6, 6 − 5	397077.70	82Ch
$6_{16}-5_{15}$	5½ − 4½	6 − 5, 5 − 4	384392.68	82Ch
	6½ − 5½	7 − 6, 6 − 5	386147.31	82Ch
$6_{24}-5_{23}$	5½ − 4½	6 − 5, 5 − 4	387333.95	82Ch
	6½ − 5½	7 − 6, 6 − 5	393690.63	82Ch
$6_{23}-5_{24}$	5½ − 4½	6 − 5, 5 − 4	387178.99	82Ch
	6½ − 5½	7 − 6, 6 − 5	393543.30	82Ch
$6_{33}-5_{32}$, $6_{34}-5_{33}$	6½ − 5½	7 − 6, 6 − 5	396034.05	82Ch

Microwave data for HO_2 (continued)				
Transition			ν MHz	Ref.
rotational $N'_{K_aK_c}-N''_{K_aK_c}$	fine $J'-J''$	hyperfine $F'-F''$		
State: electronic $\tilde{X}\ ^2A''$; vibrational (0, 0, 0) (continued)				
$6_{42}-5_{41}$ } $6_{43}-5_{42}$	5½−4½	6−5 } 5−4	380375.03	82Ch
	6½−5½	7−6 } 6−5	398470.75	82Ch
$7_{07}-6_{06}$	6½−5½	7−6 } 6−5	455990.94	82Ch
	7½−6½	8−7 } 7−6	455794.60	82Ch
$7_{16}-6_{15}$	6½−5½	7−6 } 6−5	461847.69	82Ch
$7_{17}-6_{16}$	6½−5½	7−6 } 6−5	448867.53	82Ch
	7½−6½	8−7 } 7−6	450136.82	82Ch
$7_{25}-6_{24}$	6½−5½	7−6 } 6−5	453447.97	82Ch
	7½−6½	8−7 } 7−6	458231.55	82Ch
$7_{26}-6_{25}$	6½−5½	7−6 } 6−5	453200.48	82Ch
	7½−6½	8−7 } 7−6	457994.45	82Ch
$7_{34}-6_{33}$ } $7_{35}-6_{34}$	6½−5½	7−6 } 6−5	450374.25	82Ch
	7½−6½	8−7 } 7−6	460006.04	82Ch
$7_{43}-6_{42}$ } $7_{44}-6_{43}$	6½−5½	7−6 } 6−5	447289.85	82Ch
	7½−6½	8−7 } 7−6	462034.53	82Ch
$8_{08}-7_{07}$	7½−6½	8−7 } 7−6	520972.33	82Ch
	8½−7½	9−8 } 8−7	520780.45	82Ch
$11_{0,11}-11_{1,10}$	10½−9½	11−10	184373.81	82Ch
		10−9	183378.30	82Ch
	11½−10½	12−11	189600.00	82Ch
		11−10	189596.64	82Ch
$12_{0,12}-11_{1,11}$	11½−10½	12−11	258872.75	82Ch
		11−10	258877.05	82Ch
	12½−11½	13−12	263460.52	82Ch
		12−11	263457.21	82Ch
$13_{0,13}-12_{1,12}$	12½−11½	13−12	333936.32	82Ch
		12−11	333940.27	82Ch
	13½−12½	14−13	337977.84	82Ch
		13−12	337974.55	82Ch
$17_{2,16}-18_{1,17}$	16½−17½	17−18	408522.19	82Ch
		16−17	408517.95	82Ch

Microwave data for HO_2 (continued)

Transition			ν MHz	Ref.
rotational $N'_{K_aK_c}-N''_{K_aK_c}$	fine $J'-J''$	hyperfine $F'-F''$		
State: electronic $\tilde{X}\,^2A''$; vibrational (0, 0, 0) (continued)				
$17_{2,16}-18_{1,17}$	17½−18½	18−19	326358.47	82Ch
		17−18	326354.31	82Ch
$18_{2,17}-19_{1,18}$	17½−18½	18−19	399022.02	82Ch
		17−18	399025.60	82Ch
	18½−19½	19−20	317603.71	82Ch
		18−19	317607.23	82Ch
$19_{2,18}-20_{1,19}$	18½−19½	19−20	243474.69	82Ch
		18−19	243470.60	82Ch
	19½−20½	20−21	235399.96	82Ch
		19−20	235403.51	82Ch
$25_{1,24}-24_{2,23}$	25½−24½	26−25	186594.26	82Ch
		25−24	186590.82	82Ch
$26_{1,25}-25_{2,24}$	25½−24½	26−25	268124.79	82Ch
		25−24	268128.81	82Ch
	26½−25½	27−26	273016.56	82Ch
		26−25	273013.19	82Ch

Far Infrared LMR data for HO_2

Laser			Rotational transition	Ref.
Gas	λ µm	ν GHz		
State: electronic $\tilde{X}\,^2A''$; vibrational (0, 0, 0)				
MeOH [a])	171	1757.5263	$8_{27}-8_{18}$	75Ho
$H_{2}0$ [b])	118.6	2527.9520	$4_{32}-5_{23}$	75Ho
			$4_{31}-5_{24}$	75Ho
			$13_{2,12}-12_{1,11}$	75Ho
			$11_{29}-10_{1,10}$	75Ho
			$14_{2,13}-13_{1,12}$	75Ho
			$12_{2,10}-11_{1,11}$	75Ho
D_2O [b])	108	2783.0666	$18_{4,15}-19_{3,16}$	75Ho
			$18_{4,14}-19_{3,17}$	75Ho
			$15_{2,13}-14_{1,14}$	75Ho
D_2O [b])	84	3557.1474	$11_{38}-10_{29}$	75Ho
			$11_{39}-10_{28}$	75Ho
			$10_{37}-9_{28}$	75Ho
			$10_{38}-9_{27}$	75Ho
H_2O [b])	79	3790.4745	$14_{3,12}-13_{2,11}$	75Ho
			$14_{3,11}-13_{2,12}$	75Ho
H_2O [b])	78	3821.7713	$14_{3,12}-13_{2,11}$	75Ho
			$14_{3,11}-13_{2,12}$	75Ho
			$15_{3,12}-14_{2,13}$	75Ho
			$15_{3,13}-14_{2,12}$	75Ho
D_2O [b])	73	4120.9843	$19_{3,16}-18_{2,17}$	75Ho

[a]) Optically pumped laser. [b]) Discharge laser.

EPR data for HO_2

Operating frequency GHz	Rotational transition	Ref.
State: electronic $\tilde{X}\ ^2A''$; vibrational (0, 0, 0)		
8.97 GHz	$1_{10}-1_{11}$	78Ba
	$2_{11}-2_{12}$	78Ba
	$2_{20}-2_{21}$	78Ba

Microwave data for DO_2

Transition rotational $N'_{K_aK_c}-N''_{K_aK_c}$	fine $J'-J''$	hyperfine $F'-F''$	ν MHz	Ref.
State: electronic $\tilde{X}\ ^2A''$; vibrational (0, 0, 0)				
$1_{01}-0_{00}$	1½ – ½	2½ – 1½	60370.826 (50) [a]	82Sa
	1½ – ½	1½ – ½	60367.827 (50)	82Sa
	1½ – ½	1½ – 1½	60374.102 (50)	82Sa
	1½ – ½	½ – ½	60369.836 (50)	82Sa
	½ – ½	1½ – 1½	60661.797 (50)	82Sa
	½ – ½	1½ – ½	60655.538 (50)	82Sa
	½ – ½	½ – 1½	60660.600 (50)	82Sa
$2_{12}-1_{11}$	2½ – 1½	3½ – 2½	120452.265 (50)	82Sa
	2½ – 1½	2½ – 1½	120451.600 (50) [b]	82Sa
	2½ – 1½	1½ – ½	120451.600 (50) [b]	82Sa
	2½ – 1½	2½ – 2½	120455.529 (50)	82Sa
	2½ – 1½	1½ – 1½	120453.796 (100)	82Sa
	1½ – ½	2½ – 1½	112424.500 (50)	82Sa
	1½ – ½	1½ – ½	112426.236 (50)	82Sa
$2_{02}-1_{01}$	2½ – 1½	3½ – 2½	120816.277 (50)	82Sa
	2½ – 1½	2½ – 1½	120815.670 (50) [b]	82Sa
	2½ – 1½	1½ – ½	120815.676 (50) [b]	82Sa
	2½ – 1½	2½ – 2½	120818.999 (50)	82Sa
	2½ – 1½	1½ – 1½	120817.691 (50)	82Sa
	1½ – ½	2½ – 1½	121007.661 (50)	82Sa
	1½ – ½	1½ – ½	121007.151 (50)	82Sa
	1½ – ½	1½ – 1½	121006.099 (50) [b]	82Sa
	1½ – ½	½ – ½	121006.099 (50) [b]	82Sa
$2_{11}-1_{10}$	2½ – 1½	3½ – 2½	126011.644 (50)	82Sa
	2½ – 1½	2½ – 1½	126010.698 (50) [b]	82Sa
	2½ – 1½	1½ – ½	126010.698 (50) [b]	82Sa
	2½ – 1½	2½ – 2½	126014.016 (50)	82Sa
	2½ – 1½	1½ – 1½	126012.603 (50)	82Sa
	1½ – ½	2½ – 1½	118147.898 (50) [b]	82Sa
	1½ – ½	1½ – ½	118147.898 (50) [b]	82Sa
	1½ – ½	1½ – 1½	118147.173 (50) [b]	82Sa
	1½ – ½	½ – ½	118147.173 (50) [b]	82Sa
$3_{03}-2_{02}$	3½ – 2½	4½ – 3½	181218.118 (50) [b]	82Sa
	3½ – 2½	3½ – 2½	181218.118 (50) [b]	82Sa
	3½ – 2½	2½ – 1½	181218.118 (50) [b]	82Sa
	2½ – 1½	3½ – 2½	181406.861 (50) [b]	82Sa
	2½ – 1½	2½ – 1½	181406.861 (50) [b]	82Sa
	2½ – 1½	1½ – ½	181406.861 (50) [b]	82Sa

[a]) The figures in parentheses are estimates of the experimental error, in units of the last quoted decimal place.
[b]) Hyperfine components not resolved.

Microwave data for DO_2 (continued)

Transition			ν MHz	Ref.
rotational $N'_{K_aK_c} - N''_{K_aK_c}$	fine $J' - J''$	hyperfine $F' - F''$		
State: electronic $\tilde{X}\ ^2A''$; vibrational (0, 0, 0) (continued)				
$3_{13} - 2_{12}$	3½ − 2½	4½ − 3½	178347.465 (50) [a][b]	82Sa
	3½ − 2½	3½ − 2½	178347.465 (50) [b]	82Sa
	3½ − 2½	2½ − 1½	178347.465 (50) [b]	82Sa
	2½ − 1½	3½ − 2½	174758.106 (50) [b]	82Sa
	2½ − 1½	2½ − 1½	174758.106 (50) [b]	82Sa
	2½ − 1½	1½ − ½	174758.106 (50) [b]	82Sa
$3_{12} - 2_{11}$	3½ − 2½	4½ − 3½	186756.729 (50) [b]	82Sa
	3½ − 2½	3½ − 2½	186756.729 (50) [b]	82Sa
	3½ − 2½	2½ − 1½	186756.729 (50) [b]	82Sa
	2½ − 1½	3½ − 2½	183354.967 (50) [b]	82Sa
	2½ − 1½	2½ − 1½	183354.967 (50) [b]	82Sa
	2½ − 1½	1½ − ½	183354.967 (50) [b]	82Sa
$1_{11} - 2_{02}$	1½ − 2½	2½ − 3½	175422.408 (50)	82Sa
	1½ − 2½	1½ − 2½	175423.696 (50) [b]	82Sa
	1½ − 2½	½ − 1½	175423.696 (50)	82Sa
	½ − 1½	1½ − 2½	196268.259 (50)	82Sa
	½ − 1½	½ − 1½	196265.074 (50)	82Sa
$2_{12} - 3_{03}$	2½ − 3½	3½ − 4½	114656.364 (50)	82Sa
	2½ − 3½	2½ − 3½	114657.274 (50) [b]	82Sa
	2½ − 3½	1½ − 2½	114657.274 (50) [b]	82Sa
	1½ − 2½	2½ − 3½	127285.778 (50)	82Sa
	1½ − 2½	1½ − 2½	127284.485 (50)	82Sa
$3_{13} - 4_{04}$	3½ − 4½	4½ − 5½	51449.189 (50)	82Sa
	3½ − 4½	3½ − 4½	51449.995 (50) [b]	82Sa
	3½ − 4½	2½ − 3½	51449.995 (50) [b]	82Sa
	2½ − 3½	3½ − 4½	60303.677 (50)	82Sa
	2½ − 3½	2½ − 3½	60302.723 (50)	82Sa
	2½ − 3½	1½ − 2½	60302.010 (50)	82Sa
$6_{06} - 5_{15}$	6½ − 5½	7½ − 6½	79866.391 (50)	82Sa
	6½ − 5½	6½ − 5½	79865.604 (50) [b]	82Sa
	6½ − 5½	5½ − 4½	79865.604 (50) [b]	82Sa
	5½ − 4½	6½ − 5½	74578.543 (50)	82Sa
	5½ − 4½	5½ − 4½	74579.170 (50)	82Sa
	5½ − 4½	4½ − 3½	74579.950 (50)	82Sa
$7_{07} - 6_{16}$	7½ − 6½	8½ − 7½	147452.286 (50)	82Sa
	7½ − 6½	7½ − 6½	147451.514 (50) [b]	82Sa
	7½ − 6½	6½ − 5½	147451.514 (50) [b]	82Sa
	6½ − 5½	7½ − 6½	143160.834 (50)	82Sa
	6½ − 5½	6½ − 5½	143161.548 (50)	82Sa

[a]) The figures in parentheses are estimates of the experimental error, in units of the last quoted decimal place.
[b]) Hyperfine components not resolved.

Far Infrared LMR data for $^{1}D^{16}O_2$

Laser			Rotational transition	Ref.
Gas	λ µm	ν GHz		
State: electronic $\tilde{X}\ ^2A''$; vibrational (0, 0, 0)				
CH_3OH	163	1838.8393	$13_{2,11}-12_{1,12}$	80Ba
N_2H_4	193	1554.0760		80Ba
CH_3OD	229			80Ba
N_2H_4	234	1281.6258		80Ba
CH_3OD	295	1016.8972		80Ba
HCOOH	302	991.7778	$10_{29}-10_{1,10}$	80Ba
			$9_{28}-9_{1,9}$	80Ba
N_2H_4	311	963.7314		80Ba
CH_3NH_2	315	952.1850	$7_{26}-7_{17}$	80Ba
			$6_{25}-6_{16}$	80Ba
CH_3OD	330			80Ba
N_2H_4	332	903.8894		80Ba
CH_2CHCl	386	776.8471	$11_{39}-12_{2,10}$	80Ba
HCOOH	394	761.6065		80Ba
HCOOH	433	692.9505	$7_{17}-6_{06}$	80Ba
CH_3OH	469	639.1846	$6_{16}-5_{05}$	80Ba
			$4_{22}-5_{15}$	80Ba
HCOOH	513	584.3869	$5_{13}-4_{04}$	80Ba

EPR data for $^{1}D^{16}O_2$

Operating frequency	Rotational transition	Ref.
State: electronic $\tilde{X}\ ^2A''$; vibrational (0, 0, 0)		
8.97, 9.27 GHz	$1_{10}-1_{11}$	80Ba
	$2_{11}-2_{12}$	80Ba
	$2_{20}-2_{21}$	80Ba
	$3_{21}-3_{22}$	80Ba
	$4_{22}-4_{23}$	80Ba
	$5_{05}-4_{14}$	80Ba

Molecular parameters for HO_2

Parameter [a])		Value	Method	Ref.
State: electronic $\tilde{X}\ ^2A''$; vibrational (0, 0, 0)				
A	[MHz]	610274.31 (72) [b])	Fit to S-reduced Hamiltonian MW, EPR, LMR	75Ho, 77Sa, 78Ba, 79Br
B	[MHz]	33513.838 (34)		
C	[MHz]	31671.662 (37)		
D_K	[MHz]	123.616 (17)		
D_{NK}	[MHz]	3.471 (23)		
D_N	[MHz]	0.1151 (13)		
d_1	[MHz]	$-0.698\ (50) \cdot 10^{-2}$		
d_2	[MHz]	$-0.811\ (12) \cdot 10^{-3}$		
H_K	[MHz]	$9.69\ (14) \cdot 10^{-2}$		
H_{KN}	[MHz]	$2.095\ (94) \cdot 10^{-3}$		

[a]) Parameters determined in a fit to the S-reduced Hamiltonian.
[b]) The numbers in parentheses are 1 standard deviation of the least-squares fit, in units of the last quoted decimal place.

Molecular parameters for HO_2 (continued)

Parameter[a])		Value	Method	Ref.
State: electronic $\tilde{X}\ ^2A''$; vibrational (0, 0, 0) (continued)				
h_1	[MHz]	$4.50(29)\cdot 10^{-5}$	Fit to S-reduced Hamiltonian MW, EPR, LMR	75Ho, 77Sa, 78Ba, 79Br
$\tilde{\varepsilon}_{aa}$	[MHz]	−49571.15 (57)		
$\tilde{\varepsilon}_{bb}$	[MHz]	−422.573 (80)		
$\tilde{\varepsilon}_{cc}$	[MHz]	8.623 (89)		
$\lvert\tilde{\varepsilon}_{ab}\rvert$	[MHz]	193.95 (55)		
D^s_K	[MHz]	22.71 (16)		
$(D^s_{NK}+D^s_{KN})$	[MHz]	0.1264 (75)		
$(O)_H$	[MHz]	−27.48 (15)		
$(a\,a)_H$	[MHz]	−8.34 (22)		
$(b\,b)_H$	[MHz]	19.68 (31)		
g_S^{aa}		2.04204 (19)		
g_S^{bb}		2.00790 (24)		
g_S^{cc}		2.00152 (24)		
$10^3\,g_r^{aa}$		−9.857 (41)		
$10^4\,(g_r^{bb}+g_r^{cc})$		−1.85 (41)		
μ_a	[D]	1.412 (33)	MW	80Sa
μ_b	[D]	1.541 (16)		
μ_{tot}	[D]	2.090 (34)		
r_o(H–O)	[Å]	0.9974		
r_o(O–O)	[Å]	1.3339		
$\measuredangle_o$ (HOO)	[degrees]	104.15°		
Results of fit to A-reduced Hamiltonian				
A	[MHz]	610273.223 (56) [b])	MW	82Ch
B	[MHz]	33517.816 (51)		
C	[MHz]	31667.654 (51)		
Δ_K	[MHz]	123.572 (28)		
Δ_{NK}	[MHz]	3.44552 (77)		
Δ_N	[MHz]	0.11693 (5)		
δ_K	[MHz]	2.017 (26)		
δ_N	[MHz]	0.00613 (2)		
Φ_K	[MHz]	$9.69\cdot 10^{-2}$ [c])		
Φ_{KN}	[MHz]	$1.051(21)\cdot 10^{-3}$		
Φ_{NK}	[MHz]	$2.29(57)\cdot 10^{-5}$		
$\tilde{\varepsilon}_{aa}$	[MHz]	−49571.409 (143)		
$\tilde{\varepsilon}_{bb}$	[MHz]	−422.755 (60)		
$\tilde{\varepsilon}_{cc}$	[MHz]	8.605 (60)		
$\lvert\tilde{\varepsilon}_{ab}\rvert$	[MHz]	193.45 (15)		
Δ^s_K	[MHz]	23.061 (20)		
Δ^s_{NK}	[MHz]	0.1261 (22)		
δ^s_K	[MHz]	0.079 (18)		

[a]) Parameters determined in a fit to the S-reduced Hamiltonian.

[b]) The numbers in parentheses are 1 standard deviation of the least-squares fit, in units of the last quoted decimal place.

[c]) Parameter constrained to this value in the fit.

Molecular parameters for DO_2

Parameter [a])		Value	Method	Ref.
State: electronic $\tilde{X}\ ^2A''$; vibrational (0, 0, 0)				
A	[MHz]	335602.94 (78) [b])	LMR, EPR	80Ba
B	[MHz]	31654.47 (12)		
C	[MHz]	28814.92 (11)		
D_K	[MHz]	38.39 (21)		
D_{NK}	[MHz]	2.511 (22)		
D_N	[MHz]	0.1047 (15)		
d_1	[MHz]	$-0.758(76)\cdot 10^{-2}$		
d_2	[MHz]	$-0.180(27)\cdot 10^{-2}$		
h_1	[MHz]	$0.296(57)\cdot 10^{-4}$		
$\tilde{\varepsilon}_{aa}$	[MHz]	−27145.8 (26)		
$\tilde{\varepsilon}_{bb}$	[MHz]	−392.03 (31)		
$\tilde{\varepsilon}_{cc}$	[MHz]	6.10 (21)		
$\lvert\tilde{\varepsilon}_{ab}\rvert$	[MHz]	300 (34)		
D_K^s	[MHz]	8.91 (70)		
$(D_{KN}^s+D_{NK}^s)$	[MHz]	0.154 (82)		
$(O)_D$	[MHz]	−4.182 (13)	MW	82Sa
$(aa)_D$	[MHz]	−0.952 (19)		
$(bb)_D$	[MHz]	2.724 (20)		
$(ab)_D$	[MHz]	3.03		
$\lvert\mu_a/\mu_b\rvert$		0.85 (10)	EPR intensity	80Ba
g_S^{aa}		2.04235 (20)		
g_S^{bb}		2.00788 (20)		
g_S^{cc}		2.00048 (32)		
g_r^{aa}		$-0.5310(84)\cdot 10^{-2}$		
g_r^{bb}		$-0.142(58)\cdot 10^{-3}$		
g_r^{cc}		$-0.63(58)\cdot 10^{-4}$		

[a]) Parameters determined in a fit to the S-reduced Hamiltonian.
[b]) The numbers in parentheses are 1 standard deviation of the least-squares fit, in units of the last quoted decimal place.

References for HO_2

75Ho Hougen, J.T., Radford, H.E., Evenson, K.M., Howard, C.J.: J. Mol. Spectrosc. **56** (1975) 210.
77Sa Saito, S.: J. Mol. Spectrosc. **65** (1977) 229.
78Ba Barnes, C.E., Brown, J.M., Carrington, A., Pinkstone, J., Sears, T.J., Thistlethwaite, P.J.: J. Mol. Spectrosc. **72** (1978) 86.
79Br Brown, J.M., Sears, T.J.: J. Mol. Spectrosc. **75** (1979) 111.
80Sa Saito, S., Matsumara, C.: J. Mol. Spectrosc. **80** (1980) 34.
80Ba Barnes, C.E., Brown, J.M., Radford, H.E.: J. Mol. Spectrosc. **84** (1980) 179.
82Ch Charo, A., De Lucia, F.C.: J. Mol. Spectrosc. **94** (1982) 426.
82Sa Saito, S.: J. Mol. Spectrosc., to be published.

7 HSO

Microwave data for $^1H^{32}S^{16}O$

Transition			ν MHz	Ref.
rotational $N'_{K_aK_c}-N''_{K_aK_c}$	fine $J'-J''$	hyperfine $F'-F''$		
State: electronic $\tilde{X}\ ^2A''$; vibrational (0, 0, 0)				
$2_{02}-1_{01}$	1½← ½	1←0	79373.524 [a])	81En
$2_{02}-1_{01}$	1½← ½	2←0	79378.688	
$2_{02}-1_{01}$	2½←1½	2←1	79162.482	

[a]) Experimental precision less than 0.05 MHz.

Microwave data for $^1H^{32}S^{16}O$ (continued)

Transition rotational $N'_{K_aK_c}-N''_{K_aK_c}$	fine $J'-J''$	hyperfine $F'-F''$	ν MHz	Ref.
State: electronic $\tilde{X}\,^2A''$; vibrational (0, 0, 0) (continued)				
$2_{02}-1_{01}$	2½←1½	3←2	79164.923 [a])	81En
$3_{03}-2_{02}$	2½←1½	3←2	118997.266	
$3_{03}-2_{02}$	3½←2½	4←3	118786.810	
$3_{13}-2_{12}$	2½←1½	2←1	116019.400	
$3_{13}-2_{12}$	2½←1½	3←2	116019.400	
$3_{13}-2_{12}$	3½←2½	3←2	117258.597	
$3_{13}-2_{12}$	3½−2½	4−3	117260.309	
$3_{12}-2_{11}$	2½−1½	2−1	120227.178	
$3_{12}-2_{11}$	2½−1½	3−2	120227.716	
$3_{12}-2_{11}$	3½−2½	3−2	121254.318	
$3_{12}-2_{11}$	3½−2½	4−3	121256.207	
$3_{21}-2_{20}$	2½−1½	2−1	116069.147	
$3_{21}-2_{20}$	2½−1½	3−2	116065.814	
$3_{21}-2_{20}$	3½−2½	3−2	120465.037	
$3_{21}-2_{20}$	3½−2½	4−3	120468.134	
$3_{22}-2_{21}$	2½−1½	2−1	116047.515	
$3_{22}-2_{21}$	2½−1½	3−2	116047.173	
$3_{22}-2_{21}$	3½−2½	3−2	120446.404	
$3_{22}-2_{21}$	3½−2½	4−3	120449.551	
$4_{04}-3_{03}$	3½−2½	3−2	158597.989	
$4_{04}-3_{03}$	3½−2½	4−3	158598.747	
$4_{04}-3_{03}$	4½−3½	4−3	158391.353	
$4_{04}-3_{03}$	4½−3½	5−4	158391.984	
$4_{13}-3_{12}$	3½−2½	3−2	160962.286	
$4_{13}-3_{12}$	3½−2½	4−3	160962.817	
$4_{13}-3_{12}$	4½−3½	4−3	161385.603	
$4_{13}-3_{12}$	4½−3½	5−4	161386.598	
$4_{14}-3_{13}$	4½−3½	4−3	156021.814	
$4_{14}-3_{13}$	4½−3½	5−4	156022.759	
$4_{22}-3_{21}$	3½−2½	3−2	157009.280	
$4_{22}-3_{21}$	3½−2½	4−3	157008.312	
$4_{22}-3_{21}$	4½−3½	4−3	159505.794	
$4_{22}-3_{21}$	4½−3½	5−4	159507.528	
$4_{23}-3_{22}$	3½−2½	3−2	156956.395	
$4_{23}-3_{22}$	4½−3½	4−3	159458.454	
$4_{23}-3_{22}$	4½−3½	5−4	159460.017	
$4_{31}-3_{30}$	3½−2½	3−2	155355.641	
$4_{31}-3_{30}$	4½−3½	4−3	160536.683	
$4_{31}-3_{30}$	4½−3½	5−4	160539.528	
$4_{32}-3_{31}$	3½−2½	3−2	155355.641	
$4_{32}-3_{31}$	4½−3½	4−3	160536.683	
$4_{32}-3_{31}$	4½−3½	5−4	160539.529	
$9_{09}-8_{18}$	8½−7½	8−7	100131.718	
$9_{09}-8_{18}$	8½−7½	9−8	100130.411	
$9_{09}-8_{18}$	9½−8½	9−8	100412.611	
$9_{09}-8_{18}$	9½−8½	10−9	100413.829	
$10_{0,10}-9_{19}$	9½−8½	9−8	145329.634	
$10_{0,10}-9_{19}$	9½−8½	10−9	145328.434	
$10_{0,10}-9_{19}$	10½−9½	10−9	145397.168	
$10_{0,10}-9_{19}$	10½−9½	11−10	145398.303	

[a]) Experimental precision less than 0.05 MHz.

Microwave data for $^{2}D^{32}S^{16}O$

Transition rotational $N'_{K_aK_c}-N''_{K_aK_c}$	fine $J'-J''$	hyperfine $F'-F''$	ν MHz	Ref.
State: electronic $\tilde{X}\ ^2A''$; vibrational (0, 0, 0)				
$3_{03}-2_{02}$	2½−1½		112197.580 [a]	81En
$3_{03}-2_{02}$	3½−2½		112012.794	
$3_{13}-2_{12}$	3½−2½		108990.649	
$3_{12}-2_{11}$	3½−2½		115621.195	
$3_{22}-2_{11}$	2½−1½		110621.065	
$3_{22}-2_{11}$	3½−2½		113005.422	
$3_{21}-2_{20}$	2½−1½		110735.485	
$3_{21}-2_{20}$	3½−2½		113111.186	
$4_{04}-3_{03}$	3½−2½		149498.382	
$4_{04}-3_{03}$	4½−3½		149254.856	
$4_{13}-3_{12}$	3½−2½		153994.155	
$4_{13}-3_{12}$	4½−3½		154132.999	
$4_{14}-3_{13}$	3½−2½		144841.900	
$4_{14}-3_{13}$	4½−3½		145130.076	
$4_{22}-3_{21}$	3½−2½		149055.594	
$4_{22}-3_{21}$	4½−3½		150304.793	
$4_{23}-3_{22}$	3½−2½		148772.953	
$4_{23}-3_{22}$	4½−3½		150038.224	
$3_{12}-3_{03}$	2½−2½		148093.186	
$3_{12}-3_{03}$	3½−3½		145982.736	
$4_{13}-4_{04}$	3½−3½		152588.906	
$4_{13}-4_{04}$	4½−4½		150860.831	
$5_{14}-5_{05}$	4½−4½		158629.326	
$5_{14}-5_{05}$	5½−5½		156979.039	

[a]) Experimental precision less than 0.05 MHz.

Molecular parameters for $^{1}H^{32}S^{16}O$

Parameter [a])		Value	Method	Ref.
State: electronic $\tilde{X}\ ^2A''$; vibrational (0, 0, 0)				
A	[MHz]	299484.63 (49) [b])	MW	81En
B	[MHz]	20504.56 (59)		
C	[MHz]	19133.93 (58)		
Δ_N	[MHz]	$0.3070(52)\cdot 10^{-1}$		
Δ_{NK}	[MHz]	0.8960 (26)		
Δ_K	[MHz]	27.2 [c])		
δ_N	[MHz]	$0.193(22)\cdot 10^{-2}$		
δ_K	[MHz]	0.89 (29)		
$\tilde{\varepsilon}_{aa}$	[MHz]	−10365.99 (55)		
$\tilde{\varepsilon}_{bb}$	[MHz]	−426.65 (109)		
$\tilde{\varepsilon}_{cc}$	[MHz]	0.226 (161)		
$\lvert\tilde{\varepsilon}_{ab}\rvert$	[MHz]	378.0 (27)		
Δ^s_{NK}	[MHz]	$0.38(43)\cdot 10^{-1}$		
Δ^s_K	[MHz]	2.953 (115)		

[a]) The parameters have been determined in a fit to the A-reduced Hamiltonian.
[b]) The numbers in parentheses are 2½ standard deviations of the least-squares fit, in units of the last quoted decimal place.
[c]) Parameter constrained to this value in the fit, obtained from the optical spectrum [80Oh].

Molecular parameters for $^1H^{32}S^{16}O$ (continued)

Parameter[a])		Value	Method	Ref.
State: electronic $\tilde{X}\ ^2A''$; vibrational (0, 0, 0) (continued)				
$(O)_H$	[MHz]	−36.37 (142) [b])	MW	81En
$(aa)_H$	[MHz]	−11.96 (24)		
$(bb)_H$	[MHz]	10.44 (38)		
$\|(ab)_H\|$	[MHz]	−7.8 [d])		

[a]) The parameters have been determined in a fit to the A-reduced Hamiltonian.
[b]) The numbers in parentheses are 2½ standard deviations of the least-squares fit, in units of the last quoted decimal place.
[d]) Parameter constrained in the fit.

Molecular parameters for $^2D^{32}S^{16}O$

Parameter [a])		Value	Method	Ref.
State: electronic $\tilde{X}\ ^2A''$; vibrational (0, 0, 0)				
A	[MHz]	158726.938 (71) [b])	MW	81En
B	[MHz]	19836.533 (89)		
C	[MHz]	17570.206 (87)		
Δ_N	[MHz]	$0.2646(31)\cdot 10^{-1}$		
Δ_{NK}	[MHz]	0.66982 (134)		
Δ_K	[MHz]	8.9 [c])		
δ_N	[MHz]	$0.323(23)\cdot 10^{-2}$		
δ_K	[MHz]	0.558 (42)		
$\tilde{\varepsilon}_{aa}$	[MHz]	−5646.83 (42)		
$\tilde{\varepsilon}_{bb}$	[MHz]	−393.997 (37)		
$\tilde{\varepsilon}_{cc}$	[MHz]	$0.53(56)\cdot 10^{-1}$		
$\|\tilde{\varepsilon}_{ab}\|$	[MHz]	322.667 (78)		
Δ_K^s	[MHz]	0.667 (101)		
$(O)_D$	[MHz]	−5.58 [c])		
$(aa)_D$	[MHz]	−1.84 [c])		
$(bb)_D$	[MHz]	1.60 [c])		
$(ab)_D$	[MHz]	−1.20 (3)		

[a]) The parameters have been determined in a fit to the A-reduced Hamiltonian.
[b]) The numbers in parentheses are 2½ standard deviations of the least-squares fit in units of the last quoted decimal place.
[c]) Parameter constrained in the least-squares fit.

References for HSO

80Oh Ohashi, N., Kasimoto, M., Saito, S., Hirota, E.: J. Mol. Spectrosc. **84** (1980) 204.
81En Endo, Y., Saito, S., Hirota, E.: J. Chem. Phys. **75** (1981) 4379.

8 FSO

Microwave data for $^{19}F^{32}S^{16}O$

Transition rotational $N'_{K_aK_c}-N''_{K_aK_c}$	fine $J'-J''$	hyperfine $F'-F''$	ν MHz	Ref.
State: electronic $\tilde{X}\ ^2A''$; vibrational (0, 0, 0)				
$1_{10}-1_{01}$	½ − ½	1−1	31446.076 [a])	81En
$1_{10}-1_{01}$	1½ − ½	1−1	31141.038	
$1_{10}-1_{01}$	1½−1½	1−1	31102.405	
$1_{10}-1_{01}$	1½−1½	2−2	31078.227	
$2_{11}-2_{02}$	1½−1½	1−1	33117.305	
$2_{11}-2_{02}$	1½−1½	2−2	33241.550	
$2_{11}-2_{02}$	2½−1½	2−2	33098.837	
$2_{11}-2_{02}$	1½−2½	2−2	33198.074	
$2_{11}-2_{02}$	2½−2½	2−2	33055.330	
$2_{11}-2_{02}$	2½−2½	3−3	33039.466	
$3_{12}-3_{03}$	2½−2½	2−2	36135.602	
$3_{12}-3_{03}$	2½−2½	3−3	36250.058	
$3_{12}-3_{03}$	2½−3½	3−3	36191.099	
$3_{12}-3_{03}$	3½−2½	3−3	36158.083	
$3_{12}-3_{03}$	3½−3½	3−3	36099.201	
$3_{12}-3_{03}$	3½−3½	4−4	36091.783	
$4_{13}-4_{04}$	3½−3½	3−3	40435.842	
$4_{13}-4_{04}$	3½−3½	4−4	40481.303	
$4_{13}-4_{04}$	4½−4½	4−4	40500.886	
$4_{13}-4_{04}$	4½−4½	5−4	40426.168	
$5_{14}-5_{05}$	4½−4½	4−4	46257.410	
$5_{14}-5_{05}$	4½−4½	5−5	46308.835	
$5_{14}-5_{05}$	5½−5½	5−5	46326.785	
$5_{14}-5_{05}$	5½−5½	6−6	46253.546	
$6_{15}-6_{06}$	5½−5½	5−5	53744.953	
$6_{15}-6_{06}$	5½−5½	6−6	53806.236	
$6_{15}-6_{06}$	6½−6½	6−6	53848.251	
$6_{15}-6_{06}$	6½−6½	7−7	53779.283	
$7_{16}-7_{07}$	6½−6½	6−6	63096.759	
$7_{16}-7_{07}$	6½−6½	7−7	63163.014	
$7_{16}-7_{07}$	7½−7½	7−7	63225.909	
$7_{16}-7_{07}$	7½−7½	8−8	63156.215	
$8_{17}-8_{08}$	7½−7½	7−7	74394.001	
$8_{17}-8_{08}$	7½−7½	8−8	74442.250	
$8_{17}-8_{08}$	8½−8½	8−8	74506.365	
$8_{17}-8_{08}$	8½−8½	9−9	74435.399	
$6_{24}-6_{15}$	5½−5½	5−5	77973.092	
$6_{24}-6_{15}$	5½−5½	6−6	77988.340	
$6_{24}-6_{15}$	6½−6½	6−6	77767.542	
$6_{24}-6_{15}$	6½−6½	7−7	77802.948	
$7_{25}-7_{16}$	6½−6½	6−6	76313.277	
$7_{25}-7_{16}$	6½−6½	7−7	76350.491	
$7_{25}-7_{16}$	7½−7½	7−7	76139.964	
$7_{25}-7_{16}$	7½−7½	8−8	76184.008	
$8_{26}-8_{17}$	7½−7½	7−7	75757.753	
$8_{26}-8_{17}$	7½−7½	8−8	75756.700	
$8_{26}-8_{17}$	8½−8½	8−8	75712.595	
$8_{26}-8_{17}$	8½−8½	9−9	75703.780	

[a]) Experimental precision less than or equal to 0.08 MHz.

Microwave data for $^{19}F^{32}S^{16}O$ (continued)

Transition			ν MHz	Ref.
rotational $N'_{K_aK_c}-N''_{K_aK_c}$	fine $J'-J''$	hyperfine $F'-F''$		
State: electronic $\tilde{X}\ ^2A''$; vibrational (0, 0, 0) (continued)				
$9_{27}-9_{18}$	8½ − 8½	8 − 8	76767.316 [a])	81En
$9_{27}-9_{18}$	8½ − 8½	9 − 9	76769.634	
$9_{27}-9_{18}$	9½ − 9½	9 − 9	76704.396	
$9_{27}-9_{18}$	9½ − 9½	10 − 10	76684.508	
$1_{11}-0_{00}$	½ − ½	1 − 0	46398.243	
$1_{11}-0_{00}$	½ − ½	0 − 1	46430.286	
$1_{11}-0_{00}$	½ − ½	1 − 1	46330.770	
$1_{11}-0_{00}$	1½ − ½	1 − 0	46129.324	
$1_{11}-0_{00}$	1½ − ½	1 − 1	46062.185	
$1_{11}-0_{00}$	1½ − ½	2 − 1	46123.203	
$2_{12}-1_{01}$	1½ − ½	1 − 0	61306.586	
$2_{12}-1_{01}$	1½ − ½	2 − 1	61298.708	
$2_{12}-1_{01}$	2½ − 1½	2 − 1	61133.181	
$2_{12}-1_{01}$	2½ − 1½	3 − 2	61164.780	
$3_{13}-2_{02}$	2½ − 1½	2 − 1	75416.785	
$3_{13}-2_{02}$	2½ − 1½	3 − 2	75381.815	
$3_{13}-2_{02}$	3½ − 2½	3 − 2	75268.177	
$3_{13}-2_{02}$	3½ − 2½	4 − 3	75303.333	
$4_{04}-3_{13}$	3½ − 2½	3 − 2	41323.300	
$4_{04}-3_{13}$	3½ − 2½	4 − 3	41359.704	
$4_{04}-3_{13}$	4½ − 3½	4 − 3	41497.778	
$4_{04}-3_{13}$	4½ − 3½	5 − 4	41476.613	
$5_{05}-4_{14}$	4½ − 3½	4 − 3	60400.740	
$5_{05}-4_{14}$	4½ − 3½	5 − 4	60432.257	
$5_{05}-4_{14}$	5½ − 4½	5 − 4	60542.058	
$5_{05}-4_{14}$	5½ − 4½	6 − 5	60552.198	
$6_{06}-5_{15}$	5½ − 4½	5 − 4	79443.904	
$6_{06}-5_{15}$	5½ − 4½	6 − 5	79469.102	
$6_{06}-5_{15}$	6½ − 5½	6 − 5	79557.882	
$6_{06}-5_{15}$	6½ − 5½	7 − 6	79540.298	
$7_{16}-6_{25}$	6½ − 5½	6 − 5	50569.919	
$7_{16}-6_{25}$	6½ − 5½	7 − 6	50685.226	
$7_{16}-6_{25}$	7½ − 6½	7 − 6	51300.001	
$7_{16}-6_{25}$	7½ − 6½	8 − 7	50560.874	
$8_{17}-7_{26}$	7½ − 6½	7 − 6	73525.956	
$8_{17}-7_{26}$	7½ − 6½	8 − 7	73540.309	
$8_{17}-7_{26}$	8½ − 7½	8 − 7	73762.673	
$8_{17}-7_{26}$	8½ − 7½	9 − 8	73738.367	
$5_{33}-6_{24}$	4½ − 5½	4 − 5	45991.234	
$5_{33}-6_{24}$	4½ − 5½	5 − 6	45981.171	
$5_{33}-6_{24}$	5½ − 6½	5 − 6	45585.824	
$5_{33}-6_{24}$	5½ − 6½	6 − 7	45576.155	
$11_{29}-10_{38}$	10½ − 9½	10 − 9	65752.061	
$11_{29}-10_{38}$	10½ − 9½	11 − 10	65769.611	
$11_{29}-10_{38}$	11½ − 10½	11 − 10	66083.955	
$11_{29}-10_{38}$	11½ − 10½	12 − 11	66044.228	
$7_{44}-8_{35}$	6½ − 7½	6 − 7	75001.983	
$7_{44}-8_{35}$	6½ − 7½	7 − 8	75009.880	
$7_{44}-8_{35}$	7½ − 8½	7 − 8	74609.657	

[a]) Experimental precision less than or equal to 0.08 MHz.

Microwave data for $^{19}F^{32}S^{16}O$ (continued)

Transition			ν MHz	Ref.
rotational $N'_{K_aK_c}-N''_{K_aK_c}$	fine $J'-J''$	hyperfine $F'-F''$		
State: electronic $\tilde{X}\ ^2A''$; vibrational (0, 0, 0) (continued)				
$7_{44}-8_{35}$	7½−8½	8−9	74600.723 [a])	81En
$7_{43}-8_{36}$	6½−7½	6−7	76072.484	
$7_{43}-8_{36}$	6½−7½	7−8	76084.522	
$7_{43}-8_{36}$	7½−8½	7−8	75623.782	
$7_{43}-8_{36}$	7½−8½	8−9	75610.955	
$8_{45}-9_{36}$	7½−8½	7−8	56795.096	
$8_{45}-9_{36}$	7½−8½	8−9	56795.096	
$8_{45}-9_{36}$	8½−9½	8−9	56435.444	
$8_{45}-9_{36}$	8½−9½	9−10	56429.196	
$8_{44}-9_{37}$	7½−8½	7−8	58917.030	
$8_{44}-9_{37}$	7½−8½	8−9	58926.366	
$8_{44}-9_{37}$	8½−9½	8−9	58612.691	
$8_{44}-9_{37}$	8½−9½	9−10	58603.941	
$11_{56}-12_{49}$	10½−11½	10−11	67250.726	
$11_{56}-12_{49}$	10½−11½	11−12	67255.355	
$11_{56}-12_{49}$	11½−12½	11−12	66980.658	
$11_{56}-12_{49}$	11½−12½	12−13	66976.424	
$11_{57}-12_{48}$	10½−11½	10−11	66555.164	
$11_{57}-12_{48}$	10½−11½	11−12	66557.189	
$11_{57}-12_{48}$	11½−12½	11−12	66232.229	
$11_{57}-12_{48}$	11½−12½	12−13	66227.604	
$12_{57}-13_{4,10}$	11½−12½	11−12	49641.113	
$12_{57}-13_{4,10}$	11½−12½	12−13	49642.999	
$12_{57}-13_{4,10}$	12½−13½	12−13	49358.119	
$12_{57}-13_{4,10}$	12½−13½	13−14	49352.925	
$12_{58}-13_{49}$	11½−12½	11−12	48341.284	
$12_{58}-13_{49}$	11½−12½	12−13	48339.298	
$12_{58}-13_{49}$	12½−13½	12−13	48031.921	
$12_{58}-13_{49}$	12½−13½	13−14	48027.760	
State: electronic $\tilde{X}\ ^2A''$; vibrational (0, 0, 1)				
$4_{13}-4_{04}$	3½−3½	3−3	41092.359 [a])	81En
$4_{13}-4_{04}$	3½−3½	4−4	41140.383	
$4_{13}-4_{04}$	4½−4½	4−4	41160.787	
$4_{13}-4_{04}$	4½−4½	5−5	41088.937	
$5_{14}-5_{05}$	4½−4½	4−4	46941.065	
$5_{14}-5_{05}$	4½−4½	5−5	46991.997	
$5_{14}-5_{05}$	5½−5½	5−5	47005.532	
$5_{14}-5_{05}$	5½−5½	6−6	46931.782	
$6_{15}-6_{06}$	5½−5½	5−5	54436.096	
$6_{15}-6_{06}$	5½−5½	6−6	54497.976	
$6_{15}-6_{06}$	6½−6½	6−6	54542.484	
$7_{16}-7_{07}$	6½−6½	6−6	63809.754	
$7_{16}-7_{07}$	6½−6½	7−7	63876.402	
$7_{16}-7_{07}$	7½−7½	7−7	63942.381	
$7_{16}-7_{07}$	7½−7½	8−8	63873.648	
$8_{17}-8_{08}$	7½−7½	7−7	75105.621	
$8_{17}-8_{08}$	7½−7½	8−8	75172.660	

[a]) Experimental precision less than or equal to 0.08 MHz.

Microwave data for $^{19}F^{32}S^{16}O$ (continued)

Transition rotational $N'_{K_aK_c}-N''_{K_aK_c}$	fine $J'-J''$	hyperfine $F'-F''$	ν MHz	Ref.
State: electronic $\tilde{X}$ $^2A''$; vibrational (0, 0, 1) (continued)				
$8_{17}-8_{08}$	8½−8½	8−8	75255.622 [a])	81En
$8_{17}-8_{08}$	8½−8½	9−9	75185.582	
$6_{24}-6_{15}$	5½−5½	5−5	79661.940	
$6_{24}-6_{15}$	5½−5½	6−6	79678.100	
$6_{24}-6_{15}$	6½−6½	6−6	79465.108	
$6_{24}-6_{15}$	6½−6½	7−7	79502.212	
$7_{25}-7_{16}$	6½−6½	6−6	77934.210	
$7_{25}-7_{16}$	6½−6½	7−7	77907.514	
$7_{25}-7_{16}$	7½−7½	7−7	77831.918	
$7_{25}-7_{16}$	7½−7½	8−8	77812.968	
$8_{26}-8_{17}$	7½−7½	7−7	77314.638	
$8_{26}-8_{17}$	7½−7½	8−8	77312.172	
$8_{26}-8_{17}$	8½−8½	8−8	77258.468	
$8_{26}-8_{17}$	8½−8½	9−9	77249.532	
$9_{27}-9_{18}$	8½−8½	8−8	78219.208	
$9_{27}-9_{18}$	8½−8½	9−9	78222.180	
$9_{27}-9_{18}$	9½−9½	9−9	78154.510	
$9_{27}-9_{18}$	9½−9½	10−10	78136.316	
$2_{12}-1_{01}$	1½− ½	1−0	61820.999	
$2_{12}-1_{01}$	1½− ½	2−1	61812.831	
$2_{12}-1_{01}$	2½−1½	2−1	61659.059	
$2_{12}-1_{01}$	2½−1½	3−2	61690.233	
$3_{13}-2_{02}$	2½−1½	3−2	75838.024	
$3_{13}-2_{02}$	3½−2½	3−2	75735.142	
$3_{13}-2_{02}$	3½−2½	4−3	75771.126	
$5_{05}-4_{14}$	4½−3½	4−3	59653.103	
$5_{05}-4_{14}$	4½−3½	5−4	59684.714	
$5_{05}-4_{14}$	5½−4½	5−4	59790.096	
$5_{05}-4_{14}$	5½−4½	6−5	59770.088	
$6_{06}-5_{15}$	5½−4½	5−4	78697.348	
$6_{06}-5_{15}$	5½−4½	6−5	78721.840	
$6_{06}-5_{15}$	6½−5½	6−5	78809.608	
$6_{06}-5_{15}$	6½−5½	7−6	78790.852	
$8_{17}-7_{26}$	7½−6½	7−6	71568.341	
$8_{17}-7_{26}$	7½−6½	8−7	71589.407	
$8_{17}-7_{26}$	8½−7½	8−7	71833.112	
$8_{17}-7_{26}$	8½−7½	9−8	71797.447	
$11_{29}-10_{38}$	10½−9½	10−9	62159.911	
$11_{29}-10_{38}$	10½−9½	11−10	62177.121	
$11_{29}-10_{38}$	11½−10	11−10	62484.212	
$11_{29}-10_{38}$	11½−10½	12−11	62444.235	
$7_{44}-8_{35}$	6½−7½	6−7	79658.452	
$7_{44}-8_{35}$	6½−7½	7−8	79666.298	
$7_{44}-8_{35}$	7½−8½	7−8	79299.900	
$7_{44}-8_{35}$	7½−8½	8−9	79290.790	
$8_{45}-9_{36}$	7½−8½	7−8	61514.479	
$8_{45}-9_{36}$	7½−8½	8−9	61513.691	
$8_{45}-9_{36}$	8½−9½	8−9	61181.801	
$8_{45}-9_{36}$	8½−9½	9−10	61174.580	

[a]) Experimental precision less than or equal to 0.08 MHz.

Microwave data for $^{19}F^{32}S^{16}O$ (continued)

Transition rotational $N'_{K_aK_c}-N''_{K_aK_c}$	fine $J'-J''$	hyperfine $F'-F''$	ν MHz	Ref.
State: electronic $\tilde{X}\ ^2B$; vibrational (0, 0, 1) (continued)				
$8_{44}-9_{37}$	7½−8½	7−8	63588.214 [a]	81En
$8_{44}-9_{37}$	7½−8½	8−9	63597.834	
$8_{44}-9_{37}$	8½−9½	8−9	63339.878	
$8_{44}-9_{37}$	8½−9½	9−10	63333.014	
$11_{57}-12_{48}$	10½−11½	10−11	72617.664	
$11_{57}-12_{48}$	10½−11½	11−12	72619.526	
$11_{57}-12_{48}$	11½−12½	11−12	72326.264	
$11_{56}-12_{49}$	10½−11½	10−11	73287.563	
$11_{56}-12_{49}$	10½−11½	11−12	73292.837	
$11_{56}-12_{49}$	11½−12½	11−12	73128.516	
$11_{56}-12_{49}$	11½−12½	12−13	73128.516	

[a]) Experimental precision less than or equal to 0.08 MHz.

Microwave data for $^{19}F^{34}S^{16}O$

Transition rotational $N'_{K_aK_c}-N''_{K_aK_c}$	fine $J'-J''$	hyperfine $F'-F''$	ν MHz	Ref.
State: electronic $\tilde{X}\ ^2A''$; vibrational (0, 0, 0)				
$5_{14}-5_{05}$	4½−4½	4−4	45622.260 [a]	81En
$5_{14}-5_{05}$	4½−4½	5−5	45675.843	
$5_{14}-5_{05}$	5½−5½	5−5	45699.390	
$5_{14}-5_{05}$	5½−5½	6−6	45628.168	
$6_{15}-6_{06}$	5½−5½	5−5	53393.158	
$6_{15}-6_{06}$	5½−5½	6−6	53455.770	
$6_{15}-6_{06}$	6½−6½	6−6	53501.625	
$6_{15}-6_{06}$	6½−6½	7−7	53432.570	
$3_{13}-2_{02}$	2½−1½	2−1	73993.440	
$3_{13}-2_{02}$	2½−1½	3−2	73958.279	
$3_{13}-2_{02}$	3½−2½	3−2	73844.886	
$4_{14}-3_{03}$	3½−2½	3−2	87206.393	
$4_{14}-3_{03}$	3½−2½	4−3	87168.692	
$4_{14}-3_{03}$	4½−3½	4−3	87075.006	
$4_{14}-3_{03}$	4½−3½	5−4	87107.811	
$5_{05}-4_{14}$	4½−3½	4−3	67250.820	
$5_{05}-4_{14}$	4½−3½	5−4	61281.303	
$5_{05}-4_{14}$	5½−4½	5−4	61388.847	
$5_{05}-4_{14}$	5½−4½	6−5	61369.824	
$6_{06}-5_{15}$	5½−4½	5−4	80155.828	
$6_{06}-5_{15}$	5½−4½	6−5	80179.616	
$6_{06}-5_{15}$	6½−5½	6−5	80265.852	
$6_{06}-5_{15}$	6½−5½	7−6	80249.004	

[a]) Experimental precision less than or equal to 0.08 MHz.

Molecular parameters for $^{19}F^{32}S^{16}O$				
Parameter [a])		Value	Method	Ref.
State: electronic $\tilde{X}\ ^2A''$; vibrational (0, 0, 0)				
A	[MHz]	38698.179 (19) [b])	MW	81En
B	[MHz]	9340.810 (4)		
C	[MHz]	7505.064 (5)		
Δ_c	[amu Å^2]	0.17449 (5)		
Δ_N	[MHz]	$0.9970 (45) \cdot 10^{-2}$		
Δ_{NK}	[MHz]	$-0.8286 (42) \cdot 10^{-1}$		
Δ_K	[MHz]	0.88177 (48)		
δ_N	[MHz]	$0.2899 (18) \cdot 10^{-2}$		
δ_K	[MHz]	$0.2711 (88) \cdot 10^{-1}$		
$\tilde{\varepsilon}_{aa}$	[MHz]	−339.543 (83)		
$\tilde{\varepsilon}_{bb}$	[MHz]	34.896 (30)		
$\tilde{\varepsilon}_{cc}$	[MHz]	1.861 (26)		
$\lvert\tilde{\varepsilon}_{ab}\rvert$	[MHz]	207.952 (45)		
Δ_N^s	[MHz]	$0.47 (49) \cdot 10^{-3}$		
$(\Delta_{NK}^s + \Delta_{KN}^s)$	[MHz]	$-0.131 (31) \cdot 10^{-1}$		
Δ_K^s	[MHz]	$0.426 (63) \cdot 10^{-1}$		
$(O)_F$	[MHz]	67.228 (57)		
$(aa)_F$	[MHz]	−118.16 (19)		
$(bb)_F$	[MHz]	−117.06 (13)		
$\lvert(ab)_F\rvert$	[MHz]	10.29 (74)		
State: electronic $\tilde{X}\ ^2A''$; vibrational (0, 0, 1)				
A	[MHz]	39295.969 (28)	MW	81En
B	[MHz]	9325.356 (6)		
C	[MHz]	7479.494 (8)		
Δ_c	[amu Å^2]	0.51427 (8)		
Δ_N	[MHz]	$0.9987 (70) \cdot 10^{-2}$		
Δ_{NK}	[MHz]	$-0.8526 (69) \cdot 10^{-1}$		
Δ_K	[MHz]	0.96169 (67)		
δ_N	[MHz]	$0.2915 (24) \cdot 10^{-2}$		
δ_K	[MHz]	$0.291 (14) \cdot 10^{-2}$		
$\tilde{\varepsilon}_{aa}$	[MHz]	−313.228 (40)		
$\tilde{\varepsilon}_{bb}$	[MHz]	35.695 (26)		
$\tilde{\varepsilon}_{cc}$	[MHz]	1.915 (32)		
$\lvert\tilde{\varepsilon}_{ab}\rvert$	[MHz]	210.82 (11)		
Δ_N^s	[MHz]	$0.47 \cdot 10^{-3}$ [c])		
$(\Delta_{NK}^s + \Delta_{KN}^s)$	[MHz]	$-0.131 \cdot 10^{-1}$ [c])		
Δ_K^s	[MHz]	$0.426 \cdot 10^{-1}$ [c])		
$(O)_F$	[MHz]	66.86 (11)		
$(aa)_F$	[MHz]	−118.24 (33)		
$(bb)_F$	[MHz]	−116.85 (19)		
$\lvert(ab)_F\rvert$	[MHz]	10.5 (15)		

[a]) The parameters have been determined in a fit to the A-reduced Hamiltonian.

[b]) The numbers in parentheses are 2½ standard deviations of the least-squares fit, quoted in units of the last decimal place.

[c]) Parameter constrained in the least-squares fit to value for (0, 0, 0) level.

Molecular parameters for $^{19}F^{32}S^{16}O$ (continued)

Parameter a)		Value	Method	Ref.
State: electronic $\tilde{X}$ $^2A''$; vibrational (0, 0, 0)				
μ_a	[D]	0.374 (12) b)	MW	81En
μ_b	[D]	1.624 (9)		
μ_{tot}	[D]	1.666 (13)		
r_o(S-F)	[Å]	1.602 (3)		
r_o(S-O)	[Å]	1.452 (3)		
$\angle_o$(FSO)	[degrees]	108.32 (6)		

a) The parameters have been determined in a fit to the A-reduced Hamiltonian.
b) The numbers in parentheses are 2½ standard deviations of the least-squares fit, quoted in units of the last decimal place.

Molecular parameters for $^{19}F^{34}S^{16}O$

Parameter a)		Value	Method	Ref.
State: electronic $\tilde{X}$ $^2A''$; vibrational (0, 0, 0)				
A	[MHz]	37536.604 (63) b)	MW	81En
B	[MHz]	9334.689 (7)		
C	[MHz]	7456.138 (6)		
Δ_N	[MHz]	$0.9970 \cdot 10^{-2}$ c)		
Δ_{NK}	[MHz]	$-0.8286 \cdot 10^{-1}$ c)		
Δ_K	[MHz]	0.88177 c)		
δ_N	[MHz]	$0.2899 \cdot 10^{-2}$ c)		
δ_K	[MHz]	$0.2711 \cdot 10^{-1}$ c)		
$\tilde{\varepsilon}_{aa}$	[MHz]	−333.48 (66)		
$\tilde{\varepsilon}_{bb}$	[MHz]	35.89 (16)		
$\tilde{\varepsilon}_{cc}$	[MHz]	1.82 (15)		
$\|\tilde{\varepsilon}_{ab}\|$	[MHz]	214.2 (22)		
Δ_N^s	[MHz]	0.47 c) $\cdot 10^{-3}$		
$(\Delta_{NK}^s + \Delta_{KN}^s)$	[MHz]	−0.131 c) $\cdot 10^{-1}$		
Δ_K^s	[MHz]	0.426 c) $\cdot 10^{-1}$		
$(O)_F$	[MHz]	66.82 (53)		
$(aa)_F$	[MHz]	−115.7 (20)		
$(bb)_F$	[MHz]	−117.7 (9)		

a) The parameters have been determined in a fit to the A-reduced Hamiltonian.
b) The numbers in parentheses are 2½ standard deviations of the least-squares fit, in units of the last quoted decimal place.
c) Parameter constrained to value for $F^{32}SO$ in (0, 0, 0) level in the least-squares fit.

FSO

Observations by Far Infrared LMR

The radical was first detected by Wayne and Radford [77Wa], using the LMR technique. Spectra were observed laser lines at 513 μm, 419 μm and 394 μm. They have not yet been analysed.

References for FSO

77Wa Wayne, F.D., Radford, H.E.: unpublished results.
81En Endo, Y., Saito, S., Hirota, E.: J. Chem. Phys. **74** (1981) 1568.

9 PH_2

Microwave data for $^{31}P^{1}H_2$

Transition				ν MHz	Ref.
rotational $N'_{K_aK_c}-N''_{K_aK_a}$	$J'-J''$	$F_1'-F_1''$ [a]	$F'-F''$		
State: electronic $\tilde{X}\,^2B_1$; vibrational (0, 0, 0)					
$1_{10}-1_{01}$	½ − ½		1−1	151800.837 (25) [b]	83En
			0−1	151450.678 (25)	83En
			1−0	151534.074 (25)	83En
	1½ − ½		2−1	143558.083 (25)	83En
	½ − 1½		1−2	153517.854 (25)	83En
	1½ − 1½		1−1	145395.609 (25)	83En
			2−2	145275.113 (25)	83En
			2−1	145455.348 (25)	83En
			1−2	145215.341 (25)	83En
$2_{20}-2_{11}$	1½ − 1½	1−1	1−1	188154.845 (25)	83En
			2−2	188156.120 (25)	83En
			1−2	188129.449 (25)	83En
			2−1	188181.562 (25)	83En
			1−0	188167.302 (25)	83En
			0−1	188141.670 (25)	83En
		1−2	2−3	188100.245 (25)	83En
		2−1	3−2	188322.417 (25)	83En
		2−2	1−1	188265.264 (25)	83En
			2−2	188264.350 (25)	83En
			3−3	188266.471 (25)	83En
	2½ − 2½	2−2	1−1	181506.501 (25)	83En
			2−2	181505.578 (50) [c]	83En
			3−3	181505.578 (50) [c]	83En
			1−2	181530.002 (25)	83En
			2−1	181482.168 (25)	83En
		2−2	2−3	181540.451 (25)	83En
			3−2	181470.862 (25)	83En
		3−3	2−2	181457.725 (50) [c]	83En
			3−3	181457.725 (50) [c]	83En
			4−4	181457.035 (25)	83En
$3_{30}-3_{21}$	2½ − 2½		2−2	252867.979 (25)	83En
			3−3	252906.473 (25)	83En
	3½ − 3½		3−3	244036.269 (25)	83En
			4−4	244013.727 (25)	83En

[a]) The intermediate angular momentum quantum number F_1 refers to the coupling of $\mathbf{I}_P$ to $\mathbf{J}$.

[b]) The figures in parentheses represent the experimental uncertainty, in units of the last quoted decimal place.

[c]) Line blended with at least one other hyperfine component.

Far Infrared LMR data for PH_2

Laser			Rotational transition	Ref.
Gas	λ µm	ν GHz		
State: electronic $\tilde{X}\,^2B_1$; vibrational (0, 0, 0)				
HCOOH	458.5	653.821	$2_{12}-1_{01}$	79Da
N_2H_4	331.7	903.889	$4_{13}-4_{04}$	
			$5_{33}-5_{24}$	

Far Infrared LMR data for PH_2 (continued)

Laser			Rotational transition	Ref.
Gas	λ μm	ν GHz		
State: electronic $\tilde{X}\ ^2B_1$; vibrational (0, 0, 0) (continued)				
HCOOH	432.6	692.950	$5_{42}-5_{33}$	79Da
H_2O	118.6	2527.952	$6_{43}-5_{32}$	
CH_2F_2	407.3	736.059	$6_{52}-6_{43}$	
HCOOH	418.6	716.156	$6_{61}-6_{52}$	
CH_3OH	96.5	3105.937	$7_{53}-6_{42}$	

Molecular parameters for $^{32}P^1H_2$

Parameter [a])		Value	Method	Ref.
State: electronic $\tilde{X}\ ^2B_1$; vibrational (0, 0, 0)				
A	[MHz]	273782.0 (10) [b])	LMR, MW	79Da, 79Hi, 81Mc, 83En
B	[MHz]	242346.6 (10)		
C	[MHz]	126342.5 (10)		
Δ_N	[MHz]	16.554 (16)		
Δ_{NK}	[MHz]	−54.41 (17)		
Δ_K	[MHz]	87.27 (48)		
δ_N	[MHz]	7.470 (10)		
δ_K	[MHz]	−2.984 (76)		
Φ_N	[kHz]	5.56 (63)		
Φ_{NK}	[kHz]	31.8 (42)		
Φ_{KN}	[kHz]	25.1 (85)		
Φ_K	[kHz]	27.9 (52)		
ϕ_N	[kHz]	2.94 (23)		
ϕ_{NK}	[kHz]	−11.3 (12)		
ϕ_K	[kHz]	23.5 (23)		
ε_{aa}	[MHz]	−8427.48 (63)		
ε_{bb}	[MHz]	−2458.62 (18)		
ε_{cc}	[MHz]	−7.12 (42)		
Δ_N^S	[MHz]	0.44 (12)		
$\Delta_{NK}^S + \Delta_{KN}^S$	[MHz]	−2.11 (62)		
Δ_K^S	[MHz]	6.75 (61)		
δ_N^S	[MHz]	0.183 (50)		
δ_K^S	[MHz]	0.13 (32)		
$((0)_P$	[MHz]	207.25 (5)		
$(aa)_P$	[MHz]	−300.24 (5)		
$(bb)_P$	[MHz]	−321.86 (12)		
$(cc)_P$	[MHz]	622.10 (12)		
$(0)_H$	[MHz]	−48.85 (5)		
$(aa)_H$	[MHz]	−1.00 (4)		
$(bb)_H$	[MHz]	−4.46 (10)		
$(cc)_H$	[MHz]	5.46 (10)		
C_{aa}	[MHz]	0.960 (21)		
C_{bb}	[MHz]	0.523 (30)		
C_{cc}	[MHz]	0.13 (6)		

[a]) The parameters have been determined in a fit to the A-reduced Hamiltonian.

[b]) The numbers in parentheses are 1 standard deviation of the least-squares fit, in units of the last quoted decimal place.

For References, see next page.

References for PH_2

79Da Davies, P.B., Russell, D.K., Thrush, B.A., Radford, H.E.: Chem. Phys. **44** (1979) 421.
79Hi Hills, G.W., McKellar, A.R.W.: J. Chem. Phys. **71** (1979) 1141.
81Mc McKellar, A.R.W.: Discuss. Faraday Soc. **71** (1981) 63.
83En Endo, Y., Saito, S., Hirota, E.: J. Mol. Spectrosc., **97** (1983) 204.

10 CH_2

Microwave data for CH_2

Transition			ν MHz	Ref.
rotational $N'_{K_aK_c} - N''_{K_aK_c}$	fine $J' - J''$	hyperfine $F' - F''$		
State: electronic $\tilde{X}\ ^3B_1$; vibrational (0, 0, 0)				
$4_{04} - 3_{13}$	5−4	6−5	68371.278 (41) [a]	83Lo
		5−4	68375.875 (39)	
		4−3	68380.873 (41)	
	4−3	5−4	70680.720 (38)	
		4−3	70679.543 (45)	
		3−2	70678.633 (42)	
	3−2	4−3	69019.187 (44)	
		3−2	69014.202 (37)	
		2−1	69007.179 (37)	

[a]) The figure in parenthesis are the authors' estimate of experimental error.

Far Infrared LMR data for $^{12}C^1H_2$

Laser			Rotational transition	Ref.
Gas	λ µm	ν GHz		
State: electronic $\tilde{X}\ ^3B_1$; vibrational (0, 0, 0)				
$^{13}CH_3OH$	63.1	4751.3414	$5_{23} - 5_{14}$	82Se
CH_3OH	63.4	4730.8605	$5_{23} - 5_{14}$	
CH_2DOH	102.0	2938.4651	$3_{22} - 4_{13}$	
			$4_{22} - 5_{15}$	
CH_2DOH	108.8	2754.9957	$2_{12} - 1_{01}$	
CD_3OH	128.0	2341.5104	$1_{11} - 0_{00}$	
			$6_{15} - 6_{06}$	
CD_3OH	144.1	2080.1893	$4_{13} - 4_{04}$	
$^{13}CH_3OH$	149.3	2008.3601	$2_{11} - 2_{02}$	
			$3_{12} - 3_{03}$	
CD_2F_2	149.5	2005.7834	$3_{12} - 3_{03}$	
CH_2DOH	150.8	1987.7990	$2_{11} - 2_{02}$	
$^{13}CH_3OH$	157.9	1898.2799	$1_{10} - 1_{01}$	
CH_2F_2	158.5	1891.2743	$1_{10} - 1_{01}$	
			$2_{11} - 2_{02}$	
CD_2F_2	317.0	945.5625	$1_{11} - 2_{02}$	
CD_2F_2	500.6	598.8937	$5_{05} - 4_{14}$	
CH_2CHBr	680.5	440.5205	$2_{12} - 3_{03}$	
State: electronic $\tilde{X}\ ^3B_1$; vibrational (1, 0, 0)				
CH_2F_2	158.5	1891.2743	$3_{12} - 3_{03}$	82Se
CH_3OH	163.0	1838.8393	$2_{11} - 2_{02}$	

Molecular parameters for $^{12}C^{1}H_2$

Parameter [a]		Value	Method	Ref.
State: electronic $\tilde{X}\ ^3B_1$; vibrational (0, 0, 0)				
A	[MHz]	2190216.2 (33) [b]	MW, LMR	82Se, 83Lo
B	[MHz]	252280.5 (23)		
C	[MHz]	216428.3 (13)		
Δ_K	[MHz]	59690.1 (14)		
Δ_{NK}	[MHz]	−589.39 (81)		
Δ_N	[MHz]	9.03 (10)		
δ_K	[MHz]	0.0 [c]		
δ_N	[MHz]	3.034 (36)		
Φ_{KN}	[MHz]	−58.211 (63)		
Φ_{NK}	[MHz]	0.384 (26)		
Φ_N	[MHz]	$0.75 (17) \cdot 10^{-2}$		
ϕ_N	[MHz]	$0.585 (90) \cdot 10^{-2}$		
α	[MHz]	7778.8 (42)		
β	[MHz]	1196.4 (11)		
ε_{aa}	[MHz]	13.3 (23)		
ε_{bb}	[MHz]	−154.33 (54)		
ε_{cc}	[MHz]	−123.09 (81)		
$(O)_H$	[MHz]	−20.26 (51)		
$(aa)_H$	[MHz]	39.7 (17)		
$(bb)_H$	[MHz]	−20.2 (19)		
g_S^{aa}		2.002315 [c]		
g_S^{bb}		2.002624 [c]		
g_S^{cc}		2.002604 [c]		
g_r^{aa}		$-0.20 \cdot 10^{-4}$ [c]		
g_r^{bb}		$-2.57 \cdot 10^{-4}$ [c]		
g_r^{cc}		$-1.47 \cdot 10^{-4}$ [c]		
State: electronic $\tilde{X}\ ^3B_1$; vibrational (1, 0, 0)				
A	[MHz]	2072065 (15) [d]		82Se
B	[MHz]	251817.8 (54)		
α	[MHz]	7853 (11)		
β	[MHz]	1313.4 (93)		
ε_{bb}	[MHz]	−126.8 (42)		
ε_{cc}	[MHz]	83.3 (36)		

[a]) Parameters determined in a fit to the A-reduced Hamiltonian.
[b]) The figures in parenthesis represent one standard deviation of the least-squares fit, in units of the last quoted decimal place.
[c]) Parameter constrained to this value in the fit.
[d]) All other parameters constrained to their ground state (0, 0, 0) values.

References for CH_2

82Se Sears, T.J., Bunker, P.R., Mc Kellar, A.R.W., Evenson, K.M., Jennings, D.A., Brown, J.M.: J. Chem. Phys., **77** (1982) 5348.

83Lo Lovas, F.J., Suenram, R.D., Evenson, K.M.: to be published.

4.2.3 Non-linear larger radicals

4.2.3.1 Preliminary remarks

1. Introduction

Since the publication of the last supplement, an important development in free radical spectroscopy has been the detection of a molecule with 5 atoms in an open-shell state, the methoxy radical CH_3O by Radford and Russell [77Ra]. This molecule has a non-degenerate ground state (^2E) which is subject to a weak Jahn-Teller effect [66He]. The Hamiltonian used to describe the energy levels of the molecule in its ground vibrational state can be written as:

$$\mathcal{H} = \mathcal{H}_{so} + \mathcal{H}_{rot} + \mathcal{H}_{cd} + \mathcal{H}_{sr} + \mathcal{H}_Z + \mathcal{H}'.$$

The first term on the right hand side represents the spin-orbit coupling,

$$\mathcal{H}_{so} = \xi L_z S_z$$

where L_z and S_z are the components of the orbital and spin angular momenta respectively along the symmetric top (z) axis and ξ is the spin-orbit coupling parameter. $\mathcal{H}_{rot}$ is the rotational Hamiltonian, written for a prolate symmetric top as

$$\mathcal{H}_{rot} = (A-B)[J_z - L_z - G_z - S_z]^2 + B[\boldsymbol{J} - \boldsymbol{L} - \boldsymbol{G} - \boldsymbol{S}]^2$$

where A and B are the two rotational constants, related to a good approximation to the inverse moments of inertia:

$$A \simeq \hbar^2/2hI_{aa} \quad \text{and} \quad B \simeq \hbar^2/2hI_{bb}.$$

The vector operators $\boldsymbol{J}$ and $\boldsymbol{G}$ represent the total and vibrational angular momenta respectively. The next term in the Hamiltonian, $\mathcal{H}_{cd}$, represents the quartic centrifugal distortion corrections:

$$\mathcal{H}_{cd} = -D_K[J_z - L_z - G_z - S_z]^4 - D_{JK}[J_z - L_z - G_z - S_z]^2[\boldsymbol{J} - \boldsymbol{L} - \boldsymbol{G} - \boldsymbol{S}]^2$$
$$-D_J[\boldsymbol{J} - \boldsymbol{L} - \boldsymbol{S} - \boldsymbol{G}]^4 + \eta_e J_z L_z [\boldsymbol{J} - \boldsymbol{L} - \boldsymbol{G} - \boldsymbol{S}]^2.$$

The last term in this equation describes the centrifugal distortion of the electronic Coriolis coupling term. $\mathcal{H}_{sr}$ represents the spin-rotation coupling term and is taken in the isotropic form

$$\mathcal{H}_{sr} = \gamma(\boldsymbol{J} - \boldsymbol{S}) \cdot \boldsymbol{S}.$$

The effects of the external magnetic field in the LMR experiment are included in the Zeeman Hamiltonian, $\mathcal{H}_Z$, taken as

$$\mathcal{H}_Z = \mu_B \boldsymbol{B} \cdot \{g_L \boldsymbol{L} + g_S \boldsymbol{S} - g_r(\boldsymbol{J} - \boldsymbol{L} - \boldsymbol{G} - \boldsymbol{S})\} + \mu_B g_l (B_+ S_- + B_- S_+)$$

In this equation $\boldsymbol{B}$ is the applied flux density, μ_B is the Bohr magneton, g_L, g_S, g_r are the orbital, spin and rotational g-factors and $B_\pm$ is defined as $B_x \pm \mathrm{i} B_y$ (with a similar definition for $S_\pm$). The final term in the Hamiltonian, $\mathcal{H}'$, contains perturbation terms defined below. The involvement of these terms has been suggested by Hougen [80Ho].

The matrix representation of the effective Hamiltonian has been constructed by Russell and Radford [80Ru] in a Hund's case (a) basis set; see also the earlier paper by Brown [71Br]z A member of this basis set is written as $|\eta\Lambda\rangle |vl\rangle |S\Sigma\rangle |JM_J P\rangle$ where the four factors are respectively, the electron orbital wavefunction, the (degenerate) vibrational wavefunction, the electron spin and the rotational wavefunction. The quantum numbers Λ, l, Σ and P are associated with z component of the orbital, vibrational, spin and rotational angular momenta ($\boldsymbol{L}, \boldsymbol{G}, \boldsymbol{S}$ and $\boldsymbol{J}$ respectively). The various internal interactions in the molecule mean that strictly speaking, Λ, l and Σ are not good quantum numbers. For CH_3O in its ground state, such departures from goodness are not great and the quantum numbers can still be used unambiguously. Lambda-type doubling effects are not resolved and Λ is used as an (unsigned) quantum number of unit magnitude. The states can be referred to in short by the ket $|K\Sigma; JM_J\rangle$ where $K = P - \Sigma$. The expectation value of L_z in this basis set is $\zeta_e \Lambda$. Russell and Radford [80Ru] found it necessary to include small centrifugal distortion corrections to the parameters ξ and ζ_e in order to fit their data.

The perturbations included in $\mathcal{H}'$ above can now be defined in terms of their matrix elements in the chosen basis set. These perturbations all have matrix elements off-diagonal in K and have been defined by analogy with the corresponding terms in degenerate vibration-rotation theory. The major perturbation was found to be the 2, -1 perturbation which has matrix elements of the form

$$\langle -K-1, -\Sigma; JM_J | \mathcal{H}'_{21} | K\Sigma; JM_J\rangle = X_{21}(K+1)[J(J+1) - K^2 + \tfrac{1}{4}]^{\frac{1}{2}}.$$

The second perturbation is of the 2, 2 type, with matrix elements

$$\langle -K+2, -\Sigma; JM_J | \mathcal{H}'_{22} | K\Sigma; JM_J\rangle = X_{22}[J(J+1) - K^2 + \tfrac{1}{4}]^{\frac{1}{2}}[J(J+1) - (K-1)^2 + \tfrac{1}{4}]^{\frac{1}{2}}.$$

The final perturbation is of the $\Delta K=3$ type, with matrix elements

$$\langle K+3,\Sigma;JM_J|\mathcal{H}'_{33}|K,\Sigma;JM_J\rangle = X_{33}\Lambda[J(J+1)-(K+\Sigma)(K+\Sigma+1)]^{\frac{1}{2}}[J(J+1)-(K+\Sigma+1)(K+\Sigma+2)]^{\frac{1}{2}}\cdot[J(J+1)-(K+\Sigma+2)(K+\Sigma+3)]^{\frac{1}{2}}.$$

2. List of tabulated parameters

ξ, ξ_D	spin-orbit coupling parameter and its centrifugal distortion correction coefficient (effective parameter $=\xi+\xi_D K^2$)
A, B	rotational constants for a prolate symmetric top molecule. The subscripts v have been omitted in the tables because the vibrational level is shown explicitly
ζ_e, ζ_K	electron orbital angular momentum parameter and centrifugal distortion correction (defined so that $\langle L_z\rangle=\zeta_e\Lambda+\zeta_K K$)
ζ_v	vibrational angular momentum coefficient ($\langle G_z\rangle=\zeta_v\ell$)
D_{JK}, D_J	quartic centrifugal distortion parameters
η_e	centrifugal distortion correct to the electronic Coriolis coupling term
γ	spin-rotation coupling parameter
g_S, g_K	electron spin g-factor and K-dependent correction term
g_l	anisotropic correction terms to the electron spin Zeeman interaction
X_{21}, X_{22}, X_{33}	coefficients describing perturbations of levels off-diagonal in the quantum number K
$(O)_H$	fermi contact parameter used to describe the proton hyperfine interaction

3. Arrangement of molecules

1. CH_3O
2. CF_3

4. References

66He Herzberg, G.: "Molecular Spectra and Molecular Structure III. Electronic Spectra and Electronic Structure of Polyatomic Molecules", Van Nostrand Reinhold Company, New York, **1966**,
71Br Brown, J.M.: Mol. Phys. **20** (1971) 817.
77Ra Radford, H.E., Russell, D.K.: J. Chem. Phys. **66** (1977) 222.
80Ho Hougen, J.T.: J. Mol. Spectrose. **81** (1980) 73.
80Ru Russell, D.K., Radford, H.E.: J. Chem. Phys. **72** (1980) 2750.

4.2.3.2 Data and references

1 CH_3O
Far Infrared LMR data for CH_3O

Laser			Rotational transition $J'-J''$, K [a])	Ref.
Gas	λ µm	ν GHz		
State: electronic $\tilde{X}\,^2E$; vibrational: ground				
CH_2CHCl	634	472.5078	8½ − 7½, −5	80Ru
DCOOD	568	527.9260	9½ − 8½, −5	
HCOOH	513	584.3729	10½ − 9½, −5	
CH_3CH_2F	502	596.884	10½ − 9½, −5	
			10½ − 9½, −4	
			10½ − 9½, 3	

[a]) The rotational selection rule on the transitions is $\Delta J=1$, $\Delta K=0$.

Far Infrared LMR data for CH_3O (continued)

Laser			Rotational transition $J'-J''$, K [a])	Ref.
Gas	λ µm	ν GHz		
State: electronic $\tilde{X}$ ^2E; vibrational: ground (continued)				
HCOOH	458	653.8222	11½−10½, −3	80Ru
			11½−10½, −4	
			11½−10½, −5	
CH_2CHCl	442	678.006	12½−11½, −10	
			12½−11½, −11	
N_2H_4	436	687.9574	12½−11½, −6	
			12½−11½, −7	
			12½−11½, −8	
			12½−11½, 2	
			12½−11½, 3	
			12½−11½, 4	
			12½−11½, 5	
			12½−11½, 6	
			12½−11½, 7	
			12½−11½, 8	
			12½−11½, 9	
CD_3OH	419	715.9876	12½−11½, −8	
			12½−11½, −9	
CF_2CH_2	407	736.060	13½−12½, −9	
			13½−12½, −10	
			13½−12½, −11	
HCOOH	406	739.1610	13½−12½, −3	
			13½−12½, −8	
			13½−12½, −9	
			13½−12½, 9	
			13½−12½, 10	
			13½−12½, 11	
HCOOH	394	761.6077	13½−12½, −6	
CH_3OH	392	764.6426	13½−12½, −6	
			13½−12½ [b])	
			13½−12½ [b])	
CH_3OH	386	775.9824	13½−12½, −10	
			13½−12½, −11	
CF_2CH_2	375	798.287	14½−13½, −7	
			14½−13½, 2	
			14½−13½, 3	
			14½−13½, 4	
			14½−13½, 5	
			14½−13½, 5	
			14½−13½, 6	
			14½−13½, 7	
			14½−13½ [b])	
N_2H_4	332	903.8894	16½−15½, −4	
			16½−15½, −9	
			16½−15½, −10	
			16½−15½, 11	
N_2H_4	311	963.7314	17½−16½, −7	
HCOOH	302	991.7769	17½−16½, −3	
CH_3OH	251	1193.7273	21½−20½, 9	

[a]) The rotational selection rule on the transitions is $\Delta J=1$, $\Delta K=0$.
[b]) Not assigned.

Molecular parameters for CH_3O

Parameter		Value	Method	Ref.
State: electronic $\tilde{X}\ ^2E$; vibrational: ground				
A	[MHz]	159729 (22) [a])	LMR	80Ru
B	[MHz]	27933.9 (51)		
ξ	[GHz]	−4281 (20)		
ξ_D	[MHz]	794 (30)		
ζ_e		0.4397 (20)		
ζ_K		$0.882 (5) \cdot 10^{-2}$		
ζ_v		−0.1035 (20)		
D_{JK}	[MHz]	1.0058 (33)		
D_J	[MHz]	$0.794 (30) \cdot 10^{-1}$		
η_e	[MHz]	11.33 (10)		
γ	[MHz]	−1634 (48)		
X_{21}	[MHz]	3096.9 (22)		
X_{22}	[MHz]	43.2 (60)		
X_{33}	[MHz]	0.18 (15)		
g_S		2.077 (5)		
g_K		$-0.29 (4) \cdot 10^{-2}$		
g_l		$-0.91 (10) \cdot 10^{-1}$		

[a]) The figures in parenthesis are 1 standard deviation of the least-squares fit, in units of the last quoted decimal place.

Reference for CH_3O

80Ru Russell, D.K., Radford, H.E.: J. Chem. Phys., **72** (1980) 2750.

2 CF_3
Microwave data for $^{12}C^{19}F_3$

Transition			ν MHz	Ref.
rotational $N', K' - N'', K''$	fine $J' - J''$	hyperfine $F' - F''$		
State: electronic $\tilde{X}\ ^2A_1$; vibrational: ground				
4,0 – 3,0	3½ – 2½	3 – 2	87240.670 (50) [a])	82En
	3½ – 2½	4 – 3	87218.404 (50)	
	3½ – 2½	5 – 4	87191.348 (50)	
	4½ – 3½	3 – 2	87194.217 (50)	
	4½ – 3½	4 – 3	87202.228 (50)	
	4½ – 3½	6 – 5	87178.291 (50)	
4,1 – 3,1	3½ – 2½	3 – 2	87223.184 (50)	
	3½ – 2½	4 – 3	87217.580 (50)	
	4½ – 3½	4 – 3	87198.837 (50)	
4,2 – 3,2	3½ – 2½	4 – 3	87244.208 (50)	
	4½ – 3½	4 – 3	87199.531 (50)	
	4½ – 3½	5 – 4	87170.663 (50)	
5,0 – 4,0	4½ – 3½	3 – 2	109031.822 (50)	
	4½ – 3½	5 – 4	109011.177 (50)	
	4½ – 3½	6 – 5	108993.674 (50)	
	5½ – 4½	4 – 3	108992.980 (50)	
	5½ – 4½	5 – 4	108999.423 (50)	
	5½ – 4½	6 – 5	109004.073 (50)	
	5½ – 4½	7 – 6	108978.948 (50)	

[a]) Estimated experimental uncertainty in units of the last quoted decimal place.

Microwave data for $^{12}C^{19}F_3$ (continued)				
Transition			ν MHz	Ref.
rotational $N', K' - N'', K''$	fine $J' - J''$	hyperfine $F' - F''$		
State: electronic $\tilde{X}\ ^2A_1$; vibrational: ground (continued)				
5,1 – 4,1	4½ – 3½	4 – 3	109022.212 (50) [a]	82En
	4½ – 3½	5 – 4	109012.926 (50)	
	5½ – 4½	5 – 4	108996.192 (50)	
	5½ – 4½	6 – 5	108980.540 (50)	
5,2 – 4,2	4½ – 3½	5 – 4	109026.628 (50)	
	5½ – 4½	5 – 4	108996.911 (50)	
	5½ – 4½	6 – 5	108975.396 (50)	
5,3 – 4,3	4½ – 3½	3 – 2	108955.406 (50)	
	4½ – 3½	5 – 4	109055.860 (50)	
	4½ – 3½	6 – 5	109016.353 (50)	
	5½ – 4½	4 – 3	109007.208 (50)	
	5½ – 4½	5 – 4	108984.018 (50)	
	5½ – 4½	6 – 5	109050.151 (50)	
	5½ – 4½	7 – 6	108940.406 (50)	
5,4 – 4,4	4½ – 3½	4 – 3	108992.147 (50)	
	5½ – 4½	5 – 4	109001.555 (50)	
	5½ – 4½	6 – 5	108954.778 (50)	
6,0 – 5,0	5½ – 4½	4 – 3	130824.429 (50)	
	5½ – 4½	5 – 4	130817.922 (50)	
	5½ – 4½	6 – 5	130806.049 (50)	
	5½ – 4½	7 – 6	130793.101 (50)	
	6½ – 5½	5 – 4	130789.637 (50)	
	6½ – 5½	6 – 5	130795.334 (50)	
	6½ – 5½	7 – 6	130799.973 (50)	
	6½ – 5½	8 – 7	130777.044 (50)	
6,1 – 5,1	5½ – 4½	5 – 4	130818.885 (50)	
	5½ – 4½	6 – 5	130808.957 (50)	
	6½ – 5½	6 – 5	130791.967 (50)	
	6½ – 5½	7 – 6	130778.493 (50)	
6,2 – 5,2	5½ – 4½	5 – 4	130816.598 (50)	
	5½ – 4½	6 – 5	130817.299 (50)	
	6½ – 5½	6 – 5	130792.748 (50)	
	6½ – 5½	7 – 6	130775.527 (50)	
6,3 – 5,3	5½ – 4½	4 – 3	130787.029 (50)	
	5½ – 4½	5 – 4	130818.552 (50)	
	5½ – 4½	6 – 5	130835.408 (50)	
	5½ – 4½	7 – 6	130806.049 (50)	
	6½ – 5½	5 – 4	130796.608 (50)	
	6½ – 5½	6 – 5	130787.886 (50)	
	6½ – 5½	7 – 6	130826.932 (50)	
	6½ – 5½	8 – 7	130753.990 (50)	
6,4 – 5,4	5½ – 4½	5 – 4	130807.594 (50)	
	5½ – 4½	6 – 5	130849.761 (50)	
	6½ – 5½	6 – 5	130796.608 (50)	
	6½ – 5½	7 – 6	130763.708 (50)	
6,5 – 5,5	5½ – 4½	5 – 4	130800.884 (50)	
	5½ – 4½	6 – 5	130872.425 (50)	
	6½ – 5½	6 – 5	130801.526 (50)	
	6½ – 5½	7 – 6	130754.874 (50)	

[a]) Estimated experimental uncertainty in units of the last quoted decimal place.

Molecular parameters for CF_3

Parameter		Value	Method	Ref.
State: electronic $\tilde{X}\ ^2A_1$; vibrational: ground				
B	[MHz]	10900.9118 (52) [a]	MW	82En
D_N	[MHz]	0.013882 (87)		
D_{NK}	[MHz]	−0.02343 (17)		
ε_{bb}	[MHz]	−36.500 (42)		
ε_{cc}	[MHz]	3.35 (15)		
$(O)_F$	[MHz]	408.5 (16)		
$(cc)_F$	[MHz]	320.01 (20)		
$\lvert(aa)_F-(bb)_F\rvert$	[MHz]	40.1 (49)		

[a]) The figures in parentheses represent 2½ standard deviations of the least squares fit, in units of the last quoted decimal place.

Reference for CF_3

82En Endo, Y., Yamada, C., Saito, S., Hirota, E.: J. Chem. Phys. **77** (1982) 3376.

5 Appendix

Appendix to 2.2 Diatomic molecules

For preliminary remarks see p. 5 of Vol. II/14a Landolt-Börnstein, N.S. The literature was updated to September 1982.

The molecules are included in the order as given in the survey of p. 8 of Vol. II/14a. The molecular numbers are extended by a small letter.

Additional molecules:

Nr.	Molecule	Ground state	Data in
13a	HCl^+	$^2\Pi_i$	4.1.5
22a	KF	$^1\Sigma^+$	2.2 of Appendix
33a	AgCl	$^1\Sigma^+$	2.2 of Appendix
33b	AgI	$^1\Sigma^+$	2.6.1 and 2.7.1 of Appendix
40a	CaBr	$^2\Sigma^+$	2.7.1 of Appendix and 4.1.2
40b	CaI	$^2\Sigma^+$	2.7.1 of Appendix and 4.1.2
47a	BCl	$^1\Sigma^+$	2.2 of Appendix
77a	NO^+	$^1\Sigma^+$	2.2 and 2.7.1 of Appendix
81a	AsO	$^2\Pi_r$	2.7.1 of Appendix and 4.1.5
90a	Te_2	$^3\Sigma_g^-$, O_g^+	2.9.1 of Appendix

Nr.	Molecule	B [1]) MHz	D [1]) kHz	α, γ [1]) MHz	Ref.	Meth.	Derived constants	r_e [1]) Å	See Tab.
3	$^{23}Na^1H$	$Y_{01}=146999.10\ (30)$	$Y_{02}^{der}=-10.307\ (20)$ MHz	$Y_{11}=-4108.99\ (80)$	81SAS	MW		1.88654 (10)	2.6.1
				$Y_{21}=32.83\ (50)$	81SAS	MW			
				$Y_{31}=-0.96\ (7)$	81SAS	MW			
	$^{23}Na^2D$	$Y_{01}=76659.59\ (20)$	$Y_{02}=-2.802\ (6)$ MHz	$Y_{11}=-1546.94\ (40)$	81SAS	MW			
				$Y_{21}=8.76\ (20)$	81SAS	MW			
				$Y_{31}=-0.16\ (4)$	81SAS	MW			
20 *6–18*	LiCl				82LEF		$d_{Li}=-0.181\ (7)$ $d_{Cl}=0.53\ (6)$		2.9.1
22a *6–25*	$^{39}K^{19}F$	$Y_{01}=8392.3166\ (19)$	$Y_{02}=-14.486\ (15)$	$Y_{11}=-70.0104\ (29)$	82MAK2	IR/MW	$Y_{10}=426.26119\ (84)\ cm^{-1}$		
			$Y_{12}^{der}=1.9$ Hz	$Y_{21}=211.57\ (75)$ kHz	82MAK2	IR/MW	$Y_{20}=-2.44951\ (40)\ cm^{-1}$		
			$Y_{22}=1.118\ (74)$ Hz		82MAK2	IR/MW	$Y_{30}=0.009345\ (56)\ cm^{-1}$		
	KF				82MAK2	[2])	$a_1=-3.11676\ (21)$		
					82MAK2	[2])	$a_2=6.3075\ (24)$		
					82MAK2	[2])	$a_3=-9.363\ (19)$		
					82MAK2	[2])	$a_4=9.63\ (40)$		
					82MAK2	[2])	$a_5=-5.1\ (22)$		
	$^{39}K^{19}F$				82MAK2	[2])	$\omega_e=426.26204\ (52)\ cm^{-1}$		
					82MAK2	[2])	$B_e=8392.2936\ (25)$ MHz		
31	CuCl				82LEF		$d_{Cu}=0.71\ (15)$ $d_{Cl}=0.10\ (5)$		2.7.1 2.9.1
32	CuBr				82LEF		$d_{Cu}=0.41\ (13)$ $d_{Br}=0.28\ (15)$		2.7.1
33a *6–39*	AgCl				82LEF		$d_{Cl}=0.143\ (32)$		
47 *6–45*	$^{27}Al^{19}F$	$Y_{01}=16562.9360\ (45)$	$Y_{02}=-31.378\ (13)$	$Y_{11}=-149.4281\ (74)$	82MAK2	IR/MW	$Y_{10}=802.3243\ (15)\ cm^{-1}$		2.7.1 2.9.1
			$Y_{12}=50.98\ (87)$ Hz	$Y_{21}=517.9\ (37)$ kHz	82MAK2	IR/MW	$Y_{20}=-4.84945\ (90)\ cm^{-1}$		
			$Y_{22}=2.07\ (11)$ Hz	$Y_{31}=0.99\ (53)$ kHz	82MAK2	IR/MW	$Y_{30}=0.01931\ (16)\ cm^{-1}$		
Cont.			$Y_{03}^{der}=-10.9\cdot 10^{-3}$ Hz		82MAK2	IR/MW			

[1]) Or related terms.
[2]) The spectral data were directly fitted to the potential coefficients $a_1, a_2, a_3, \ldots$, and ω_e, B_e; [82MAK2] presents an additional set of Dunham-parameters Y_{lk} calculated from the a_i's and ω_e, B_e.

Nr.	Molecule	B [1]) MHz	D [1]) kHz	α, γ [1]) MHz	Ref.	Meth.	Derived constants	r_e [1]) Å	See Tab.
47 Cont.	$^{27}Al^{19}F$		$Y_{13}=0.361\ (42)\ \cdot 10^{-3}$ Hz		82MAK2	IR/MW			
	AlF				82MAK2	[2])	$a_1=-3.183496\ (81)$		
					82MAK2	[2])	$a_2=6.81574\ (63)$		
					82MAK2	[2])	$a_3=-11.579\ (18)$		
					82MAK2	[2])	$a_4=16.63\ (16)$		
					82MAK2	[2])	$a_5=-22.51\ (45)$		
					82MAK2	[2])	$a_6=29.9\ (93)$		
	$^{27}Al^{19}F$				82MAK2	[2])	$\omega_e=802.3253\ (14)$ cm^{-1}		
					82MAK2	[2])	$B_e=16562.9300\ (27)$ MHz		
47a	$^{11}B^{35}Cl$	$Y_{01}=20514.26\ (74)$	$Y_{02}=-55.2\ (13)$	$Y_{11}=-204.229\ (85)$	82MAK1	IR	$Y_{10}=840.2947\ (13)$ cm^{-1}		
			$Y_{12}=184\ (59)$ Hz	$Y_{21}=0.986\ (10)$	82MAK1	IR	$Y_{20}=-5.49170\ (66)$ cm^{-1}		
					82MAK1	IR	$Y_{30}=0.02995\ (13)$ cm^{-1}		
	BCl				82MAK1		$a_0=257969.3\ (85)$ cm^{-1}	1.715283 (31)	
					82MAK1		$a_1=-3.0376\ (10)$		
					82MAK1		$a_2=6.183\ (7)$		
					82MAK1		$a_3=-9.43\ (10)$		
					82MAK1		$a_4=10.2\ (12)$		
50 *6-46*	GaF				82TIE	MW	$d_{Ga}=0.30\ (15)$	$r_e^{cor}=1.7743351\ (25)$	2.7.1 2.9.1
51	GaBr				82LEF	MW	$d_{Ga}=0.33\ (10)$ $d_{Br}=0.67\ (13)$		2.7.1
52	GaI				82TIE	MW	$d_{Ga}=0.353\ (48)$	$r_e^{cor}=2.5746263\ (40)$	2.7.1 2.9.1
54 *6-49*	InI				82TIE	MW	$d_{In}=1.34\ (14)$	$r_e^{cor}=2.753618\ (10)$	
55 *6-50*	TlF [3])				82TIE	MW	$d_{Tl}=9.38\ (55)$	$r_e^{cor}=2.084386\ (10)$	

[1]) Or related terms.
[2]) The spectral data were directly fitted to the potential coefficients $a_1, a_2, a_3, \ldots$, and ω_e, B_e; [82MAK2] presents an additional set of Dunham-parameters Y_{lk} calculated from the a_i's and ω_e, B_e.
[3]) For reinterpretation of the d-value of the group-III elements involving isotopic field shift, see [82SCH, 82KNÖ1].

Nr.	Molecule	B [1]) MHz	D [1]) kHz	α, γ [1]) MHz	Ref.	Meth.	Derived constants	r_e [1]) Å	See Tab.
56 *6-51*	TlCl [2])				82TIE	MW	$d_{Tl}=9.48\,(100)$ $d_{Cl}=0.622\,(25)$	$r_e^{cor}=2.484739\,(10)$	
57 *6-52*	TlBr [2])				82TIE	MW	$d_{Tl}=7.81\,(23)$ $d_{Br}=0.569\,(32)$	$r_e^{cor}=2.6181114\,(15)$	
58 *6-53*	TlI [2])				82TIE	MW	$d_{Tl}=7.34\,(24)$	$r_e^{cor}=2.813614\,(10)$	
59 *6-54*	CO				82TIE	MW	$d_C=1.031\,(17)$ $d_O=1.059\,(24)$	$r_e^{cor}=1.1282291\,(14)$	2.6.1 2.9.1
60 *6-55*	$^{12}C^{32}S$	$Y_{01}=24584.2833\,(22)$	$Y_{02}^{ev}=-40.06$	$Y_{11}=-177.4235\,(21)$	82BOG	MW/IR	$d_C=1.293\,(13)$		2.6.1
				$Y_{21}=\ \ -0.03039\,(50)$	82BOG	MW/IR	$d_S=1.262\,(32)$		2.9.1
			$Y_{12}^{ev}=-40$ Hz	$Y_{31}=-298\,(49)$ Hz	82BOG	MW/IR	$Y_{10}=1285.1526\,(13)\ cm^{-1}$		
				$Y_{41}=\ -35.1\,(14)$ Hz	82BOG	MW/IR	$Y_{20}=-6.50185\,(63)\ cm^{-1}$		
					82BOG	MW/IR	$Y_{30}=\ \ 0.00377\,(10)\ cm^{-1}$		
	CS				82BOG		$a_0=503433.8\,(13)\ cm^{-1}$	$r_e^{cor}=1.5348192\,(12)$	
					82BOG		$a_1=-2.884430\,(25)$		
					82BOG		$a_2=\ \ 5.11503\,(18)$		
					82BOG		$a_3=-7.1060\,(14)$		
					82BOG		$a_4=\ \ 8.539\,(41)$		
	$^{12}C^{33}S$	$Y_{01}=24380.9866\,(70)$	$D_0=39.41\,(15)$	$Y_{11}=-175.2903\,(60)$	81BOG	MW			
		$B_0=24293.3420\,(28)$			81BOG	MW			
		$B_1=24118.0513\,(72)$	$D_1=39.20\,(46)$		81BOG	MW			
	$^{13}C^{33}S$	$B_0=22921.611\,(12)$	$D_0=35.33\,(71)$		81BOG	MW			
	$^{13}C^{34}S$	$B_0=22731.7811\,(32)$	$D_0=34.14\,(14)$		81BOG	MW			
	$^{12}C^{36}S$	$B_0=23754.4776\,(37)$	$D_0=37.60\,(14)$		81BOG	MW			
63 *6-57*	SiS				82TIE	MW	$d_{Si}=0.696\,(30)$ $d_S\ =0.935\,(33)$	$r_e^{cor}=1.9292639\,(19)$	2.9.1
65 *6-60*	GeS				82TIE	MW	$d_{Ge}=0.732\,(35)$ $d_S\ =0.936\,(23)$	$r_e^{cor}=2.0120431\,(10)$	2.9.1
66 *6-61*	GeSe				82TIE	MW	$d_{Ge}=0.806\,(23)$ $d_{Se}=1.007\,(35)$	$r_e^{cor}=2.1346018\,(10)$	2.9.1

[1]) Or related terms.
[2]) For reinterpretation of the *d*-value of the group-III elements involving isotopic field shift see [82SCH, 82KNÖ1].

Nr.	Molecule	B [1]) MHz	D [1]) kHz	α, γ [1]) MHz	Ref.	Meth.	Derived constants	r_e [1]) Å	See Tab.
69 *6-64*	SnS				82TIE	MW	$d_{Sn}=0.88\,(10)$ $d_S=0.911\,(33)$	$r_e^{cor}=2.2089829\,(22)$	2.9.1
70 *6-65*	SnSe				82TIE	MW	$d_{Sn}=0.778\,(42)$ $d_{Se}=1.062\,(25)$	$r_e^{cor}=2.3255738\,(11)$	2.9.1
71 *6-66*	SnTe				82TIE	MW	$d_{Sn}=0.875\,(49)$ $d_{Te}=0.827\,(49)$	$r_e^{cor}=2.5227979\,(13)$	2.9.1
73 *6-68*	PbS [2])				82TIE	MW	$d_{Pb}=6.47\,(71)$ $d_S=0.999\,(36)$	$r_e^{cor}=2.2867844\,(40)$	2.9.1
74 *6-69*	PbSe [2])				82TIE	MW	$d_{Pb}=5.93\,(46)$ $d_{Se}=1.060\,(38)$	$r_e^{cor}=2.4021776\,(24)$	2.9.1
75 *6-70*	PbTe [2])				82TIE	MW	$d_{Pb}=5.99\,(41)$ $d_{Te}=0.897\,(55)$	$r_e^{cor}=2.5949253\,(27)$	2.9.1
77a	$^{14}N^{16}O^+$	$B_0=59597.132\,(16)$	$D_0=171\,(1)$		82BOW	MW			2.7.1

[1]) Or related terms.
[2]) For reinterpretation of the *d*-value of the group-IV elements involving isotopic field shift see [82SCH, 82KNÖ1].

References for Appendix to 2.2

81BOG Bogey, M., Demuynck, C., Destombes, J.L.: Chem. Phys. Lett. **81** (1981) 256.
81SAS Sastry, K.V.L.N., Herbst, E., DeLucia, F.C.: J. Chem. Phys. **75** (1981) 4753.
82BOG Bogey, M., Demuynck, C., Destombes, J.L.: J. Mol. Spectrosc. **95** (1982) 35.
82BOW Bowman, W.C., Herbst, E., DeLucia, F.C.: J. Chem. Phys. **77** (1982) 4261.
82KNÖ1 Knöckel, H., Tiemann, E.: Chem. Phys. **68** (1982) 13.
82LEF Lefloch, A.C., Rostas, J.: J. Mol. Spectrosc. **92** (1982) 276.
82MAK1 Maki, A.G., Lovas, F.J., Suenram, R.D.: J. Mol. Spectrosc. **91** (1982) 424.
82MAK2 Maki, A.G., Lovas, F.J.: J. Mol. Spectrosc. **95** (1982) 80.
82SCH Schlembach, J., Tiemann, E.: Chem. Phys. **68** (1982) 21.
82TIE Tiemann, E., Arnst, H., Stieda, W.U., Törring, T., Hoeft, J.: Chem. Phys. **67** (1982) 133.

Appendix to 2.3 Linear molecules

For preliminary remarks, see p. 37 of Vol. II/14a, Landolt-Börnstein, N.S.

Nr.	Molecule	v	B MHz	D kHz	Ref.	q MHz	Ref.	Varia, remarks	See also Tab., Nr.
2	$OC^{33}S$	$v=0$	6004.916202 (205)	1.2707086 (798)	81WEL			uncertainties are 2σ	
		$v_3=1$	5987.00885 (60)	1.295829 (233)				$\nu_3=853.20084\ (16)\ cm^{-1}$	
	$OC^{34}S$	$v=0$	5932.833784 (172)	1.241278 (351)					
		$v_3=1$	5915.14727 (188)	1.264708 (408)				$\nu_3=847.73936\ (10)\ cm^{-1}$	
	$O^{13}CS$	$v=0$	6061.923922 (269)	1.297790 (529)					
		$v_3=1$	6043.93360 (189)	1.327721 (566)				$\nu_3=854.46843\ (8)\ cm^{-1}$	
	^{18}OCS	$v=0$	5704.857270 (218)	1.1330001 (779)					
		$v_3=1$	5688.65259 (189)	1.154340 (235)				$\nu_3=838.00433\ (18)\ cm^{-1}$	
	$O^{13}C^{34}S$	$v=0$	5911.733163 (251)	1.237879 (101)					
		$v_3=1$	5894.14152 (401)	1.261246 (507)				$\nu_3=843.13127\ (24)\ cm^{-1}$	
16	O=C=C=C=S	$v=0$	1413.91242 (25)	0.04398 (91)	81WIN				2.6.2, 11
		$v_5=1$	1415.18691 (26)	0.04741 (96)		$q=0.32312\ (10)$	81WIN		
		$v_6=1$	1415.97451 (28)	0.04263 (102)		$q=0.38424\ (22)$		*) fixed value	
		$v_7=1$	1420.54876 (23)	0.05484 (85)		$q^{(0)}=1.75041\ (47)$		$\gamma_{11}=5.30$ kHz (fixed)	
						$q^{(1)}=1.90\ (171)\cdot 10^{-6}$			
		$v_7=2$	1427.14471 (40)	0.06595 (116)		$q^{(0)}=1.75952$ *)		$\gamma_{11}=6.346\ (103)$ kHz	
						$q^{(1)}=1.99\cdot 10^{-6}$ *)		$x_{11}=0.4007\ (41)\ cm^{-1}$	
		$v_7=3$	1433.69383 (27)	0.07319 (87)		$q^{(0)}=1.76872\ (24)$		$\gamma_{11}=7.098\ (50)$ kHz	
						$q^{(1)}=2.07\ (100)\cdot 10^{-6}$		$x_{11}=0.3793\ (99)\ cm^{-1}$	
		$v_7=4$	1440.19162 (33)	0.08575 (124)		$q^{(0)}=1.77600$ *)		$\gamma_{11}=7.816\ (18)$ kHz	
						$q^{(1)}=2.72\cdot 10^{-6}$ *)		$x_{11}=0.3742\ (9)\ cm^{-1}$	
		$v_7=5$	1446.63449 (22)	0.09526 (82)		$q^{(0)}=1.78496\ (19)$		$\gamma_{11}=8.419\ (10)$ kHz	
						$q^{(1)}=4.31\ (80)\cdot 10^{-6}$		$x_{11}=0.3799\ (19)\ cm^{-1}$	
		$v_7=6$	1453.02344 (61)	0.10323 (241)		$q^{(0)}=1.79248$ *)		$\gamma_{11}=8.800\ (17)$ kHz	
						$q^{(1)}=6.84\cdot 10^{-6}$ *)		$x_{11}=0.3587\ (9)\ cm^{-1}$	
		$v_7=7$	1459.35934 (46)	0.11469 (311)		$q^{(0)}=1.79989\ (36)$		$\gamma_{11}=8.926\ (10)$ kHz	
						$q^{(1)}=10.30\ (246)\cdot 10^{-6}$		$x_{11}=0.3637\ (12)\ cm^{-1}$	
27	HC≡CCN	$v_3=1$	$\alpha=13.95\ (4)$	$\beta=-0.0300\ (13)$	81YAM			$\omega_3^0+x_{33}^0+y_{333}^0$ $=2079.30606\ (71)\ cm^{-1}$	2.6.2, 16 2.7.2, 14
	$DC\equiv CC^{15}N$	$v=0$	4100.380 (2)		81FLI				
27a	$^{35}ClC\equiv CF$	$v=0$	2224.8442 (12)	0.1445 (6)	81GUA1				2.6.2, 16a
	$^{37}ClC\equiv CF$	$v=0$	2171.6642 (12)	0.1377 (6)					2.7.2, 23a

Nr.	Molecule	v	B MHz	D kHz	Ref.	q MHz	Ref.	Varia, remarks	See also Tab., Nr.
30a	HC≡C—C≡P	$v=0$	2656.39442 (62)	0.1948 (81)	81KRO2				2.6.2, 17a
		$v_5=1$	$\alpha=-0.9797$ (18)			$q=0.92193$ (86)	81KRO2	$\nu_5=587$ (77) cm^{-1}	
		$v_6=1$	$\alpha=-4.5022$ (50)			$q=1.3142$ (64)		$\nu_6=412$ (54) cm^{-1}	
		$v_7=1$	$\alpha=-7.17884$ (77)			$q=2.6256$ (12)		$\nu_7=206$ (27) cm^{-1}	
	DC≡C—C≡P	$v=0$	2488.9027 (17)	0.149 (16)					
		$v_6=1$	$\alpha=-3.2522$ (48)			$q=1.1701$ (26)		$\nu_6=406$ (53) cm^{-1}	
		$v_7=1$	$\alpha=-6.5114$ (12)			$q=2.4157$ (12)		$\nu_7=197$ (26) cm^{-1}	
30b	HC≡CC≡CH	$v_6=1$	4391.3230 (84)	0.582 (154)	81MAT	$q=2.4830$ (32)	81MAT	uncertainties are 2.5 σ	2.6.2, 17b
		$v_8=1$	4391.1921 (94)	0.594 (179)		$q=2.4073$ (37)		ν_6, C≡C—H symmetric bending	
								ν_8, C≡C—H antisymmetric bending	
								$\nu_8-\nu_6$ band origin	
								= 71868.336 (57) MHz	
	HC≡CC≡CD	$v=0$	4084.74 *)		81ETO			*) extrapolated value	
		$v_6=1$	4086.2165 (44)	0.408 (36)		$q=2.1316$ (28)	81ETO		
		$v_6=2$	4087.6963 (58)	0.349 (50)				for structure see also	
		$v_6=1$, $v_9=1$	4097.7537 (50)	0.358 (44)				Vol. II/7 (3, 473)	
		$v_8=2$	4094.4786 (103)	0.351 (85)					
		$v_6=1$, $v_7=1$	4092.9074 (176)	1.419 (145)					
		$v_7=1$, $v_8=1$	4095.8925 (354)	0.355 (290)					
		$v_6=2$, $v_9=1$	4098.9505 (339)	0.372 (303)					
31	HC≡C—C≡CCN	$v=0$	1331.33271	0.03010	81YAM1				2.6.2, 18
		$v_2=1$	1326.9663 (36)	0.02705 (66)				$\nu_2=2256.12406$ (13) cm^{-1}	
34a	NCCN···HF	$v=0$	1195.0616	0.34	81LEG2				2.7.2, 14b
	^{15}NCCN···HF	$v=0$	1168.2207	0.41					
	NCC^{15}N···HF	$v=0$	1194.1515	0.36					
	^{15}NCC^{15}N···HF	$v=0$	1167.2187	0.34					
35	OC · HF	$v=0$	3063.8994	9.77	81LEG1				2.7.2, 3a
	O^{13}C · HF	$v=0$	3049.8111	9.72					
	^{18}OC · HF	$v=0$	2927.9125	8.69					
	OC · DF	$v=0$	3047.8981	9.40					
	^{18}OC · DF	$v=0$	2910.2637	8.17					
37a	CO_2 · HF	$v=0$	1951.1696 (11)	10.69 (7)	81BAI				2.6.2, 19b
	CO_2 · DF	$v=0$	1935.9959 (7)	11.21 (3)					2.7.2, 3b
37b	SCO · HF	$v=0$	1307.462 (4)	2.46 (9)	81BAI				2.6.2, 19c

Nr.	Molecule	v	B MHz	D kHz	Ref.	q MHz	Ref.	Varia, remarks	See also Tab., Nr.
44	$^{83}Kr \cdot H^{35}Cl$ $^{83}Kr \cdot D^{35}Cl$	$v=0$ $v=0$	$B-8D_J$ $=1204.84742\,(40)$ $=1188.01053\,(60)$		81CAM5				2.6.2, 25 2.7.2, 29 2.9.2, 13
45a	$^{83}Kr \cdot HCN$	$v=0$	$B-8D_J$ $=1184.60700\,(20)$		81CAM5				2.7.2, 14d
46	$^{83}Kr \cdot {}^{35}ClF$	$v=0$	$B-18D_J$ $=929.2174\,(10)$		81BUX2				2.6.2, 26 2.7.2, 30
46a	$^{129}Xe \cdot HF$ $^{131}Xe \cdot HF$ $^{132}Xe \cdot HF$ $^{129}Xe \cdot DF$ $^{131}Xe \cdot DF$ $^{132}Xe \cdot DF$	$v=0$ $v=0$ $v=0$ $v=0$ $v=0$ $v=0$	2039.690 (4) 2035.563 (5) 2033.552 (5) 1995.469 (5) 1991.268 (3) 1989.217 (4)	18.32 (19) 18.19 (16) 18.19 (20) 15.87 (16) 15.77 (9) 15.73 (14)	81BAI1				2.6.2, 26a 2.7.2, 61
47	$^{131}Xe \cdot H^{35}Cl$ $^{131}Xe \cdot D^{35}Cl$ $^{132}Xe \cdot D^{35}Cl$	$v=0$ $v=0$ $v=0$	990.86302 (32) 976.11556 (40) 974.50768 (35)	3.796 (20) 3.4285 (67) 3.4209 (20)	81CAM5				2.6.2, 27 2.7.2, 31

References for Appendix to 2.3

81BAI Baiocchi, F.A., Dixon, T.A., Joyner, C.H., Klemperer, W.: J. Chem. Phys. **74** (1981) 6544.

81BAI1 Baiocchi, F.A., Dixon, T.A., Joyner, C.H., Klemperer, W.: J. Chem. Phys. **75** (1981) 2041.

81BUX2 Buxton, L.W., Campbell, E.J., Flygare, W.H.: Chem. Phys. **59** (1981) 55.

81CAM5 Campbell, E.J., Buxton, L.W., Keenan, M.R., Flygare, W.H.: Phys. Rev. A **24** (1981) 812.

81ETO Etoh, T., Matsumura, K., Tanaka, T.: J. Mol. Spectrosc. **89** (1981) 511.

81FLI Fliege, E., Bestmann, G., Andolfatto, M., Dreizler, H.: Z. Naturforsch. **36a** (1981) 1126.

81GUA1 Guarnieri, A., Andolfatto, M.: Z. Naturforsch. **36a** (1981) 899.

81KRO2 Kroto, H.W., Nixon, J.F., Ohno, K.: J. Mol. Spectrosc. **90** (1981) 512.

81LEG1 Legon, A.C., Soper, P.D., Flygare, W.H.: J. Chem. Phys. **74** (1981) 4944.

81LEG2 Legon, A.C., Soper, P.D., Flygare, W.H.: J. Chem. Phys. **74** (1981) 4936.

81MAT Matsumura, K., Etoh, T., Tanaka, T.: J. Mol. Spectrosc. **90** (1981) 106.

81WEL Wells, J.S., Petersen, F.R., Maki, A.G., Sakle, D.J.: J. Mol. Spectrosc. **89** (1981) 421.

81WIN Winnewisser, M., Peau, E.W., Yamada, K., Christiansen, J.J.: Z. Naturforsch. **36a** (1981) 819.

81YAM Yamada, K., Winnewisser, G.: Z. Naturforsch. **36a** (1981) 23.

81YAM1 Yamada, K., Winnewisser, G.: Z. Naturforsch. **36a** (1981) 1052.

Appendix to 2.4 Symmetric top molecules

For preliminary remarks see p. 63 of Vol. II/14a, Landolt-Börnstein, N.S.

Nr.	Molecule	v	B MHz	D kHz	Ref.	q MHz	Ref.	Ref. struct.	Varia, remarks	See also Tab., Nr.
103a	$Kr \cdot SO_3$									2.6.2, 69a
106	$CH_3CN \cdot HF$	$v=0$	1853.3323 (5)	D_J = 0.93 (4) D_{JK}=66.8 (5)	81SOP1					2.7.2, 14c
110	 $\cdot H^{35}Cl$	$v=0$	1237.68362 (52)	D_J = 1.223 (13) D_{JK}=13.35 (2)	81REA					2.7.2, 40a
	 $\cdot H^{37}Cl$	$v=0$	1201.92498 (170)	D_J = 1.155 (39) D_{JK}=13.41 (81)						
	 $\cdot D^{35}Cl$	$v=0$	1228.24396 (58)	D_J = 1.186 (15) D_{JK}=14.57 (3)						

References for Appendix to 2.4

81REA Read, W.G., Campbell, E.J., Henderson, G., Flygare, W.H.: J. Am. Chem. Soc. **103** (1981) 7670.

81SOP1 Soper, P.D., Legon, A.C., Read, W.G., Flygare, W.H.: J. Phys. Chem. **85** (1981) 3440.

Appendix to 2.5 Asymmetric top molecules

For preliminary remarks, see p. 128 of Vol. II/14a, Landolt-Börnstein, N.S.

Nr.	Molecule	v	A, B, C MHz	D, H kHz	Ref.	Ref. struct.	Varia, remarks	See also Tab., Nr.
53 *6–37* *4–29*	H_2O	$v_2=1$	A=933236.58 (460) B=440311.98 (270) C=273679.11 (140)	Δ_J = 41900.0 (500) Δ_{JK} = −233354 (290) Δ_K = 1749385 (570) δ_J = 17401.6 (190) δ_K =113856 (390) H_J = 18.925 (370) H_{JK}= 99.99 (850) H_{KJ}= −1674.5 (280) H_K =12321.2 (500)	80KUZ		ν_2=1594.74982 (59) cm^{-1}	2.6.3, 30 2.7.3, 3
Cont.								

Nr.	Molecule	v	A, B, C MHz	D, H kHz	Ref.	Ref. struct.	Varia, remarks	See also Tab., Nr.
53 Cont.	H_2O	$v_2=1$		h_J = 9.785 (150) h_{JK} = −30.58 (370) h_K = 3157.5 (720) L_{JJK} = −0.1015 (590) L_{JK} = −7.473 ± 1.2 L_{KKJ} = 30.40 (320) L_K = −131.32 (370) l_K = −39.81 (310) *)	80KUZ		*) higher order constants (in kHz) P_K = 1.0651 (540) P_{KKJ} = −0.03569 (700) p_K = 0.10429 (870) Q_K = −6.633 (630) · 10^{-3} R_K = 1.969 (270) · 10^{-5}	
23a *6-16* *4-15b*	N_2O_3						see appendix to 2.7.3, 18a for structure, see also Vol. II/7 (2, 333)	
310a *6-133*	$CH_2{=}CHCH_2OH$	$v=0$	A' = 25592.1371 (854) B' = 4323.3065 (146) C' = 4228.2209 (143)	D'_J = 4.4239 (491) D'_{JK} = −42.3089 (740) D'_K = 43.4043 (355) R'_6 = −0.1074 (191)	81JAM1		for structure, see also Vol. II/7 (3, 404)	
328	$CH_2{=}CHOCH_2CF_3$	$v=0$	A = 4421.2 (43) *) B = 1128.3 (196) C = 1053.7 (196)		81DEN		*) LRMW	
448a	$(CH_3)_3CCH_2C(=O)H$ skew⁺⁺) syn⁺) free⁺⁺⁺)		$B+C$ = 3230.2 (5) *) $2A-B-C$ = 4406 (2) *) $B+C$ = 3328 (3) *)		80THO		⁺) $\tau_{OCCC}=0°$ ⁺⁺) $\tau_{OCCC}\sim 120°$ ⁺⁺⁺) free rotation about τ_{OCCC} *) LRMW	
448b	$(CH_3)_3CCH_2C(=O)F$ skew syn free		$B+C$ = 2638 (2) *) $B+C$ = 2452 (1) *) $B+C$ = 2554 (2) *)		80THO		*) LRMW, for definitions see Nr. 448a	
448c	$(CH_3)_3CCH_2C(=O)^{35}Cl$ syn $(CH_3)_3CCH_2C(=O)^{37}Cl$ syn		$B+C$ = 1755.4 (5) *) $B+C$ = 1716.0 (5) *)		80THO		*) LRMW, for definitions see Nr. 448a	
448d	$(CH_3)_3CCH_2C(=O)CN$ skew syn free		$B+C$ = 2082 *) $B+C$ = 1787 (1) *) $B+C$ = 1987 (2) *)		80THO		*) LRMW, for definitions see Nr. 448a	

References for Appendix to 2.5

80KUZ Kuze, H.: Astrophys. J. **239** (1980) 1131.

80THO Thomas, L. P., True, N.S., Bohn, R.K.: J. Phys. Chem. **84** (1980) 1785.

81DEN Dennison, F.T., Gillies, C.W., Borchert, S.J.: J. Chem. Phys. **75** (1981) 3238.

81JAM1 Jaman, A.I., Ghosh, D.K.: Pramaña **17** (1981) 245.

Appendix to 2.6.1 Dipole moments: Diatomic molecules

For preliminary remarks see p. 584 of Vol. II/14a, Landolt-Börnstein, N.S.

Nr.	Molecule	v	μ D	μ_e D	μ_I D	μ_{II} D	Ref.	Meth.	State		Varia, remarks	See Tab.
1 *6-1*	$^1H^2D$	$v=0$										
		$J=2\leftarrow1$	$9.36(30)\cdot10^{-4}$				82NEL	[1])	$X\,^1\Sigma_g^+$	gr	[1]) intensity measurements of the rotational spectrum	
		$J=3\leftarrow2$	$8.00(20)\cdot10^{-4}$				82NEL	[1])	$X\,^1\Sigma_g^+$	gr		
		$J=4\leftarrow3$	$9.79(30)\cdot10^{-4}$				82NEL	[1])	$X\,^1\Sigma_g^+$	gr		
15 *6-56*	$^1H^{19}F$	$v=1$	1.872 (3)	1.803 (2)[ev]			81GOU	MB/IR	$X\,^1\Sigma^+$	gr		2.2.2 2.9.1
33b	$^{107}Ag^{127}I$	$v=0$	5.10 (15)				83HOE	MW	$X\,^1\Sigma^+$	gr		2.7.1
40	$^{40}Ca^{35}Cl$	$v=0$	3.6 (6)				82DAG	ED	$X\,^2\Sigma^+$	gr		4.1.2
87	$^{16}O_2$	$v=0$					74GUS	MW	$X\,^3\Sigma^-$	gr	$\alpha_\parallel-\alpha_\perp=1.12(7)$ Å^3	4.1.3
102	$^{127}I\,Br$	$v=0$	0.7264 (28)				82NAI	MW	$X\,^1\Sigma^+$	gr	additional error of 0.5 % due to Stark-voltage measurement; the value of μ_0 was obtained for both isotopes ^{79}Br and ^{81}Br	2.2.2 2.7.1

References for Appendix to 2.6.1

74GUS Gustafson, S., Gordy, W.: Phys. Lett. **49A** (1974) 161.

81GOU Gough, T.E., Miller, R.E., Scoles, G.: Farad. Disc. **71** (1981) 77.

82DAG Dagdigian, P.J.: Chem. Phys. Lett. **88** (1982) 225.

83HOE Hoeft, J., Nair, K.P.R.: J. Mol. Struct. **97** (1983) 347.

82NAI Nair, K.P.R., Hoeft, J.: J. Mol. Struct **79** (1982) 227.

82NEL Nelson, J.B., Tabisz, G.C.: Phys. Rev. Lett. **48** (1982) 1393, Phys. Rev. Lett. **48** (1982) 1870.

Appendix to 2.6.3 Dipole moments: Asymmetric top molecules

For preliminary remarks see p. 598 of Vol. II/14a, Landolt-Börnstein, N.S.

Nr.	Molecule	v	μ D	Components of μ	Ref.	Varia, remarks	See Tab., Nr.
30	H_2O	$v_2=1$	1.827 (5) *)	$\mu_b=1.827$ (5)	81KUZ	*) rotationless moment	2.5.3, 53

References for Appendix to 2.6.3

81KUZ Kuze, H., Amano, T., Shimizu, T.: J. Chem. Phys. **75** (1981) 4869.

Appendix to 2.7.1 Quadrupole coupling constants: Diatomic molecules

For preliminary remarks see p. 644 of Vol. II/14a, Landolt-Börnstein, N.S.

Nr.	Molecule	J, v	Nuclear quadrupole coupling constants: Nucleus	eqQ [MHz]	Ref.	Meth.	Varia, remarks	See Tab.
18 *6-51*	$^2D^{127}I$	$v=0$ $v=0$	^{127}I ^2D	−1823.36535 (91) 0.0985 (92)	81SIV 81SIV	MW/MB MW/MB		2.2.2 2.9.1
77a	$^{14}N^{16}O^+$	$v=0$	^{14}N	−6.76 (10)	82BOW	MW		5: 2.2.2
80 *6-12*	$^{14}N^{16}O$	$v=1$	^{14}N	$eq^{(a)}Q=-1.78$ (3) $eq^{(b)}Q=23.0$ (1)	81LOW 81LOW	IR/DR IR/DR	} $X\,^2\Pi$	4.1.5
81 *6-13*	$^{14}N^{32}S$ $^{14}N^{34}S$	$v=0$ $v=0$	^{14}N ^{14}N	$eq^{(a)}Q=-2.70$ (20) $eq^{(a)}Q=-2.80$ (22)	82LOV 82LOV	MW MW	$X\,^2\Pi$ $X\,^2\Pi$	4.1.5 4.1.5
59 *6-15*	$^{12}C^{17}O$	$v=0$	^{17}O	4.337 (56)	81FRE	MW	spectra taken by radio astronomy	2.2.2 2.6.1 2.9.1
87 *6-16*	$^{16}O^{17}O$ $^{17}O^{18}O$	$v=0$ $v=0$	^{17}O ^{17}O	−8.3 (3) −7.8 (5)	81CAZ 81CAZ	MW MW	} $X\,^3\Sigma_g^-$	4.1.3
88 *6-26*	$^{32}S^{17}O$	$v=0$	^{17}O	−3.5 (2)	82CAZ	MW	$X\,^3\Sigma^-$	4.1.3

Nr.	Molecule	J, v	Nuclear quadrupole coupling constants: Nucleus	eqQ [MHz]	Ref.	Meth.	Varia, remarks	See Tab.
60	$^{12}C^{33}S$	$v=0$	^{33}S	12.833 (36)	81BOG	MW		2.2.2
		$v=1$	^{33}S	12.567 (82)	81BOG	MW		2.9.1
	$^{13}C^{33}S$	$v=0$	^{33}S	13.13 (13)	81BOG	MW		2.2.2
40	$^{40}Ca^{35}Cl$		^{35}Cl		82CHI1	MB	$eq_eQ\,(^{35}Cl) = -0.905\,(9)$ MHz } $X\,^2\Sigma^+$	4.1.2
					82CHI1	MB	$eq_1Q\,(^{35}Cl) = -0.23\,(1)$ MHz } $X\,^2\Sigma^+$	
79	$^{12}C^{35}Cl$	$v=0$	^{35}Cl	$eq^{(a)}Q = -34.26\,(13)$	82END	MW	} $X\,^2\Pi$	4.1.5
				$eq^{(b)}Q = 84.6\,(95)$	82END	MW		
92 *6–30*	$^{35}Cl^{16}O$	$v=0$	^{35}Cl	$eq^{(a)}Q = -87.95\,(19)$	78KAK	MW	} $X\,^2\Pi$; the sign of $eq^{(b)}Q$ was reversed according to [82END]	4.1.5
				$eq^{(b)}Q = -126.4\,(65)$	78KAK	MW		
81 a	$^{75}As^{16}O$	$v=0, 1$	^{75}As	$eq^{(a)}Q = -170$	81UEH	LMR	$X\,^2\Pi_{3/2}$	4.1.5
40 a	$^{40}CaBr$	$v=0$	^{79}Br	20.015 (7)	81CHI1	MB	} $X\,^2\Sigma^+$	4.1.2
		$v=0$	^{81}Br	16.714 (6)	81CHI1	MB		
					82CHI2	MB	$eq_eQ\,(^{79}Br) = 19.262\,(6)$ MHz, $eq_1Q\,(^{79}Br) = 1.482\,(5)$ MHz	
		$v=0$	^{79}Br	30.4 (10)	81CHI2	LIF/MB	} $B\,^2\Sigma^+$	4.1.2
		$v=0$	^{81}Br	21.8 (23)	81CHI2	LIF/MB		
		$v=0$	^{79}Br	$\Delta eqQ = 31\,(1)$	81BER	SLS	$\Delta eqQ = eqQ\,(A\,^2\Pi) - eqQ\,(X\,^2\Sigma^+)$	4.1.5
		$v=0$	^{81}Br	$\Delta eqQ = 28\,(2)$	81BER	SLS	systematic uncertainty due to neglect of magnetic hyperfine interaction	
18	DI		^{127}I	see ^2D				
33 b *6–52*	$Ag^{127}I$	$v=0, 1, 2, 3, 4$	^{127}I		83HOE	MW	$eq_eQ\,(^{127}I) = -1061.37\,(40)$ MHz $eq_1Q\,(^{127}I) = -2.13\,(20)$ MHz the parameters are valid for both isotopes ^{107}AgI and ^{109}AgI	2.6.1
40 b	$^{40}Ca^{127}I$	$v=0$	^{127}I	$\Delta eqQ = -55\,(1)$	81BER	SLS	$\Delta eqQ = eqQ\,(A\,^2\Pi) - eqQ\,(X\,^2\Sigma^+)$ systematic uncertainty due to neglect of magnetic hyperfine interactions	4.1.5
		$v=0$	^{127}I	$\Delta eqQ = -66\,(3)$	81BER	SLS	$\Delta eqQ = eqQ\,(B\,^2\Sigma) - eqQ\,(X\,^2\Sigma^+)$	4.1.2
100 *6–54 a*	$^{127}I^{35}Cl$	$v=0$	^{127}I	$-664\,(9)$	82HAN	MODR	$B\,^3\Pi_{0^+}$	4.1.6
			^{127}I		82KNÖ2	MODR	$eq_eQ\,(^{127}I) = -657.14$ MHz } $B\,^3\Pi_{0^+}$	
					82KNÖ2	MODR	$eq_1Q\,(^{127}I) = -5.96$ MHz } $B\,^3\Pi_{0^+}$	
					82KNÖ2	MODR	$eq_{11}Q\,(^{127}I) = -15.53$ MHz } $B\,^3\Pi_{0^+}$	

Nr.	Molecule	J, v	Nuclear quadrupole coupling constants Nucleus	eqQ [MHz]	Ref.	Meth.	Varia, remarks	See Tab.
103 6–54c	$^{127}I_2$	$v=83$	^{127}I	−1554.7 (5) [1]	78KOF	[2]	$X\ ^1\Sigma_g^+$ [1] recalculated using B-state constants of [80YOK]. [2] Stimulated-emission spectroscopy	2.2.2
		$v=0$	^{127}I	1160 (250)	81CHU	LIF [3]	$a'0_g^+$ [3] Laser-induced-fluorescence Fourier-transform-spectroscopy	2.9.1
		$v=1$	^{127}I	1030 (70)	81CHU			
		$v=2$	^{127}I	880 (140)	81CHU			
		$v=3$	^{127}I	820 (110)	81CHU			
		$J=11, v=61$	^{127}I	−560 (8)	81PIQ	LIF	$B\ ^3\Pi_{0u}^+$	
		$J=14, v=61$	^{127}I	−563 (9)	81PIQ			
		$J=27, v=62$	^{127}I	−566 (9)	81PIQ			
		$J=40, v=63$	^{127}I	−574 (10)	81PIQ			
		$J=50, v=67$	^{127}I	−583 (14)	81PIQ			
	$^{129}I_2$	$J=110, v=2$	^{129}I	−1813.2 (71) [4]	81GLÄ	SLS	$X\ ^1\Sigma_g^+$ ground state } $\Delta eqQ=1365.30\,(23)$ MHz	
		$J=109, v=10$	^{129}I	−447.9 (72) [4]	81GLÄ		$B\ ^3\Pi_{0u}^+$ excited state }	
		$J=113, v=4$	^{129}I	−1887 (12) [4]	81GLÄ		$X\ ^1\Sigma_g^+$ ground state } $\Delta eqQ=1358.42\,(15)$ MHz	
		$J=114, v=14$	^{129}I	−529 (12) [4]	81GLÄ		$B\ ^3\Pi_{0u}^+$ excited state } [4] strong correlation between the parameters of the ground and excited state	
43	$^{137}Ba^{19}F$	$v=0$	^{137}Ba	−117 (12)	82RYZ	MW	$X\ ^2\Sigma^+$	4.1.2
44	$^{137}Ba^{35}Cl$	$v=0$	^{137}Ba	−134 (42)	82RYZ	MW	$X\ ^2\Sigma^+$	4.1.2

References for Appendix to 2.7.1

78KAK Kakar, R.K., Cohen, E.A., Geller, M.: J. Mol. Spectrosc. **70** (1978) 243.

78KOF Koffend, J.B., Goldstein, St., Bacis, R., Field, R.W., Ezekiel, S.: Phys. Rev. Lett. **41** (1978) 1040.

80YOK Yokozeki, A., Muenter, J.S.: J. Chem. Phys. **72** (1980) 3796.

81BER Bernath, P.F., Pinchemel, B., Field, R.W.: J. Chem. Phys. **74** (1981) 5508.

81BOG Bogey, U., Demuynck, C., Destombes, J.L.: Chem. Phys. Lett. **81** (1981) 256.

81CAZ Cazzoli, G., Degli Esposti, C., Favero, P.G., Severi, G.: Il Nuovo Cimento **62B** (1981) 243.

81CHI1 Childs, W.J., Cok, D.R., Goodman, G.L., Goodman, L.S.: J. Chem. Phys. **75** (1981) 501.

81CHI2 Childs, W.J., Cok, D.R., Goodman, L.S.: Can. J. Phys. **59** (1981) 1308.

81CHU Churassy, S., Martin, F., Bacis, R., Vergès, J., Field, R.W.: J. Chem. Phys. **75** (1981) 4863.

81FRE Frerking, M.A., Langer, W.D.: J. Chem. Phys. **74** (1981) 6990.

81GLÄ Gläser, M., Kegung, D., Foth, H.J.: Opt. Comm. **38** (1981) 119.

81LOW Lowe, R.S., McKellar, A.R.W., Veillette, P., Meerts, W.L.: J. Mol. Spectrosc. **88** (1981) 372.

Cont.

References for Appendix to 2.7.1 (continued)

81PIQ Pique, J.P., Hartmann, F., Bacis, R., Churassy, S.: Opt. Comm. **36** (1981) 354.
81SIV Sivery, B., Demaison, J., Burie, J., Dubrulle, A.: J. Mol. Spectrosc. **88** (1981) 434.
81UEH Uehara, H., Chem. Phys. Lett. **84** (1981) 539.
82BOW Bowman, W.C., Herbst, E., DeLucia, F.C.: J. Chem. Phys. **77** (1982) 4261.
82CAZ Cazzoli, G., Degli Esposti, C., Favero, P.G.: Il Nuovo Cimento, **1D** (1982) 648.
82CHI1 Childs, W.J., Cok, D.R., Goodman, L.S.: J. Chem. Phys. **76** (1982) 3993.
82CHI2 Childs, W.J., Cok, D.R., Goodman, L.S.: J. Mol. Spectrosc. **95** (1982) 153.
82END Endo, Y., Saito, S., Hirota, E.: J. Mol. Spectrosc. **94** (1982) 199.
82HAN Hansen, S.G., Howard, B.J.: Chem. Phys. Lett. **85** (1982) 249.
82KNÖ2 Knöckel, H., Tiemann, E.: Chem. Phys. **70** (1982) 345.
82LOV Lovas, F.J., Suenram, R.D.: J. Mol. Spectrosc. **93** (1982) 416.
82RYZ Ryzlewicz, Ch., Schütze-Pahlmann, H.-U., Hoeft, J., Törring, T.: Chem. Phys. **71** (1982) 389.
83HOE Hoeft, J., Nair, K.P.R.: J. Mol. Struct. **97** (1983) 347.

Appendix to 2.7.2 Quadrupole coupling constants: Linear and symmetric top molecules

For preliminary remarks see p. 654 of Vol. II/14a Landolt-Börnstein, N.S.

Nr.	Molecule	v	Nucleus	eqQ MHz	Ref.	Meth.	Varia, remarks	See Tab., Nr.
14d	$^{83}Kr \cdot HCN$	$v=0$	^{14}N ^{83}Kr	−3.2630 (60) 7.457 (50)	81CAM5	MWFT		2.3, 45a
40a	 $\cdot H^{35}Cl$	$v=0$	^{35}Cl	−52.189 (18)	81REA	MWFT		2.4, 110
	 $\cdot H^{37}Cl$	$v=0$	^{37}Cl	−41.248 (275)				
	 $\cdot D^{35}Cl$	$v=0$	^{35}Cl	−54.705 (22)				

References for Appendix to 2.7.2

81CAM5 Campbell, E.J., Buxton, L.W., Keenan, M.R., Flygare, W.H.: Phys. Rev. A **24** (1981) 812.
81REA Read, W.G., Campbell, E.J., Henderson, G., Flygare, W.H.: J. Am. Chem. Soc. **103** (1981) 7670.

Appendix to 2.7.3 Quadrupole coupling constants: Asymmetric top molecules

For preliminary remarks see p. 669 of Vol. II/14a Landolt-Börnstein, N.S.

Nr.	Molecule	v	Nucleus	Nuclear quadrupole coupling tensor components in system of principal axes of inertia χ_{gg}, [MHz]	in system of principal coupling tensor or bond axes $\chi_{\alpha\beta}$, χ_{xy} [MHz]	position of the axes θ	Ref.	Varia, remarks	See Tab., Nr.
18a	$O^{14}N_{\alpha}{}^{14}N_{\beta}O_2$	$v=0$	$^{14}N_{\alpha}$	$\chi_{aa}=-1.9(1)$ *) $\chi_{bb}=1.6(6)$ $\chi_{cc}=0.3(6)$			73COX	*) rotation of the inertial axes with ^{15}N substitution has negligible effect on the χ values	2.5, 23a
			$^{14}N_{\beta}$	$\chi_{aa}=-2.0(1)$ $\chi_{bb}=0.15(40)$ $\chi_{cc}=1.85(40)$					

References for Appendix to 2.7.3

73COX Cox, A. P., Finnigan, D. J.: J. Chem. Soc. Farad. Trans 2 **69** (1973) 49.

Appendix to 2.9.1 Magnetic constants: Diatomic molecules

For preliminary remarks see p. 745 of Vol. II/14a Landolt-Börnstein, N.S.

Nr.	Molecule	v, J	c_1 kHz	c_2 kHz	c_3 kHz	c_4 kHz	Ref.	g_J	ξ_T $\frac{Hz}{kG^2}$	Ref.	Meth.	(Θ/e) Å^2	Ref.	See Tab.
18 *6-11*	$^2D^{127}I$	$v=0$	$-8.8(17)$	170.45(54)	0.8856[ev]		81SIV				MW/MB			2.2.2 2.7.1
59 *6-36*	$^{12}C^{17}O$	$v=0$		$-30.4(12)$			81FRE				MW			2.2.2 2.6.1 2.7.1 in appendix
60 *6-37*	$^{12}C^{33}S$	$v=0$ $v=1$		13.4(38) 17.2(65)			81BOG 81BOG				MW			2.2.2 2.7.1

Nr.	Molecule v, J	c_1 kHz	c_2 kHz	c_3 kHz	c_4 kHz	Ref.	g_J	ξ_T $\frac{Hz}{kG^2}$	Ref.	Meth.	(Θ/e) Å^2	Ref.	See Tab.
90a	$^{130}Te_2$ $v=6$						(−) 0.308 (9)		82FER	QB			
	XO_g^+ $J=52$												
101	$^{81}Br_2$, $B^3\Pi_{0u}^+$												
	$v=40$						2.19 (18)		81LUY	QB			2.7.1
	$v=45$						5.35 (22)		81LUY	QB			2.9.1
	$^{79}Br_2$, $B^3\Pi_{0u}^+$												
	$v=42$						3.76 (24)		81LUY	QB			
103	$^{127}I_2$ $X^1\Sigma_g^+$												2.2.2
6-45	$J=13, v=83$	62 (5) [1]	$c_1=c_2$			78KOF				[2]			2.7.1
	$^{127}I_2$ $B^3\Pi_{0u}^+$												
	$J=11, v=61$	864 (15)	$c_1=c_2$			81PIQ				LIF			
	$J=14, v=61$	868 (20)	$c_1=c_2$			81PIQ							
	$J=27, v=62$	941 (9)	$c_1=c_2$			81PIQ							
	$J=40, v=63$	1159 (9)	$c_1=c_2$			81PIQ							
	$J=50, v=67$	1633 (11)	$c_1=c_2$			81PIQ							
	$J=52, v=68$ [3]	1844 (12)	$c_1=c_2$			81PIQ							
	$J=55, v=70$ [3]	2333 (20)	$c_1=c_2$			81PIQ							
	$J=56, v=71$ [3]	2743 (28)	$c_1=c_2$			81PIQ							
	$J=55, v=73$ [3]	4073 (45)	$c_1=c_2$			81PIQ							
	$^{129}I_2$ [4]	$c_1'-c_1''$ $=17.62$ (6) [4]		$c_3'-c_3''$ $=-34$ (12) [4]	$c_4'-c_4''$ $=-7.5$ (14) [4]	81GLÄ				SLS			2.2.2 2.7.1
	$^{129}I_2$ [5]	$c_1'-c_1''$ $=21.22$ (4) [5]		$c_3'-c_3''$ $=14$ (9) [5]	$c_4'-c_4''$ $=13$ (1) [5]	81GLÄ				SLS			

[1]) Recalculated using B-state constants of [80YOK].
[2]) Stimulated-emission spectroscopy.
[3]) $eqQ(B^3\Pi_{0u}^+)$ kept fixed at −570 MHz for fitting the spectra.
[4]) Difference between the molecular parameters of the states: $' \mathrel{\hat=} B^3\Pi_{0u}^+$, $v'=10$, $J'=109$; $'' \mathrel{\hat=} X^1\Sigma_g^+$, $v''=2$, $J''=110$
[5]) Difference between the molecular parameters of the states: $' \mathrel{\hat=} B^3\Pi_{0u}^+$, $v'=14$, $J'=114$; $'' \mathrel{\hat=} X^1\Sigma_g^+$, $v''=4$, $J''=113$

For References, see next page.

References for Appendix to 2.9.1

78KOF Koffend, J.B., Goldstein, St., Bacis, R., Field, R.W., Ezekiel, S.: Phys. Rev. Lett. **41** (1978) 1040.

80YOK Yokozeki, A., Muenter, J.S.: J. Chem. Phys. **72** (1980) 3796.

81BOG Bogey, M., Demuynck, Ċ., Destombes, J.L.: Chem. Phys. Lett. **81** (1981) 256.

81FRE Frerking, M.A., Langer, W.D.: J. Chem. Phys. **74** (1981) 6990.

81GLÄ Gläser, M., Kegung, D., Foth, H.J.: Opt. Comm. **38** (1981) 119.

81LUY Luypaert, R., Van Craen, J., Coremans, J., De Vlieger, G.: J. Phys. **B14** (1981) 2575.

81PIQ Pique, J.P., Hartmann, F., Bacis, R., Churassy, S.: Opt. Comm. **36** (1981) 354.

81SIV Sivery, B., Demaison, J., Burie, J., Dubrulle, A.: J. Mol. Spectrosc. **88** (1981) 434.

82FER Ferber, R.S., Okunevich, A.I., Shmit, Ȯ.A., Tamanis, M.Ya.: Chem. Phys. Lett. **90** (1982) 476.

Appendix to 2.9.2 Magnetic constants: Linear and symmetric top molecules

For preliminary remarks see p. 755 of Vol. II/14a Landolt-Börnstein, N.S.

Nr.	Molecule	g	$\Delta\xi=\xi_{\parallel}-\xi_{\perp}$ 10^{-6} erg/Gauss^2 mol	Meth.	Ref.	Varia, remarks
a) Linear molecules						
1	$H_2\cdots H_2$ (ortho-ortho) $H_2\cdots H_2$ (ortho-para)			MBM	80V1, 81V1, 81W1	magnetic hfs measurements are reported and analyzed; the dimer spin-rotation and spin-spin interaction energies are expressed in terms of the H_2 molecular constants, assuming negligible variations on complex formation
2	$H_2\cdots Ne$ [1]			MBM	81W2, 81W3	see Nr. 1
3	$H_2\cdots Ar$ [1]			MBM	81W2, 81W3	see Nr. 1
4	$H_2\cdots Kr$ [1]			MBM	81W2, 81W3	see Nr. 1
5	Ar···HCl Ar···DCl			MBE	81H1	$M_{\perp}^{35Cl}=0.132\,(16)$ kHz $M_{\perp}^{35Cl}=0.169\,(37)$ kHz
6	Ar···HF Ar···DF			MW[2]	81K1	$D_{\perp}^{HF}=49.0\,(19)$ kHz $D_{\perp}^{DF}=12.5\,(11)$ kHz
7	ClC≡CD			BMS	82T1	$C^{35Cl}(J=1)=1.3\,(5)$ kHz

[1]) Ortho-H_2.
[2]) Fabry-Perot Fourier transform technique.

Nr.	Molecule	g	$\Delta\xi=\xi_{\parallel}-\xi_{\perp}$ 10^{-6} erg/Gauss^2 mol	Meth.	Ref.	Varia, remarks
8	OC···HF $O^{13}C$···HF ^{18}OC···HF OC···DF ^{18}OC···DF			MW [2]	81L1	$D_{\perp}^{HF}=113.4\,(2)$ kHz $D_{\perp}^{HF}=114.9\,(2)$ kHz $D_{\perp}^{HF}=114.1\,(2)$ kHz $D_{\perp}^{DF}=19.3\,(2)$ kHz $D_{\perp}^{DF}=20.9\,(2)$ kHz
9	OCO···HF OCO···DF			MBE	81B1	$D_{\perp}^{HF}=104.7\,(6)$ kHz $D_{\perp}^{DF}=17.39\,(24)$ kHz
10	SCO···HF			MBE	81B1	$D_{\perp}^{HF}=105\,(5)$ kHz
11	NCCN···HF $^{15}NCCN$···HF $NCC^{15}N$···HF $^{15}NCC^{15}N$···HF			MW [2]	81L2	$D_{\perp}^{HF}=110\,(10)$ kHz $D_{\perp}^{HF}=106\,(11)$ kHz $D_{\perp}^{HF}=117\,(8)$ kHz $D_{\perp}^{HF}=117\,(2)$ kHz
b) Symmetric top molecules						
12	NH_3	see Subvolume a, 2.9.2, Nr. 15		MBE	81M1	$(J, K)=(1, 0)$ state: $C^{H}=-17.859\,(7)$ kHz, $C^{N}=6.774\,(9)$ kHz, $r_{NH}=1.0178\,(4)$ Å, ∢HNH$=106.75\,(5)$ deg. $(J, K)=(1, 1)$ state: $C^{H}=-3.834\,(4)$ kHz [1]), $C^{N}=6.737\,(18)$ kHz [1]), $r_{NH}=1.025\,(2)$ Å, ∢HNH$=107.3$ deg. held fixed [1]) compare earlier BMS results [67K1, 68K1, 72H1]
				Theory	81S1	dependence of the rotational g-tensor on the inversional ground states, and estimation of the effective g-factors in the excite inversion state
	$^{15}NH_3$					dependence of the rotational g-tensor on the inversional ground states
13	AsH_3			RFIRDR	81S2	$M_{\perp}^{As}=111.7\,(11)$ kHz $M_{\parallel}^{As}=69.1\,(11)$ kHz compare Subvolume a, 2.9.2, Nr. 18

[2]) Fabry-Perot Fourier transform technique.

Nr.	Molecule	g	$\Delta\xi = \xi_{\parallel} - \xi_{\perp}$ 10^{-6} erg/Gauss^2 mol	Meth.	Ref.	Varia, remarks
14	$CH_3{}^{79}Br$ $CH_3{}^{81}Br$			MW	81D1	$M_{\perp}^{79\,Br} = 12.26\,(52)$ kHz [1] $M_{\parallel}^{79\,Br} = 36.4\,(114)$ kHz [1] $M_{\perp}^{81\,Br} = 13.46\,(61)$ kHz [1] $M_{\parallel}^{81\,Br} = 31.1\,(87)$ kHz [1]
15	CH_3I			MW	81D1	$M_{\perp}^{I} = 17.54\,(42)$ kHz [1] $M_{\parallel}^{I} = 31.3\,(19)$ kHz [1]
16	CHF_3 CDF_3	$g_{\perp} = -0.03627\,(2)$ $g_{\parallel} = -0.03333\,(7)$ $g_{\perp} = -0.03485\,(2)$ $g_{\parallel} = -0.0325\,(8)$		MBE	81M2	the signs of the g-values were obtained in a hfs avoided-crossing experiment compare earlier results, Subvolume a, 2.9.2, Nr. 24
17	SiH_3CH_3	$g_{\perp} = -0.036391\,(21)$ $g_{\parallel} = -0.10667\,(13)$	−2.70 (14)	MBE	82O1 82M3	the sign of $g_{\perp}/g_{\parallel}$ disagrees with an earlier MW result, see Vol. II/6, 2.9.2, Nr. 34a the signs of the g-values were obtained in a hfs avoided-crossing experiment
18	$CH_3CN \cdots HF$			MW [2]	81S3	$D_{\perp}^{HF} = 138.1\,(34)$ kHz

[1]) In the $v_6 = 1$ vib. state.
[2]) Fabry-Perot Fourier transform technique.

References for Appendix to 2.9.2

67K1 Kukolich, S.G.: Phys. Rev. **156** (1967) 83.
68K1 Kukolich, S.G.: Phys. Rev. **172** (1968) 59.
72H1 Hougen, J.: J. Chem. Phys. **57** (1972) 4207.
80V1 Verberne, J., Reuss, J.: Chem. Phys. **50** (1980) 137.
81B1 Baiocchi, F.A., Dixon, T.A., Joyner, C.H., Klemperer, W.: J. Chem. Phys. **74** (1981) 6544.
81D1 Dubrulle, A., Burie, J., Boucher, D., Herlemont, F., Demaison, J.: J. Mol. Spectrosc. **88** (1981) 394.
81H1 Hutson, J.M., Howard, B.J.: J. Chem. Phys. **74** (1981) 6520.
81K1 Keenan, M.R., Buxton, L.W., Campbell, E.J., Legon, A.C., Flygare, W.H.: J. Chem. Phys. **74** (1981) 2133.
81L1 Legon, A.C., Soper, P.D., Flygare, W.H.: J. Chem. Phys. **74** (1981) 4944.
81L2 Legon, A.C., Soper, P.D., Flygare, W.H.: J. Chem. Phys. **74** (1981) 4936.
81M1 Marshall, M.D., Muenter, J.S.: J. Mol. Spectrosc. **85** (1981) 322.
81M2 Meerts, M.L., Ozier, I.: J. Chem. Phys. **75** (1981) 596.
81S1 Špirko, V.: Mol. Phys. **43** (1981) 723.
81S2 Scappini, F., Oka, T.: J. Mol. Spectrosc. **85** (1981) 390.
81S3 Soper, P.D., Legon, A.C., Read, W.G., Flygare, W.H.: J. Am. Chem. Soc. **85** (1981) 3440.
81V1 Verberne, J., Reuss, J.: Chem. Phys. **54** (1981) 189.
81W1 Waaijer, M., Jacobs, M., Reuss, J.: Chem. Phys. **63** (1981) 257.
81W2 Waaijer, M., Reuss, J.: Chem. Phys. **63** (1981) 263.
81W3 Waaijer, M., Jacobs, M., Reuss, J.: Chem. Phys. **63** (1981) 247.
82M3 Meerts, W.L., Ozier, I.: J. Mol. Spectrosc. **94** (1982) 38.
82O1 Ozier, I., Meerts, W.L.: J. Mol. Spectrosc. **93** (1982) 164.
82T1 Tack, L.M., Kukolich, S.G.: J. Mol. Spectrosc. **94** (1982) 95.

Appendix to 2.9.3 Magnetic constants: Asymmetric top molecules

For preliminary remarks see p. 772 of Vol. II/14a Landolt-Börnstein, N.S.

Nr.	Molecule	g_{aa}, g_{bb}, g_{cc}	$\Delta\xi_1, \Delta\xi_2$ 10^{-6} erg/G^2 mol	$\Theta_{aa}, \Theta_{bb}, \Theta_{cc}$ 10^{-26} esu cm^2	$\langle a^2\rangle, \langle b^2\rangle, \langle c^2\rangle$ 10^{-16} cm^2	Meth.	Ref.	Varia, remarks
1	NO···NO					MBE	81W4	$M_{aa}^{14N}=10.43\,(48)$ kHz $M_{bb}^{14N}=13.84\,(48)$ kHz $M_{cc}^{14N}=0.75\,(46)$ kHz
2	HC≡CH···HF HC≡CH···DF $H^{13}C$≡CH···HF					MW [1]	82R1	$D_{aa}^{HF}=-237.2\,(44)$ kHz $D_{aa}^{DF}=-38.3\,(57)$ kHz $\frac{1}{2}(M_{bb}+M_{cc})^F=7.7\,(30)$ kHz $D_{aa}^{HF}=-237\,(12)$ kHz
3	C_2H_4···HF C_2H_4···DF					MW [1]	82S1	$D_{aa}^{HF}=-237.4\,(82)$ kHz $D_{aa}^{DF}=-38.3\,(87)$ kHz $\frac{1}{2}(M_{bb}+M_{cc})^F=7.3\,(38)$ kHz
4	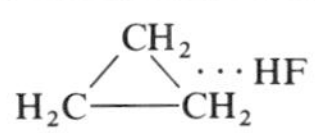					MW [1]	81B2	$D_{aa}^{HF}=121\,(43)$ kHz
5	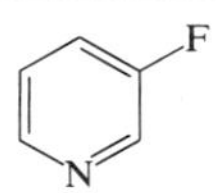	$g_{aa}=-0.0917\,(5)$ [2] $g_{bb}=-0.0476\,(5)$ [2] $g_{cc}=-0.0194\,(5)$ [2]	$\Delta\xi_1=53.3\,(8)$ $\Delta\xi_2=60.5\,(7)$	$\Theta_{aa}=-1.2\,(12)$ $\Theta_{bb}=3.2\,(13)$ $\Theta_{cc}=-2.0\,(18)$	$\langle a^2\rangle=108.0\,(24)$ $\langle b^2\rangle=58.0\,(24)$ $\langle c^2\rangle=8.4\,(24)$	MW	81H2, 82H1	
6	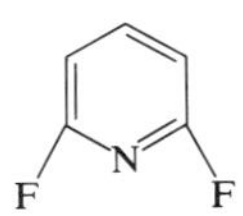	$g_{aa}=-0.0573\,(6)$ [2] $g_{bb}=-0.0311\,(6)$ [2] $g_{cc}=-0.0102\,(6)$ [2]	$\Delta\xi_1=48.4\,(11)$ $\Delta\xi_1=51.7\,(11)$	$\Theta_{aa}=-0.4\,(20)$ $\Theta_{bb}=1.5\,(22)$ $\Theta_{cc}=-1.0\,(30)$	$\langle a^2\rangle=140.9\,(24)$ $\langle b^2\rangle=81.6\,(24)$ $\langle c^2\rangle=8.7\,(24)$	MW	81H2, 82H1	

[1]) Fabry-Perot Fourier transform technique

[2]) The opposite choice of the sign of the g-values yields $\langle c^2\rangle<0$.

References for Appendix to 2.9.3

81B2 Buxton, L.W., Aldrich, P.D., Shea, J.A., Legon, A.C., Flygare, W.H.: J. Chem. Phys. **75** (1981) 2681.

81H2 Hübner, D., Stolze, M., Sutter, D.H.: Z. Naturforsch. **36a** (1981) 332.

81W4 Western, C.M., Langridge-Smith, P.R.R., Howard, B.J., Novick, S.E.: Mol. Phys. **44** (1981) 145.

82H1 Hübner, D., Stolze, M., Sutter, D.H.: Z. Naturforsch. **37a** (1982) 95 (Erratum).

82R1 Reed, W.G., Flygare, W.H.: J. Chem. Phys. **76** (1982) 2238.

82S1 Shea, J.A., Flygare, W.H.: J. Chem. Phys. **76** (1982) 4857.

6 Index of substances for volumes II/4, II/6 and II/14

The substances are listed with their individual number in the tables. The names of the substances which already have been discussed in Vol. II/6 and/or in Vol. II/4 are taken from the Index of substances of Vol. II/6. The names of the remaining substances are formed according to the latest IUPAC-rules.

The location of the substances discussed in the volumes II/4, II/6 and/or II/14a, b is indicated by two numbers:

a bold-faced number which refers to the respective table (**2.1**···**2.5**: Constants of diamagnetic molecules; **2.6**: Dipole moments; **2.7**: Quadrupole coupling constants; **2.8**: Hindered rotation; **2.9**: Magnetic constants; **4** (of II/14) or **5** (of II/6): Constants of radicals; **5** (of II/14) or **6** (of II/6): Appendix);

a light-faced number which represents the individual number of the substances in the respective table.

Substances from volume II/14a, b: the individual number in each table with data on the respective substance is listed.

Substances from volumes II/4 and II/6: only the individual number in the table with the first entry is listed. There, references are given to the following tables.

6.1 Inorganic substances

All substances without any C are included.

The substances are listed alphabetically according to their alphabetically ordered gross formula.

*) Including deuterated compounds.

Alphabetic formula	Chemical formula	Name	Table Nr., Substance Nr.		
			II/14	II/6	II/4
ArO_3	$Ar \cdot O_3$	Argon ozone complex	**2.5,** 760; **2.6.3,** 368		
ArO_2S	$Ar \cdot SO_2$	Argon sulfur dioxide complex	**2.5,** 759; **2.6.3,** 367		
ArO_3S	$Ar \cdot SO_3$	Argon sulfur trioxide complex	**2.4,** 103; **2.6.2,** 69;		
$AsBr_3$	$AsBr_3$	Arsenous tribromide	**2.4,** 9		
$AsCl_3$	$AsCl_3$	Trichloroarsine	**2.4,** 8		**2.4,** 9
AsF_3	AsF_3	Trifluoroarsine	**2.4,** 7; **2.7.2,** 43	**2.4,** 8	**2.4,** 8
AsH_3	AsH_3	Arsine	**2.4,** 6; **2.6.2,** 31; **2.7.2,** 42; **2.9.2,** 18; **5: 2.9.2,** 13	**2.4,** 7	**2.4,** 7
AsO	AsO	Arsenic monoxide	**4.1.5,** 81a; **5: 2.7.1,** 81a		
B_2BrH_5	B_2H_5Br	Bromodiborane		**2.5,** 173	**2.5,** 147
BCl	BCl	Chloroborane (1)	**5: 2.2,** 47a		
$BClF_2$	$BClF_2$	Chlorodifluoroborane	**2.5,** 45; **2.7.3,** 92		
BClS	ClBS	Chlorothioxoborane	**2.3,** 13; **2.6.2,** 9; **2.7.2,** 22		
BF	BF	Boron monofluoride		**2.2,** 44a	
BF_2H	HBF_2	Difluoroborane		**2.5,** 34	
BF_2H_2N	BF_2NH_2 *)	Aminodifluoroborane	**2.5,** 46; **2.6.3,** 23		
BFH_2O	BFHOH	Fluorohydroxyborane	**2.5,** 49; **2.6.3,** 26		
BFH_2O_2	$BF(OH)_2$ *)	Fluorodihydroxyborane	**2.5,** 50; **2.6.3,** 27		
BF_2HO	BF_2OH	Difluorohydroxiborine		**2.5,** 34a	
BF_2H_4P	$BH_3 \cdot PHF_2$	Difluorophosphine borane		**2.5,** 35	
BF_3H_3P	$H_3P \cdot BF_3$ *)	Phosphine(P—B) trifluoroborane	**2.4,** 80; **2.6.2,** 61; **2.8,** 133		
BF_3H_3P	$BH_3 \cdot PF_3$	Trifluorophosphine borane		**2.4,** 63	**2.4,** 91a
BF_3N_2	$N_2 \cdot BF_3$	Dinitrogen(N—B)trifluoroborane	**2.4,** 104; **2.6.2,** 70		
BF_5Si	SiF_3BF_2	Difluoro-(trifluorosilyl) borane	**2.5,** 400; **2.6.3,** 189; **2.8,** 127		
BH	BH	Boron monohydride		**2.6.1,** 2aa	
B_4H_{10}	B_4H_{10} *)	Tetraborane (10)	**2.5,** 418; **2.6.3,** 204		
B_5H_9	B_5H_9 *)	Pentaborane (9)	**2.4,** 98		**2.4,** 96
B_6H_{10}	B_6H_{10}	Hexaborane (10)	**2.5,** 419; **2.6.3,** 205		
BH_4N	BH_2NH_2	Aminoborane	**2.5,** 47; **2.6.3,** 24		
BH_5N_2	$BH(NH_2)_2$ *)	Diaminoborane	**2.5,** 51; **2.6.3,** 28		
BH_6N	$BH_3 \cdot NH_3$	Ammonia(N—B) borane	**2.4,** 79; **2.6.2,** 60		
B_2H_7N	$H_2NB_2H_5$ *)	Aminodiborane (6)	**2.5,** 415; **2.6.3,** 201; **2.7.3,** 44		
BH_3O	BH_2OH *)	Borinic acid-Hydroxyborane	**2.5,** 47a; **2.6.3,** 24a; **2.7.3,** 9b		

*) Including deuterated compounds.

Alphabetic formula	Chemical formula	Name	**Table Nr.**, Substance Nr.		
			II/14	II/6	II/4
BH_3O_2	$BH(OH)_2$	Boronic acid-Dihydroxyborane	**2.5**, 48; **2.6.3**, 25		
$B_2H_2O_3$	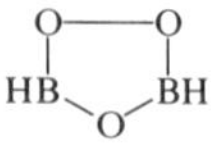	1,3,4-Trioxa-2,5-borocyclopentane		**2.5**, 223	
BH_6P	BH_3PH_3 *)	Phosphine borane	**2.4**, 81; **2.6.2**, 62; **2.8**, 132	**2.4**, 62a	
BHS	HBS	Thioxoborane	**2.3**, 12; **2.6.2**, 8; **2.7.2**, 5; **2.9.2**, 1		
BO_2	BO_2	Boron dioxide	**4.2.1**, 5		
BaCl	BaCl	Barium monochloride	**2.2**, 44; **2.7.1**, 44; **4.1.2**, 44; **5: 2.7.1**, 44		
BaF	BaF	Barium monofluoride	**2.2**, 43; **4.1.2**, 43; **5: 2.7.1**, 43		
BaO	BaO	Bariumoxide	**2.2**, 36; **2.6.1**, 36	**2.2**, 42	
BaS	BaS	Bariumsulfide	**2.2**, 37; **2.6.1**, 37	**2.2**, 43	
BiBr	BiBr	Bismuth monobromide	**2.2**, 85; **4.1.3**, 85		
BiCl	BiCl	Bismuth monochloride	**2.2**, 84; **2.7.1**, 84; **4.1.3**, 84		
$BiCl_3$	$BiCl_3$	Bismuth trichloride			**2.4**, 12
BiF	BiF	Bismuth monofluoride	**2.2**, 83; **2.7.1**, 83; **4.1.3**, 83		
BiI	BiI	Bismuth monoiodide	**2.2**, 86; **2.7.1**, 86; **4.1.3**, 86		
Br_2	Br_2	Bromine	**2.2**, 101; **2.7.1**, 101; **2.9.1**, 101; **4.1.6**, 101; **5: 2.9.1**, 101		
BrCa	CaBr	Calcium monobromide	**4.1.2**, 40a; **4.1.5**, 40a; **5: 2.7.1**, 40a		
BrCl	BrCl	Bromine chloride	**2.2**, 99; **2.6.1**, 99; **2.7.1**, 99; **2.9.1**, 99		**2.2**, 39
BrCs	CsBr	Caesium bromide	**2.2**, 28; **2.6.1**, 28	**2.2**, 35	**2.2**, 13
BrCu	CuBr	Copper(I)bromide	**2.2**, 32; **2.7.1**, 32; **5: 2.2**, 32		
BrF	BrF	Bromine fluoride	**2.2**, 97; **2.6.1**, 97; **2.7.1**, 97; **2.9.1**, 97	**2.2**, 84a	
BrF_3	BrF_3	Bromine trifluoride			**2.5**, 26
BrF_5	BrF_5	Bromine pentafluoride	**2.4**, 89; **2.7.2**, 54	**2.4**, 66	
$BrFO_2S$	SO_2BrF	Sulfuryl bromidefluoride	**2.5**, 33; **2.7.3**, 147		
BrF_5S	SF_5Br	Bromopentafluorosulfur(VI)	**2.4**, 85; **2.7.2**, 53		**2.4**, 95
BrF_3Si	SiF_3Br	Bromotrifluorosilane			**2.4**, 31
Br_3FSi	$SiFBr_3$	Fluorotribromosilane		**2.4**, 29	
BrGa	GaBr	Gallium(I)bromide	**2.2**, 51; **2.7.1**, 51; **5: 2.2**, 51		**2.2**, 18
$BrGeH_3$	GeH_3Br *)	Bromogermane	**2.4**, 29; **2.4**, 45c; **2.7.2**, 51a	**2.4**, 30b	**2.4**, 34
BrH	HBr *)	Hydrogen bromide	**2.2**, 17; **2.6.1**, 17; **2.7.1**, 17; **2.9.1**, 17	**2.2**, 15	**2.7**, 51

*) Including deuterated compounds.

Alphabetic formula	Chemical formula	Name	Table Nr., Substance Nr.		
			II/14	II/6	II/4
BrH	HBr^+	Hydrogen bromide (1+)ion	**2.2,** 14, **4.1.5,** 14		
BrHKr	Kr · HBr*)	Krypton hydrogen-bromide complex	**2.3,** 45; **2.7.2,** 48		
BrH_3Si	SiH_3Br*)	Bromosilane	**2.4,** 24; **2.6.2,** 45; **2.7.2,** 51; **2.9.2,** 27	**2.4,** 22	**2.4,** 26
Br_3HSi	$SiHBr_3$	Tribromosilane		**2.4,** 25	
BrH_3Sn	SnH_3Br	Bromostannane		**2.4,** 33	
BrI	IBr	Iodine bromide	**2.2,** 102; **2.6.1,** 102; **2.7.1,** 102; **5: 2.6.1,** 102		**2.2,** 41
BrIn	InBr	Indium(I)bromide	**2.2,** 53; **2.7.1,** 53		**2.2,** 21
BrK	KBr	Potassium bromide		**2.2,** 27	
BrLi	LiBr	Lithium bromide	**2.2,** 21; **2.9.1,** 21	**2.2,** 19	**2.2,** 2
BrNO	BrNO	Nitrosyl bromide	**2.5,** 15	**2.5,** 7	**2.5,** 11
BrNa	NaBr	Sodium bromide		**2.2,** 23	**2.2,** 5
BrO	OBr	Bromine oxide radical	**2.2,** 93; **2.7.1,** 93; **4.1.5,** 93	**2.2,** 79	
Br_3P	PBr_3	Phosphorus tribromide		**2.4,** 6	**2.4,** 6
BrRb	RbBr	Rubidium bromide	**2.2,** 24; **2.6.1,** 24; **2.7.1,** 24	**2.2,** 31	**2.2,** 9
BrTl	TlBr	Thallium(I)bromide	**2.2,** 57; **5: 2.2,** 57	**2.2,** 52	
CaCl	CaCl	Calcium monochloride	**2.2,** 40; **4.1.2,** 40; **4.1.5,** 40; **5: 2.6.1,** 40; **5: 2.7.1,** 40		
CaF	CaF	Calcium monofluoride	**2.2,** 39; **4.1.2,** 39		
CaI	CaI	Calcium monoiodide	**4.1.2,** 40b; **4.1.5,** 40b; **5: 2.7.1,** 40b		
CaO	CaO	Calcium oxide	**2.2,** 34		
ClCs	CsCl	Caesium chloride	**2.2,** 27; **2.7.1,** 27; **2.9.1,** 27	**2.2,** 34	**2.2,** 12
ClCu	CuCl	Copper(I)chloride	**2.2,** 31; **2.7.1,** 31; **2.9.1,** 31; **5: 2.2,** 31		
ClF	ClF	Chlorine fluoride	**2.2,** 96	**2.2,** 84	
ClF_3	ClF_3	Chlorine trifluoride			**2.5,** 25
ClF_5	ClF_5	Chlorine pentafluoride	**2.4,** 88; **2.6.2,** 64; **2.7.2,** 39		
ClF_3Ge	GeF_3Cl	Chlorotrifluorogermane			**2.4,** 36
$ClFH_2$	HFHCl*)	Hydrogen chloride hydrogen fluoride	**2.5,** 40; **2.6.3,** 18; **2.7.3,** 90		
ClF_2H	HFClF*)	Hydrogen fluoride chlorine fluoride complex	**2.5,** 39; **2.6.3,** 17; **2.7.3,** 89		
ClFK	Kr · ClF	Krypton chloride fluoride	**2.3,** 46; **2.6.2,** 26; **2.7.2,** 30; **5: 2.3,** 46		
$ClFO_2$	$FClO_2$	Chloryl fluoride	**2.5,** 42; **2.6.3,** 20; **2.7.3,** 91	**2.5,** 27a	
$ClFO_3$	ClO_3F	Perchloryl fluoride	**2.4,** 35; **2.6.2,** 48; **2.7.2,** 35		**2.4,** 44a

*) Including deuterated compounds.

Alphabetic formula	Chemical formula	Name	Table Nr., Substance Nr. II/14	II/6	II/4
$ClFO_2S$	SO_2ClF	Sulfuryl chloride fluoride		**2.5**, 21 a	
ClF_2P	PF_2Cl	Chlorodifluoro-phosphine		**2.5**, 29	
ClF_5S	SF_5Cl	Chloropentafluoro-sulfur (VI)	**2.4**, 84; **2.7.2**, 38	**2.4**, 64	**2.4**, 94
Cl_3FSi	$SiFCl_3$	Fluorotrichlorosilane		**2.4**, 28	
ClF_3Si	SiF_3Cl	Chlorotrifluorosilane			**2.4**, 30
ClF_5Te	TeF_5Cl	Tellurium chloride pentafluoride	**2.4**, 86		
ClF_5W	WF_5Cl	Chloropentafluoro-tungsten (VI)		**2.4**, 65	
ClGa	GaCl	Gallium(I)chloride		**2.2**, 46a	**2.2**, 17
$ClGeH_3$	GeH_3Cl*)	Chlorogermane	**2.4**, 45b	**2.4**, 30a	**2.4**, 33
Cl_3GeH	$GeHCl_3$	Trichlorogermane			**2.4**, 35
ClH	HCl	Hydrogen chloride	**2.2**, 16; **2.6.1**, 16; **2.7.1**, 16; **2.9.1**, 16	**2.2**, 14	
Cl^+H	HCl^+	Hydrogen chloride (1+)ion	**4.1.5**, 13a		
ClHKr	Kr · HCl	Krypton hydrogen chloride complex	**2.3**, 44; **2.6.2**, 25; **2.7.2**, 29; **2.9.2**, 13; **5**: **2.3**, 44		
ClH_2N	NH_2Cl	Monochloramine	**2.5**, 1 b	**2.5**, 1 b	
ClHNe	Ne · HCl*)	Neon hydrogen chloride (complex)	**2.3**, 38; **2.6.2**, 20; **2.7.2**, 26		
ClHO	HOCl	Hypochlorous acid		**2.5**, 27	
ClH_3Si	SiH_3Cl	Chlorosilane		**2.4**, 21	**2.4**, 25
Cl_2H_2Si	SiH_2Cl_2	Dichlorosilane	**2.5**, 379; **2.6.3**, 176; **2.7.3**, 129		
Cl_3HSi	$HSiCl_3$	Trichlorosilane	**2.4**, 27; **2.7.2**, 34	**2.4**, 24	**2.4**, 29
ClH_2Sn	SnH_3Cl	Chlorostannane		**2.4**, 32	
ClHXe	Xe · HCl	Xenon hydrogen chloride complex	**2.3**, 47; **2.6.2**, 27; **2.7.2**, 31; **5**: **2.3**, 47		
ClI	ICl	Iodine chloride	**2.2**, 100; **2.6.1**, 100; **2.7.1**, 100; **4.1.6**, 100; **5**: **2.7.1**, 100	**2.2**, 85	
ClIn	InCl	Indium(I)chloride		**2.2**, 48	**2.2**, 20
ClK	KCl	Potassium chloride		**2.2**, 26	**2.2**, 7
ClLi	LiCl	Lithium chloride	**2.2**, 20; **2.9.1**, 20; **5**: **2.2**, 20	**2.2**, 18	**2.2**, 1
Cl_3N	NCl_3	Nitrogen trichloride	**2.4**, 2; **2.6.2**, 29; **2.7.2**, 32		
ClN_3	ClN_3	Chlorazide		**2.5**, 14	
ClNO	ClNO	Nitrosyl chloride	**2.5**, 14; **2.7.3**, 83	**2.5**, 6	**2.5**, 10
$ClNO_2$	ClONO	Nitryl chloride	**2.5**, 18; **2.7.3**, 85		
$ClNO_2$	NO_2Cl		**2.5**, 17; **2.7.3**, 84	**2.5**, 11	**2.5**, 14
$ClNO_3$	$ClNO_3$	Chlorine nitrate	**2.5**, 24; **2.6.3**, 11; **2.7.3**, 86		
ClNS	NSCl	Thionitrosyl chloride		**2.5**, 9	
ClNa	NaCl	Sodium chloride		**2.2**, 22	**2.2**, 4
ClO	OCl	Chlorine oxide radical	**2.2**, 92; **2.7.1**, 92; **4.1.5**, 92; **5**: **2.7.1**, 92	**2.2**, 78	
ClO_2	ClO_2	Chlorine dioxide	**4.2.2**, 2		

*) Including deuterated compounds.

Alphabetic formula	Chemical formula	Name	**Table Nr.**, Substance Nr.		
			II/14	II/6	II/4
Cl_2O	Cl_2O	Chlorine monoxide	**2.5**, 58	**2.5**, 41	**2.5**, 33
Cl_3OP	$POCl_3$	Phosphoryl chloride		**2.4**, 36	**2.4**, 38
ClO_3Re	ReO_3Cl	Rhenium oxychloride			**2.4**, 44
Cl_2OS	$SOCl_2$	Thionyl chloride	**2.5**, 31; **2.7.3**, 87	**2.5**, 20	
Cl_2O_2S	SO_2Cl_2	Sulfuryl chloride	**2.5**, 32	**2.5**, 21	**2.5**, 22
Cl_3OV	$VOCl_3$	Vanadyl(V)chloride		**2.4**, 36b	
Cl_3P	PCl_3	Phosphorus trichloride	**2.4**, 5	**2.4**, 5	**2.4**, 5
Cl_3PS	$PSCl_3$	Thiophosphoryl chloride			**2.4**, 40
ClRb	RbCl	Rubidium chloride		**2.2**, 30	
Cl_2S	SCl_2	Sulfur dichloride	**2.5**, 37; **2.6.3**, 15; **2.7.3**, 88	**2.5**, 22a	
Cl_2S_2	S_2Cl_2	Disulfur dichloride	**2.5**, 59; **2.7.3**, 93		
Cl_3Sb	$SbCl_3$	Trichlorostibine	**2.4**, 10		**2.4**, 11
ClSr	SrCl	Strontium monochloride	**2.2**, 42; **4.1.2**, 42		
ClTl	TlCl	Thallium(I)chloride	**2.2**, 56; **5: 2.2**, 56	**2.2**, 51	
CrF_2O_2	CrO_2F_2	Chromyl fluoride			**2.5**, 27
Cs_2	Cs_2	molecular Caesium	**2.2**, 6	**2.2**, 8	
CsF	CsF	Caesium fluoride	**2.2**, 26; **2.6.1**, 26	**2.2**, 33	**2.2**, 11
CsHO	CsOH	Caesium hydroxide	**2.3**, 22	**2.3**, 13	
CsI	CsI	Caesium iodide	**2.2**, 29; **2.6.1**, 29	**2.2**, 36	**2.2**, 14
CsNa	CsNa	Sodium caesium		**2.6.1**, 1f	
CuF	CuF	Copper(I)fluoride	**2.2**, 30; **2.7.1**, 30; **2.9.1**, 30	**2.2**, 37	
CuI	CuI	Copper(I)iodide	**2.2**, 33; **2.7.1**, 33		
D_2	D_2	Deuterium	**2.2.2**, 1; **2.9.1.** 1; **4.1.3**, 1; **4.1.6**, 1; **5: 2.6.1**, 1	**2.2**, 2; **2.9.1**, 1; **2.6.1**, 1	
F_2	F_2	molecular Fluorine		**2.9.1**, 41	
FGa	GaF	Gallium(I)fluoride	**2.2.**, 50; **2.7.1**, 50; **2.9.1**, 50; **5: 2.2**, 50	**2.2**, 46	
F_2Ge	GeF_2	Germanium difluoride		**2.5**, 1a	
F_4Ge	GeF_4	Tetrafluorogermane	**2.4**, 30; **2.9.2**, 30	**2.4**, 31	
$FGeH_3$	GeH_3F	Fluorogermane	**2.4**, 45a	**2.4**, 30	**2.4**, 32
FH	HF	Hydrogen fluoride	**2.2**, 15; **2.9.1**, 15; **5: 2.6.1**, 15	**2.2**, 13	
F_2H_2	$(HF)_2$ *)	Hydrogen fluoride dimer	**2.5**, 38; **2.6.3**, 16; **2.7.3**, 2	**2.5**, 35a	
FHKr	Kr · HF	Krypton hydrogen fluoride complex	**2.3**, 43; **2.7.2**, 55		
F_2HN	NHF_2	Difluoroamine	**2.5**, 3		**2.5**, 3
FHN_2O	$N_2O \cdot HF$	Nitrous oxide hydrogen fluoride complex	**2.5**, 763a; **2.6.3**, 369a		
F_2HNOS	$HN{=}SOF_2$ *)	Imidosulfuryl fluoride	**2.5**, 4; **2.6.3**, 1; **2.7.3**, 11		
FHO	HOF*)	Hypofluorous acid	**2.5**, 41; **2.6.3**, 19	**2.5**, 26a	

*) Including deuterated compounds.

*) Including deuterated compounds.

Alphabetic formula	Chemical formula	Name	**Table Nr.**, Substance Nr.		
			II/14	II/6	II/4
F_2O_2S	SO_2F_2	Sulfuryl fluoride		**2.9.3**, 3	**2.5**, 21
F_4OS	SOF_4	Thionyl tetrafluoride	**2.5**, 34	**2.5**, 18a	
F_2OSe	$SeOF_2$	Selenyl fluoride		**2.5**, 25	
F_4OXe	$XeOF_4$	Xenon oxyfluoride		**2.4**, 69	**2.4**, 98
F_3P	PF_3	Trifluorophosphine	**2.4**, 4; **2.9.2**, 17	**2.4**, 4	**2.4**, 4
F_3PS	SPF_3	Thiophosphoryl fluoride	**2.4**, 32; **2.7.2**, 21		**2.4**, 39
FRb	RbF	Rubidium fluoride		**2.2**, 29	
FS	SF	Sulfur monofluoride	**2.2**, 95; **2.6.1**, 95; **4.1.5**, 95	**2.2**, 81	
F_2S	SF_2	Sulfur difluoride	**2.5**, 35	**2.5**, 22	
F_4S	SF_4	Sulfur tetrafluoride	**2.5**, 36	**2.5**, 23	**2.5**, 23
F_6S	SF_6	Sulfur(VI)fluoride	**2.4**, 83; **2.9.2**, 36		
F_2S_2	$S{=}SF_2$	Sulfursulfur difluoride			**2.5**, 37
F_2S_2	FSSF	Disulfur difluoride			**2.5**, 36
FSe	SeF	Selenium monofluoride		**2.6.1**, 58a	
F_4Se	SeF_4	Selenium tetrafluoride		**2.5**, 26	
F_2Si	SiF_2	Silicon difluoride	**2.5**, 1a	**2.5**, 1	
F_4Si	SiF_4	Tetrafluorosilane	**2.4**, 25; **2.9.2**, 28	**2.9.2**, 26	
FSr	SrF	Strontium monofluoride	**2.2**, 41; **4.1.2**, 41		
FTl	TlF	Thallium(I)fluoride	**2.2**, 55; **5: 2.2**, 55	**2.2**, 50	
GaI	GaI	Gallium monoiodide	**2.2**, 52; **2.7.1**, 52; **2.9.1**, 52; **5: 2.2**, 52		**2.2**, 19
GeH_4	GeH_4	Germane	**2.4**, 28; **2.6.2**, 46; **2.9.2**, 29	**2.9.2**, 27	
GeH_3I	GeH_3I *)	Iodogermane	**2.4**, 45d	**2.4**, 30c	
GeH_6Si	SiH_3GeH_3	Silylgermane		**2.4**, 62	**2.4**, 92
GeO	GeO	Germanium monoxide	**2.2**, 64; **2.6.1**, 64; **2.9.1**, 64	**2.2**, 59	
GeS	GeS	Germanium monosulfide	**2.2**, 65; **2.9.1**, 65; **5: 2.2**, 65	**2.2**, 60	
GeSe	GeSe	Germanium monoselenide	**2.2**, 66; **2.9.1**, 66; **5: 2.2**, 66	**2.2**, 61	
GeTe	GeTe	Germanium monotelluride	**2.2**, 67; **2.9.1**, 67	**2.2**, 62	
H_2^+	H_2^+	Hydrogen molecular ion		**5.1.2**, 1	
H_2	H_2	molecular Hydrogen		**2.6.1**, 1	
H_4	$H_2\cdots H_2$	Hydrogen dimer (ortho-ortho and ortho-para)	**5: 2.9.2**, 1		
HI	HI *)	Hydrogen iodide	**2.2**, 18; **2.7.1**, 18; **5: 2.7.1**, 18; **5: 2.9.1**, 18	**2.2**, 16	**2.2**, 36
H_3ISi	SiH_3I	Iodosilane		**2.4**, 23	**2.4**, 27
H_3ISn	SnH_3I	Iodostannane		**2.4**, 34	
HKO	KOH	Potassium hydroxide	**2.3**, 20; **2.7.2**, 41	**2.3**, 11	
H_2Kr	$H_2\cdots Kr$	ortho Hydrogen-krypton complex	**5: 2.9.2**, 4		

*) Including deuterated compounds.

Alphabetic formula	Chemical formula	Name	Table Nr., Substance Nr.		
			II/14	II/6	II/4
HLi	LiH	Lithium hydride	**2.2**, 2; **2.6.1**, 2; **2.9.1**, 2	**2.2**, 3	
HLiO	LiOH	Lithium hydroxide		**2.3**, 10	
HN	NH	Nitrogen monohydride	**2.2**, 9; **2.6.1**, 9; **4.1.3**, 9	**2.2**, 8c	
HN_3	HN_3*)	Hydrogen azide	**2.5**, 21; **2.6.3**, 10	**2.5**, 13	**2.5**, 16
H_2N	NH_2*)	Aminyl	**4.2.2**, 4		
H_3N	NH_3*)	Ammonia	**2.4**, 1; **2.6.2**, 28; **2.7.2**, 15; **2.9.2**, 15	**2.4**, 1	
H_4N_2	N_2H_4*)	Hydrazine	**2.5**, 25; **2.6.3**, 12; **2.7.3**, 19; **2.8**, 137		**2.5**, 4
HNO	HNO*)	Nitroxyl	**2.5**, 12; **2.6.3**, 6; **2.7.3**, 15	**2.5**, 4a	
HNO_2	HNO_2	Nitrous acid	**2.5**, 19; **2.6.3**, 9	**2.5**, 12	**2.5**, 15a
HNO_3	HNO_3*)	Nitric acid	**2.5**, 20		**2.5**, 15
$H_2N_2O_2$	NH_2NO_2	Nitramide			**2.5**, 2
H_3NO	NH_2OH	Hydroxylamine		**2.5**, 2a	
HNOS	HNSO*)	Thionylimine	**2.5**, 23; **2.7.3**, 18	**2.5**, 15	**2.5**, 17
HNa	NaH	Sodium hydride	**2.2**, 3; **2.6.1**, 3; **5: 2.2**, 3		
HNaO	NaOH	Sodium hydroxide	**2.3**, 19		
H_2Ne	$H_2\cdots Ne$	ortho Hydrogen-neon complex	**5: 2.9.2**, 2		
HO	OH	Hydroxyl	**2.2**, 11; **2.6.1**, 11; **2.7.1**, 11; **4.1.2**, 11; **4.1.5**, 11	**2.2**, 9	
HO_2	HO_2*)	Perhydroxyl	**4.2.2**, 6		
H_2O	H_2O*)	Water	**2.5**, 53; **2.6.3**, 30; **2.7.3**, 3; **2.9.3**, 6; **5: 2.5**, 53; **5: 2.6.3**, 30	**2.5**, 37	**2.5**, 29
H_2O_2	H_2O_2	Hydrogen peroxide	**2.5**, 62; **2.6.3**, 32a; **2.9.3**, 7		**2.5**, 34
H_4O_2	$(H_2O)_2$*)	Water dimer	**2.5**, 60; **2.6.3**, 31		
HORb	RbOH	Rubidium hydroxide	**2.3**, 21	**2.3**, 12	
HOS	HSO*)	Hydrosulfinyl radical	**4.2.2**, 7		
H_2O_4S	H_2SO_4*)	Sulfuric acid	**2.5**, 37a; **2.6.3**, 15a		
H_6OSi_2	$(SiH_3)_2O$	Disiloxane	**2.5**, 387		
HP	PH	Phosphorus monohydride	**2.2**, 10; **4.1.3**, 10; **4.1.7**, 10		
H_2P	PH_2		**4.2.2**, 9		
H_3P	PH_3*)	Phosphine	**2.4**, 3; **2.6.2**, 30; **2.9.2**, 16	**2.4**, 3	**2.4**, 3
H_4P_2	P_2H_4*)	Diphosphine	**2.5**, 44; **2.6.3**, 22		
H_5PSi	SiH_3PH_2*)	Silylphosphine	**2.5**, 401; **2.6.3**, 190; **2.8**, 131	**2.5**, 166a	
HS	SH	Sulfurmonohydride	**2.2**, 12; **2.6.1**, 12; **2.7.1**, 12; **4.1.5**, 12	**2.2**, 10	
H^+S	SH^+	Sulfur monohydride (1+)ion	**4.1.3**, 10a		
H_2S	H_2S	Hydrogen sulfide	**2.5**, 54	**2.5**, 38	**2.5**, 30
H_2S_2	HSSH	Hydrogenpersulfide		**2.5**, 42	

*) Including deuterated compounds.

Alphabetic formula	Chemical formula	Name	Table Nr., Substance Nr.		
			II/14	II/6	II/4
H_6SSi_2	$(SiH_3)_2S$	Disilylsulfide	**2.5**, 388; **2.6.3**, 179; **2.8**, 106		
H_3Sb	SbH_3	Stibine		**2.4**, 9	**2.4**, 10
HSe	SeH	Selenium monohydride	**2.2**, 13; **4.1.5**, 13	**2.2**, 11	
H_2Se	H_2Se*)	Hydrogen selenide	**2.5**, 55	**2.5**, 39	**2.5**, 31
H_4Si	SiH_4	Silane	**2.4**, 22; **2.6.2**, 43; **2.9.2**, 26	**2.9.2**, 24	
HTe	TeH	Tellurium hydride		**5.1.5**, 5	
H_2Te	H_2Te	Hydrogen telluride	**2.5**, 56		
He_2	He_2	Dihelium molecule	**2.2**, 7; **4.1.3**, 7; **4.1.7**, 7		
I_2	I_2	molecular Iodine	**2.2**, 103; **2.6.1**, 103; **2.7.1**, 103; **2.9.1**, 103; **4.1.6**, 103; **5: 2.7.1**, 103; **5: 2.9.1**, 103	**2.2**, 86	
IIn	InI	Indium (I)iodide	**2.2**, 54; **5: 2.2**, 54	**2.2**, 49	
IK	KI	Potassium iodide	**2.2**, 23; **2.6.1**, 23; **2.7.1**, 23	**2.2**, 28	**2.2**, 8
ILi	LiI	Lithium iodide	**2.2**, 22; **2.7.1**, 22; **2.9.1**, 22	**2.2**, 20	**2.2**, 3
INa	NaI	Sodium iodide		**2.2**, 24	**2.2**, 6
IO	OI	Iodine oxide radical	**2.2**, 94; **2.7.1**, 94; **4.1.5**, 94	**2.2**, 80	
IRb	RbI	Rubidium iodide	**2.2**, 25; **2.6.1**, 25; **2.7.1**, 25	**2,2**, 63	**2.2**, 10
ITl	TlI	Thallium (I)iodide	**2.2**, 58; **5: 2.2**, 58	**2.2**, 53	
K_2	K_2	molecular Potassium		**2.9.1**, 5	
KLi	KLi	Lithium potassium		**2.6.1**, 1 b	
KNa	NaK	Sodium potassium	**2.2**, 5; **2.6.1**, 5	**2.2**, 6 b	
KrO_3S	$Kr \cdot SO_3$	Krypton sulfur trioxide complex	**2.4**, 103 a; **2.6.2**, 69 a; **5: 2.4**, 103 a		
Li_2	Li_2	molecular Lithium		**2.7.1**, 4	
LiNa	NaLi	Lithium sodium		**2.6.1**, 1 a	
LiO	LiO	Lithium monoxide		**2.6.1**, 5 a	
LiRb	RbLi	Rubidium lithium		**2.6.1**, 1 d	
N_2	N_2	molecular Nitrogen	**2.2**, 76; **2.7.1**, 76; **4.1.3**, 76	**2.2**, 71	
NO	NO	Nitrogen monoxide	**2.2**, 80; **2.6.1**, 80; **2.7.1**, 80; **2.9.1**, 80; **4.1.2**, 80; **4.1.5**, 80; **5: 2.7.1**, 80	**2.2**, 73	
NO^+	NO^+	Nitric oxide	**5: 2.2**, 77 a; **5: 2.7.1**, 77 a		
NO_2	NO_2	Nitrogen dioxide	**4.2.2**, 1	**5.2.2**, 2	
N_2O	N_2O	Nitrous oxide	**2.3**, 18; **2.6.2**, 12; **2.7.2**, 13; **2.9.2**, 7	**2.3**, 14	**2.3**, 15
N_2O_2	NO···NO	Nitric oxid dimer	**5:2.9.3**, 1		

*) Including deuterated compounds.

6.2 Organic substances

All substances containing C are included.

The substances are listed according to increasing C number, increasing H number, and then alphabetically. All molecules not containing any H are listed before those with one H atom.

Alphabetic formula	Chemical formula	Name	**Table Nr.,** Substance Nr.		
			II/14	II/6	II/4
C_1					
CArClN	Ar · ClCN	Argon cyanogen chloride complex	**2.5,** 756a; **2.7.3,** 144a		
$CArO_2$	Ar · CO_2	Argon carbon dioxide complex	**2.5,** 755; **2.6.3,** 363		
CArOS	Ar · OCS	Argon carbonyl sulfide	**2.5,** 756; **2.6.3,** 364		
CBF_3O	BF_3CO	Trifluoroborane carbonyl	**2.4,** 52; **2.6.2,** 55; **2.7.2,** 8		
$CBrF_3$	CF_3Br	Bromotrifluoromethane	**2.4,** 20; **2.6.2,** 41; **2.7.2,** 50		**2.4,** 21
CBrN	BrCN	Cyanogen bromide	**2.3,** 7	**2.9.2,** 6a	**2.3,** 8
CBr_2O	$OCBr_2$	Carbonic dibromide	**2.5,** 127; **2.7.3,** 150		
CCl	CCl	Carbon monochloride (radical)	**2.2,** 79; **4.1.5,** 79; **5: 2.7.1,** 79		
$CClF_3$	CF_3Cl	Chlorotrifluoromethane			**2.4,** 20
CCl_2F_2	CF_2Cl_2	Difluorodichloromethane	**2.5,** 107; **2.7.3,** 101	**2.5,** 61	
CClFO	OCFCl	Fluoro carbonyl chloride	**2.5,** 128; **2.7.3,** 104; **2.9.3,** 19	**2.6.3,** 31	**2.5,** 67
$CClF_3O_2$	CF_3OOCl	Trifluoromethyl peroxyhypochlorite	**2.5,** 79; **2.7.3,** 97		
CClFS	SCFCl	Carbonthioic chloride fluoride	**2.5,** 148; **2.7.3,** 108		
CClN	ClCN	Cyanogen chloride	**2.3,** 6; **2.6.2,** 5; **2.7.2,** 10	**2.6.2,** 2a	**2.3,** 7
CClNO	ClNCO	Chlorine isocyanate	**2.5,** 8; **2.7.3,** 81	**2.5,** 4	
CClNS	ClSCN	Chlorothiocyanate	**2.5,** 11; **2.7.3,** 82		
CCl_2O	$OCCl_2$	Carbonyl chloride	**2.5,** 126; **2.7.3,** 103	**2.5,** 67a	**2.5,** 66
CCl_2S	$SCCl_2$	Carbonthioic dichloride	**2.5,** 147; **2.7.3,** 107		
CF	CF	Carbon monofluoride	**2.2,** 78; **2.6.1,** 78; **4.1.5,** 78	**2.2,** 72	
CF_2	CF_2	Carbon difluoride	**2.5,** 64; **2.6.3,** 33a	**2.5,** 42a	
CF_3	CF_3	Trifluoromethyl radical	**4.2.3,** 2		
CF_4	CF_4	Carbon tetrafluoride	**2.4,** 19; **2.6.2,** 40; **2.9.2,** 25	**2.9.2,** 23	
CF_3I	CF_3I	Trifluoromethyliodide	**2.4,** 21; **2.6.2,** 42; **2.7.2,** 59		**2.4,** 22
CFN	FCN	Cyanogen fluoride	**2.3,** 5; **2.6.2,** 4	**2.3,** 7	**2.3,** 6
CF_2N_2	NF_2CN	Cyanodifluoroamine		**2.5,** 2b	
CF_3NO	CF_3NO	Trifluoronitrosomethane	**2.5,** 73; **2.6.3,** 38; **2.8,** 113		
CF_3NO_2	CF_3NO_2	Trifluoronitromethane			**2.5,** 44a
CF_3NOS	$F_2S{=}NC(=O)F$	Fluoroformyliminosulfur difluoride	**2.5,** 130		

Alphabetic formula	Chemical formula	Name	Table Nr., Substance Nr.		
			II/14	II/6	II/4
CF_2NP	PF_2CN	Cyanodifluoro-phosphine		**2.5,** 29a	
CF_2O	OCF_2	Carbonyl fluoride	**2.5,** 125; **2.9.3,** 18	**2.5,** 67	**2.5,** 65
CF_4O	CF_3OF	Trifluoromethyl hypofluorite	**2.5,** 77; **2.6.3,** 41; **2.8,** 74	**2.5,** 43a	
CF_4O_2	CF_3OOF	Trifluoromethyl peroxohypofluorite	**2.5,** 78; **2.8,** 75		
CFP	FCP	(Fluoromethylidyne) phosphine	**2.3,** 11; **2.6.2,** 7		
CF_7P	CF_3PF_4	Trifluoromethyl phosphorus tetra-fluoride		**2.4,** 61	
CF_2S	F_2CS	Carbonothioic difluoride	**2.5,** 146; **2.6.3,** 66		
CF_8S	CF_3SF_5	Trifluoromethyl sulfurpentafluoride			**2.4,** 91
CF_6Si	CF_3SiF_3	Hexafluoromethyl-silane		**2.4,** 60a	
CIN	ICN	Cyanogen iodide	**2.3,** 8; **2.7.2,** 56; **2.9.2,** 5	**2.3,** 8b	**2.3,** 9
CKN	KCN	Potassium cyanide	**2.5,** 63		
CN	CN	Carbon mononitride	**2.2,** 46; **2.6.1,** 46; **2.7.1,** 46; **4.1.2,** 46; **4.1.5,** 46	**2.7.1,** 10	
CN_4	NCN_3	Cyanogen azide	**2.5,** 22; **2.7.3,** 17	**2.5,** 14a	
CNO	NCO	NCO-radical	**4.2.1,** 4	**5.2.1,** 1	
CN_2O	NOCN	Nitrosyl cyanide	**2.5,** 16; **2.6.3,** 8; **2.7.3,** 16		
CNS	NCS	NCS-radical		**5.2.1,** 2	
CO	CO	Carbon monoxide	**2.2,** 59; **2.6.1,** 59; **2.9.1,** 59; **4.1.6,** 59; **5: 2.2,** 59; **5: 2.7.1,** 59; **5: 2.9.1,** 59	**2.2,** 54	**2.2,** 27
CO	CO^+	Carbon monoxide (1^+) ion	**2.2,** 45; **4.1.2,** 45; **4.1.5,** 45		
CO_2	OCO	Carbon dioxide	**2.3,** 1; **2.6.2,** 1	**2.9.2,** 1	
COS	OCS	Carbonyl sulfide	**2.3,** 2; **2.6.2,** 2; **2.7.2,** 20; **2.9.2,** 2; **5: 2.3,** 2	**2.3,** 2	**2.3,** 1
COSe	OCSe	Carbonyl selenide	**2.3,** 3	**2.3,** 4	**2.3,** 2
CS	CS	Carbon monosulfide	**2.2,** 60; **2.6.1,** 60; **2.9.1,** 60; **5: 2.2,** 60; **5: 2.7.1,** 60; **5: 2.9.1,** 60	**2.6.1,** 35	**2.2,** 28
CS_2	CS_2	Carbon disulfide		**2.9.2,** 3	
CSSe	SCSe	Carbon sulfoselenide		**2.3,** 5	
CSe	CSe	Carbon selenide	**2.2,** 61; **2.6.1,** 61; **2.9.1,** 61		
CSTe	SCTe	Carbon sulfotelluride			**2.3,** 4
CH	CH	Carbon monohydride	**2.2,** 8; **2.6.1,** 8; **4.1.5,** 8	**2.6.1,** 2a	
$CHBr_3$	$CHBr_3$*)	Bromoform			**2.4,** 19

*) Including deuterated compounds.

Alphabetic formula	Chemical formula	Name	**Table Nr.,** Substance Nr.		
			II/14	II/6	II/4
CHBrO	OC···HBr	Carbon monoxide hydrogen bromide complex	**2.3,** 37; **2.7.2,** 46; **2.9.2,** 14		
$CHCl_3$	$CHCl_3$*)	Chloroform		**2.4,** 16	**2.4,** 18
$CHClF_2$	CHF_2Cl	Chlorodifluoromethane			**2.5,** 55
$CHCl_2F$	$CHCl_2F$	Fluorodichloromethane			**2.5,** 56
CHClO	HCOCl*)	Formyl chloride	**2.5,** 109; **2.6.3,** 55; **2.7.3,** 102; **2.9.3,** 17		
CHClO	OC···HCl	Carbon monoxide hydrogen chloride complex	**2.3,** 36; **2.7.2,** 25		
CHF_3	CF_3H*)	Fluoroform	**2.4,** 18; **2.6.2,** 39; **2.7.2,** 4; **2.9.2,** 24; **5: 2.9.2,** 16	**2.4,** 15	**2.4,** 17
CHFN	HCN···HF	Hydrogen cyanide hydrogen fluoride adduct	**2.3,** 33; **2.6.2,** 19		
CHFO	HCOF	Formyl fluoride	**2.5,** 108a	**2.7.3,** 4a	**2.5,** 59
CHFO	OC···HF*)	Carbon monoxide hydrogen fluoride complex	**2.3,** 35; **5: 2.3,** 35; **5: 2.9.2,** 8		
$CHFO_2$	CO_2···HF*)	Carbon dioxide hydrogen fluoride complex	**2.3,** 37a; **2.6.2,** 19b; **5: 2.3,** 37a; **5: 2.9.2,** 9		
CHFOS	SCO · HF	Carbon sulfoxide hydrogen fluoride complex	**2.3,** 37b; **2.6.2,** 19c; **5: 2.3,** 37b; **5: 2.9.2,** 10		
CHF_2P	CF_2=PH	(Difluoromethylene)-phosphine	**2.5,** 431		
CHKrN	Kr · HCN	Krypton hydrogen cyanide	**2.3,** 45a; **5: 2.3,** 45a; **5: 2.7.2,** 14d		
CHN	HCN	Hydrogen cyanide	**2.3,** 4; **2.6.2,** 3; **2.7.2,** 1	**2.3,** 6	**2.3,** 5
CHN	HNC	Hydrogen isocyanide	**2.3,** 9; **2.6.2,** 6; **2.7.2,** 11; **2.9.2,** 3		
CHNO	HCNO*)	Fulminic acid	**2.3,** 17; **2.9.2,** 8; **2.7.2,** 12	**2.3,** 15	
CHNO	HNCO*)	Hydrogen isocyanate	**2.5,** 5; **2.6.3,** 2; **2.7.3,** 12	**2.5,** 3	**2.5,** 7
CHN_3O_6	$HC(NO_2)_3$*)	Trinitromethane	**2.4,** 87		
CHNS	HNCS*)	Hydrogen isothiocyanate	**2.5,** 6; **2.6.3,** 3; **2.7.3,** 13		**2.5,** 8
CHNSe	HNCSe*)	Isoseleno cyanate	**2.5,** 7; **2.6.3,** 4; **2.7.3,** 14		
CHO	HCO*)	HCO-radical	**4.2.2,** 5	**5.2.2,** 1	
CHP	HCP*)	Methinophosphide	**2.3,** 10; **2.7.2,** 2; **2.9.2,** 6	**2.3,** 9	**2.3,** 10
CH_2	CH_2	Methylene radical	**4.2.2,** 10		
CH_2Br_2	CH_2Br_2*)	Dibromomethane	**2.5,** 95	**2.5,** 57	
CH_2BrF	CH_2BrF	Bromofluoromethane			**2.5,** 52
CH_2Cl_2	CH_2Cl_2*)	Dichloromethane	**2.5,** 94		**2.5,** 49

*) Including deuterated compounds.

Alphabetic formula	Chemical formula	Name	**Table Nr.,** Substance Nr.		
			II/14	II/6	II/4
CH_2ClF	CH_2FCl	Chlorofluoromethane	**2.5,** 97; **2.6.3,** 48; **2.7.3,** 100		**2.5,** 51
CH_2ClP	$CH_2{=}PCl$	Chloromethylene-phosphine	**2.5,** 430		
CH_2F_2	CH_2F_2 *)	Difluoromethane	**2.5,** 93; **2.6.3,** 47; **2.7.3,** 5; **2.9.3,** 12	**2.5,** 56	**2.5,** 48
CH_2FI	CH_2FI	Fluoroiodomethane			**2.5,** 52a
CH_2FNO	$FC({=}O){-}NH_2$	Carbamyl fluoride		**2.5,** 70	
CH_2F_3P	CF_3PH_2 *)	Trifluoromethyl-phosphine		**2.5,** 177	
CH_2N_2	CH_2NN *)	Diazomethane			**2.5,** 57
CH_2N_2	NH_2NC	Isocyanamide	**2.5,** 2a		
CH_2N_2	NH_2CH *)	Cyanamide	**2.5,** 2	**2.5,** 2	**2.5,** 1
CH_2N_2	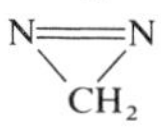	Diazirine	**2.5,** 590; **2.7.3,** 55		**2.5,** 167
CH_2N_4	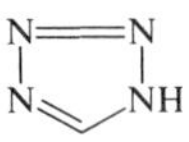 *)	1-H-Tetrazole	**2.5,** 644; **2.6.3,** 317		
CH_2O	$HCOH$ *)	Formaldehyde	**2.5,** 108; **2.6.3,** 54; **2.7.3,** 6; **2.9.3,** 14	**2.5,** 62	**2.5,** 58
CH_2O_2	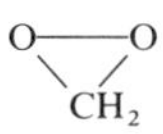 *)	Dioxirane	**2.5,** 488; **2.6.3,** 235		
CH_2O_2	$HCOOH$ *)	Formic acid	**2.5,** 111; **2.6.3,** 56; **2.7.3,** 7; **2.9.3,** 15	**2.5,** 64	**2.5,** 61
CH_2OS	$HCOSH$ *)	Methanethioic S-acid	**2.5,** 112; **2.6.3,** 57		
CH_2OS	$CH_2{=}SO$ *)	Thioformaldehyde oxide	**2.5,** 102; **2.6.3,** 50		
CH_2S	$HC({=}S)H$ *)	Thioformaldehyde	**2.5,** 145; **2.6.3,** 65; **2.9.3,** 20	**2.5,** 73	
CH_2S_2	$HC({=}S)SH$	Methane (dithioic) acid	**2.5,** 150; **2.6.3,** 68		
CH_3ArCl	$Ar \cdot CH_3Cl$	Argon chloro-methane complex	**2.5,** 761		
CH_3AsF_2	CH_3AsF_2	Methyldifluoroarsine			**2.5,** 145
CH_3BF_2	CH_3BF_2 *)	Methyldifluoroborine	**2.5,** 412; **2.6.3,** 198; **2.7.3,** 10	**2.5,** 171	**2.5,** 146
CH_3BO	BH_3CO *)	Borine carbonyl	**2.4,** 51; **2.6.2,** 54; **2.7.2,** 7		**2.4,** 53a
CH_3BS	CH_3BS *)	Methyl thioborane	**2.4,** 40; **2.6.2,** 49; **2.7.2,** 6		
CH_3Br	CH_3Br *)	Bromomethane	**2.4,** 14; **2.6.2,** 35; **2.7.2,** 49; **2.9.2,** 22; **5: 2.9.2,** 14	**2.4,** 13	**2.4,** 15
CH_3BrHg	CH_3HgBr	Methylmercuric bromide	**2.4,** 37		**2.4,** 46
CH_3Br_3Si	CH_3SiBr_3	Methyltribromo-silane		**2.4,** 59	
CH_3Cl	CH_3Cl *)	Chloromethane	**2.4,** 13; **2.6.2,** 34; **2.7.2,** 33; **2.9.2,** 21	**2.4,** 12	**2.4,** 14
CH_3Cl_3Ge	CH_3GeCl_3	Trichloro (methyl)-germane	**2.4,** 76; **2.8,** 67		

*) Including deuterated compounds.

Alphabetic formula	Chemical formula	Name	**Table Nr.**, Substance Nr.		
			II/14	II/6	II/4
CH_3ClHg	CH_3HgCl *)	Methylmercuric chloride	**2.4**, 36	**2.4**, 37	**2.4**, 45
CH_3ClO	CH_3OCl *)	Methyl hypochlorite	**2.5**, 75; **2.6.3**, 39; **2.7.3**, 95; **2.9.3**, 11		**2.5**, 46
CH_3ClO_2S	CH_3SO_2Cl	Methane sulfonyl-chloride	**2.5**, 91; **2.7.3**, 99		
CH_3ClS	CH_3SCl *)	Methane sulfenyl-chloride	**2.5**, 76; **2.6.3**, 40; **2.7.3**, 96	**2.5**, 54	
CH_3Cl_3Si	CH_3SiCl_3	Methyltrichloro-silane	**2.4**, 75	**2.4**, 58	**2.4**, 86
CH_3F	CH_3F *)	Fluoromethane	**2.4**, 12; **2.6.2**, 33; **2.9.2**, 20	**2.4**, 11	**2.4**, 13
CH_3F_3Ge	CH_3GeF_3	Methyltrifluoro-germane	**2.4**, 77; **2.8**, 66		**2.4**, 88
CH_3F_2N	CH_3NF_2 *)	Methyldifluoroamine		**2.5**, 46	
CH_3F_2OP	CH_3OPF_2 *)	Methoxydifluoro-phosphine	**2.5**, 426; **2.8**, 90	**2.5**, 177a	
CH_3F_2OP	CH_3POF_2	Methylphosphine difluoride	**2.5**, 427; **2.6.3**, 212; **2.8**, 125		
CH_3FOS_2	CH_3SO_2F	Methane sulfonyl		**2.5**, 54a	
CH_3FO_3S	CH_3OSO_2F	Methylfluorosulfate	**2.5**, 92; **2.6.3**, 46		
CH_3F_3OSi	CH_3OSiF_3	Trifluoromethoxy-silane	**2.5**, 386		
CH_3F_2P	CH_3PF_2	Methyldifluoro-phosphine	**2.5**, 424; **2.6.3**, 210; **2.8**, 124	**2.5**, 176a	
CH_3F_3Si	CH_3SiF_3 *)	Methyltrifluorosilane		**2.4**, 57a	**2.4**, 85
CH_3GeN	GeH_3CN	Germyl cyanide		**2.4**, 44	
CH_3GeNO	GeH_3NCO	Germyl isocyanate		**2.5**, 169	
CH_3GeNS	GeH_3NCS *)	Isothio cyanatogermane	**2.4**, 46		
CH_3HgI	CH_3HgI *)	Methylmercuric iodide	**2.4**, 38	**2.4**, 38	**2.4**, 47
CH_3I	CH_3I *)	Iodomethane	**2.4**, 15; **2.6.2**, 36; **2.7.2**, 58; **2.9.2**, 23; **5: 2.9.2**, 15	**2.4**, 14	**2.4**, 16
CH_3N	CH_2NH *)	Methylene imine	**2.5**, 103; **2.6.3**, 53; **2.7.3**, 25	**2.5**, 58c	
CH_3N_3	CH_3N_3	Methyl azide		**2.5**, 50	**2.5**, 43b
CH_3NO	CH_3NO *)	Nitrosomethane	**2.5**, 72; **2.6.3**, 37; **2.8**, 112	**2.5**, 51	
CH_3NO	$HCONH_2$ *)	Formamide	**2.5**, 110	**2.5**, 63	**2.5**, 60
CH_3NO	H_2CNOH *)	Formaldoxime	**2.5**, 121		**2.5**, 62
CH_3NO_2	CH_3ONO *)	Methyl nitrite	**2.5**, 82; **2.7.3**, 21; **2.8**, 76	**2.5**, 52	**2.5**, 44b
CH_3NO_2	CH_3NO_2 *)	Nitromethane	**2.5**, 74; **2.8**, 114; **2.9.3**, 10	**2.5**, 51a	**2.5**, 44
CH_3NO_3	CH_3ONO_2 *)	Methyl nitrate		**2.5**, 53	**2.5**, 45
CH_3NOS	CH_3NSO	Methylthionylamine			**2.5**, 43a
CH_3NOSi	SiH_3NCO *)	Silyl isocyanate	**2.4**, 44; **2.7.2**, 18	**2.4**, 43	
CH_3NS	$HC(=S)NH_2$ *)	Methanethioamide	**2.5**, 149; **2.6.3**, 67; **2.7.3**, 28		
CH_3NSSi	SiH_3NCS *)	Isothiocyanatosilane	**2.4**, 45; **2.6.2**, 52; **2.7.2**, 19		**2.4**, 53

*) Including deuterated compounds.

Alphabetic formula	Chemical formula	Name	Table Nr., Substance Nr.		
			II/14	II/6	II/4
CH_3NSi	SiH_3CN *)	Silyl cyanide	**2.4**, 43; **2.7.2**, 17		**2.4**, 52
CH_3O	CH_3O	Hydroxymethyl radical	**4.2.3**, 1		
CH_4	CH_4 *)	Methane	**2.4**, 11; **2.6.2**, 32; **2.9.2**, 19	**2.6.2**, 12	
CH_4ClN	CH_3NHCl *)	N-Chloromethane-amine	**2.5**, 69; **2.7.3**, 94; **2.8**, 109		
CH_4Cl_2Si	CH_3SiHCl_2	Dichloromethylsilane	**2.5**, 384; **2.7.3**, 131; **2.8**, 61		
CH_4F_2Si	CH_3SiHF_2 *)	Methyldifluorosilane			**2.5**, 140
CH_4N_2	$CH_3{-}N{=}NH$ *)	Methyldiazine	**2.5**, 105; **2.6.3**, 51	**2.5**, 60	
CH_4N_2O	$OC(NH_2)_2$	Urea	**2.5**, 129; **2.6.3**, 63; **2.7.3**, 27		
CH_4O	CH_3OH *)	Methanol	**2.5**, 65; **2.6.3**, 33; **2.7.3**, 4; **2.8**, 73; **2.9.3**, 9	**2.5**, 43	**2.5**, 38
CH_4S	CH_3SH *)	Methanethiol	**2.5**, 66; **2.8**, 93	**2.8**, 86	**2.5**, 39
CH_4Se	CH_3SeH *)	Methaneselenol	**2.5**, 67; **2.6.3**, 34; **2.8**, 94	**2.5**, 45	
CH_5As	CH_3AsH_2	Methylarsine		**2.8**, 72	
CH_5BrSi	SiH_3CH_2Br *)	(Bromomethyl)silane	**2.5**, 380; **2.7.3**, 155		
CH_5ClGe	CH_3GeH_2Cl *)	Chloromethyl-germane	**2.5**, 411		
CH_5ClGe	GeH_3CH_2Cl *)	(Chloromethyl)-germane	**2.5**, 407; **2.7.3**, 134; **2.8**, 69		
CH_5ClSi	CH_3SiH_2Cl *)	Methylchlorosilane	**2.5**, 383; **2.7.3**, 130	**2.5**, 167c	
CH_5ClSi	SiH_3CH_2Cl *)	Chloromethylsilane			**2.5**, 138
CH_5FGe	CH_3GeH_2F *)	Fluoromethyl-germane	**2.5**, 410; **2.6.3**, 197; **2.8**, 71		
CH_5FGe	GeH_3CH_2F	(Fluoromethyl)-germane	**2.5**, 406; **2.6.3**, 194; **2.8**, 68		
CH_5FSi	CH_3SiH_2F *)	Methylfluorosilane			**2.5**, 139
CH_5N	CH_3NH_2 *)	Methylamine	**2.5**, 68	**2.5**, 45a	**2.5**, 40
CH_5NO	CH_3NHOH *)	N-Hydroxy-methanamine	**2.5**, 71; **2.6.3**, 36; **2.7.3**, 20; **2.8**, 111		
CH_5NO	CH_3ONH_2 *)	O-Methylhydroxyl-amine	**2.5**, 84; **2.7.3**, 22; **2.8**, 77		
CH_5P	CH_3PH_2	Methylphosphine			**2.5**, 144
CH_6BF_2P	$CH_3PF_2BH_3$ *)	Difluoromethyl-phosphine (P—B)borane	**2.5**, 433; **2.6.3**, 215; **2.8**, 123		
CH_6Ge	CH_3GeH_3 *)	Methylgermane		**2.6.2**, 34	**2.4**, 87
CH_6N_2	$CH_3NH{-}NH_2$	Methylhydrazine		**2.5**, 55	
CH_6OSi	CH_3OSiH_3 *)	Methoxysilane	**2.5**, 385; **2.6.3**, 178; **2.8**, 87	**2.6.3**, 80b	
CH_6SSi	CH_3SSiH_3 *)	(Methylthio)silane	**2.5**, 389; **2.6.3**, 180; **2.8**, 101		
CH_6Si	CH_3SiH_3 *)	Methylsilane	**2.9.2**, 34a; 5: **2.9.2**, 17	**2.4**, 57	**2.4**, 84
CH_6Sn	CH_3SnH_3	Methylstannane			**2.4**, 90
CH_7B_5	CB_5H_7	2-Carbahexa-borane (7)	**2.5**, 420; **2.6.3**, 206	**2.5**, 174a	
CH_7PSi	$CH_3SiH_2PH_2$	(Methylsilyl)-phosphine	**2.5**, 402		

*) Including deuterated compounds.

Alphabetic formula	Chemical formula	Name	**Table Nr.,** Substance Nr.		
			II/14	II/6	II/4
CH_8B_2	$CH_3B_2H_5$ *)	1-Methyldiborane(6)	**2.5,** 417; **2.6.3,** 203; **2.8,** 129		
CH_8BP	$CH_3PH_2BH_3$ *)	Methylphosphine-borane	**2.5,** 432; **2.8,** 122	**2.5,** 176	
CH_8Si	$(SiH_3)_2CH_2$ *)	Disilylmethane	**2.5,** 392; **2.6.3,** 183; **2.8,** 62		
$CH_{11}B_5$	$CH_3B_5H_8$	1-Methylpenta-borane		**2.4,** 78	
C_2					
C_2BrCl	$ClC{\equiv}CBr$	Chlorobromo-acetylene		**2.3,** 21	
C_2ClF	$ClC{\equiv}CF$	Chlorofluoroethyne	**2.3,** 27a; **5: 2.3,** 27a		
C_2ClF_3	$CF_2{=}CFCl$	Chlorotrifluoro-ethylene		**2.5,** 122	
$C_2Cl_3F_3$	CCl_3CF_3	2,2,2-Trifluoro-1,1,1-trichloroethane		**2.4,** 56	
C_2ClI	$ClC{\equiv}CI$	Chloroiodoacetylene		**2.3,** 22	
C_2Cl_3N	CCl_3CN	Trichloroacetonitrile			**2.4,** 51
C_2F_3N	CF_3CN	Trifluoroacetonitrile	**2.4,** 42; **2.6.2,** 51	**2.4,** 42	**2.4,** 50
C_2N_2O	NCNCO	Cyanogen isocyanate	**2.5,** 9; **2.6.3,** 5		
C_2NP	$N{\equiv}C{-}C{\equiv}P$	(Cyanomethylidyne)-phosphine	**2.3,** 15; **2.6.2,** 10		
C_2N_2S	NCNCS	Sulfurdicyanide	**2.5,** 10		**2.5,** 24
C_2O_2	$(CO)_2$	Carbon monoxide dimer	**2.5,** 763		
C_2H	C_2H	Ethynyl radical	**4.2.1,** 1		
C_2HBF_2	$HC{\equiv}C{-}BF_2$ *)	Ethynyldifluoro-borane		**2.5,** 172	
C_2HBr	$HC{\equiv}CBr$ *)	Bromoacetylene	**2.3,** 26; **2.6.2,** 15; **2.7.2,** 44; **2.9.2,** 11	**2.3,** 19	
C_2HCl	$HC{\equiv}CCl$ *)	Chloroacetylene	**2.3,** 25; **2.7.2,** 23; **5: 2.9.2,** 7	**2.3,** 18	**2.3,** 13
C_2HClF_2	$CF_2{=}CHCl$	1,1-Difluoro-2-chloroethylene		**2.5,** 121	**2.5,** 104
C_2HF	$HC{\equiv}CF$ *)	Fluoroacetylene	**2.3,** 24; **2.6.2,** 14; **2.9.2,** 10	**2.7.2,** 1	**2.3,** 12
C_2HF_3	$CF_2{=}CHF$	1,1,2-Trifluoroethane	**2.5,** 304; **2.6.3,** 136		**2.5,** 103
C_2HF_5	CF_3CHF_2	Pentafluoroethane		**2.5,** 89	
C_2HFN_2	NCCN···HF	Cyanogen-hydrogen fluoride complex	**2.3,** 34a; **5: 2.3,** 34a; **5: 2.9.2,** 11		
C_2HF_3O	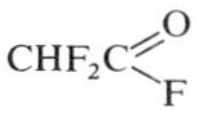 *)	Difluoroacetyl fluoride	**2.5,** 211; **2.6.3,** 104		
C_2HF_3O	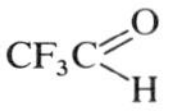 *)	Trifluoroacet-aldehyde		**2.8,** 8	**2.5,** 83b
C_2HN_3	$NN{=}CHCN$ *)	Diazoacetonitrile		**2.5,** 59	**2.5,** 57a
C_2HNO_2S	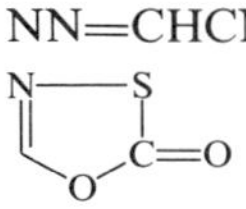	1,3,4-Oxathiazol-2-one	**2.5,** 641		
C_2H_2	$HC{\equiv}CH$ *)	Acetylene	**2.3,** 23; **2.6.2,** 13; **2.9.2,** 9	**2.9.2,** 8	**2.3,** 11

*) Including deuterated compounds.

Alphabetic formula	Chemical formula	Name	Table Nr., Substance Nr. II/14	II/6	II/4
C_2H_2Ar	$Ar \cdot HC{\equiv}CH$ *)	Argon ethyne complex	**2.5**, 762; **2.6.3**, 369; **2.7.3**, 9		
C_2H_2BrN	CH_2BrCN	Bromoacetonitrile	**2.5**, 98; **2.7.3**, 149	**2.5**, 58	
$C_2H_2Cl_2$	$CHCl{=}CHCl$	1,2-Dichloroethylene		**2.5**, 106	
$C_2H_2Cl_2$	$CH_2{=}CCl_2$	1,1-Dichloroethylene			**2.5**, 101
C_2H_2ClF	$CHF{=}CHCl$	1-Chloro-2-fluoroethylene			**2.5**, 107
C_2H_2ClF	$CH_2{=}CFCl$	1-Fluoro-1-chloroethylene		**2.5**, 118	**2.5**, 102
C_2H_2ClFO	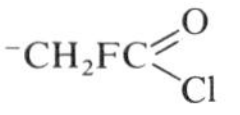	Fluoroacetyl chloride	**2.5**, 202; **2.7.3**, 114		
C_2H_2ClN	CH_2ClCN *)	Chloroacetonitrile			**2.5**, 54
$C_2H_2Cl_2O$		Chloroacetyl chloride	**2.5**, 204; **2.7.3**, 116		
$C_2H_2F_2$	$CH_2{=}CF_2$	1,1-Difluoroethylene	**2.5**, 295; **2.6.3**, 132; **2.9.3**, 23	**2.5**, 117	**2.5**, 100
$C_2H_2F_4$	CHF_2CHF_2	1,1,2,2-Tetrafluoroethane		**2.5**, 88	
$C_2H_2F_2O$	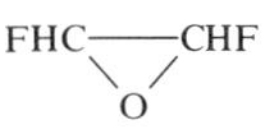 *)	2,3-Difluorooxirane	**2.5**, 482; **2.6.3**, 231		
$C_2H_2F_2O$	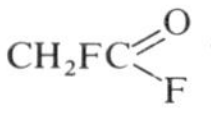 *)	Fluoroacetyl fluoride	**2.5**, 201 a	**2.5**, 95	**2.5**, 85
$C_2H_2F_2O_3$	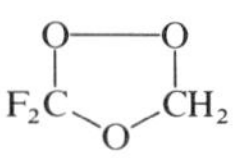	3,3-Difluoro-1,2,4-trioxolane	**2.5**, 517; **2.6.3**, 251		
C_2H_2IN	CH_2ICN	Iodoacetonitrile	**2.5**, 99; **2.7.3**, 158		
$C_2H_2N_2$	$HCN \cdots HCN$	Hydrogen cyanide dimer	**2.3**, 34; **2.7.2**, 14a		
$C_2H_2N_2$	$CH_2{=}N{-}CN$ *)	N-Cyanoformimine	**2.5**, 104		
$C_2H_2N_2O$	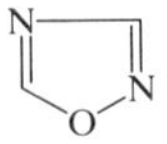	1,2,4-Oxadiazole		**2.5**, 261	
$C_2H_2N_2O$	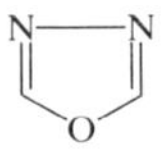 *)	1,3,4-Oxadiazole		**2.5**, 260 a	**2.5**, 174 a
$C_2H_2N_2O$	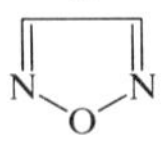 *)	1,2,5-Oxadiazole			**2.5**, 175
$C_2H_2N_2S$	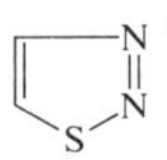 *)	1,2,3-Thiadiazole	**2.5**, 640; **2.6.3**, 315		
$C_2H_2N_2S$	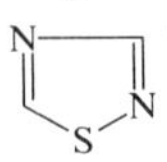 *)	1,2,4-Thiadiazole	**2.5**, 639		
$C_2H_2N_2S$	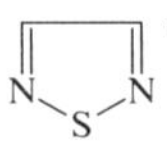 *)	1,2,5-Thiadiazole	**2.5**, 638; **2.6.3**, 314; **2.7.3**, 68		**2.5**, 174
$C_2H_2N_2S$	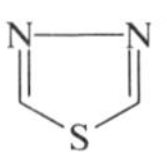 *)	1,3,4-Thiadiazole		**2.5**, 265	**2.5**, 173
$C_2H_2N_2Se$	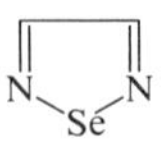	1,2,5-Selenadiazole		**2.5**, 267	

*) Including deuterated compounds.

Alphabetic formula	Chemical formula	Name	Table Nr., Substance Nr. II/14	II/6	II/4
$C_2H_2N_2Se$	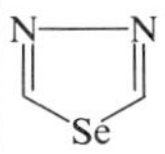	1,3,4-Selenadiazole		**2.5,** 266	
C_2H_2O	$CH_2{=}C{=}O$ *)	Ketene	**2.5,** 299; **2.6.3,** 133; **2.7.3,** 8; **2.9.3,** 25	**2.5,** 119	**2.5,** 119
$C_2H_2O_2$	$HC(=O){-}C(=O){-}H$ *)	Glyoxal	**2.5,** 212	**2.5,** 97b	
$C_2H_2O_3$	$HC(=O){-}C(=O)OH$ *)	Glyoxylic acid	**2.5,** 213; **2.6.3,** 105	**2.5,** 97c	
$C_2H_2O_3$	$O{=}C(H){-}O{-}C(=O){-}H$ *)	Formic anhydride	**2.5,** 123; **2.6.3,** 62		
C_2H_2S	$CH_2{=}C{=}S$ *)	Ethenethione	**2.5,** 301; **2.6.3,** 134		
C_2H_2Se	$CH_2{=}C{=}Se$	Etheneselenide	**2.5,** 303		
$C_2H_3BF_2$	$CH_2{=}CHBF_2$ *)	Ethenyldifluoro-borane	**2.5,** 414; **2.6.3,** 200; **2.8,** 130		
C_2H_3Br	$CH_2{=}CHBr$ *)	Vinyl bromide	**2.5,** 283; **2.7.3,** 152	**2.5,** 110	**2.5,** 95
$C_2H_3Br_3$	CH_3CBr_3	1,1,1-Tribromo-ethane		**2.4,** 55a	
$C_2H_3BrN_2$	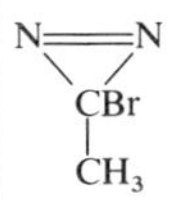	Methylbromo-diazirine		**2.5,** 237	
C_2H_3BrO	$CH_3C(=O){-}Br$	Acetyl bromide		**2.5,** 90b	**2.5,** 81
$C_2H_3BrO_2$	$CH_2BrC(=O)OH$ *)	Bromoacetic acid	**2.5,** 201; **2.7.3,** 151		
C_2H_3Cl	$CH_2{=}CHCl$ *)	Vinyl chloride	**2.5,** 282; **2.7.3,** 118	**2.5,** 109	**2.5,** 94
C_2H_3Cl	$HC{\equiv}CH \cdot HCl$	Ethyne hydrogen chloride adduct	**2.5,** 762a; **2.7.3,** 144b		
$C_2H_3Cl_3$	CH_3CCl_3*)	1,1,1-Trichloro-ethane	**2.4,** 74; **2.8,** 3	**2.4,** 55	**2.4,** 83
$C_2H_3ClF_2$	CH_3CF_2Cl	1-Chloro-1,1-difluoroethane		**2.5,** 81	
$C_2H_3ClN_2$	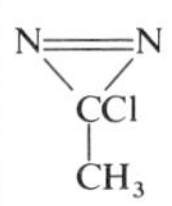	Methylchloro-diazirine		**2.5,** 236	
C_2H_3ClO	$CH_3C(=O){-}Cl$ *)	Acetyl chloride		**2.5,** 90a	**2.5,** 80
C_2H_3ClO	$CH_2ClC(=O)H$	Chloroacetaldehyde	**2.5,** 203; **2.7.3,** 115		
$C_2H_3ClO_2$	$ClC(=O)OCH_3$ *)	Methylcarbono-chloridate	**2.5,** 131; **2.7.3,** 105; **2.8,** 82		
$C_2H_3ClO_2$	$CH_2ClC(=O)OH$ *)	Chloroacetic acid	**2.5,** 200; **2.7.3,** 113; **2.6.3,** 99		
C_2H_3ClOS	$ClC(=O)SCH_3$	Methyl thiolcyanoformate	**2.5,** 132; **2.7.3,** 106; **2.8,** 98		
C_2H_3F	$HC{\equiv}CH \cdots HF$ *)	Acetylene-hydrogen fluoride complex	**5: 2.9.3,** 2		
C_2H_3F	$CH_2{=}CHF$*)	Vinyl fluoride	**2.5,** 281		**2.5,** 93
$C_2H_3F_3$	CH_3CF_3*)	1,1,1-Trifluoroethane	**2.4,** 73; **2.6.2,** 58		**2.4,** 82

*) Including deuterated compounds.

Alphabetic formula	Chemical formula	Name	Table Nr., Substance Nr.		
			II/14	II/6	II/4
$C_2H_3F_3Hg$	CF_3HgCH_3	Methyl (trifluoro-methyl) mercury	**2.4,** 39		
C_2H_3FO	$CH_3C(=O)F$ *)	Acetyl fluoride			**2.5,** 79
$C_2H_3FO_2$	$CH_2FC(=O)OH$ *)	Fluoroacetic acid	**2.5,** 199	**2.5,** 96	**2.5,** 84
$C_2H_3FO_2$	$HC(=O)OCH_2F$ *)	Fluoromethyl-formate	**2.5,** 120; **2.6.3,** 61		
$C_2H_3FO_2$	$FC(=O)—OCH_3$	Methylfluoroformate		**2.5,** 71	
$C_2H_3FO_3$	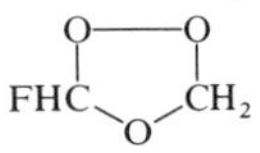	3-Fluoro-1,2,4-trioxolane	**2.5,** 516; **2.6.3,** 250		
$C_2H_3F_2O_2$	$CHF_2C(=O)OH$ *)	Difluoroacetyl acid	**2.5,** 210		
C_2H_3FOS	$FC(=O)SCH_3$	Methyl thiofluoro-formate	**2.5,** 131a; **2.6.3,** 63a; **2.8,** 97a		
$C_2H_3F_2P$	$CH_2{=}CHPF_2$	Ethenyl difluoro-phosphine	**2.5,** 428; **2.6.3,** 213		
$C_2H_3F_3Si$	$CH_2{=}CHSiF_3$	Vinyl trifluorosilane		**2.5,** 167b	
C_2H_3I	$CH_2{=}CHI$	Vinyl iodide		**2.5,** 111	**2.5,** 96
C_2H_3IO	$CH_3C(=O)I$	Acetyl iodide		**2.5,** 91	
C_2H_3N	CH_3CN *)	Methyl cyanide	**2.4,** 16; **2.6.2,** 37; **2.7.2,** 16; **2.9.2,** 32	**2.4,** 39	**2.4,** 48
C_2H_3N	CH_3NC *)	Methyl isocyanide	**2.4,** 17; **2.6.2,** 38	**2.4,** 40	**2.4,** 49
C_2H_3N	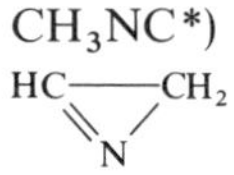	2H-Azirine	**2.5,** 589; **2.6.3,** 290; **2.7.3,** 54		
$C_2H_3N_3$	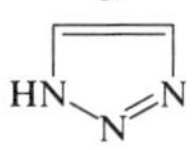 *)	1,2,3-Triazole	**2.5,** 642; **2.7.3,** 69	**2.5,** 268	
$C_2H_3N_3$	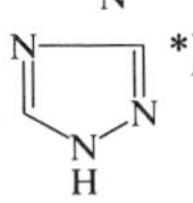 *)	1,2,4-Triazole	**2.5,** 643; **2.6.3,** 316; **2.7.3,** 70	**2.5,** 267a	
$C_2H_3N_3$	$CH_2{=}CHN_3$	Azidoethene	**2.5,** 288; **2.6.3,** 127		
C_2H_3NO	CH_3CNO	Acetonitrile-N-oxide		**2.4,** 41	
C_2H_3NO	CH_2OHCN *)	Hydroxyacetonitrile	**2.5,** 100; **2.7.3,** 23	**2.5,** 58a	
C_2H_3NO	CH_3NCO	Methyl isocyanate		**2.5,** 49	**2.5,** 43
$C_2H_3NO_2$	$HN(CHO)_2$ *)	N-Formylformamide	**2.5,** 122; **2.7.3,** 26		
$C_2H_3NO_2$	$CH_2{=}CHNO_2$ *)	Nitroethylene	**2.5,** 286	**2.5,** 113	**2.5,** 98
$C_2H_3N_3O_2$	N_3COOCH_3 *)	Methylazidoformate	**2.5,** 119; **2.8,** 83	**2.5,** 72a	
C_2H_3NS	CH_3NCS	Methyl isothiocyanate		**2.5,** 48	**2.5,** 42
C_2H_3NS	CH_3SCN *)	Methyl thiocyanate	**2.5,** 80; **2.8,** 95	**2.5,** 47	**2.5,** 41
C_2H_3NSe	CH_3SeCN *)	Selenocyanato-methane	**2.5,** 81; **2.6.3,** 42; **2.8,** 96		
C_2H_3P	$CH_2{=}PH$ *)	Methylenephosphine	**2.5,** 429; **2.6.3,** 214		
C_2H_3P	$CH_3C{\equiv}P$ *)	Ethylidynephosphine	**2.4,** 41; **2.6.2,** 50		
C_2H_4	$CH_2{=}CH_2$ *)	Ethene	**2.5,** 280a; **2.6.3,** 125b		
$C_2H_4Cl_2$	CH_3CHCl_2	1,1-Dichloroethane			**2.5,** 75

*) Including deuterated compounds.

Alphabetic formula	Chemical formula	Name	**Table Nr.**, Substance Nr.		
			II/14	II/6	II/4
C_2H_4ClF	CH_3CHFCl	1-Chloro-1-fluoro-methane	**2.5**, 170; **2.6.3**, 83; **2.7.3**, 112; **2.8**, 13		
C_2H_4ClF	CH_2FCH_2Cl	1-Chloro-2-fluoro-ethane		**2.5**, 83	**2.5**, 76
C_2H_4ClN	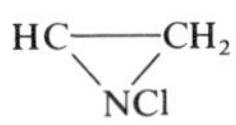 *)	1-Chloroaziridine		**2.5**, 203	
$C_2H_4F_2$	CH_3CHF_2	1,1-Difluoroethane	**2.5**, 169; **2.8**, 12		**2.5**, 74
$C_2H_4F_2$	CH_2FCH_2F	1,2-Difluoroethane	**2.5**, 175a	**2.5**, 82	
C_2H_4FN	$CH_3CN \cdots HF$	Acetonitrile-hydrogen fluoride complex	**2.4**, 106; **5**: **2.4**, 106; **5**: **2.9.2**, 18		
$C_2H_4F_3N$	$CF_3CH_2NH_2$ *)	1,1,1-Trifluoroethyl-amine		**2.5**, 89a	
C_2H_4FNO	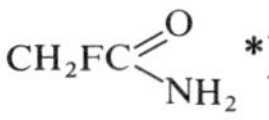 *)	2-Fluoroacetamide	**2.5**, 205; **2.6.3**, 100; **2.7.3**, 35		
$C_2H_4F_2O$	CHF_2CH_2OH	2,2-Difluoroethanol	**2.5**, 184; **2.6.3**, 94		
C_2H_4Ge	$HC{\equiv}CGeH_3$	Germylacetylene		**2.4**, 47	
$C_2H_4N_2$	NH_2CH_2CN *)	Amino acetonitrile	**2.5**, 101; **2.6.3**, 49; **2.7.3**, 24	**2.5**, 58b	
$C_2H_4N_2$	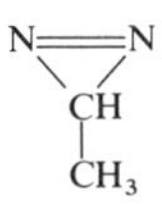 *)	Methyldiazirine		**2.5**, 235	
$C_2H_4N_2$	CH_3NHCN *)	Imino-acetonitrile	**2.5**, 70; **2.6.3**, 35; **2.8**, 110		
C_2H_4O	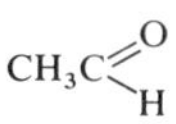 *)	Acetaldehyde	**2.5**, 185; **2.6.3**, 95; **2.8**, 17	**2.5**, 90	**2.5**, 78
C_2H_4O	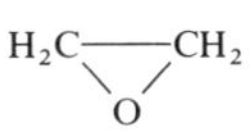 *)	Ethylene oxide	**2.5**, 478; **2.6.3**, 228; **2.7.3**, 80; **2.9.3**, 32		**2.5**, 152
C_2H_4O	$CH_2{=}CHOH$	Ethenol	**2.5**, 290; **2.6.3**, 129		
$C_2H_4O_2$	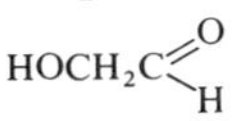 *)	Glycolaldehyde	**2.5**, 207; **2.6.3**, 102	**2.5**, 97	
$C_2H_4O_2$	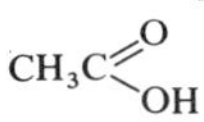 *)	Acetic acid	**2.5**, 188; **2.8**, 19a	**2.5**, 91a	**2.5**, 83
$C_2H_4O_2$	$HCOOCH_3$ *)	Methyl formate	**2.5**, 113; **2.8**, 81	**2.9.3**, 15	**2.5**, 63
$C_2H_4O_3$	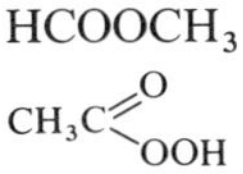	Hydroxyacetic acid	**2.5**, 189; **2.6.3**, 97; **2.8**, 20		
$C_2H_4O_3$	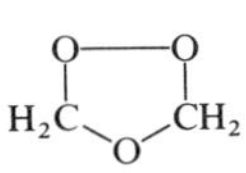	1,2,4-Trioxolane	**2.5**, 515a; **2.6.3**, 249		
$C_2H_4O_3$	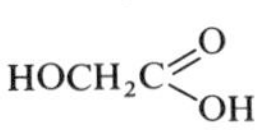	Glycolic acid	**2.5**, 207a; **2.6.3**, 102a	**2.5**, 97a	
C_2H_4OS	$HCOSCH_3$	S-Methyl methane-thioate	**2.5**, 114; **2.6.3**, 58; **2.8**, 97		
C_2H_4OS	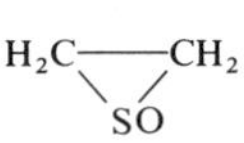	Ethylene episulfoxide		**2.5**, 199	
C_2H_4OS	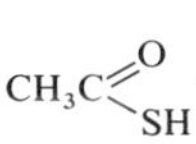 *)	Ethanethioic-S-acid	**2.5**, 190; **2.6.3**, 98; **2.8**, 21		
C_2H_4OS	$CH_3CH{=}SO$	Ethanethial S-oxide	**2.5**, 173; **2.6.3**, 85; **2.8**, 16		

*) Including deuterated compounds.

Alphabetic formula	Chemical formula	Name	Table Nr., Substance Nr.		
			II/14	II/6	II/4
$C_2H_4OS_2$	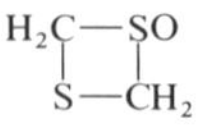	1,3-Dithiotane-1-oxide	**2.5,** 508; **2.6.3,** 246		
$C_2H_4O_2S$	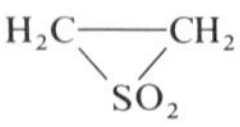	Ethylene episulfone		**2.5,** 200	
C_2H_4S	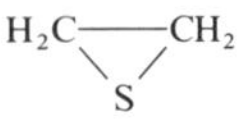	Ethylene sulfide	**2.5,** 489	**2.5,** 198	**2.5,** 154
C_2H_4S	$CH_2{=}CHSH$ *)	Ethenethiol	**2.5,** 294; **2.6.3,** 131		
C_2H_4S	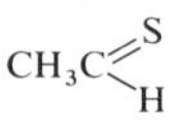 *)	Ethenethial	**2.5,** 217; **2.6.3,** 109; **2.8,** 25		
C_2H_4Se	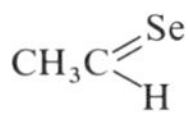	Ethaneselenal	**2.5,** 218; **2.8,** 26		
C_2H_4Si	$SiH_3C{\equiv}CH$ *)	Silylacetylene	**2.4,** 50		**2.4,** 61
$C_2H_5B_4Cl$	$C_2B_4H_5Cl$	2-Chloro-1,6-dicarba-hexaborane (6)		**2.5,** 175a	
$C_2H_5BF_2$	$CH_3CH_2BF_2$	Ethyldifluoroborane	**2.5,** 413; **2.6.3,** 199; **2.8,** 128		
$C_2H_5BO_2$	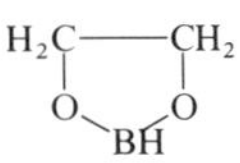	1,3,2-Dioxaborolane		**2.5,** 224	
C_2H_5Br	CH_3CH_2Br *)	Ethyl bromide	**2.5,** 153		**2.5,** 70
C_2H_5BrO	CH_3OCH_2Br *)	Bromomethoxy-methane	**2.5,** 87; **2.7.3,** 148		
C_2H_5BrO	CH_2BrCH_2OH *)	2-Bromoethanol		**2.5,** 86	
C_2H_5Cl	$CH_2{=}CH_2 \cdot HCl$	Ethene hydrogen chloride adduct	**2.5,** 766; **2.7.3,** 145a		
C_2H_5Cl	CH_3CH_2Cl *)	Ethyl chloride	**2.5,** 152; **2.7.3,** 109; **2.9.3,** 13		**2.5,** 69
C_2H_5ClO	CH_3OCH_2Cl	Monochloro-dimethylether	**2.5,** 86; **2.7.3,** 98; **2.8,** 79		
C_2H_5ClO	CH_3CH_2OCl	Ethyl hypochloride	**2.5,** 166; **2.6.3,** 81; **2.7.3,** 110		
C_2H_5ClO	CH_2ClCH_2OH *)	2-Chloroethanol		**2.5,** 85	
C_2H_5F	CH_3CH_2F *)	Ethyl fluoride	**2.5,** 151; **2.8,** 4	**2.5,** 74	**2.5,** 68
C_2H_5F	$C_2H_4 \cdots HF$ *)	Ethylene-hydrogen fluoride complex	**5: 2.9.3,** 3		
C_2H_5FO	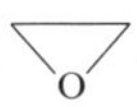 ···HF	Oxirane hydrogen fluoride adduct	**2.5,** 764a; **2.6.3,** 369b		
C_2H_5FO	CH_2FCH_2OH *)	2-Fluoroethanol		**2.5,** 84	
C_2H_5FO	CH_3OCH_2F	Fluoromethoxy-methane	**2.5,** 85; **2.6.3,** 43; **2.8,** 78		
C_2H_5I	CH_3CH_2I	Ethyl iodide	**2.5,** 154; **2.7.3,** 159; **2.8,** 5		**2.5,** 71
C_2H_5N	$CH_2{=}CHNH_2$	Ethenamine	**2.5,** 287; **2.6.3,** 126; **2.7.3,** 37		
C_2H_5N	$CH_3CH{=}NH$	Ethanimine	**2.5,** 171; **2.6.3,** 84; **2.7.3,** 31; **2.8,** 14		
C_2H_5N	$CH_3N{=}CH_2$ *)	N-Methyl methylenimine	**2.5,** 90		**2.5,** 47
C_2H_5N	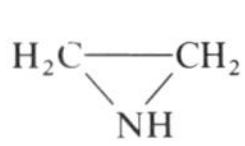 *)	Ethylenamine		**2.5,** 201	**2.5,** 156

*) Including deuterated compounds.

Alphabetic formula	Chemical formula	Name	Table Nr., Substance Nr.		
			II/14	II/6	II/4
C_2H_5NO	$CH_3C(=NOH)H$ *)	Acetaldoxime	**2.5**, 216; **2.6.3**, 108; **2.8**, 24	**2.5**, 94	
C_2H_5NO	$HCONH(CH_3)$	N-Methylformamide	**2.5**, 115; **2.6.3**, 59; **2.8**, 116		
$C_2H_5NO_2$	$CH_3CH_2NO_2$	Nitroethane		**2.5**, 79	
$C_2H_5NO_2$	CH_3CH_2ONO *)	Ethyl nitrite	**2.5**, 163; **2.6.3**, 78		
$C_2H_5NO_2$	$NH_2CH_2C(=O)OH$	Glycine	**2.5**, 206; **2.6.3**, 101		
$C_2H_5NO_3$	$CH_3CH_2ONO_2$	Ethyl nitrate	**2.5**, 164; **2.6.3**, 79	**2.5**, 79b	
C_2H_5P	H_2C—CH_2 (ring, PH) *)	Ethylene phosphine		**2.5**, 205	
C_2H_6	CH_3CH_3 *)	Ethane	**2.4**, 72; **2.6.2**, 57; **2.8**, 1	**2.4**, 54	
$C_2H_6B_4$	$C_2B_4H_6$	2,3-Dicarbahexa-borane (6)		**2.5**, 175	
C_2H_6BN	$CH_3NC \cdot BH_3$ *)	Methyl isocyanide borane adduct	**2.4**, 107; **2.6.2**, 72; **2.8**, 134		
$C_2H_6Cl_2Si$	$(CH_3)_2SiCl_2$	Dimethyldichloro-silane	**2.5**, 393; **2.7.3**, 132		
C_2H_6FN	$CH_2NH_2CH_2F$	2-Fluoroethylamine	**2.5**, 177; **2.6.3**, 89; **2.7.3**, 32		
$C_2H_6F_2NP$	$(CH_3)_2NPF_2$ *)	Dimethylphosphor-amidous difluoride	**2.5**, 439		
C_2H_6Ge	$GeH_3CH{=}CH_2$ *)	Ethenylgermane	**2.5**, 409; **2.6.3**, 196; **2.7.3**, 146a; **2.8**, 70		
$C_2H_6N_2$	$CH_3N{=}NCH_3$ *)	Azomethane	**2.5**, 106; **2.6.3**, 52; **2.8**, 115		
$C_2H_6N_2O$	CH_3CH_3NNO *)	Dimethylnitrosamine	**2.5**, 440; **2.7.3**, 43	**2.5**, 164a	
C_2H_6O	CH_3CH_2OH *)	Ethyl alcohol	**2.5**, 156; **2.6.3**, 70; **2.8**, 7	**2.5**, 77	**2.5**, 73
C_2H_6O	$(CH_3)_2O$ *)	Dimethyl ether	**2.5**, 434; **2.8**, 89	**2.5**, 161	**2.5**, 128
$C_2H_6O_2$	CH_2OHCH_2OH *)	1,2-Ethanediol	**2.5**, 174; **2.6.3**, 86		
C_2H_6OS	$(CH_3)_2SO$ *)	Dimethyl sulfoxide		**2.5**, 165	**2.5**, 133
C_2H_6OS	$HSCH_2CH_2OH$ *)	2-Mercaptoethanol	**2.5**, 175; **2.6.3**, 87		
$C_2H_6O_2S$	$(CH_3)_2SO_2$	Dimethyl sulfone		**2.5**, 165a	
C_2H_6S	CH_3CH_2SH *)	Ethyl mercaptan	**2.5**, 157; **2.6.3**, 71; **2.8**, 8	**2.5**, 78	
C_2H_6S	$(CH_3)_2S$ *)	Dimethyl sulfide	**2.5**, 435; **2.6.3**, 166; **2.8**, 102	**2.5**, 162	**2.5**, 129
$C_2H_6S_2$	CH_3SSCH_3 *)	Dimethyl disulfide	**2.5**, 444; **2.8**, 104		**2.5**, 133a
C_2H_6Se	CH_3CH_2SeH *)	Ethaneselenol	**2.5**, 158; **2.6.3**, 73; **2.8**, 9		
C_2H_6Se	$(CH_3)_2Se$ *)	Dimethyl selenide	**2.5**, 436; **2.8**, 103	**2.5**, 163	
C_2H_6Si	$CH_2{=}CHSiH_3$ *)	Vinylsilane			**2.5**, 142
C_2H_7As	$(CH_3)_2AsH$	Dimethylarsine		**2.8**, 73	
$C_2H_7B_5$	$C_2B_5H_7$ *)	2,4-Dicarbahepta-borane (7)		**2.5**, 174	
C_2H_7ClSi	$CH_3CH_2SiH_2Cl$ *)	1-Chloro-1-silapropane	**2.5**, 396; **2.7.3**, 133	**2.5**, 167d	
C_2H_7FSi	$CH_3CH_2SiH_2F$ *)	Ethylfluorosilane	**2.5**, 395; **2.6.3**, 185		
C_2H_7N	$(CH_3)_2NH$ *)	Dimethylamine		**2.5**, 164	**2.5**, 131
C_2H_7NO	$CH_2NH_2CH_2OH$ *)	2-Aminoethanol	**2.5**, 178	**2.5**, 86a	

*) Including deuterated compounds.

*) Including deuterated compounds.

Alphabetic formula	Chemical formula	Name	**Table Nr.**, Substance Nr.		
			II/14	II/6	II/4
C_3HP	$HC{\equiv}C{-}C{\equiv}P$	Phosphabutadiyne	**2.3,** 30a; **2.6.2,** 17a; **5: 2.3,** 30a		
$C_3H_2Cl_2$	$ClC{\equiv}CCH_2Cl$	1,3-Dichloro-1-propyne	**2.5,** 371; **2.7.3,** 125		
C_3H_2ClF	$ClC{\equiv}CCH_2F$	1-Chloro-3-fluoro-propyne		**2.5,** 156	
C_3H_2ClN	$CH_2{=}CClCN$	2-Chloro-2-propenenitrile	**2.5,** 297; **2.7.3,** 119		
$C_3H_2F_2$	$CH_2{=}C{=}CF_2$	1,1-Difluoro-1,2-propadiene	**2.5,** 350; **2.6.3,** 153		
$C_3H_2F_2$	$CHF{=}C{=}CHF$	1,3-Difluoro-1,2-propadiene	**2.5,** 351; **2.6.3,** 154		
$C_3H_2F_2$	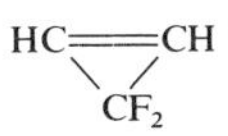 *)	3,3-Difluorocyclo-propene	**2.5,** 586; **2.6.3,** 287		
$C_3H_2F_4$	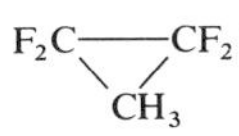 *)	1,1,2,2-Tetrafluoro-cyclopropane	**2.5,** 467		
$C_3H_2N_2$	$CH_2(CN)_2$ *)	Malononitrile	**2.5,** 96		**2.5,** 50
C_3H_2O	$CH_2{=}C{=}C{=}O$ *)	1,2-Propadien-1-one	**2.5,** 353; **2.6.3,** 155		
C_3H_2O	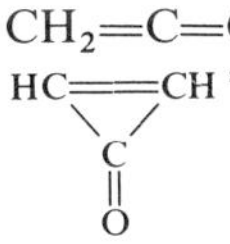 *)	2-Cyclopropen-1-one	**2.5,** 588; **2.6.3,** 289; **2.9.3,** 36		
C_3H_2O	$HC{\equiv}C{-}CHO$ *)	Propiolaldehyde	**2.5,** 364	**2.5,** 154	**2.5,** 124
$C_3H_2O_2$	$HC{\equiv}C{-}COOH$ *)	Propargylic acid	**2.5,** 367	**2.5,** 154a	
$C_3H_2O_3$	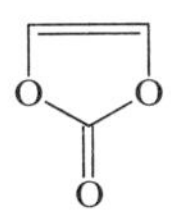	1,3-Dioxol-2-one	**2.5,** 599; **2.9.3,** 40	**2.5,** 246	**2.5,** 177
$C_3H_2OS_2$	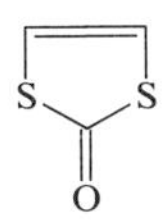	1,3-Dithiol-2-one	**2.5,** 600; **2.6.3,** 296		
$C_3H_2O_2S$	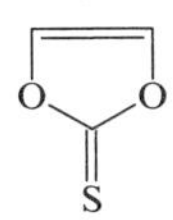	1,3-Dioxol-2-thione	**2.5,** 601; **2.6.3,** 297		
C_3H_3Br	$HC{\equiv}CCH_2Br$	Propargyl bromide			**2.5,** 123
C_3H_3Br	$CH_2{=}C{=}CHBr$	Bromoallene	**2.5,** 346; **2.7.3,** 154	**2.5,** 146	
C_3H_3Br	$CH_3C{\equiv}CBr$	Methylbromo-acetylene			**2.4,** 56
C_3H_3Cl	$HC{\equiv}CCH_2Cl$ *)	Propargylchloride			**2.5,** 122
C_3H_3Cl	$CH_2{=}C{=}CHCl$	Chloroallene			**2.5,** 117
C_3H_3Cl	$CH_3C{\equiv}CCl$ *)	Methylchloro-acetylene			**2.4,** 55
C_3H_3ClO	$CH_2{=}CH{-}C(O){-}Cl$	Acryloyl chloride		**2.5,** 138a	
C_3H_3F	$HC{\equiv}CCH_2F$	Propargyl fluoride			**2.5,** 121
C_3H_3F	$CH_2{=}C{=}CHF$	3-Fluoro-1-propyne	**2.5,** 345; **2.6.3,** 151		
$C_3H_3F_3$	$CH_3CF{=}CF_2$	1,1,2-Trifluoro-propylene		**2.5,** 129b	

*) Including deuterated compounds.

Alphabetic formula	Chemical formula	Name	Table Nr., Substance Nr. II/14	II/6	II/4
$C_3H_3F_3$	$CF_3CH{=}CH_2$ *)	3,3,3-Trifluoro-propylene	**2.5**, 307; **2.6.3**, 139; **2.8**, 45	**2.5**, 129c	
$C_3H_3F_3$	$\overline{CHFCHFCHF}$ *)	1,2,3-Trifluoro-cyclopropane	**2.4**, 91; **2.6.2**, 65		
C_3H_3FO	$CH_2{=}CH{-}C({=}O){-}F$	Acryloyl fluoride		**2.5**, 138	
$C_3H_3F_3O_2$	$CF_3C({=}O)OCH_3$ *)	Methyl trifluoroacetate	**2.5**, 191		
$C_3H_3F_3O_4$	$CF_3COOH\cdots HCOOH$ *)	Trifluoroacetic acid – Formic acid		**2.5**, 232d	**2.5**, 163
C_3H_3I	$CH_3C{\equiv}CI$	Methyliodoacetylene			**2.4**, 57
C_3H_3N	$CH_2{=}CHNC$	Vinyl isocyanide	**2.5**, 285	**2.5**, 112	
C_3H_3N	$CH_2{=}CHCN$ *)	Acrylonitrile	**2.5**, 284		**2.5**, 97
C_3H_3NO	$HC{\equiv}C{-}C({=}O)NH_2$	2-Propynamide	**2.5**, 368; **2.6.3**, 161; **2.7.3**, 42		
C_3H_3NO	$CH_3C({=}O)CN$ *)	Acetyl cyanide	**2.5**, 186; **2.7.3**, 34; **2.8**, 18		**2.5**, 82
C_3H_3NO	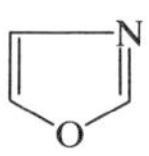 *)	Oxazole	**2.5**, 633; **2.6.3**, 312; **2.7.3**, 64; **2.9.3**, 45	**2.5**, 259	
C_3H_3NO	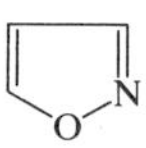 *)	Isoxazole	**2.5**, 634; **2.6.3**, 313; **2.7.3**, 65; **2.9.3**, 46	**2.5**, 260	
$C_3H_3NO_2$	$NC{-}C({=}O){-}OCH_3$	Methylcyano-formate		**2.5**, 72	
$C_3H_3NO_2$	$CH_3C({=}O)NCO$	Acetyl isocyanate	**2.5**, 187; **2.6.3**, 96; **2.8**, 19		
C_3H_3NOS	$NCC({=}O)SCH_3$	Methyl thiol-cyanoformate	**2.5**, 132a; **2.8**, 98a		
C_3H_3NS	 *)	Thiazole	**2.5**, 636; **2.7.3**, 66; **2.9.3**, 47	**2.5**, 263	**2.5**, 172
C_3H_3NS	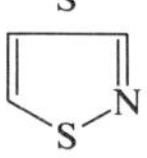	Isothiazole	**2.5**, 637; **2.7.3**, 67; **2.9.3**, 48	**2.5**, 264	
C_3H_3P	$CH_2{=}CHC{\equiv}P$	(Ethylmethylidyne)-phosphine	**2.5**, 433a; **2.6.3**, 215a		
C_3H_4	$HC{=}CH$, CH_2 (ring) *)	Cyclopropene	**2.5**, 585	**2.5**, 233	**2.5**, 166
C_3H_4	$CH_2{=}C{=}CH_2$ *)	Allene	**2.4**, 82; **2.6.2**, 63		
C_3H_4	$CH_3C{\equiv}CH$ *)	Methylacetylene	**2.4**, 47; **2.6.2**, 53; **2.9.2**, 33	**2.4**, 45	**2.4**, 54
$C_3H_4Cl_2$	$H_2C{-}CCl_2$, CH_2 (ring) *)	1,1-Dichlorocyclo-propane			**2.5**, 151
C_3H_4ClF	$CH_3CF{=}CHCl$	1-Chloro-2-fluoropropene		**2.5**, 128	
$C_3H_4Cl_2O$	$(CH_2Cl)_2C{=}O$ *)	1,3-Dichloro-2-propanone	**2.5**, 255		
$C_3H_4F_2$	$H_2C{-}CF_2$, CH_2 (ring) *)	1,1-Difluorocyclo-propane	**2.5**, 465; **2.6.3**, 220		

*) Including deuterated compounds.

Alphabetic formula	Chemical formula	Name	**Table Nr.**, Substance Nr.		
			II/14	II/6	II/4
$C_3H_4F_2$	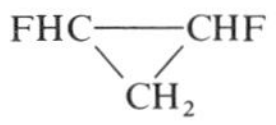	1,2-Difluorocyclo-propane	**2.5**, 468		
$C_3H_4F_2$	$CH_3CH{=}CF_2$	1,1-Difluoropropene		**2.5**, 129	
$C_3H_4F_2$	$CH_2{=}CFCH_2F$	2,3-Difluoro-1-propene	**2.5**, 316; **2.6.3**, 141		
$C_3H_4F_2$	$CH_2{=}CHCHF_2$	3,3-Difluoro-1-propene	**2.5**, 317; **2.6.3**, 142		
$C_3H_4F_2O$	$(CH_2F)_2C{=}O$	1,3-Difluoro-2-propanone	**2.5**, 254; **2.6.3**, 122		
$C_3H_4F_2O$	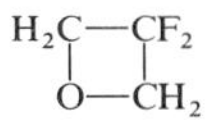	3,3-Difluoroxetane		**2.5**, 214	
$C_3H_4N_2$	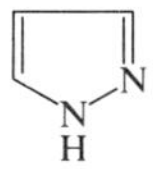 *)	Pyrazole	**2.5**, 630; **2.6.3**, 311; **2.7.3**, 62	**2.5**, 257	**2.5**, 176a
$C_3H_4N_2$	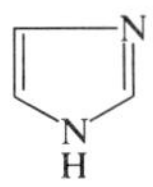 *)	Imidazole	**2.5**, 632; **2.7.3**, 63	**2.5**, 258	
$C_3H_4N_2$	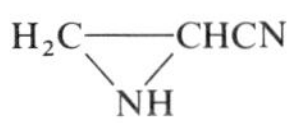	2-Aziridine-carbonitrile	**2.5**, 492; **2.7.3**, 48		
C_3H_3NO	$CH_2{=}CHNCO$	Isocyanatoethene	**2.5**, 289; **2.6.3**, 128; **2.7.3**, 38		
$C_3H_4N_2O$	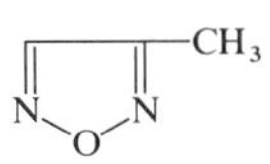	3-Methyl-1,2,5-oxadiazole		**2.5**, 262	
C_3H_4O	$HC{\equiv}COCH_3$ *)	Methoxyacetylene		**2.5**, 149	**2.5**, 125a
C_3H_4O	$HC{\equiv}CCH_2OH$ *)	Propargyl alcohol	**2.5**, 357	**2.5**, 153	
C_3H_4O	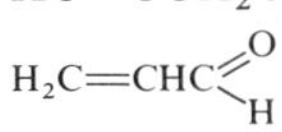 *)	Acrolein	**2.5**, 318; **2.9.3**, 26	**2.5**, 136	**2.5**, 99
C_3H_4O	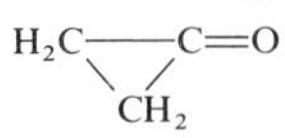 *)	Cyclopropanone		**2.5**, 184	
C_3H_4O	$CH_3CH{=}C{=}O$ *)	Methylketene			**2.5**, 120
$C_3H_4O_2$	$HCOOCH{=}CH_2$	Vinyl formate	**2.5**, 117; **2.9.3**, 16		**2.5**, 64
$C_3H_4O_2$	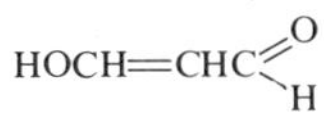	Malonaldehyde propanedial	**2.5**, 256; **2.6.3**, 123; **2.7.3**, 79		
$C_3H_4O_2$	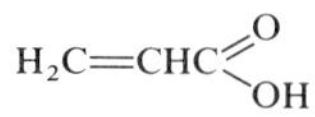 *)	Acrylic acid	**2.5**, 320; **2.6.3**, 144	**2.5**, 139	
$C_3H_4O_2$	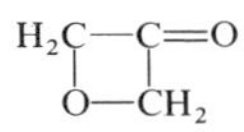	3-Oxetanone	**2.5**, 503; **2.9.3**, 33	**2.5**, 211 a	
$C_3H_4O_2$	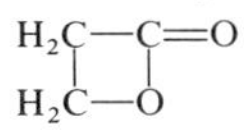	β-Propiolactone	**2.5**, 502	**2.5**, 210a	**2.5**, 160
$C_3H_4O_2$	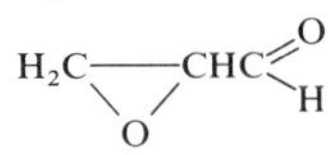	Oxirane carbox-aldehyde	**2.5**, 480; **2.6.3**, 229		
$C_3H_4O_2$	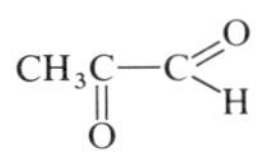	2-Oxopropanal	**2.5**, 215; **2.6.3**, 107; **2.8**, 23		
$C_3H_4O_3$	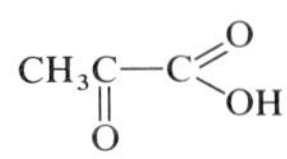	Pyruvic acid	**2.5**, 214; **2.6.3**, 106; **2.8**, 22		

*) Including deuterated compounds.

Alphabetic formula	Chemical formula	Name	Table Nr., Substance Nr. II/14	II/6	II/4
$C_3H_4O_3$	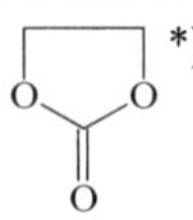 *)	Ethylene carbonate	**2.5,** 514a		**2.5,** 161a
C_3H_4OS	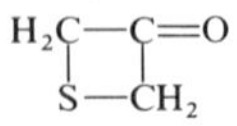	3-Thietanone	**2.5,** 506; **2.6.3,** 245		
$C_3H_4O_2S$	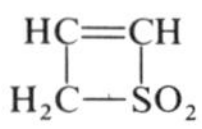	2H-Thiete-1,1-dioxide	**2.5,** 593		
C_3H_4S	$HC{\equiv}CSCH_3$ *)	Methylthioethyne		**2.5,** 150	
C_3H_4S	$CH_3CH{=}C{=}S$	Methylthioketene	**2.5,** 302; **2.6.3,** 135; **2.8,** 42		
C_3H_4S	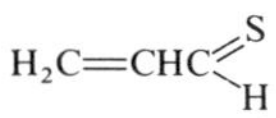	2-Propenethial	**2.5,** 319; **2.6.3,** 143		
C_3H_4S	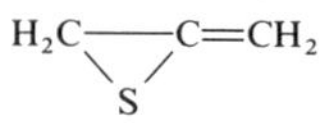	Methylenethiirane	**2.5,** 490; **2.6.3,** 236		
C_3H_4S	$HC{\equiv}CCH_2SH$ *)	Propargylmercaptan	**2.5,** 358; **2.6.3,** 158	**2.5,** 152	
$C_3H_5BF_2$	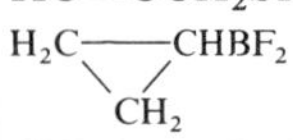	Cyclopropyldifluoroborane	**2.5,** 477; **2.6.3,** 227		
C_3H_5Br	$CH_3CH{=}CHBr$	1-Bromopropene		**2.5,** 125	
C_3H_5Br	$CH_3CBr{=}CH_2$	2-Bromopropene		**2.5,** 127	**2.5,** 112b
C_3H_5Br	$CH_2{=}CHCH_2Br$	3-Bromo-1-propene	**2.5,** 309; **2.7.3,** 153		
C_3H_5Br	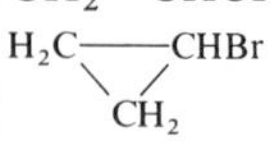 *)	Bromocyclopropane			**2.5,** 149a
C_3H_5BrO	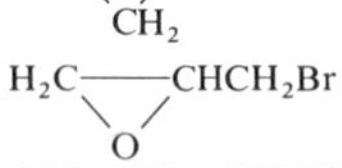	(Bromomethyl)-oxirane	**2.5,** 485		
C_3H_5Cl	$CH_3CH{=}CHCl$	1-Chloropropene			**2.5,** 110
C_3H_5Cl	$CH_3CCl{=}CH_2$ *)	2-Chloropropene	**2.5,** 305a; **2.6.3,** 138	**2.5,** 126a	**2.5,** 112a
C_3H_5Cl	$CH_2{=}CHCH_2Cl$	3-Chloropropene		**2.5,** 131	
C_3H_5Cl	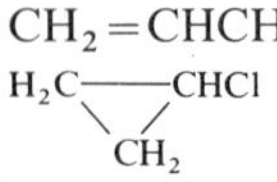 *)	Chlorocyclopropane			**2.5,** 149
C_3H_5ClO	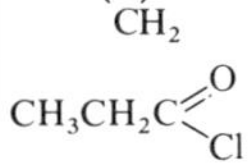	Propanoyl chloride	**2.5,** 259; **2.7.3,** 117; **2.8,** 35		
C_3H_5ClO	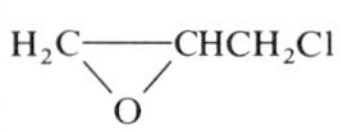	Epichlorohydrin	**2.5,** 484	**2.5,** 195a	
$C_3H_5ClO_2$	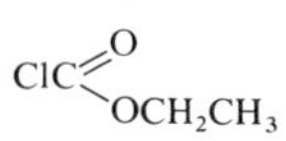	Ethyl carbonochloridate	**2.5,** 135		
C_3H_5ClOS	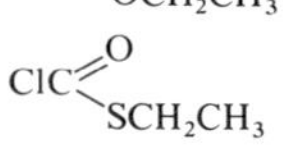	S-Methyl carbonochloridothioate	**2.5,** 137		
C_3H_5F	$CH_3CH{=}CHF$	1-Fluoropropene			**2.5,** 109
C_3H_5F	$CH_3CF{=}CH_2$	2-Fluoropropene			**2.5,** 112
C_3H_5F	$CH_2FCH{=}CH_2$ *)	3-Fluoropropene		**2.5,** 130	**2.5,** 113
C_3H_5FO	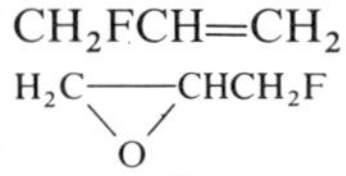	Epifluorohydrin	**2.5,** 483; **2.6.3,** 232	**2.5,** 195	
C_3H_5FO	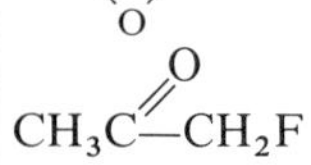	Fluoroacetone		**2.5,** 106	
C_3H_5FO	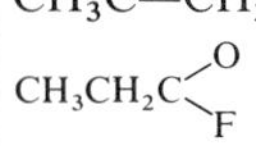	Propionyl fluoride	**2.5,** 258a; **2.8,** 34a	**2.5,** 105	
$C_3H_5FO_2$	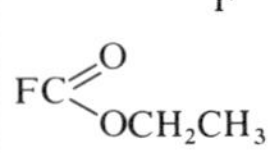	Ethyl carbonfluoridate	**2.5,** 134		

*) Including deuterated compounds.

Alphabetic formula	Chemical formula	Name	**Table Nr.**, Substance Nr.		
			II/14	II/6	II/4
C_3H_5I	$CH_3CI{=}CH_2$	2-Iodo-1-propene	**2.5**, 306; **2.7.3**, 161; **2.8**, 44		
C_3H_5I	$CH_2{=}CHCH_2I$	3-Iodo-1-propene	**2.5**, 310; **2.7.3**, 162		
C_3H_5N	CH_3CH_2NC	Ethyl isocyanide		**2.5**, 76	
C_3H_5N	CH_3CH_2CN *)	Ethyl cyanide	**2.5**, 155; **2.6.3**, 69; **2.7.3**, 29; **2.8**, 6	**2.7.3**, 21	**2.5**, 72
C_3H_5N	$CH{\equiv}CCH_2NH_2$ *)	Propargyl amine	**2.5**, 356; **2.6.3**, 157	**2.5**, 151	
C_3H_5N	$CH_2{=}CHCH{=}NH$ *)	2-Propen-1-imine	**2.5**, 322; **2.6.3**, 146; **2.7.3**, 40		
C_3H_5NO	$HOCHCH_3CN$		**2.5**, 173a		
C_3H_5NO	CH_3CH_2NCO	Isocyanatoethane	**2.5**, 162; **2.6.3**, 77		
C_3H_5NO	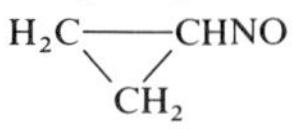	Nitroso-cyclopropane	**2.5**, 463; **2.6.3**, 219; **2.7.3**, 46		
C_3H_5NO	CH_3OCH_2CN	Methoxyacetonitrile	**2.5**, 88; **2.6.3**, 44; **2.8**, 80		
C_3H_5NO	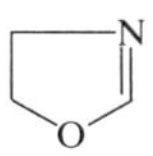	4,5-Dihydrooxazole	**2.5**, 602; **2.6.3**, 298; **2.7.3**, 56		
$C_3H_5NO_2$	$CH_3CH{=}CHNO_2$	1-Nitro-1-propene	**2.5**, 305; **2.6.3**, 137; **2.8**, 43		
$C_3H_5NO_2$	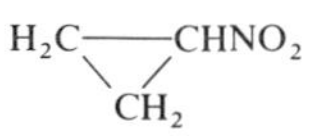 *)	Nitrocyclopropane	**2.5**, 462; **2.6.3**, 218	**2.5**, 181 b	
C_3H_5NS	CH_3SCH_2CN	(Methylthio)-acetonitrile	**2.5**, 89; **2.6.3**, 45		
C_3H_5NS	CH_3CH_2NCS	Isothiocyanatoethane	**2.5**, 161; **2.6.3**, 76		
C_3H_5NS	CH_3CH_2SCN	Ethyl thiocyanate		**2.5**, 79a	
C_3H_6	$CH_3CH{=}CH_2$ *)	Propene		**2.5**, 124	**2.5**, 108
$C_3H_6F_2$	$CH_3CF_2CH_3$	2,2-Difluoropropane	**2.5**, 252a; **2.6.3**, 121a		
$C_3H_6N_2$		4,5-Dihydro-3H-pyrazole	**2.5**, 603; **2.7.3**, 57		
$C_3H_6N_2$	$(CH_3)_2NCN$	Dimethylcyanamide	**2.5**, 438; **2.6.3**, 167		
$C_3H_6N_2$	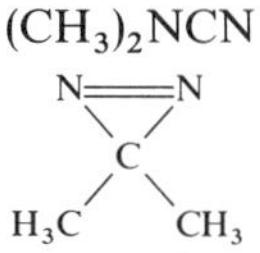	Dimethyldiazirene		**2.5**, 238	
C_3H_6O	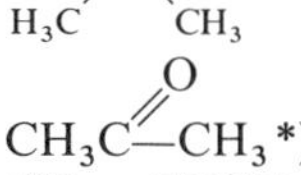 *)	Acetone		**2.5**, 161 a	**2.5**, 130
C_3H_6O	$CH_2{=}CHCH_2OH$ *)	Allyl alcohol	**5:2.5**, 310a	**2.5**, 133	
C_3H_6O	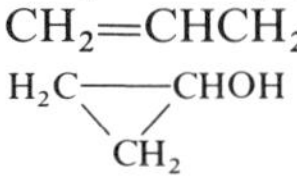 *)	Cyclopropanol	**2.5**, 461; **2.6.3**, 217		
C_3H_6O	$CH_2{=}CHOCH_3$ *)	Methylvinylether	**2.5**, 291; **2.9.3**, 24	**2.5**, 114	**2.5**, 99a
C_3H_6O	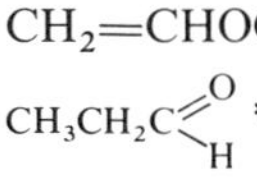 *)	Propionaldehyde	**2.5**, 258		**2.5**, 92
C_3H_6O	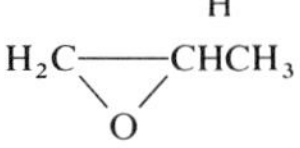 *)	Propylene oxide	**2.5**, 479		**2.5**, 153
C_3H_6O	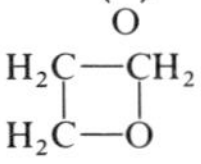 *)	Trimethylene oxide	**2.5**, 501		**2.5**, 158
$C_3H_6O_2$	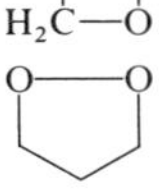	1,2-Dioxolane	**2.5**, 515		
$C_3H_6O_2$	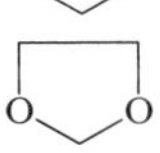	1,3-Dioxolane	**2.5**, 514; **2.6.3**, 248		

*) Including deuterated compounds.

Alphabetic formula	Chemical formula	Name	Table Nr., Substance Nr. II/14	II/6	II/4
$C_3H_6O_2$	$HC(=O)OCH_2CH_3$	Ethyl formate	**2.5**, 133	**2.5**, 66	
$C_3H_6O_2$	$CH_3C(=O)CH_2OH$ *)	1-Hydroxypropanone-(2)	**2.5**, 242 a; **2.6.3**, 121 b; **2.8**, 32 a		
$C_3H_6O_2$	$CH_3C(=O)OCH_3$ *)	Methyl acetate		**2.5**, 92	
$C_3H_6O_2$	$CH_3CH_2C(=O)OH$ *)	Propionic acid	**2.5**, 260; **2.6.3**, 125; **2.8**, 36	**2.5**, 105 a	
$C_3H_6O_3$	$HOCH_2C(=O)OCH_3$ *)	Methyl hydroxyacetate	**2.5**, 208		
$C_3H_6O_3$	$CH_3OCH_2C(=O)OH$ *)	Methoxyacetic acid	**2.5**, 209; **2.6.3**, 103		
$C_3H_6O_3$	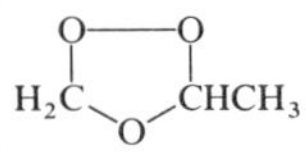 *)	3-Methyl-1,2,4-trioxolane	**2.5**, 518; **2.6.3**, 252		
$C_3H_6O_3$	$(CH_2O)_3$	1,3,5-Trioxane	**2.4**, 92; **2.9.2**, 37	**2.4**, 72	**2.4**, 99
C_3H_6OS	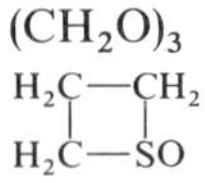 *)	Thietane 1-oxide	**2.5**, 507		
C_3H_6OS	$CH_3CH_2C(=O)SH$	Thiopropionic acid	**2.5**, 260 a; **2.6.3**, 125 a		
C_3H_6S	$CH_2{=}CHCH_2SH$ *)	Allyl mercaptan		**2.5**, 134	
C_3H_6S	$CH_2{=}CHSCH_3$	Methylthioethene		**2.5**, 115	
C_3H_6S	$(CH_3)_2C{=}S$	2-Propanethione	**2.5**, 253; **2.8**, 33		
C_3H_6S	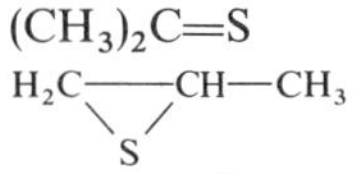	Propylene sulfide			**2.5**, 155
C_3H_6S	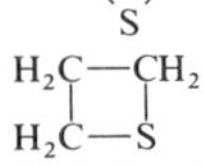 *)	Trimethylene sulfide	**2.5**, 505 a	**2.5**, 215	**2.5**, 158 a
C_3H_6Si	$CH_3C{\equiv}CSiH_3$ *)	Methylsilylacetylene			**2.4**, 93 a
C_3H_6Se	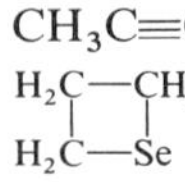	Trimethylene selenide		**2.5**, 216	
C_3H_7Br	$CH_3CH_2CH_2Br$	n-Propyl bromide	**2.5**, 221 a; **2.7.3**, 151 a	**2.5**, 99	
C_3H_7Br	$CH_3CHBrCH_3$	iso-Propyl bromide			**2.5**, 91
$C_3H_7B_3$	$C_3B_3H_7$	2,3,4-Tricarbahexaborane (7)		**2.5**, 175 b	
C_3H_7Cl	$HCl \cdot C_3H_6$	Cyclopropane hydrogen chloride adduct	**2.5**, 764; **2.7.3**, 145		
C_3H_7Cl	$CH_3CH_2CH_2Cl$	n-Propyl chloride	**2.5**, 221		**2.5**, 88
C_3H_7Cl	$CH_3CHClCH_3$	iso-Propyl chloride			**2.5**, 90
C_3H_7ClO	$(CH_3)_2CHOCl$	1-Methylethyl hypochlorite	**2.5**, 243		
C_3H_7ClO	$CH_3CH_2CH_2OCl$	Propyl hypochlorite	**2.5**, 230		
C_3H_7ClO	$CH_3CH_2OCH_2Cl$ *)	(Chloromethoxy) ethane	**2.5**, 168; **2.7.3**, 111		
C_3H_7F	$HF \cdot C_3H_6$ *)	Cyclopropane-hydrogen fluoride adduct	**2.5**, 763 b; **2.7.3**, 9 a; **5: 2.9.3**, 4		
C_3H_7F	CH_3CHFCH_3	2-Fluoropropane	**2.5**, 236; **2.6.3**, 117; **2.8**, 31		
C_3H_7F	$CH_3CH_2CH_2F$	n-Propyl fluoride	**2.5**, 220		**2.5**, 87
C_3H_7FO	$CH_3CH(OH)CH_2F$ *)	1-Fluoro-2-propanol	**2.5**, 235; **2.8**, 30		

*) Including deuterated compounds.

Alphabetic formula	Chemical formula	Name	Table Nr., Substance Nr.		
			II/14	II/6	II/4
C_3H_7FO	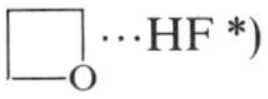	Oxetane hydrogen fluoride complex	**2.5**, 765		
C_3H_7FO	$CH_3CH_2OCH_2F$ *)	(Fluoromethoxy) ethane	**2.5**, 167; **2.6.3**, 82; **2.8**, 11		
C_3H_7I	$(CH_3)_2CHI$	2-Iodopropane	**2.5**, 237; **2.7.3**, 160		
C_3H_7I	$CH_3CH_2CH_2I$	n-Propyl iodide	**2.5**, 222	**2.5**, 100	
C_3H_7N	$CH_2{=}CHCH_2NH_2$	Allylamine	**2.5**, 311; **2.6.3**, 140; **2.7.3**, 39	**2.5**, 135	
C_3H_7N	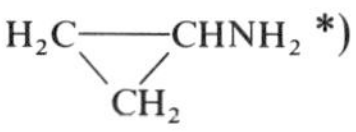	Cyclopropylamine	**2.5**, 460	**2.5**, 182	
C_3H_7N	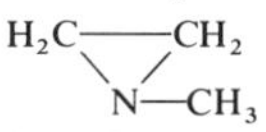	N-Methyl ethylene imine		**2.5**, 204	
C_3H_7N	$CH_3CH{=}NCH_3$ *)	N-Methylethylidene imine	**2.5**, 172; **2.8**, 15	**2.5**, 80a	
C_3H_7N	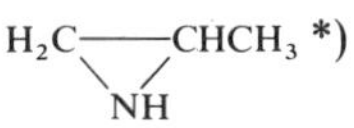	Propylene imine	**2.5**, 491; **2.7.3**, 47	**2.5**, 202	
C_3H_7NO	$HCON(CH_3)_2$	Dimethylformamide	**2.5**, 116; **2.6.3**, 60; **2.8**, 117		
$C_3H_7NO_2$	$CH_3CH_2CH_2ONO$	Propyl nitrite	**2.5**, 229		
C_3H_7NP	$(CH_3)_2PCN$	Dimethylcyano-phosphine	**2.5**, 441; **2.6.3**, 168		
C_3H_7NSi	$(CH_3)_2SiHCN$	Dimethylsilane-carbonitrile	**2.5**, 394; **2.6.3**, 184		
$C_3H_7O_2$	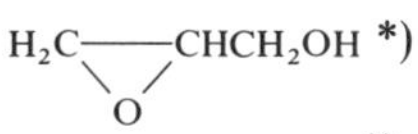	Oxiranemethanol	**2.5**, 486; **2.6.3**, 233		
C_3H_7P	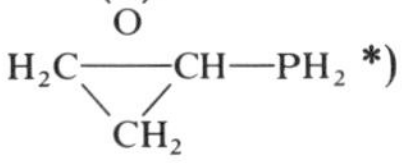	Cyclopropylphos-phine		**2.5**, 183	
C_3H_7P	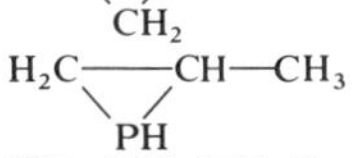	Propylenephosphine		**2.5**, 206	
C_3H_8	$CH_3CH_2CH_3$ *)	Propane	**2.5**, 219; **2.9.3**, 22	**2.5**, 98	**2.5**, 86
C_3H_8O	$CH_3CH_2OCH_3$ *)	Ethylmethyl ether	**2.5**, 159; **2.6.3**, 74; **2.8**, 10	**2.5**, 165b	**2.5**, 128a
C_3H_8O	$CH_3CH_2CH_2OH$	n-Propyl alcohol	**2.5**, 225; **2.8**, 27	**2.5**, 101	
C_3H_8O	$(CH_3)_2CHOH$ *)	iso-Propyl alcohol	**2.5**, 240; **2.6.3**, 120	**2.5**, 102	
$C_3H_8O_2$	$CH_3OCH_2CH_2OH$ *)	2-Methoxyethanol	**2.5**, 181; **2.8**, 84	**2.5**, 86b	
$C_3H_8O_2$	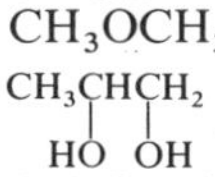	Propandiol-(1,2)	**2.5**, 232; **2.6.3**, 114		
C_3H_8S	$CH_3CH_2SCH_3$ *)	(Methylthio)ethane	**2.5**, 160; **2.6.3**, 75; **2.8**, 100		
C_3H_8S	$CH_3CH_2CH_2SH$ *)	1-Propanethiol	**2.5**, 227; **2.6.3**, 112		
C_3H_8S	$(CH_3)_2CHSH$ *)	2-Propanethiol	**2.5**, 242; **2.6.3**, 121		
C_3H_8Si	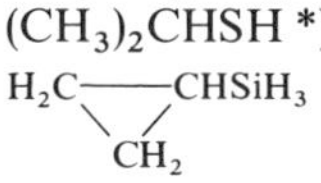	Cyclopropylsilane	**2.5**, 476; **2.6.3**, 226		
C_3H_8Si	$CH_2{=}CHSiH_2CH_3$ *)	Ethenylmethylsilane	**2.5**, 398; **2.6.3**, 187; **2.8**, 64		
C_3H_8Si	$CH_2{=}CHCH_2SiH_3$ *)	2-Propenylsilane	**2.5**, 399; **2.6.3**, 188; **2.8**, 65		
C_3H_8Si	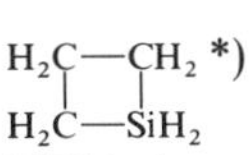	Silacyclobutane		**2.5**, 217	
C_3H_9As	$(CH_3)_3As$	Trimethylarsine			**2.4**, 78
$C_3H_9BF_3N$	$(CH_3)_3N \cdot BF_3$ *)	N,N-Dimethyl-methanamine boron trifluoride adduct	**2.4**, 69		

*) Including deuterated compounds.

Alphabetic formula	Chemical formula	Name	Table Nr., Substance Nr.		
			II/14	II/6	II/4
$C_3H_9BF_3N$	$(CH_3)_3N \cdot BF_3$	Trimethylaminetri-fluoroborine		**2.4,** 53b	**2.4,** 80
C_3H_9BrGe	$(CH_3)_3GeBr$	Trimethylbromo-germane	**2.4,** 63		
C_3H_9BrSi	$(CH_3)_3SiBr$	Trimethylbromo-silane	**2.4,** 58		**2.4,** 74
C_3H_9ClGe	$(CH_3)_3GeCl$	Trimethylchloro-germane	**2.4,** 62		
C_3H_9ClSi	$(CH_3)_3SiCl$ *)	Trimethylchloro-silane	**2.4,** 57; **2.7.2,** 37	**2.4,** 51a	**2.4,** 73
C_3H_9FSi	$(CH_3)_3SiF$	Trimethylfluorosilane			**2.4,** 72
C_3H_9ISi	$(CH_3)_3SiI$	Trimethyliodosilane			**2.4,** 75
C_3H_9N	$CH_3CH_2NHCH_3$ *)	Methylaminoethane	**2.5,** 165; **2.6.3,** 80; **2.7.3,** 30	**2.5,** 79c	
C_3H_9N	$(CH_3)_2CHNH_2$ *)	2-Propanamine	**2.5,** 239; **2.6.3,** 119; **2.7.3,** 36		
C_3H_9N	$(CH_3)_3N$ *)	Trimethylamine	**2.4,** 65; **2.8,** 108	**2.4,** 52	**2.4,** 76
C_3H_9NO	$CH_3CH(OH)CH_2NH_2$	1-Amino-2-propanol	**2.5,** 234; **2.8,** 29		
C_3H_9NO	$CH_3CH(NH_2)CH_2OH$	2-Amino-1-propanol	**2.5,** 233; **2.6.3,** 115		
C_3H_9NO	$H_2NCH_2CH_2CH_2OH$ *)	3-Amino-1-propanol	**2.5,** 231; **2.6.3,** 113		
C_3H_9NO	$CH_2NH(CH_3)CH_2OH$	2-(Methylamino) ethanol	**2.5,** 179; **2.6.3,** 90		
C_3H_9NO	$CH_3OCH_2CH_2NH_2$ *)	2-Methoxyethane-amine	**2.5,** 182; **2.6.3,** 92; **2.7.3,** 33; **2.8,** 85		
C_3H_9P	$(CH_3)_2CHPH_2$ *)	(1-Methylethyl) phosphine	**2.5,** 442; **2.6.3,** 169		
C_3H_9P	$(CH_3)_3P$ *)	Trimethylphosphine		**2.4,** 53	**2.4,** 77
$C_3H_{10}Ge$	$(CH_3)_3GeH$ *)	Trimethylgermane	**2.4,** 61		
$C_3H_{10}Si$	$CH_3CH_2SiH_2CH_3$ *)	Methyl ethylsilane	**2.5,** 397; **2.6.3,** 186	**2.5,** 167a	
$C_3H_{10}Si$	$CH_3CH_2CH_2SiH_3$ *)	Propylsilane	**2.5,** 382; **2.6.3,** 177		
$C_3H_{10}Si$	$(CH_3)_3SiH$ *)	Trimethylsilane			**2.4,** 71
$C_3H_{11}BFN$	$(CH_3)_3N \cdot BH_2F$	Trimethylamine fluoroborane adduct	**2.5,** 456		
$C_3H_{12}AsB$	$(CH_3)_3As \cdot BH_3$ *)	Trimethylarsine borane adduct	**2.4,** 71		
$C_3H_{12}BN$	$(CH_3)_3N \cdot BH_3$ *)	N,N-Dimethyl-methanamine borane adduct	**2.4,** 68; **2.6.2,** 56; **2.8,** 135		
$C_3H_{12}BN$	$(CH_3)_3N \cdot BH_3$ *)	Trimethylamine borine			**2.4,** 79
$C_3H_{12}BP$	$(CH_3)_3P \cdot BH_3$ *)	Trimethylphosphine borane		**2.4,** 53a	
$C_3H_{12}GaN$	$(CH_3)_3N \cdot GaH_3$ *)	N,N-Dimethyl-methanamine (N—Ga) gallane	**2.4,** 70; **2.8,** 136		
C_4					
C_4NiO_4	$Ni(CO)_4$	Nickel tetracarbonyl		**2.9.2,** 36	
C_4H	HC≡C—C≡C	Butadienyl radical	**4.2.1,** 3		
C_4HCl	HC≡C—C≡CCl	Chlorodiacetylene		**2.3,** 23	
C_4HN	$HC(CN)_3$	Methanetricarbo-nitrile	**2.4,** 53		

*) Including deuterated compounds.

Alphabetic formula	Chemical formula	Name	**Table Nr.**, Substance Nr. II/14	II/6	II/4
C_4H_2	HC≡C—C≡CH	1,3-Butadiyne	**2.3**, 30b; **2.6.2**, 17b; **5: 2.3**, 30b		
$C_4H_2N_2$	CH_2=C(CN)$_2$ *)	Ethenylidene-dicarbonitrile	**2.5**, 296		
C_4H_2O	CH_2=C=C=C=O *)	1,2,3-Butatrien-1-one	**2.5**, 353a; **2.6.3**, 155a		
$C_4H_2O_3$	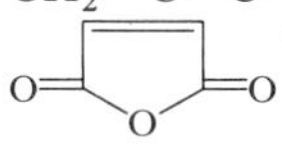 *)	Maleic acid anhydride	**2.5**, 598; **2.9.3**, 39	**2.5**, 245	
C_4H_3BrO	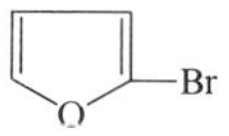	2-Bromofuran	**2.5**, 612a; **2.7.3**, 155a		
C_4H_3BrO	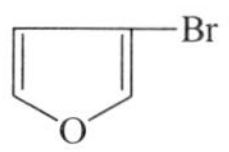	3-Bromofuran	**2.5**, 612b; **2.7.3**, 155b		
C_4H_3BrS	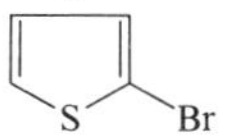	2-Bromothiophene	**2.5**, 616		
C_4H_3Cl	HC≡C—CH=CHCl *)	1-Chloro-1-buten-3-yne	**2.5**, 375; **2.7.3**, 127		
C_4H_3Cl	HC≡C—CCl=CH_2 *)	2-Chloro-1-buten-3-yne	**2.5**, 374; **2.7.3**, 126		
C_4H_3Cl	ClC≡C—CH=CH_2 *)	4-Chloro-1-buten-3-yne	**2.5**, 376; **2.7.3**, 128		
C_4H_3Cl	CH_2=C=C=CHCl	1-Chloro-1,2,3 buta-triene	**2.5**, 354; **2.7.3**, 123		
$C_4H_3ClO_2$	HC≡CCH$_2$OC(=O)Cl	2-Propynyl carbon-chloridate	**2.5**, 362		
C_4H_3ClS	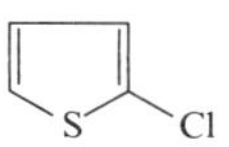	2-Chlorothiophene	**2.5**, 615; **2.7.3**, 137		
$C_4H_3F_3$	CH_3C≡CCF_3 *)	1,1,1-Trifluoro-2-butyne			**2.4**, 93
$C_4H_3FO_2$	HC≡CCH$_2$OC(=O)F	2-Propynyl carbon-fluoridate	**2.5**, 361		
C_4H_3N	CH_2=C=CHCN	Cyano allene	**2.5**, 347; **2.6.3**, 152; **2.7.3**, 41	**2.5**, 147	
C_4H_3N	CH_3C≡CCN *)	Methylcyanoacetylene			**2.4**, 58
C_4H_3N	HC≡CCH_2CN	Propargylcyanide		**2.5**, 150a	
$C_4H_3NO_2S$	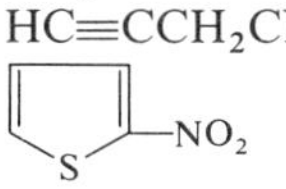	2-Nitrothiophene	**2.5**, 621		
$C_4H_3NO_2S$	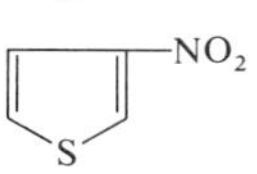	3-Nitrothiophene	**2.5**, 622		
C_4H_4	HC≡C—CH=CH_2 *)	Vinylacetylene	**2.5**, 373	**2.5**, 158	**2.5**, 126
$C_4H_4F_2$	CF_2=CH—CH=CH_2	1,1-Difluorobuta-1,3-diene			**2.5**, 115a
$C_4H_4F_4$	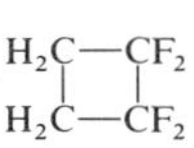 *)	1,1,2,2-Tetrafluoro-cyclobutane	**2.5**, 500; **2.6.3**, 243		
$C_4H_4F_4O_4$	CF_3COOH··· CH_2FCOOH *)	Trifluoroacetic acid-Fluoroacetic acid			**2.5**, 165
$C_4H_4N_2$	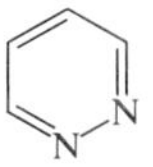	Pyridazine		**2.5**, 298	

*) Including deuterated compounds.

Alphabetic formula	Chemical formula	Name	Table Nr., Substance Nr. II/14	II/6	II/4
$C_4H_4N_2$	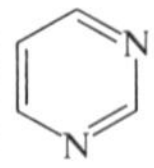 *)	Pyrimidine		**2.5**, 299	**2.5**, 191
C_4H_4O	$HC{\equiv}C{-}C(=O){-}CH_3$ *)	Acetylacetylene		**2.5**, 92a	**2.5**, 125
C_4H_4O	$CH_2{=}CH{-}CH{=}C{=}O$	1,3-Butadien-1-ene	**2.5**, 343; **2.6.3**, 150		
C_4H_4O	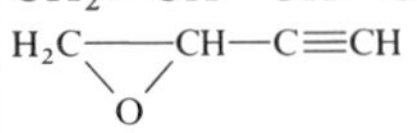 *)	Epoxybutyne		**2.5**, 196	
C_4H_4O	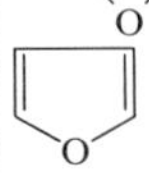 *)	Furan	**2.5**, 610; **2.9.3**, 42	**2.5**, 249	**2.5**, 170
$C_4H_4O_2$	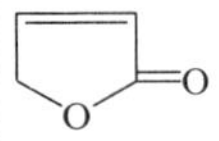	γ-Crotonolacetone	**2.5**, 597a; **2.6.3**, 295a		
$C_4H_4O_2$	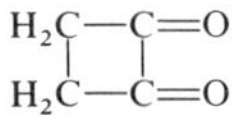	1,2-Cyclobutane-dione	**2.5**, 499; **2.6.3**, 242		
$C_4H_4O_2$	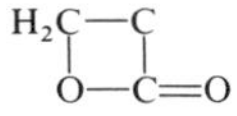	Diketene		**2.5**, 212	
$C_4H_4O_2$	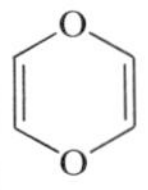	1,4-Dioxin	**2.5**, 649; **2.6.3**, 320		
$C_4H_4O_2$	$HC{\equiv}C{-}C(=O){-}OCH_3$	Methyl propiolate		**2.5**, 157	
$C_4H_4O_2$	$HCOOCH_2CCH$	Propargyl formate	**2.5**, 118		
$C_4H_4O_3$	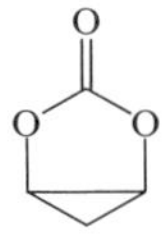	2,4-Dioxabicyclo-(3.1.0)hexan-3-one	**2.5**, 559a		
$C_4H_4O_3$	$HC(C(=O)H)_3$ *)	Methanetricarbox-aldehyde	**2.5**, 124		
C_4H_4S	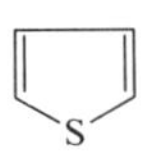 *)	Thiophene		**2.6.3**, 134b	**2.5**, 171
C_4H_4Se	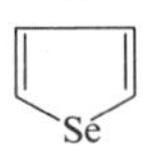 *)	Selenophene	**2.5**, 623	**2.5**, 254	
C_4H_4Te	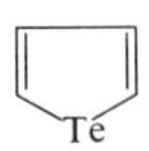	Tellurophene	**2.5**, 626; **2.6.3**, 309		
C_4H_5Cl	$CH_2{=}CH{-}CH{=}CHCl$ *)	1-Chloro-1,3 buta-diene	**2.5**, 341; **2.7.3**, 120		
C_4H_5Cl	$CH_2{=}CH{-}CCl{=}CH_2$ *)	2-Chloro-1,3-buta-diene	**2.5**, 344; **2.7.3**, 121		
C_4H_5Cl	$CH_2{=}C{=}CHCH_2Cl$ *)	3-Chloro-1,2-buta-diene	**2.5**, 348; **2.7.3**, 122		
C_4H_5Cl	$CH_3C{\equiv}CCH_2Cl$	1-Chloro-2-butyne			**2.5**, 127
C_4H_5ClO	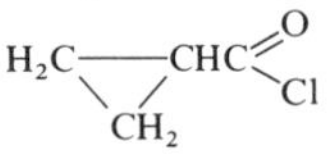	Cyclopropylcarboxy-lic chloride	**2.5**, 464		
C_4H_5ClO	$CH_3HC{=}CHC(=O)Cl$	2-Butenoyl chloride	**2.5**, 327		
$C_4H_5ClO_2$	$H_2C{=}CHCH_2OC(=O)Cl$	2-Propenyl carbon-chloridate	**2.5**, 313		

*) Including deuterated compounds.

Alphabetic formula	Chemical formula	Name	**Table Nr.,** Substance Nr. II/14	II/6	II/4
C_4H_5F	$CH_2=CH-CH=CHF$	1-Fluoro-1,3-butadiene	**2.5,** 340; **2.6.3,** 149		
C_4H_5F	$CH_2=CH-CF=CH_2$	Fluoroprene			**2.5,** 115
C_4H_5FO	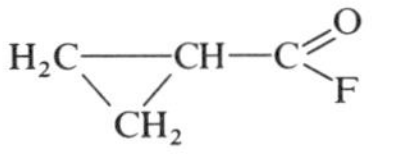	Cyclopropanecarboxylic acid fluoride		**2.5,** 191	
C_4H_5FO	$H_2C=C(CH_3)C(=O)F$	2-Methyl-2-propenoyl fluoride	**2.5,** 321; **2.6.3,** 145; **2.8,** 47		
$C_4H_5FO_2$	$H_2C=CHCH_2OC(=O)F$	2-Propenyl carbonfluoridate	**2.5,** 312a		
$C_4H_5F_3O$	$CH_2=CHOCH_2CF_3$	(2,2,2-Trifluoroethoxy)ethene	**2.5,** 328		
$C_4H_5F_3O_2$	$CF_3C(=O)OCH_2CH_3$	Ethyl trifluoroacetate	**2.5,** 192		
$C_4H_5F_3O_4$	$CH_3COOH \cdot CF_3COOH$*)	Acetic acid-Trifluoroacetic acid	**2.5,** 573; **2.6.3,** 280		**2.5,** 164
$C_4H_5F_3OS$	$CF_3C(=O)SCH_2CH_3$	S-Ethyl trifluoroethanethioate	**2.5,** 193		
C_4H_5N	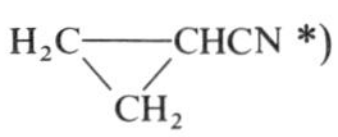	Cyanocyclopropane	**2.5,** 458; **2.7.3,** 45		**2.5,** 150
C_4H_5N	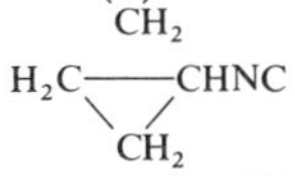	Isocyanocyclopropane	**2.5,** 459; **2.6.3,** 216		
C_4H_5N	$CH_2=CHCH_2CN$	Allyl cyanide		**2.5,** 132	
C_4H_5N	$CH_3CH=CHCN$	Crotonitrile		**2.5,** 126	**2.5,** 111
C_4H_5N	$CH_2=C(CH_3)CN$	Methacrylonitrile		**2.5,** 116a	
C_4H_5N	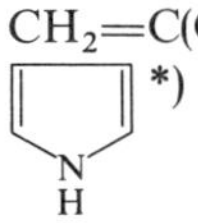	Pyrrole	**2.5,** 627; **2.7.3,** 61	**2.5,** 255	**2.5,** 176
$C_4H_5N_3$	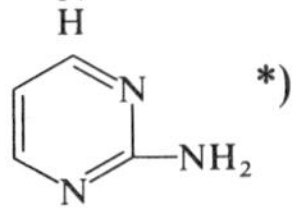	2-Pyrimidinamine	**2.5,** 754		
C_4H_5NO	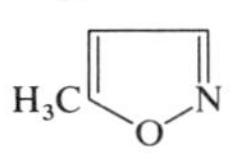	5-Methylisoxazole	**2.5,** 635; **2.8,** 55		
$C_4H_5NO_2$	$CNC(=O)OCH_2CH_3$	Ethyl carboncyanidate	**2.5,** 136; **2.6.3,** 64		
C_4H_6	$CH_3CH=C=CH_2$	Buta-1,2-diene			**2.5,** 118
C_4H_6	$HC\equiv CCH_2CH_3$ *)	Ethylacetylene		**2.5,** 155	**2.5,** 123a
C_4H_6	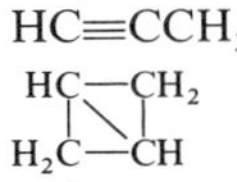	Bicyclo[1.1.0]butane		**2.5,** 231	
C_4H_6	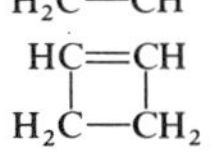	Cyclobutene			
C_4H_6	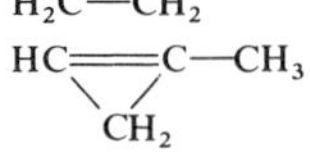	1-Methylcyclopropene		**2.5,** 234	
C_4H_6	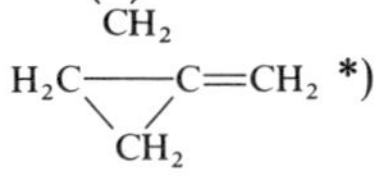	Methylenecyclopropane		**2.5,** 185	
$C_4H_6B_2$	2,3,4,5-$C_4B_2H_6$	2,3,4,5-Tetracarbahexaborane(6)	**2.5,** 423; **2.6.3,** 209		
$C_4H_6F_2$	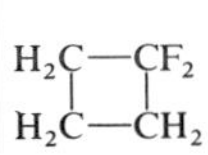	1,1-Difluorocyclobutane		**2.5,** 208	

*) Including deuterated compounds.

Alphabetic formula	Chemical formula	Name	Table Nr., Substance Nr.		
			II/14	II/6	II/4
$C_4H_6F_2$	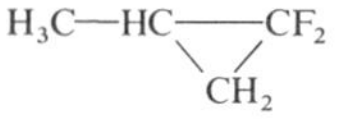	2,2-Difluoro-1-methylcyclopropane		**2.5**, 192	
$C_4H_6F_2Si$		1,1-Difluoro-1-silacyclopent-3-ene	**2.5**, 404; **2.6.3**, 192		
$C_4H_6F_3NO_3$	$CF_3COOH \cdot CH_3CONH_2$	Trifluoroacetic acid-Acetamid	**2.5**, 582		
$C_4H_6N_2$	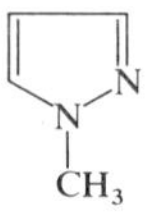	N-methylpyrazol	**2.5**, 631; **2.8**, 120		
C_4H_6O	$HC{\equiv}CCH_2CH_2OH$ *)	3-Butyn-1-ol	**2.5**, 372; **2.6.3**, 164		
C_4H_6O	$CH_3CH{=}CHC(O)H$	Crotonaldehyde		**2.5**, 142	
C_4H_6O	$H_2C{=}C(CH_3)C(O)H$	Methacrolein		**2.5**, 137	
C_4H_6O	$(CH_3)_2C{=}C{=}O$ *)	Dimethylketene	**2.5**, 300	**2.5**, 119a	**2.5**, 120a
C_4H_6O	$CH_2{=}CH{-}C({=}O){-}CH_3$	Methyl vinyl ketone			**2.5**, 99b
C_4H_6O	$HC{\equiv}CCH_2OCH_3$ *)	3-Methoxy-1-propyne	**2.5**, 359; **2.6.3**, 159; **2.8**, 86		
C_4H_6O	$HC{\equiv}C{-}OCH_2CH_3$	Ethoxyethyne	**2.5**, 369; **2.6.3**, 162		
C_4H_6O	$CH_2{=}CHOCH{=}CH_2$	Divinylether	**2.5**, 293	**2.5**, 116	
C_4H_6O	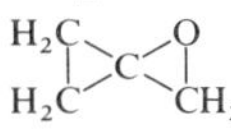 *)	1-Oxaspiro(2.2) pentane	**2.5**, 547; **2.6.3**, 264		
C_4H_6O	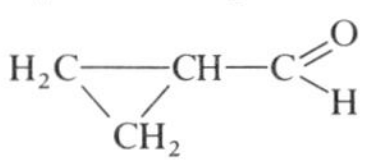	Cyclopropane carboxaldehyde		**2.5**, 190	
C_4H_6O	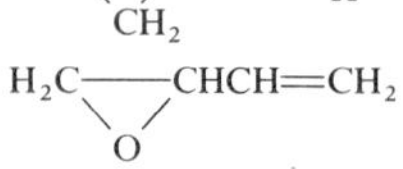	Ethenyloxirane	**2.5**, 481; **2.6.3**, 230		
C_4H_6O	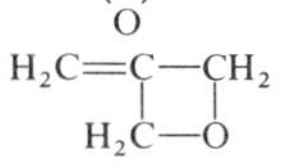	3-Methyleneoxetane	**2.5**, 505; **2.9.3**, 34	**2.5**, 213	
C_4H_6O	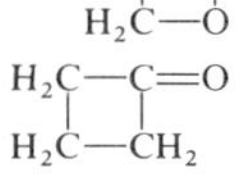 *)	Cyclobutanone	**2.5**, 496	**2.5**, 209	
C_4H_6O		2,3-Dihydrofuran		**2.5**, 244a	
C_4H_6O		2,5-Dihydrofuran		**2.5**, 244	
$C_4H_6O_2$	$H_2C{=}CHCH_2OC(O)H$	2-Propenyl formate	**2.5**, 312		
$C_4H_6O_2$	$H_2C{=}CHC(O)OCH_3$	Methyl acrylate		**2.5**, 139a	
$C_4H_6O_2$	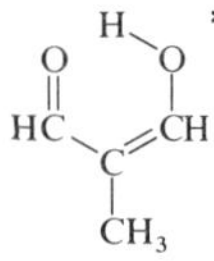 *)	Methylmalon-aldehyde methylpropanedial	**2.5**, 257; **2.6.3**, 124; **2.8**, 34		
$C_4H_6O_2$	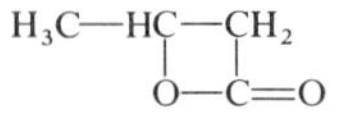	β-Butyrolacetone		**2.5**, 211b	

*) Including deuterated compounds.

Alphabetic formula	Chemical formula	Name	Table Nr., Substance Nr. II/14	II/6	II/4
$C_4H_6O_2$	$CH_3HC—CH_2$ / $O—C{=}O$ (ring)	4-Methyl-2-oxetanone	**2.5**, 504; **2.6.3**, 244		
$C_4H_6O_2$	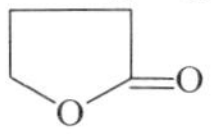	Dihydro-2(3H)-furanone	**2.5**, 513; **2.6.3**, 247		
$C_4H_6O_2$	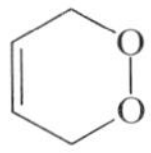	3,6-Dihydro-1,2-dioxin	**2.5**, 650; **2.6.3**, 321		
$C_4H_6O_2$	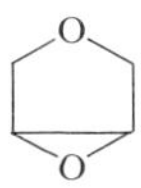	3,6-Bioxabicyclo (3.1.0) hexane	**2.5**, 559; **2.6.3**, 270		
C_4H_6S	$HC{\equiv}C—SCH_2CH_3$	(Ethylthio)ethyne	**2.5**, 370; **2.6.3**, 163		
C_4H_6S	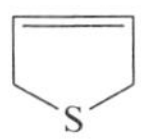	Dihydrothiophene	**2.5**, 597; **2.6.3**, 295		
C_4H_7Br	$CH_2{=}CHCH_2CH_2Br$	4-Bromo-1-butene	**2.5**, 331		
C_4H_7Br	$CH_3CH{=}CHCH_2Br$	1-Bromo-2-butene	**2.5**, 330		
C_4H_7Br	$(CH_3)_2CHCBr(CH_3)_2$	2-Bromo-2,3-dimethylbutane	**2.5**, 244		
C_4H_7Br	$H_2C—CHCH_2Br$ / CH_2 (ring)	(Bromomethyl)-cyclopropane	**2.5**, 474		
C_4H_7Br	$H_2C—CHBr$ / $H_2C—CH_2$ (ring) *)	Cyclobutyl bromide			**2.5**, 157
C_4H_7BrO	$CH_3CH_2CH_2C(=O)Br$	Butanoylbromide	**2.5**, 266		
C_4H_7Cl	$CH_3CH{=}CHCH_2Cl$	1-Chloro-2-butene	**2.5**, 329		
C_4H_7Cl	$(CH_3)_2C{=}CHCl$	1-Chloro-2-methyl-1-propene	**2.5**, 324a; **2.7.3**, 119a; **2.8**, 48a		
C_4H_7Cl	$H_2C—CHCH_2Cl$ / CH_2 (ring) *)	Cyclopropyl carbinylchloride	**2.5**, 473	**2.5**, 181a	
C_4H_7Cl	$H_2C—CHCl$ / $H_2C—CH_2$ (ring) *)	Cyclobutyl chloride		**2.7.3**, 70	**2.5**, 156b
C_4H_7ClO	$CH_3CH_2CH_2C(=O)Cl$	Butanoyl chloride	**2.5**, 265		
$C_4H_7ClO_2$	$(CH_3)_2CHOC(=O)Cl$	1-Methylethyl carbonchloridate	**2.5**, 248		
$C_4H_7ClO_2$	$ClC(=O)OCH_2CH_2CH_3$	Propyl carbon-chloridate	**2.5**, 140		
C_4H_7ClOS	$(CH_3)_2CHSC(=O)Cl$	S-(1-Methylethyl)-carbonchlorido-thioate	**2.5**, 251		
C_4H_7ClOS	$ClC(=O)SCH_2CH_2CH_3$	S-Propyl carbon-chloridothioate	**2.5**, 143		
C_4H_7F	$H_2C—CHF$ / $H_2C—CH_2$ (ring)	Cyclobutyl fluoride			**2.5**, 156a
C_4H_7FO	$CH_3CH_2CH_2C(=O)F$	Butanoyl fluoride	**2.5**, 264		
C_4H_7FO	$(CH_3)_2CHC(=O)F$	iso-Butyryl fluoride		**2.5**, 107b	
$C_4H_7FO_2$	$(CH_3)_2CHOC(=O)F$	1-Methylethyl carbonfluoridate	**2.5**, 247		

*) Including deuterated compounds.

Alphabetic formula	Chemical formula	Name	**Table Nr.,** Substance Nr.		
			II/14	II/6	II/4
$C_4H_7FO_2$	$FC(=O)OCH_2CH_2CH_3$	Propyl carbon-fluoridate	**2.5,** 139		
C_4H_7FOS	$FC(=O)SCH_2CH_2CH_3$	S-Propyl carbon-fluoridothioate	**2.5,** 142		
C_4H_7FOS	$(CH_3)_2CHSC(=O)F$	S-(1-Methylethyl)-carbonfluorido-thioate	**2.5,** 250		
C_4H_7N	$CH_3CH_2CH_2CN$	n-Butyronitrile	**2.5,** 223		**2.5,** 89
C_4H_7N	$(CH_3)_2CHCN$	iso-Butyronitrile	**2.5,** 238; **2.6.3,** 118; **2.8,** 32	**2.5,** 103	
C_4H_7N	$CH_3CH_2CH_2NC$	1-Isocyanopropane	**2.5,** 224; **2.6.3,** 110		
C_4H_7N		2,5-Dihydropyrrole		**2.5,** 246a	
C_4H_7NO	$(CH_3)_2C(OH)CN$	2-Hydroxy-2-methyl-propanenitrile	**2.5,** 241		
C_4H_7NO	$CH_3CH_2CH_2NCO$	1-Isocyanatopropane	**2.5,** 228		
C_4H_7NO	$CH_3OCH_2CH_2CN$	3-Methoxy-propanenitrile	**2.5,** 183; **2.6.3,** 93		
$C_4H_7O_3P$	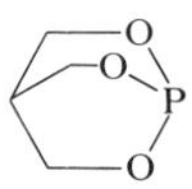	2,5,7-Trioxa-1-phosphabicyclo-(2.2.2)octane	**2.4,** 93		
C_4H_7P	 *)	2,5-Dihydro-1 H-phosphole	**2.5,** 604; **2.6.3,** 299		
C_4H_8	$CH_2{=}CHCH_2CH_3$ *)	1-Butene	**2.5,** 323; **2.9.3,** 28	**2.5,** 141	
C_4H_8	$CH_3CH{=}CHCH_3$ *)	2-Butene	**2.5,** 324; **2.8,** 48; **2.9.3,** 27	**2.5,** 140	
C_4H_8	$CH_2{=}C(CH_3)_2$ *)	Isobutene	**2.5,** 298; **2.8,** 41; **2.9.3,** 31		**2.5,** 135
C_4H_8	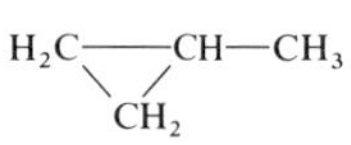	Methylcyclopropane		**2.5,** 181	
C_4H_8O	$CH_3C(=O)CH_2CH_3$ *)	Butanone		**2.5,** 107	**2.5,** 83a
C_4H_8O	$CH_3CH{=}CHCH_2OH$ *)	2-Buten-1-ol	**2.5,** 332		
C_4H_8O	$H_2C{=}CHCH_2CH_2OH$	3-Buten-1-ol	**2.5,** 323a; **2.6.3,** 146a		
C_4H_8O	$(CH_3)_2CHC(=O)H$	iso-Butyraldehyde		**2.5,** 107a	
C_4H_8O	$CH_2{=}CHOCH_2CH_3$	Ethoxyethene	**2.5,** 292; **2.6.3,** 130; **2.8,** 37		
C_4H_8O	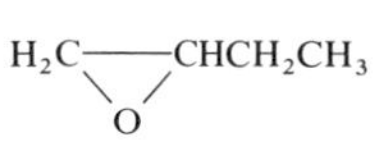	Ethyloxirane	**2.5,** 487; **2.6.3,** 234; **2.8,** 50		
C_4H_8O	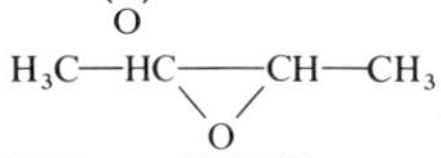	2,3-Epoxybutane		**2.5,** 197	**2.5,** 136
C_4H_8O	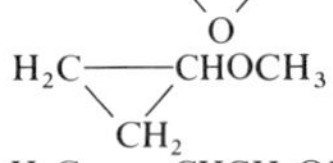	Cyclopropylmethyl ether	**2.5,** 470; **2.6.3,** 222; **2.8,** 91	**2.5,** 184a	
C_4H_8O	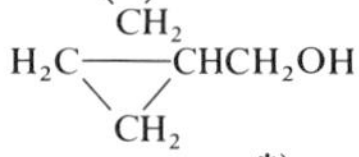 *)	Cyclopropyl carbinol	**2.5,** 475; **2.6.3,** 225	**2.5,** 187	
C_4H_8O	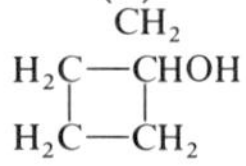 *)	Cyclobutanol	**2.5,** 495; **2.6.3,** 239		

*) Including deuterated compounds.

Alphabetic formula	Chemical formula	Name	Table Nr., Substance Nr. II/14	II/6	II/4
C_4H_8O		Tetrahydrofuran		**2.5**, 219	
$C_4H_8O_2$	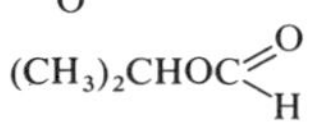	1-Methylethyl formate	**2.5**, 246		
$C_4H_8O_2$		Propyl formate	**2.5**, 138		
$C_4H_8O_2$	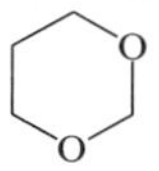	1,3-Dioxan	**2.5**, 542	**2.5**, 227a	
$C_4H_8O_3$	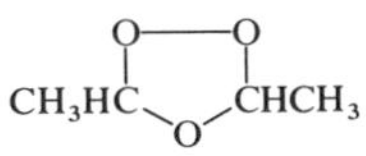	3,5-Dimethyl-1,2,4-trioxolane	**2.5**, 519; **2.6.3**, 253		
$C_4H_8O_3$	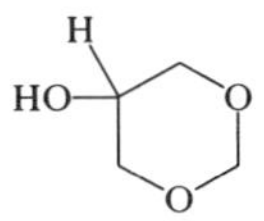	1,3-Dioxan-5-ol	**2.5**, 543; **2.6.3**, 262		
C_4H_8OS	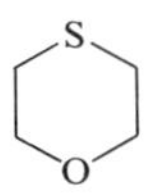	1,4-Oxathiane	**2.5**, 544; **2.6.3**, 263		
$C_4H_8O_2S$	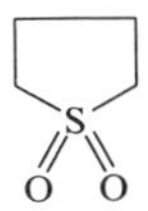	Tetrahydrothiophene 1,1-dioxide	**2.5**, 521 a		
C_4H_8S	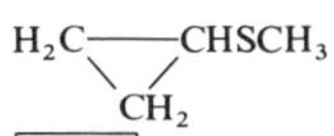	(Methylthio) cyclopropane	**2.5**, 471; **2.6.3**, 223; **2.8**, 107		
C_4H_8S	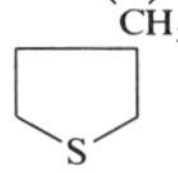	Tetrahydrothiophene	**2.5**, 521; **2.6.3**, 254	**2.5**, 220	
C_4H_8Se	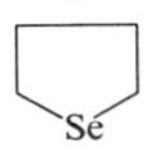	Tetrahydroselenophene	**2.5**, 522	**2.5**, 221	
C_4H_9Br	$CH_3CH_2CH_2CH_2Br$	1-Bromobutane	**2.5**, 262		
C_4H_9Br	$(CH_3)_3CBr$	tert.-Butyl bromide			**2.4**, 65
C_4H_9Br	$(CH_3)_3CBr$	2-Bromo-2-methylpropane	**2.4**, 56; **2.7.2**, 52		
C_4H_9Cl	$CH_3CH_2CH_2CH_2Cl$	1-Chlorobutane	**2.5**, 261		
C_4H_9Cl	$(CH_3)_2CH_3CCl$ *)	tert.-Butyl chloride	**2.4**, 55; **2.7.2**, 36		**2.4**, 64
C_4H_9Cl	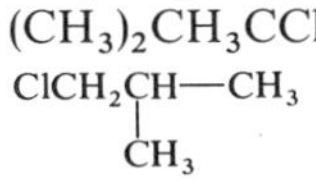	1-Chloro-2-methylpropane	**2.5**, 267		
C_4H_9F	$(CH_3)_3CF$	tert.-Butyl fluoride			**2.4**, 63
$C_4H_9F_3IN$	$(CH_3)_3N \cdot ICF_3$	N,N-Dimethyl-methanamine trifluoro iodo-methane adduct	**2.4**, 66		
C_4H_9GeN	$(CH_3)_3GeCN$	Trimethylgermane carbonitrile	**2.4**, 64		
C_4H_9I	$CH_3CH_2CH_2CH_2I$	1-Iodobutane	**2.5**, 263		
C_4H_9I	$(CH_3)_3CI$	tert.-Butyliodide		**2.4**, 49	**2.4**, 66
C_4H_9NO	$(CH_3)_3CNO$	2-Methyl-2-nitrosopropane	**2.5**, 454; **2.6.3**, 173; **2.8**, 118		
C_4H_9NO	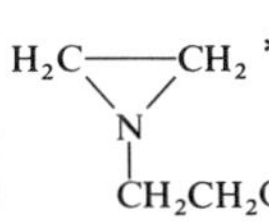 *)	1-Aziridineethanol	**2.5**, 493; **2.6.3**, 237		

*) Including deuterated compounds.

Alphabetic formula	Chemical formula	Name	Table Nr., Substance Nr. II/14	II/6	II/4
C_4H_9NO	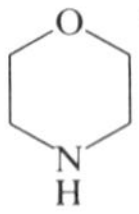 *)	Morpholine	**2.5**, 539; **2.6.3**, 259	**2.5**, 229	
$C_4H_9NO_2$	$(CH_3)_3CONO$ *)	2-Methyl-2-nitritopropane	**2.5**, 455a		
$C_4H_9NO_2$	$(CH_3)_3CNO_2$	2-Methyl-2-nitropropane	**2.5**, 455; **2.6.3**, 174; **2.8**, 119		
C_4H_9NOSi	$(CH_3)_3SiNCO$	Trimethylsilyl isocyanate		**2.4**, 51b	
C_4H_9NS	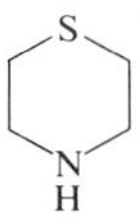 *)	Tetrahydro-4H-1,4-thiazine	**2.5**, 540; **2.6.3**, 260; **2.7.3**, 52		
C_4H_9NSi	$(CH_3)_3SiCN$	Trimethylsilane-carbonitrile	**2.4**, 59		
C_4H_9NSi	$(CH_3)_3SiNC$	Isocyanotrimethyl-silane	**2.4**, 60		
C_4H_{10}	$(CH_3)_3CH$	Isobutane	**2.4**, 54; **2.9.2**, 35		**2.4**, 62
$C_4H_{10}Ge$	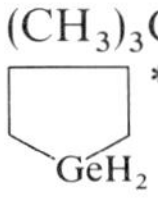 *)	Germacyclopentane		**2.5**, 222	**2.5**, 161b
$C_4H_{10}N_2$	$H_2C—NCH_3$ / $H_2C—NCH_3$	1,2-Dimethyl-1,2-diazetidine	**2.5**, 509; **2.7.3**, 51		
$C_4H_{10}O$	$(CH_3CH_2)_2O$ *)	Diethyl ether	**2.5**, 445; **2.6.3**, 171	**2.5**, 165c	
$C_4H_{10}O$	$CH_3CH_2CH_2OCH_3$ *)	Methylpropyl ether	**2.5**, 226; **2.6.3**, 111; **2.8**, 28		
$C_4H_{10}S$	$(CH_3)_3CSH$	2-Methyl-2-propanethiol	**2.5**, 446; **2.6.3**, 172; **2.8**, 105		
$C_4H_{10}Si$		Silacyclopentane	**2.5**, 403; **2.6.3**, 191		
$C_4H_{11}NO$	$CH_2N(CH_3)_2CH_2OH$ *)	2-(Dimethylamino)-ethanol	**2.5**, 180; **2.6.3**, 91		
$C_4H_{11}P$	$(CH_3)_3CPH_2$ *)	(1,1-Dimethylethyl) phosphine	**2.5**, 457; **2.6.3**, 175		
C_5					
C_5F_5N	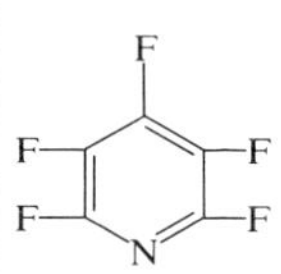	Pentafluoropyridine	**2.5**, 749; **2.6.3**, 359; **2.7.3**, 78		
C_5FeO_5	$Fe(CO)_5$	Iron pentacarbonyl		**2.9.2**, 35	
C_5HN	$HC{\equiv}C—C{\equiv}CCN$	2,4-Pentadiynenitrile	**2.3**, 31; **2.6.2**, 18; **5: 2.3**, 31		
$C_5H_3F_2N$	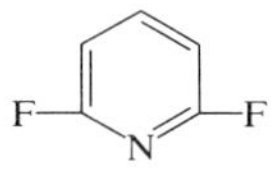	2,6-Difluoropyridine	**2.5**, 748; **2.6.3**, 358; **2.7.3**, 77; **5: 2.9.3**, 6		
$C_5H_3F_3O_2$	$HC{\equiv}CCH_2OC(=O)CF_3$	2-Propynyl trifluoro-acetate	**2.5**, 360		
C_5H_3NO	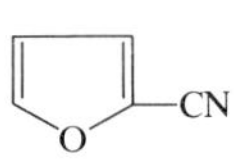	2-Furancarbonitrile	**2.5**, 613; **2.6.3**, 303; **2.7.3**, 58		

*) Including deuterated compounds.

Alphabetic formula	Chemical formula	Name	**Table Nr.**, Substance Nr.		
			II/14	II/6	II/4
$C_5H_3NO_2$	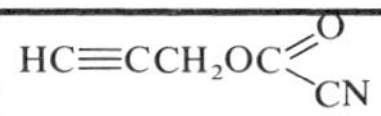	2-Propynylcarbon-cyanidate	**2.5**, 363		
$C_5H_3NO_2$	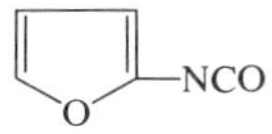	2-Isocyanatofuran	**2.5**, 613a		
C_5H_3NS	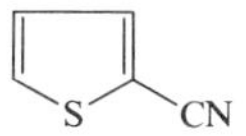	2-Thiophenecarbo-nitrile	**2.5**, 617; **2.6.3**, 304; **2.7.3**, 59		
C_5H_3NS	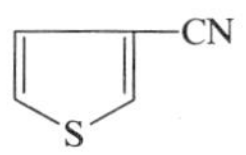	3-Thiophenecarbo-nitrile	**2.5**, 618; **2.6.3**, 305; **2.7.3**, 60		
C_5H_4	$HC\equiv CC\equiv CCH_3$ *)	Methyldiacetylene		**2.9.2**, 34	**2.4**, 60
C_5H_4	$HC\equiv CCH_2C\equiv CH$ *)	1,4-Pentadiyne	**2.5**, 377; **2.6.3**, 165		
C_5H_4BrN	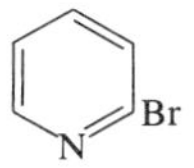	2-Bromopyridine		**2.5**, 294e	
C_5H_4BrN	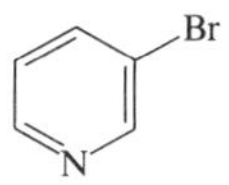	3-Bromopyridine	**2.5**, 738a; **2.7.3**, 156a		
C_5H_4BrN	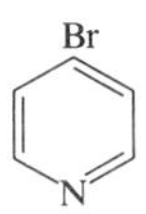	4-Bromopyridine		**2.5**, 295a	
C_5H_4ClN	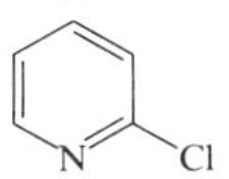	2-Chloropyridine	**2.5**, 733; **2.7.3**, 143	**2.5**, 294d	
C_5H_4ClN	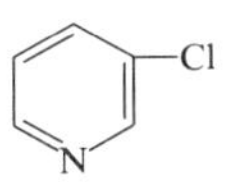	3-Chloropyridine	**2.5**, 738; **2.7.3**, 75		
C_5H_4ClN	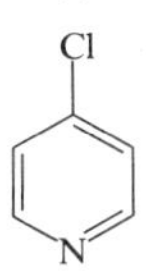	4-Chloropyridine	**2.5**, 742; **2.6.3**, 355; **2.7.3**, 144		
C_5H_4FN	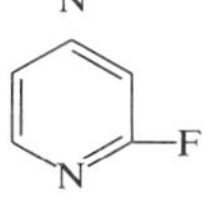	2-Fluoropyridine	**2.5**, 732	**2.5**, 294a	
C_5H_4FN	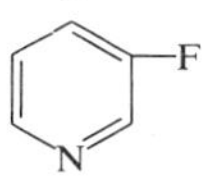	3-Fluoropyridine	**2.5**, 737; **2.6.3**, 352; **2.7.3**, 74; **5: 2.9.3**, 5		
C_5H_4IN	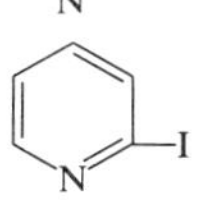	2-Iodopyridine	**2.5**, 734; **2.7.3**, 163		
C_5H_4IN	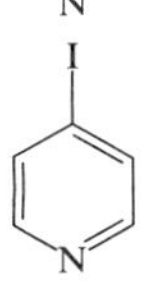	4-Iodopyridine	**2.5**, 743; **2.7.3**, 164		
$C_5H_4N_2$	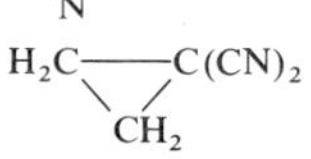	1,1-Dicyanocyclo-propane	**2.5**, 466		
$C_5H_4N_2$	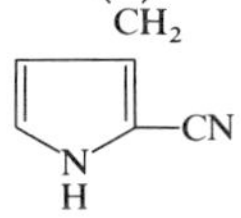 *)	1H-Pyrrole-2-carbo-nitrile	**2.5**, 629		
C_5H_4O	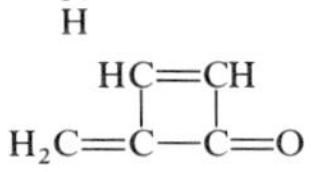	4-Methylene-2-cyclo-buten-1-one	**2.5**, 591; **2.6.3**, 291; **2.9.3**, 38		

*) Including deuterated compounds.

*) Including deuterated compounds.

Alphabetic formula	Chemical formula	Name	Table Nr., Substance Nr.		
			II/14	II/6	II/4
C_5H_5P	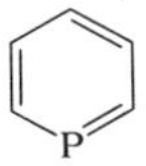	Phosphabenzene		**2.5**, 297a	
C_5H_5Sb	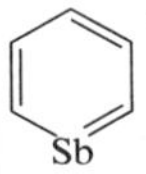 *)	Antimonin	**2.5**, 751; **2.7.3**, 157		
C_5H_5Tl	C_5H_5Tl *)	Cyclopentadienyl thallium		**2.4**, 79	**2.4**, 104
C_5H_6	$CH_3C{\equiv}C{-}CH{=}CH_2$	1-Penten-3-yne		**2.5**, 159a	
C_5H_6	$HC{\equiv}C{-}CH{=}CH{-}CH_3$	3-Penten-1-yne		**2.5**, 160	
C_5H_6	$HC{\equiv}C{-}C(CH_3){=}CH_2$	2-Methyl-1-butene-3-yne		**2.5**, 159	
C_5H_6	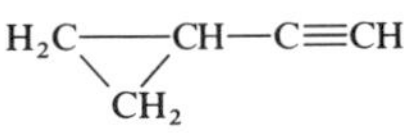 *)	Cyclopropyl acetylene		**2.5**, 189	
C_5H_6	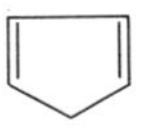 *)	Cyclopentadiene	**2.5**, 605; **2.6.3**, 300		**2.5**, 169
C_5H_6	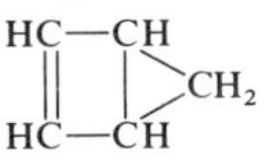	Bicyclo[2.1.0]-pent-2-ene		**2.5**, 273	
C_5H_6Be	C_5H_5BeH	(μ^5-2,4-Cyclopentadien-1-yl)hydroberyllium	**2.4**, 96; **2.6.2**, 67		
$C_5H_6F_2$	$CH_2{=}C(CH_3)CH{=}CF_2$	1,1-Difluoro-3-methylbutadiene		**2.5**, 144	
$C_5H_6N_2$	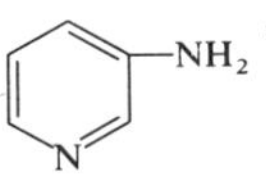 *)	3-Prydinamine	**2.5**, 740		
$C_5H_6N_2$	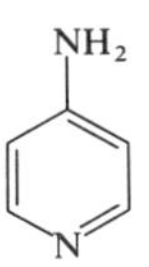 *)	4-Pyridinamine	**2.5**, 745		
$C_5H_6N_2$	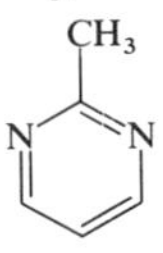	2-Methylpyrimidine	**2.5**, 753; **2.6.3**, 362; **2.8**, 59		
$C_5H_6N_2$	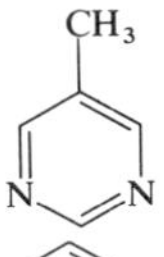	5-Methylpyrimidine	**2.5**, 752; **2.6.3**, 361; **2.8**, 58		
$C_5H_6N_2$	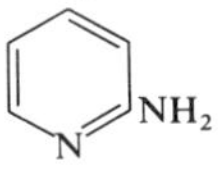	2-Aminopyridine		**2.5**, 294c	
C_5H_6O	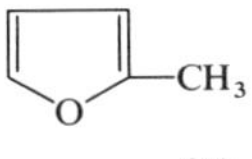	2-Methylfuran	**2.5**, 611; **2.9.3**, 43	**2.5**, 251	
C_5H_6O	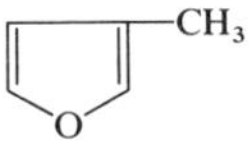	3-Methylfuran	**2.5**, 612; **2.9.3**, 44	**2.5**, 251a	
C_5H_6O	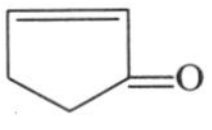	Cyclopent-2en-1-one	**2.5**, 595; **2.6.3**, 293	**2.5**, 243	

*) Including deuterated compounds.

Alphabetic formula	Chemical formula	Name	Table Nr., Substance Nr.		
			II/14	II/6	II/4
C_5H_6O	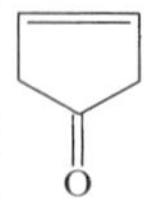 *)	Cyclopent-3-en-1-one	**2.5,** 594; **2.6.3,** 292	**2.5,** 242a	
C_5H_6S	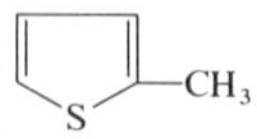	2-Methylthiophene	**2.5,** 619; **2.8,** 51	**2.5,** 253	
C_5H_6S	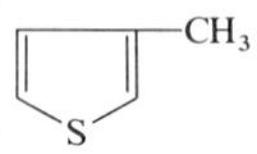	3-Methylthiophene	**2.5,** 620; **2.6.3,** 306; **2.8,** 52	**2.5,** 253a	
C_5H_6Se	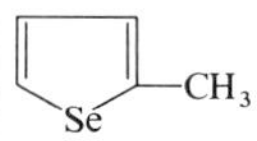	2-Methylselenophene	**2.5,** 625; **2.6.3,** 308; **2.8,** 54		
C_5H_6Se	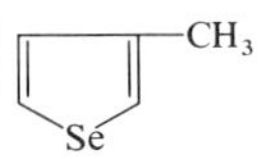	3-Methylselenophene	**2.5,** 624; **2.6.3,** 307; **2.8,** 53		
C_5H_7Cl	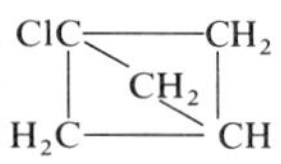	1-Chlorobicyclo-[1.1.1]-pentane		**2.4,** 76	
C_5H_7Cl	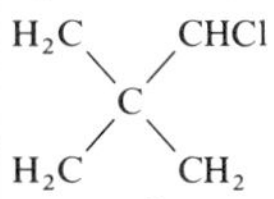	Spiropentyl chloride		**2.5,** 230	
$C_5H_7F_3O_2$	$CF_3C(=O)OCH(CH_3)_2$	1-Methylethyl trifluoroacetate	**2.5,** 196		
$C_5H_7F_3O_2$	$CF_3C(=O)OCH_2CH_2CH_3$	Propyl trifluoroacetate	**2.5,** 194		
$C_5H_7F_3O_4$	$CF_3COOH \cdot CH_3CH_2COOH$	Trifluoroacetic acid-Propanoic acid	**2.5,** 574		
$C_5H_7F_3OS$	$CF_3C(=O)SCH_2CH_2CH_3$	Ethyl trifluorothioacetate	**2.5,** 195		
$C_5H_7F_3OS$	$CF_3C(=O)SCH(CH_3)_2$	S-(1-Methylethyl) trifluoroethanethioate	**2.5,** 197		
C_5H_7N	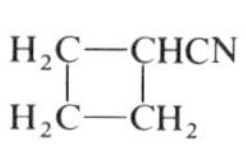	Cyanocyclobutane	**2.5,** 494; **2.6.3,** 238; **2.7.3,** 49		
C_5H_7N	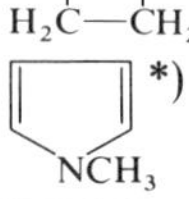 *)	N-Methylpyrrole		**2.5,** 256	
C_5H_7NO	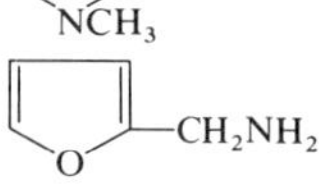	2-Furanmethanamine	**2.5,** 614		
$C_5H_7NO_2$	$(CH_3)_2CHOC(=O)CN$	1-Methylethyl carboncyanidate	**2.5,** 249		
$C_5H_7NO_2$	$NC{-}C(=O)OCH_2CH_2CH_3$	n-Propylcyanoformate	**2.5,** 141		
C_5H_7NOS	$NC{-}C(=O)SCH_2CH_2CH_3$	S-Propyl carbonthiocyanidate	**2.5,** 144		
C_5H_7NOS	$(CH_3)_2CHSC(=O)CN$	S-(1-Methylethyl)-carbonthiocyanidate	**2.5,** 252		
C_5H_8	$CH_2{=}C{=}C(CH_3)_2$	1,1-Dimethylallene	**2.5,** 352	**2.5,** 148	

*) Including deuterated compounds.

Alphabetic formula	Chemical formula	Name	**Table Nr.,** Substance Nr.		
			II/14	II/6	II/4
C_5H_8	$CH_2{=}C(CH_3)CH{=}CH_2$	Isoprene		**2.8,** 24	**2.5,** 116
C_5H_8	$HC{\equiv}CCH(CH_3)_2$ *)	3-Methyl-1-butyne	**2.5,** 355; **2.6.3,** 156; **2.8,** 38		
C_5H_8	$CH_2{=}C{=}CHCH_2CH_3$	1,2-Pentadiene	**2.5,** 349		
C_5H_8	$CH_3CH{=}CHCH{=}CH_2$	1,3-Pentadiene		**2.5,** 145	
C_5H_8	$HC{\equiv}CCH_2CH_2CH_3$	1-Pentyne		**2.5,** 157a	
C_5H_8	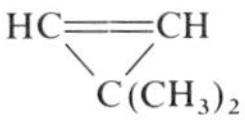	3,3-Dimethylcyclo-propene	**2.5,** 587; **2.6.3,** 288		
C_5H_8	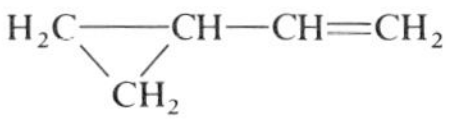	Vinylcyclopropane	**2.5,** 472; **2.6.3,** 224	**2.5,** 186	
C_5H_8	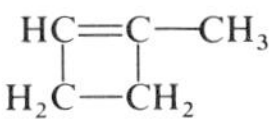	1-Methylcyclobutene		**2.5,** 241	
C_5H_8	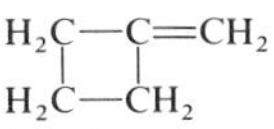	Methylene cyclobutane		**2.5,** 210	
C_5H_8		Cyclopentene		**2.5,** 242	**2.5,** 168
C_5H_8	 *)	Bicyclo[2.1.0]-pentane	**2.5,** 548	**2.5,** 231a	
$C_5H_8F_2$	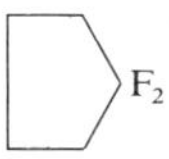 *)	1,1-Difluorocyclo-pentane			**2.5,** 160a
$C_5H_8F_3NO_3$	CF_3COOH $CH_3CH_2CONH_2$	Trifluoroacetic acid-Propanamide	**2.5,** 583; **2.6.3,** 285		
$C_5H_8N_2$	 *)	2,3-Diazabicyclo-(2.2.1) hept-2-ene	**2.5,** 572; **2.6.3,** 279		
C_5H_8O	$CH_2{=}CH{-}CH{=}CHOCH_3$	1-Methoxy-1,3-buta-diene	**2.5,** 342		
C_5H_8O	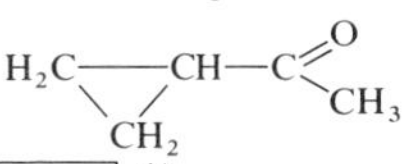	Methyl cyclopropyl ketone		**2.5,** 188	
C_5H_8O	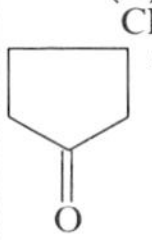 *)	Cyclopentanone	**2.5,** 510	**2.5,** 218	**2.5,** 161
C_5H_8O	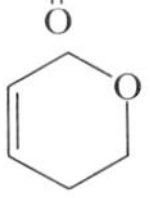	5,6-Dihydro-2H-pyran	**2.5,** 645; **2.6.3,** 318		
C_5H_8O	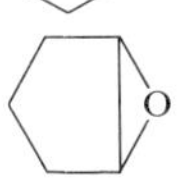	Cyclopentene oxide		**2.5,** 232	**2.5,** 162
C_5H_8O	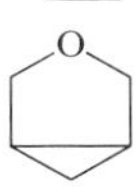	3-Oxabicyclo(3.1.0)-hexane	**2.5,** 552; **2.6.3,** 268		
$C_5H_8O_2$	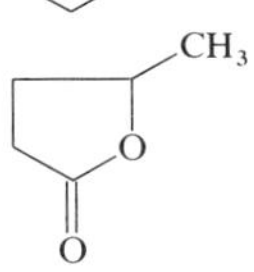	5-Methyl tetrahydro-2-furanone	**2.5,** 520; **2.8,** 50a		
$C_5H_8O_2$	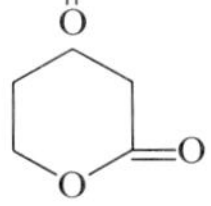	Tetrahydro-2H-pyran-2-one	**2.5,** 534; **2.6.3,** 256a		

*) Including deuterated compounds.

Alphabetic formula	Chemical formula	Name	**Table Nr.**, Substance Nr.		
			II/14	II/6	II/4
$C_5H_8O_2$	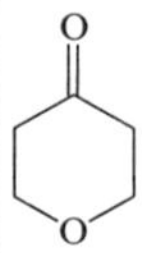	Tetrahydro-4H-pyran-4-one	**2.5**, 535; **2.6.3**, 257		
C_5H_8OS	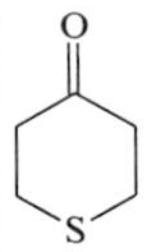	Tetrahydro-4H-thio-pyran-4-one	**2.5**, 536a; **2.6.3**, 258a		
C_5H_8S	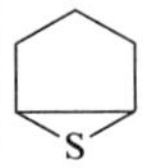	6-Thiabicyclo(3.1.0)-hexane	**2.5**, 550; **2.6.3**, 266		
C_5H_9Br	$CH_2{=}CH(CH_2)_2CH_2Br$	5-Bromo-1-pentene	**2.5**, 335		
C_5H_9Br	$CH_3CH{=}CHCH_2CH_2Br$	5-Bromo-2-pentene	**2.5**, 336		
C_5H_9Cl	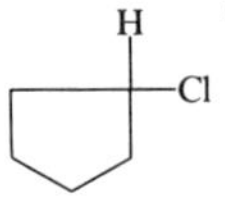 *)	Chlorocyclopentane	**2.5**, 511; **2.7.3**, 135		
$C_5H_9ClO_2$	$CH_3CH_2CHCH_3$ with $OC(=O)Cl$	1-Methylpropyl carbonchloridate	**2.5**, 271		
$C_5H_9ClO_2$	$(CH_3)_2CHCH_2OC(=O)Cl$	2-Methylpropyl carbonchloridate	**2.5**, 275		
$C_5H_9F_2N$	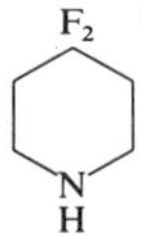 *)	4,4-Difluoro-piperidine	**2.5**, 538		
$C_5H_9FO_2$	$CH_3CH_2CHCH_3$ with $OC(=O)F$	1-Methylpropyl carbonfluoridate	**2.5**, 270		
$C_5H_9FO_2$	$(CH_3)_2CHCH_2OC(=O)F$	2-Methylpropyl carbonfluoridate	**2.5**, 274		
C_5H_9N	$(CH_3)_3CCN$	tert.-Butyl cyanide			**2.4**, 67
C_5H_9N	$(CH_3)_3CNC$	tert.-Butyl isocyanide			**2.4**, 68
C_5H_9N	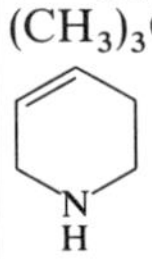	1,2,3,6-Tetrahydro-pyridine	**2.5**, 651; **2.6.3**, 322; **2.7.3**, 71		
C_5H_9NO	$(CH_3)_3C{-}CNO$	Pivalonitril-N-oxide		**2.4**, 50	
C_5H_{10}	$CH_3CH{=}CHCH_2CH_3$	2-Pentene	**2.5**, 326; **2.6.3**, 148; **2.8**, 49		
C_5H_{10}	$CH_2{=}CHCH(CH_3)_2$	3-Methyl-1-butene	**2.5**, 325; **2.6.3**, 147		
C_5H_{10}	$H_2C{-}C(CH_3)_2$ with CH_2 (ring)	1,1-Dimethylcyclo-propane	**2.5**, 469; **2.6.3**, 221		
$C_5H_{10}FN$	$(CH_3)_3CCN\cdots HF$ *)	2,2-Dimethylpropane-nitrile hydrogen fluoride adduct	**2.4**, 108		
$C_5H_{10}O$	$(CH_3)_3C{-}C(=O)H$	Pivaldehyde		**2.5**, 166	
$C_5H_{10}O$	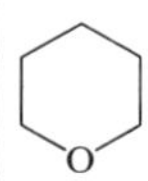	Tetrahydropyran	**2.5**, 531; **2.6.3**, 256	**2.5**, 227	

*) Including deuterated compounds.

Alphabetic formula	Chemical formula	Name	**Table Nr.,** Substance Nr.		
			II/14	II/6	II/4
$C_5H_{10}O_2$	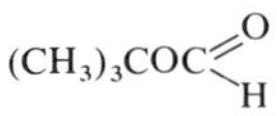	1,1-Dimethylethyl formate	**2.5,** 447		
$C_5H_{10}O_2$		2-Methylpropyl formate	**2.5,** 273		
$C_5H_{10}S$		Tetrahydro-2H-thio-pyran	**2.5,** 536; **2.6.3,** 258		
$C_5H_{11}Br$	$CH_3(CH_2)_4Br$	1-Bromopentane	**2.5,** 277		
$C_5H_{11}Br$	$(CH_3)_2CHCH_2CH_2Br$	1-Bromo-3-methyl-butane	**2.5,** 269		
$C_5H_{11}Cl$	$(CH_3)_2CHCH_2CH_2Cl$	1-Chloro-3-methyl-butane	**2.5,** 268		
$C_5H_{11}I$	$CH_3(CH_2)_4I$	1-Iodopentane	**2.5,** 278		
$C_5H_{11}N$	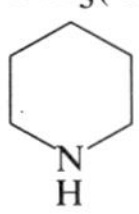 *)	Piperidine	**2.5,** 537; **2.6.3,** 258b; **2.7.3,** 51a	**2.5,** 228	**2.5,** 162a
$C_5H_{11}NO$	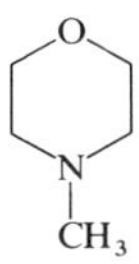	4-Methylmorpholine	**2.5,** 541; **2.6.3,** 261; **2.7.3,** 53; **2.8,** 121		
$C_5H_{12}BN$	$(CH_3)_3CNC \cdot BH_3$	2-Isocyano-2-methyl-propane borane adduct	**2.4,** 109		
C_6					
C_6HF_5	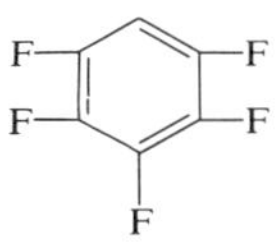	Pentafluorobenzene	**2.5,** 728; **2.6.3,** 348	**2.5,** 292	
$C_6H_2F_4$	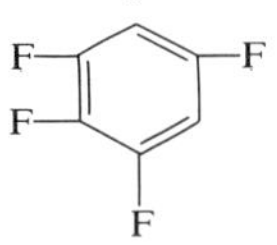	1,2,3,5-Tetrafluoro-benzene	**2.5,** 726; **2.6.3,** 347		
$C_6H_2F_4$	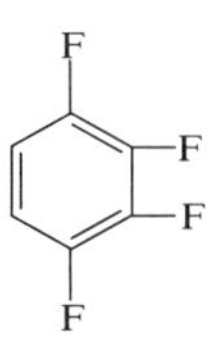	1,2,3,4-Tetrafluoro-benzene	**2.5,** 725; **2.6.3,** 346		
$C_6H_2F_3NO_2$	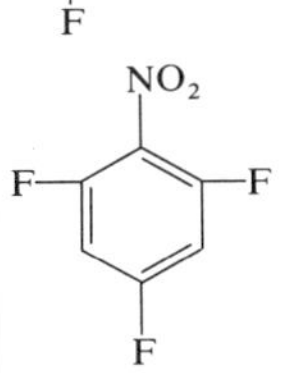	1,3,5-Trifluoro-2-nitrobenzene	**2.5,** 727		
$C_6H_3F_3$	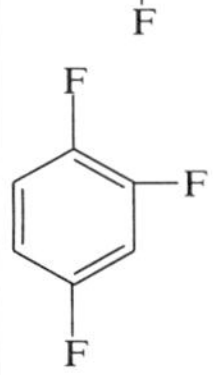	1,2,4-Trifluoro-benzene	**2.5,** 724; **2.6.3,** 345		

*) Including deuterated compounds.

Alphabetic formula	Chemical formula	Name	Table Nr., Substance Nr.		
			II/14	II/6	II/4
C_6H_3N	$CH_3(C{\equiv}C)_2CN$	Hexa-2,4-diynenitrile	**2.4**, 48		
$C_6H_4Cl_2$	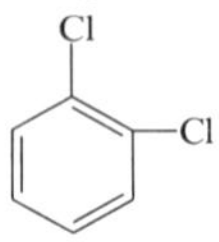	1,2-Dichlorobenzene	**2.5**, 698; **2.7.3**, 139		
$C_6H_4Cl_2$	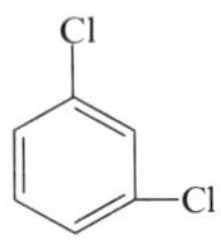	1,3-Dichlorobenzene	**2.5**, 707; **2.7.3**, 141		
C_6H_4ClF	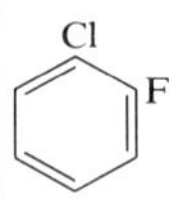	o-Chlorofluoro-benzene		**2.5**, 286	**2.5**, 186a
C_6H_4ClF	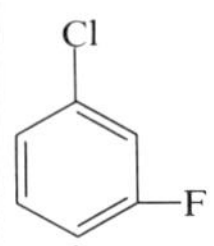	m-Chlorofluoro-benzene	**2.5**, 708; **2.7.3**, 142	**6**: **2.5**, 286a	**2.5**, 187
$C_6H_4F_2$	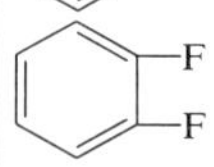	o-Difluorobenzene	**2.5**, 693; **2.9.3**, 49	**2.5**, 284	
$C_6H_4F_2$	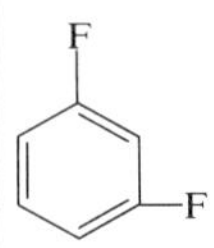	m-Difluorobenzene	**2.5**, 702; **2.9.3**, 50	**2.5**, 285	
$C_6H_4FNO_2$	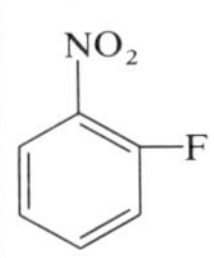	2-Fluoronitrobenzene	**2.5**, 696		
$C_6H_4FNO_2$	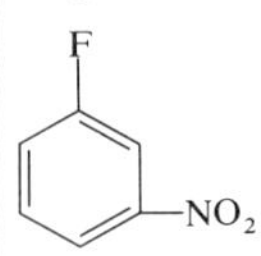	1-Fluoro-3-nitro-benzene	**2.5**, 706		
$C_6H_4FNO_2$	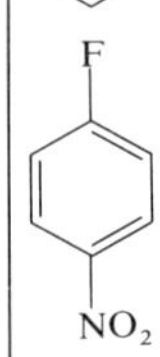	1-Fluoro-4-nitro-benzene	**2.5**, 717		
$C_6H_4N_2$	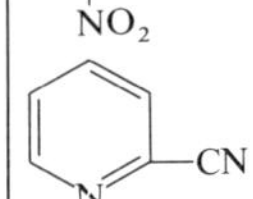	2-Cyanopyridine	**2.5**, 735; **2.6.3**, 351	**2.5**, 294b	
$C_6H_4N_2$	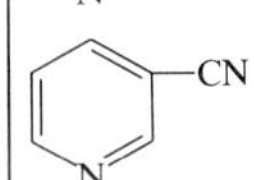	3-Cyanopyridine	**2.5**, 739; **2.6.3**, 353		
$C_6H_4N_2$	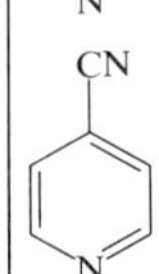	4-Cyanopyridine	**2.5**, 744; **2.6.3**, 356		
$C_6H_4N_2O$	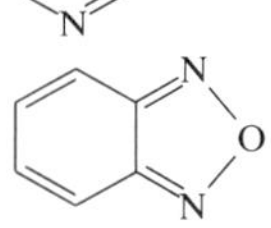	Benzo-2,1,3-oxa-diazole		**2.5**, 300	

Alphabetic formula	Chemical formula	Name	**Table Nr.,** Substance Nr.		
			II/14	II/6	II/4
$C_6H_4N_2S$	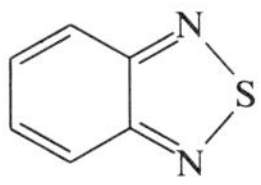	Benzo-2,1,3-thia-diazole		**2.5,** 301	
$C_6H_4O_2$	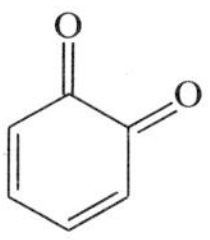	3,5-Cyclohexadiene-1,2-dione	**2.5,** 701; **2.6.3,** 344		
$C_6H_5BCl_2$	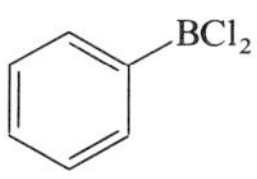	Dichlorophenyl-borane	**2.5,** 688; **2.8,** 140		
$C_6H_5BF_2$	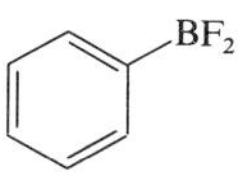	Difluorophenyl-borane	**2.5,** 687; **2.8,** 139		
C_6H_5Br	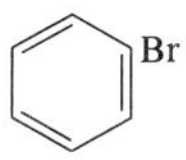 *)	Bromobenzene		**2.5,** 274b	**2.5,** 181
C_6H_5Cl	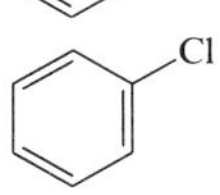 *)	Chlorobenzene	**2.5,** 675; **2.7.3,** 138	**2.5,** 274a	**2.5,** 180
C_6H_5ClO	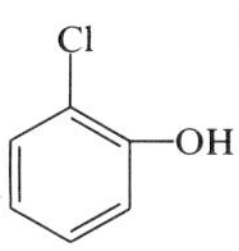 *)	2-Chlorophenol	**2.5,** 699; **2.7.3,** 140		
$C_6H_5ClO_3$	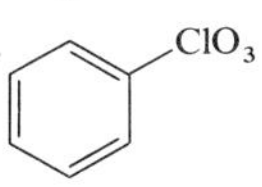	Perchlorylbenzene	**2.5,** 686		
$C_6H_5ClO_2S$	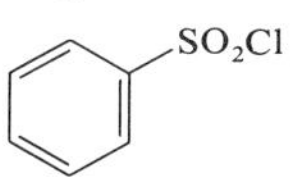	Benzenesulfonyl chloride	**2.5,** 678		
C_6H_5F	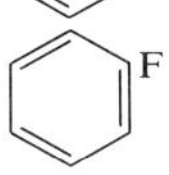 *)	Fluorobenzene		**2.5,** 274	**2.5,** 179
C_6H_5FO	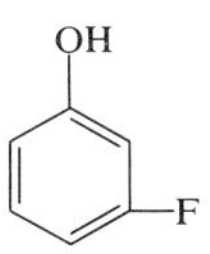	3-Fluorophenol	**2.5,** 703		
C_6H_5I	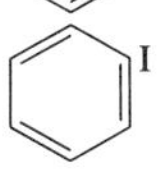	Iodobenzene		**2.5,** 275	**2.5,** 181a
C_6H_5N	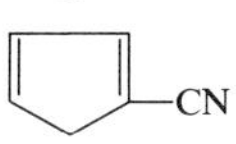	3-Cyanocyclopentene	**2.5,** 606; **2.6.3,** 301		
C_6H_5NO	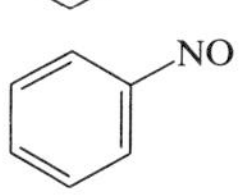	Nitrosobenzene	**2.5,** 680a; **2.7.3,** 72a	**2.5,** 280	**2.5,** 183a
C_6H_5NO	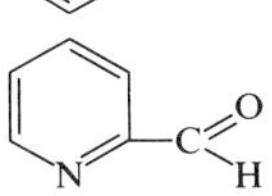	2-Pyridinecarbox-aldehyde	**2.5,** 736; **2.6.3,** 350		
C_6H_5NO	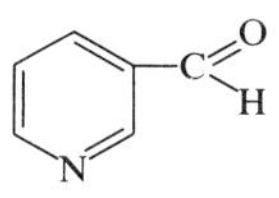	3-Pyridinecarbox-aldehyde	**2.5,** 741; **2.6.3,** 354		

*) Including deuterated compounds.

Alphabetic formula	Chemical formula	Name	**Table Nr.**, Substance Nr.		
			II/14	II/6	II/4
C_6H_5NO	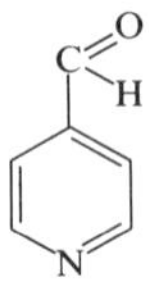	4-Pyridinecarboxaldehyde	**2.5**, 746; **2.6.3**, 357; **2.7.3**, 76		
$C_6H_5NO_3$	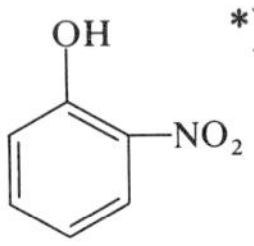 *)	2-Nitrophenol	**2.5**, 700		
C_6H_5NOS	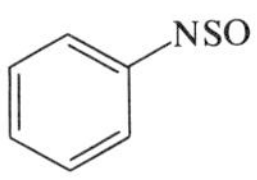	N-Sulfinylbenzeneamine	**2.5**, 683; **2.6.3**, 341		
C_6H_6	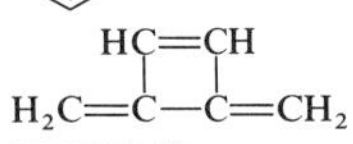	3,4-Dimethylenecyclobutene	**2.5**, 592; **2.9.3**, 37	**2.5**, 240	
C_6H_6	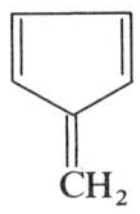 *)	Fulvene	**2.5**, 608; **2.9.3**, 41	**2.4**, 248	
C_6H_6	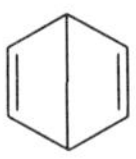	Bicyclo(2.2.0)hexa-2,5-diene	**2.5**, 660; **2.6.3**, 327		
C_6H_6		Benzvalene		**2.5**, 273a	
C_6H_6		Tricyclo[3.1.0.0^{2,6}]-hex-3-ene	**2.5**, 661		
C_6H_6FN	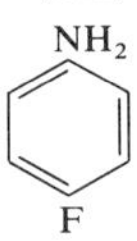	p-Fluoroaniline		**2.5**, 291	
C_6H_6FN	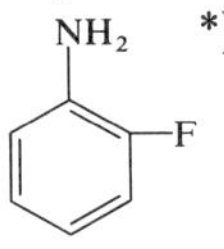 *)	2-Fluorobenzeneamine	**2.5**, 695		
C_6H_6FN	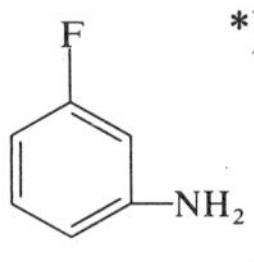 *)	3-Fluorobenzeneamine	**2.5**, 704		
C_6H_6O	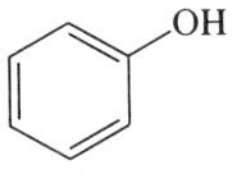 *)	Phenol	**2.5**, 677; **2.6.3**, 338; **2.8**, 92	**2.5**, 277	**2.5**, 184
C_6H_6O	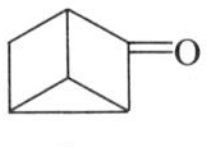	Tricyclo(2.2.0.0(2,6))-hexan-3-one	**2.5**, 569; **2.6.3**, 276		
C_6H_6S	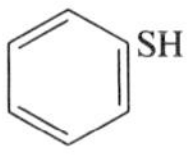	Thiophenol		**2.5**, 278	
C_6H_7Cl	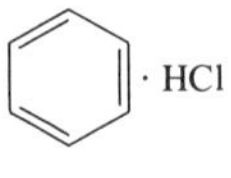 *)	Benzene hydrogen chloride complex	**2.4**, 110; **5**: **2.7.2**, 40a		
$C_6H_7F_3O$	$H_2C{=}CH(CH_2)_2OC(=O)CF_3$	3-Butenyl trifluoroacetate	**2.5**, 334		

*) Including deuterated compounds.

Alphabetic formula	Chemical formula	Name	**Table Nr.**, Substance Nr.		
			II/14	II/6	II/4
$C_6H_7F_3O_2$	$CH_3HC{=}CHCH_2OC(=O)CF_3$	2-Butenyl trifluoro-acetate	**2.5,** 333		
$C_6H_7F_3O_4$		Trifluoroacetic acid cyclopropane-carboxylic acid complex	**2.5,** 579; **2.6.3,** 284		
C_6H_7N	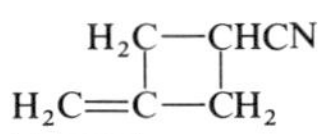	3-Methylenecyclo-butane carbonitrile	**2.5,** 498; **2.6.3,** 241; **2.7.3,** 50		
C_6H_7N		4-Cyanocyclopentane	**2.5,** 593a; **2.6.3,** 291a		
C_6H_7N	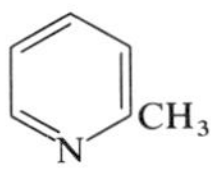	2-Methylpyridine		**2.5,** 295	
C_6H_7N	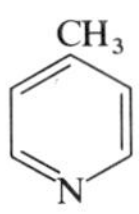	4-Methylpyridine		**2.5,** 296	**2.5,** 190a
C_6H_7N	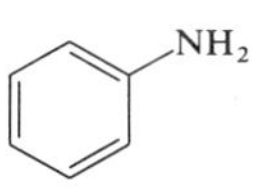 *)	Aniline	**2.5,** 679; **2.6.3,** 339; **2.7.3,** 72	**2.5,** 279	
C_6H_8	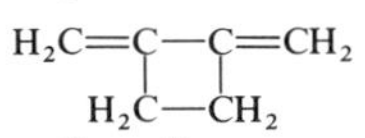	1,2-Dimethylene-cyclobutane	**2.5,** 497; **2.6.3,** 240		
C_6H_8	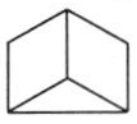 *)	Tricyclo(2.2.0.0(2,6))-hexane	**2.5,** 568		
C_6H_8		1,3-Cyclohexadiene		**2.9.3,** 46	**2.5,** 178
C_6H_8	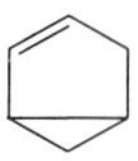	Bicyclo(3.1.0)hex-2-ene	**2.5,** 669; **2.6.3,** 333		
C_6H_8		Bicyclo(2.1.1.)hex-2-ene	**2.5,** 668; **2.6.3,** 332		
$C_6H_8F_3NO_3$	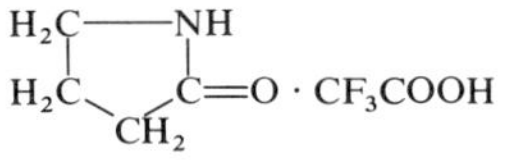	2-Pyrrolidinone-Trifluoroacetic acid	**2.5,** 584; **2.6.3,** 286		
C_6H_8O	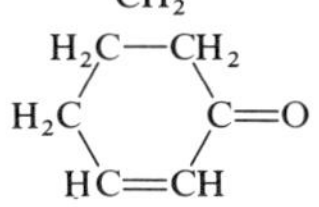	Cyclohex-2-en-1-one		**2.5,** 270	
C_6H_8O	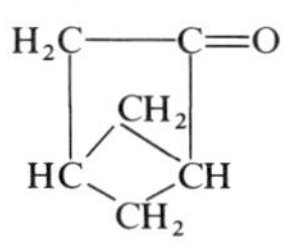	Bicyclo(2.1.1)-hexan-2-one		**2.5,** 232a	
C_6H_8O	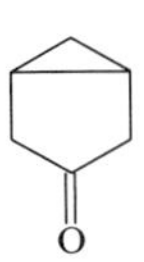	Bicyclo(3.1.0)hexan-3-one	**2.5,** 553; **2.6.3,** 269		

*) Including deuterated compounds.

Alphabetic formula	Chemical formula	Name	**Table Nr.,** Substance Nr.		
			II/14	II/6	II/4
C_6H_8O	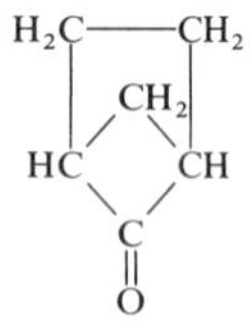	Bicyclo(2.1.1)-hexan-5-one		**2.5,** 232 b	
C_6H_8O	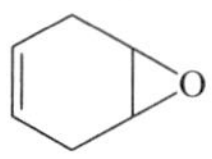	7-Oxabicyclo (4.1.0) hept-3-ene	**2.5,** 672; **2.6.3,** 336		
C_6H_8Si	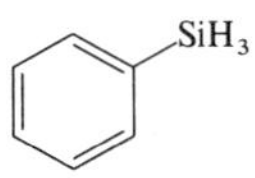 *)	Phenylsilane	**2.5,** 689; **2.6.3,** 342; **2.8,** 72		
C_6H_9Cl	$(CH_3)_3CC{\equiv}CCl$	tert.-Butylchloro-acetylene		**2.4,** 51	**2.4,** 70
$C_6H_9F_3O_2$	$CF_3C(=O)OCH_2CH(CH_3)_2$	1-Methoxypropane	**2.5,** 198		
$C_6H_9F_3O_2$	$(CH_3)_3COC(=O)CF_3$	2-Methylpropyl trifluoroacetate	**2.5,** 448		
$C_6H_9F_3O_4$	$CF_3COOH \cdot (CH_3)_2CHCOOH$	Trifluoroacetic acid-2-Methylpropanoic acid	**2.5,** 575		
C_6H_9GaN	$(CH_3)_3N \cdot Ga(CH_3)_3$	(N,N-Dimethyl-methanamine) trimethylgallium	**2.4,** 70a		
C_6H_9N	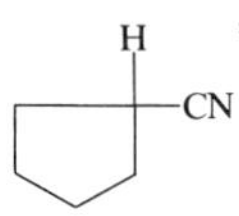 *)	Cyclopentacarbo-nitrile	**2.5,** 512		
$C_6H_9NO_2$	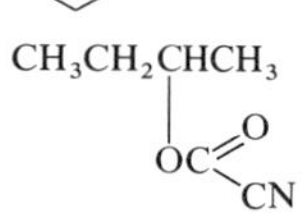	1-Methylpropyl carboncyanidate	**2.5,** 272		
$C_6H_9NO_2$	$(CH_3)_2CHCH_2OC(=O)CN$	2-Methylpropyl carboncyanidate	**2.5,** 276		
C_6H_{10}	$(CH_3)_3CC{\equiv}CH$ *)	tert.-Butylacetylene			**2.4,** 69
C_6H_{10}	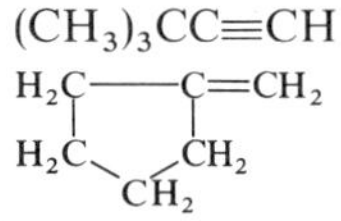	Methylene cyclopentane		**2.5,** 217a	
C_6H_{10}	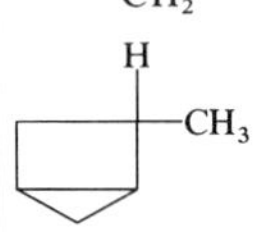	2-Methylbicyclo (2.1.0)pentane	**2.5,** 549; **2.6.3,** 265		
C_6H_{10}	 *)	Cyclohexene		**2.5,** 269	
C_6H_{10}	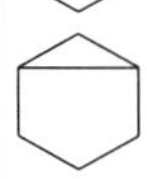	Bicyclo (3.1.0) hexane	**2.5,** 551; **2.6.3,** 267		
$C_6H_{10}F_2$	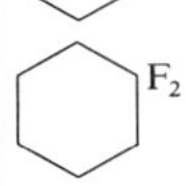	1,1-Difluorocyclo-hexane	**2.5,** 523; **2.6.4,** 255		
$C_6H_{10}O$	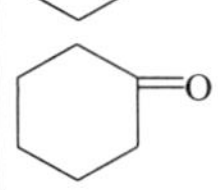	Cyclohexanone	**2.5,** 526	**2.5,** 226	

*) Including deuterated compounds.

Alphabetic formula	Chemical formula	Name	**Table Nr.,** Substance Nr.		
			II/14	II/6	II/4
$C_6H_{10}O$	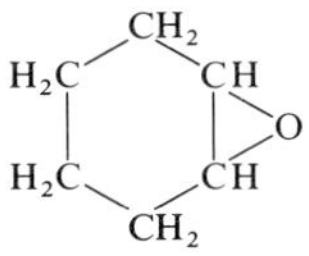	Cyclohexene oxide		**2.5,** 232c	
$C_6H_{10}O$	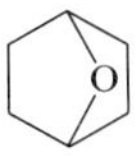	7-Oxabicyclo (2.2.1)heptane	**2.5,** 570; **2.6.3,** 277		
$C_6H_{10}S$	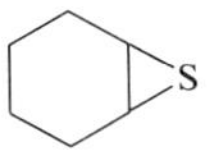	7-Thiabicyclo (4.1.0)heptane	**2.5,** 560; **2.6.3,** 271		
$C_6H_{11}Cl$	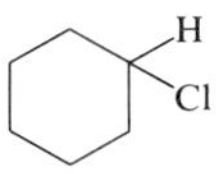	Chlorocyclohexane	**2.5,** 524; **2.7.3,** 136		
$C_6H_{11}ClO_2$	$(CH_3)_3CCH_2OC(O)Cl$	2,2-Dimethylpropyl carbonchloridate	**2.5,** 451		
$C_6H_{11}F$	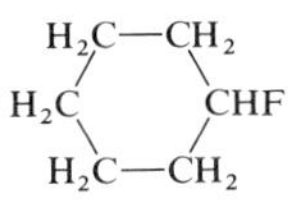	Cyclohexylfluoride		**2.5,** 225	
$C_6H_{11}FO_2$	$(CH_3)_3CCH_2OC(O)F$	2,2-Dimethylpropyl carbonfluoridate	**2.5,** 450		
$C_6H_{12}O_2$	$(CH_3)_3CCH_2OC(O)H$	2,2-Dimethylpropyl formate	**2.5,** 449		
$C_6H_{12}O_2$	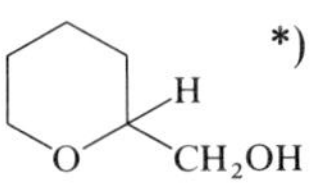 *)	Tetrahydro-2H-pyran-2-methanol	**2.5,** 532		
$C_6H_{12}O_3$	$(OCH{-}CH_3)_3$	Paraldehyde		**2.4,** 77	
$C_6H_{12}S$	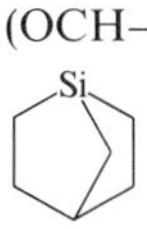	1-Silabicyclo (2.2.1)heptane	**2.5,** 405; **2.6.3,** 193		
$C_6H_{13}F$	$CH_3(CH_2)_5F$	1-Fluorohexane	**2.5,** 279		
$C_6H_{13}I$	$CH_3(CH_2)_5I$	1-Iodohexane	**2.5,** 280		
$C_6H_{13}NO$	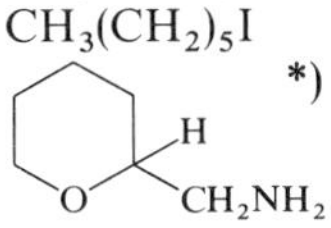 *)	Tetrahydro-2H-pyran-2-methane-amine	**2.5,** 533		
$C_6H_{18}BN$	$(CH_3)_3 \cdot N(CH_3)_3$ *)	N,N-Dimethyl-methanamine (N—B) trimethyl-borane	**2.4,** 67		
$C_6H_{18}BN$	$(CH_3)_3N \cdot B(CH_3)_3$	Trimethylamine-trimethylborine			**2.4,** 81
C_7					
C_7F_5N	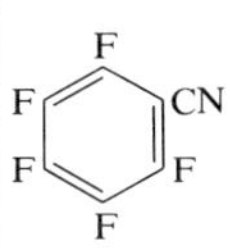	Pentafluorobenzo-nitrile		**2.5,** 293	
C_7HN	$H(C{\equiv}C)_3CN$	Hepta-2,4,6-triyne-nitrile	**2.3,** 32		
C_7H_4	$CH_3(C{\equiv}C)_3H$	1,3,5-Heptatriyne	**2.4,** 49		

*) Including deuterated compounds.

Alphabetic formula	Chemical formula	Name	Table Nr., Substance Nr.		
			II/14	II/6	II/4
$C_7H_4F_2$	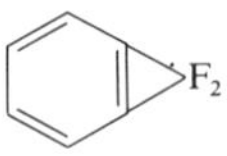	7,7-Difluorobicyclo-(4.1.0)hepta-1,3,5-triene	**2.5,** 671; **2.6.3,** 335		
C_7H_4FNO	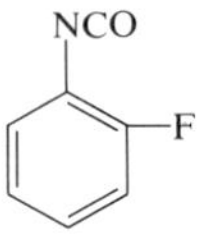	1-Fluoro-2-iso-cyanatobenzene	**2.5,** 697		
$C_7H_4FeO_3$	$C_4H_4Fe(CO)_3$	Tricarbonyl(μ^4-1,3-cyclobutadiene)iron	**2.4,** 97		
C_7H_5ClO	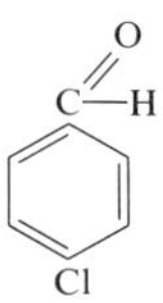	p-Chlorobenz-aldehyde		**2.5,** 287	
$C_7H_5F_3$	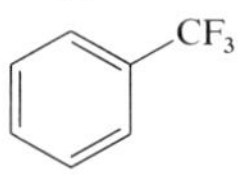	(Trifluoromethyl)-benzene	**2.5,** 685; **2.8,** 56		
C_7H_5FO	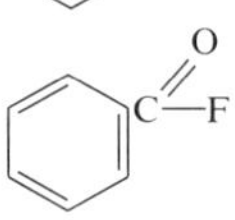	Benzoyl fluoride		**2.5,** 282a	
C_7H_5N	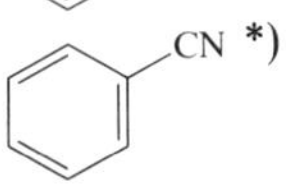	Benzonitrile	**2.5,** 675a; **2.7.3,** 71a	**2.5,** 276	**2.5,** 182
C_7H_5N	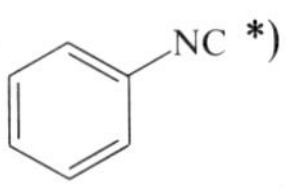	Isocyanobenzene	**2.5,** 676		
C_7H_5NO	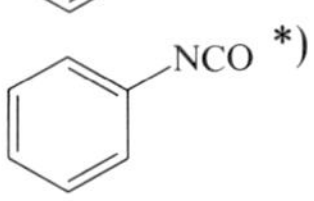	Isocyanatobenzene	**2.5,** 681; **2.6.3,** 340		
C_7H_5NS	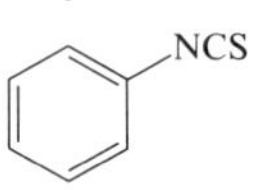	Isothiocyanato-benzene	**2.5,** 682		
C_7H_6O	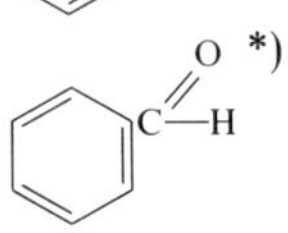	Benzaldehyde		**2.5,** 282	**2.5,** 185
C_7H_6O	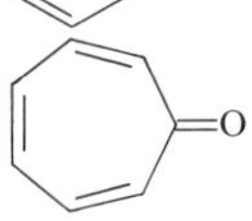	2,4,6-Cycloheptatrien-1-one	**2.5,** 657; **2.6.3,** 324; **2.9.3,** 54		
C_7H_6O	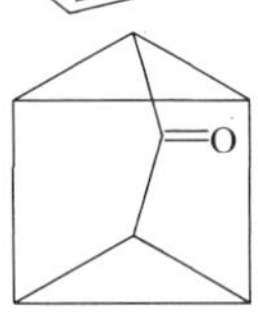	Tetracyclo(3.2.0.0(2,7).0(4,6))heptan-3-one	**2.5,** 561; **2.6,3,** 272		
$C_7H_6O_2$	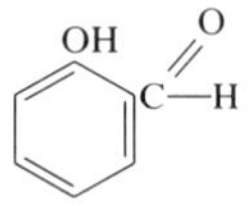	Salicyl aldehyde		**2.5,** 287a	
$C_7H_6O_2$	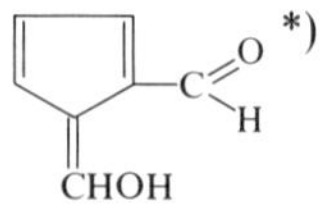	5(Hydroxy-methylene)-1,3-cyclopentadiene	**2.5,** 609		

*) Including deuterated compounds.

Alphabetic formula	Chemical formula	Name	Table Nr., Substance Nr.		
			II/14	II/6	II/4
C_7H_7Cl		p-Chlorotoluene		**2.5**, 290	
C_7H_7F	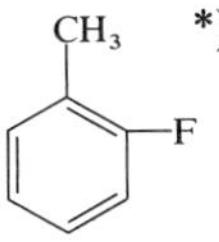 *)	o-Fluorotoluene	**2.5**, 694; **2.8**, 57	**2.5**, 288	
C_7H_7F	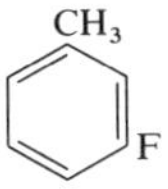	m-Fluorotoluene		**2.5**, 289	
C_7H_7F		p-Fluorotoluene			**2.5**, 187a
C_7H_7FO	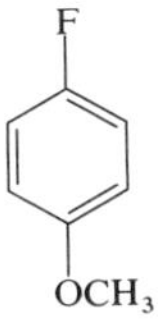	1-Fluoro-4-methoxy-benzene	**2.5**, 713		
C_7H_7FS	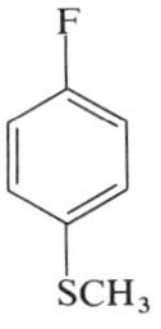	1-Fluoro-4-(methylthio)benzene	**2.5**, 714		
C_7H_7N	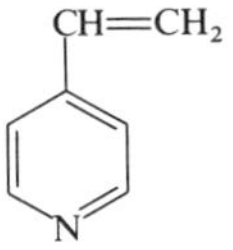	4-Ethenylpyridine	**2.5**, 747		
$C_7H_7NO_3$	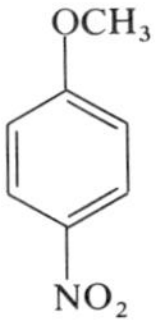	1-Methoxy-4-nitro-benzene	**2.5**, 723		
C_7H_8	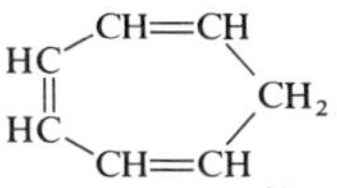	1,3,5-Cyclo-heptatriene			**2.5**, 188
C_7H_8	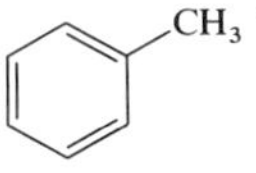 *)	Toluene	**2.5**, 684	**2.5**, 283	
C_7H_8		Bicyclo(2.2.1)hept-2-ene	**2.5**, 663; **2.6.3**, 328		
C_7H_8	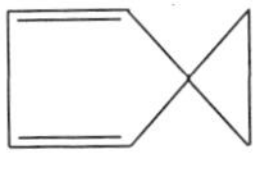 *)	Spiro(2.4)hepta-4,6-diene	**2.5**, 670; **2.6.3**, 334		

*) Including deuterated compounds.

Alphabetic formula	Chemical formula	Name	Table Nr., Substance Nr.		
			II/14	II/6	II/4
C_7H_8FN	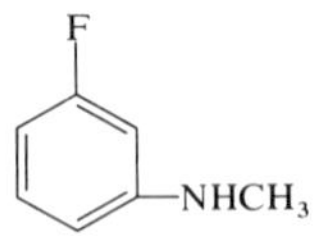	3-Fluoro-N-methyl-benzenamine	**2.5,** 705		
C_7H_8FN	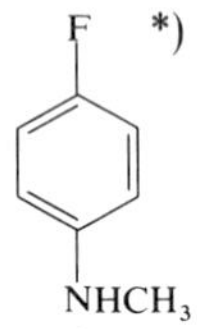 *)	4-Fluoro-N-methyl-benzenamine	**2.5,** 715		
C_7H_8O	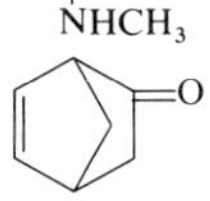	Bicyclo(2.2.1)hept-2-en-5-one	**2.5,** 664; **2.6.3,** 329		
C_7H_8O	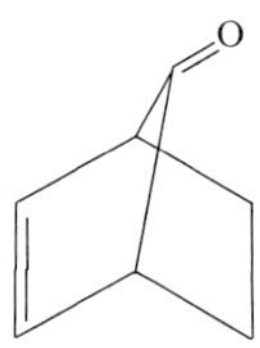	Bicyclo(2.2.1)hept-2-en-7-one	**2.5,** 667; **2.6.3,** 331		
C_7H_8O	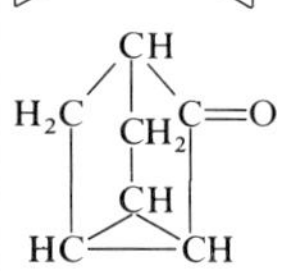	Tricyclo(2.2.1.0(2,6))-heptan-3-one	**2.5,** 566		
C_7H_9Cl	C_7H_9Cl	4-Chlorotricyclo-(2.2.1.0^{2,6})heptane	**2.4,** 95; **2.7.2,** 40		
C_7H_9Cl	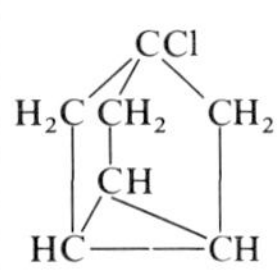	4-Chloronortricyclene		**2.4,** 83a	
$C_7H_9F_3O_2$	$CH_3HC{=}CH(CH_2)_2OC({=}O)CF_3$	3-Pentenyltrifluoro-acetate	**2.5,** 337		
$C_7H_9F_3O_4$	$CF_3COOH \cdot (CH_3)_2C{=}CHCOOH$	Trifluoroacetic acid 3-methyl-2-butenoic acid complex	**2.5,** 578; **2.6.3,** 284		
$C_7H_9F_3O_4$	$H_2C{-}CHCOOH \cdot CF_3COOH$ $H_2C{-}CH_2$	Trifluoroacetic acid cyclopentane carb-oxylic acid complex	**2.5,** 580		
C_7H_9N	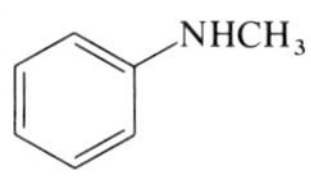 *)	N-Methylbenzene-amine	**2.5,** 680		
C_7H_{10}		1,3-Cycloheptadiene	**2.5,** 658		
C_7H_{10}	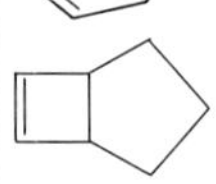	Δ^6-Bicyclo(3.2.0)-heptene	**2.5,** 666; **2.6.3,** 330		
$C_7H_{10}O$	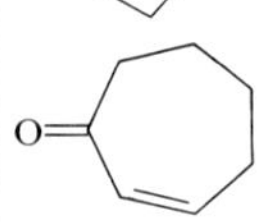	2-Cyclohepten-1-one	**2.5,** 656; **2.6.3,** 323		

*) Including deuterated compounds.

Alphabetic formula	Chemical formula	Name	Table Nr., Substance Nr.		
			II/14	II/6	II/4
$C_7H_{10}O$	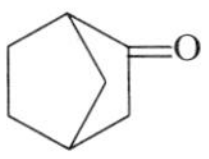	Bicyclo(2.2.1)heptan-2-one	**2.5**, 563; **2.6.3**, 274		
$C_7H_{10}O$	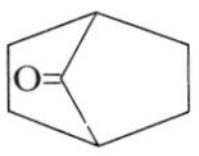	Bicyclo(2.2.1)heptan-7-one	**2.5**, 571; **2.6.3**, 278		
$C_7H_{10}O$	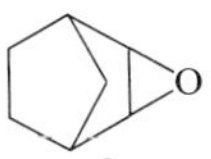	3-Oxatricyclo-(3.2.1.0)(2,4))octane	**2.5**, 564		
$C_7H_{10}O_3$	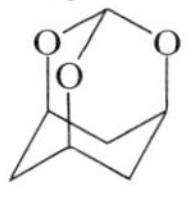	2,8,9-Trioxatricyclo-(3.3.1.1(3,7))decane	**2.4**, 94; **2.6.2**, 66		
$C_7H_{11}F_3O_2$	$(CH_3)_3CCH_2OC(=O)CF_3$	2,2-Dimethylpropyl trifluoroacetate	**2.5**, 453		
$C_7H_{11}F_3O_4$	$CF_3COOH \cdot (CH_3)_3CCOOH$	Trifluoroacetic acid 2,2-dimethylpropanoic acid complex	**2.5**, 576; **2.6.3**, 281		
$C_7H_{11}N$	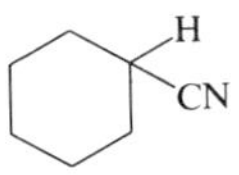	Cyclohexanecarbonitrile	**2.5**, 525		
$C_7H_{11}NO$	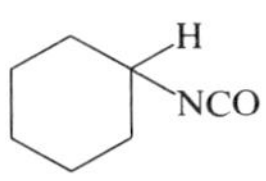	Isocyanatocyclohexane	**2.5**, 527		
$C_7H_{11}NO_2$	$(CH_3)_3CCH_2OC(=O)CN$	2,2-Dimethylpropyl carboncyanidate	**2.5**, 452		
$C_7H_{11}NS$	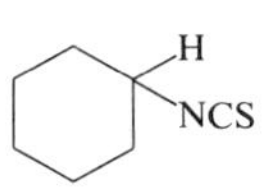	Isothiocyanatocyclohexane	**2.5**, 528		
C_7H_{12}	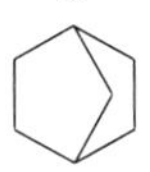	Bicyclo(2.2.1)heptane	**2.5**, 562; **2.6.3**, 273		
$C_7H_{12}O$	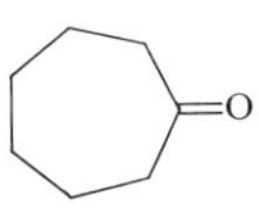	Cycloheptanone	**2.5**, 545		
$C_7H_{12}O$	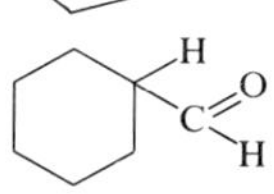	Cyclohexane carboxaldehyde	**2.5**, 529		
$C_7H_{12}O$	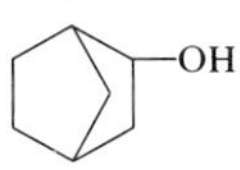 *)	2-Hydroxybicyclo-(2.2.1)-heptane	**2.5**, 562a; **2.6.3**, 273a		
$C_7H_{12}O_2$	$CH_3CH_2HC{=}CH(CH_2)_2OC(=O)H$	3-Hexenyl formate	**2.5**, 339		
$C_7H_{12}O_2$	$H_2C{=}CH(CH_2)_4OC(=O)H$	5-Hexenyl formate	**2.5**, 338		
$C_7H_{13}I$	$(CH_3)_2CHCI(CH_3)_2$	2-Iodo-2,3-dimethylbutane	**2.5**, 245		
$C_7H_{13}N$	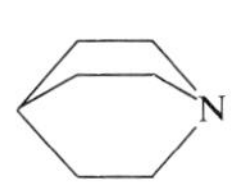	Quinuclidine		**2.4**, 83	

*) Including deuterated compounds.

Alphabetic formula	Chemical formula	Name	**Table Nr.,** Substance Nr.		
			II/14	II/6	II/4
C_8					
$C_8H_5MnO_3$	$C_5H_5Mn(CO)_3$	Cyclopentadienyl manganese tricarbonyl			**2.4,** 105
C_8H_6	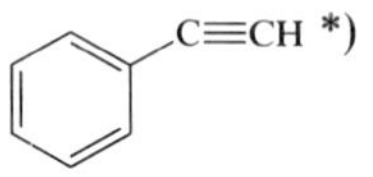	Phenylacetylene	**2.5,** 690; **2.6.3,** 343		**2.5,** 186
C_8H_7Br	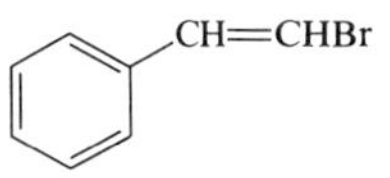	(2-Bromoethenyl)-benzene	**2.5,** 691		
C_8H_7Br	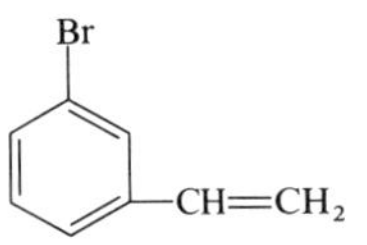	1-Bromo-3-ethenyl-benzene	**2.5,** 710		
C_8H_7Br	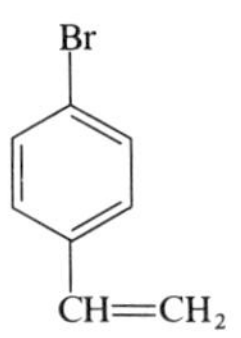	1-Bromo-4-ethenyl-benzene	**2.5,** 719		
C_8H_7Cl	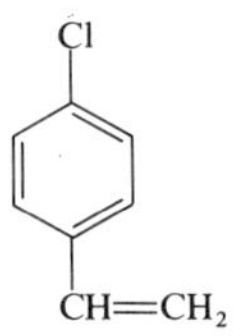	1-Chloro-4-ethenyl-benzene	**2.5,** 718		
C_8H_7F	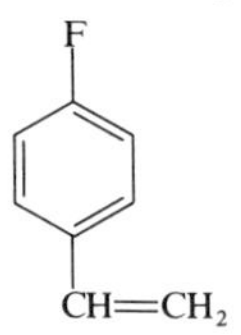	1-Ethenyl-4-fluoro-benzene	**2.5,** 716		
C_8H_7I	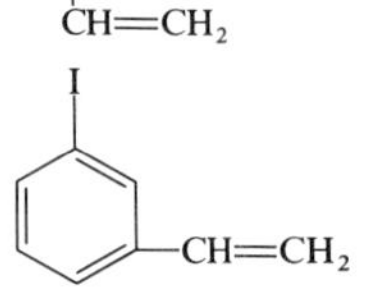	1-Ethenyl-3-iodo-benzene	**2.5,** 711		
C_8H_7NO	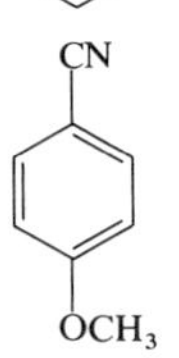	4-Methoxybenzo-nitrile	**2.5,** 720		
C_8H_8	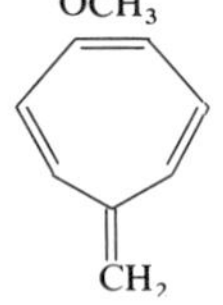	7-Methylene-1,3,5-cycloheptatriene	**2.5,** 659; **2.6.3,** 326		
$C_8H_8O_2$	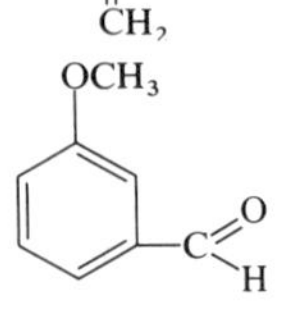	3-Methoxybenz-aldehyde	**2.5,** 712		

*) Including deuterated compounds.

Alphabetic formula	Chemical formula	Name	Table Nr., Substance Nr.		
			II/14	II/6	II/4
$C_8H_8O_2$	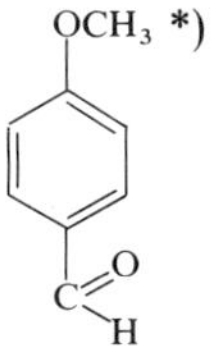	4-Methoxybenz-aldehyde	**2.5**, 721		
C_8H_9Cl	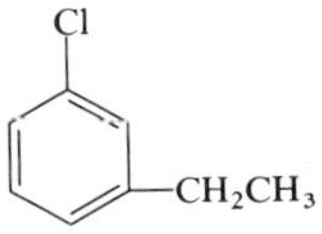	1-Chloro-3-ethyl-benzene	**2.5**, 709		
C_8H_{10}	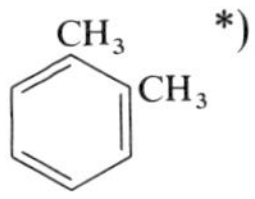	o-Xylene		**2.5**, 287 b	
$C_8H_{10}O$	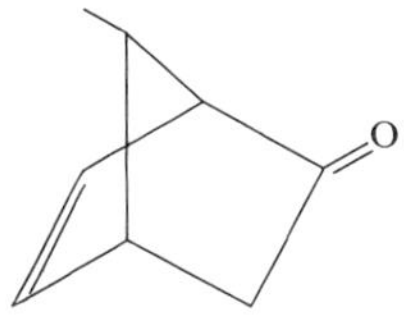	7-Methyl bicyclo-(2.2.1)hept-2-en-5-one	**2.5**, 665		
$C_8H_{11}F_3O_4$	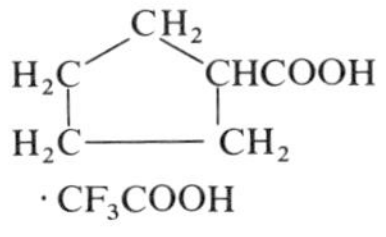	Trifluoroacetic acid cyclopentane carboxylic acid complex	**2.5**, 581		
C_8H_{12}	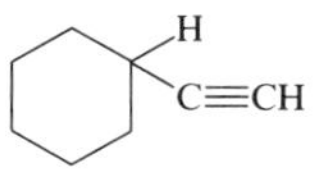	Ethynylcyclohexane	**2.5**, 530		
$C_8H_{12}O$	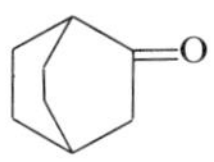	Bicyclo (2.2.2) octan-2-one	**2.5**, 565		
$C_8H_{13}Br$	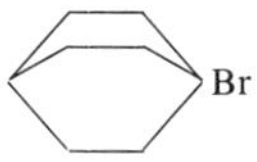	1-Bromobicyclo (2.2.2)-octane			**2.4**, 101
$C_8H_{13}Cl$	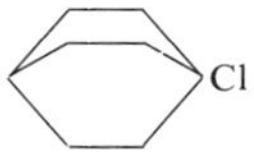	1-Chlorobicyclo (2.2.2)-octane		**2.4**, 82	**2.4**, 100
$C_8H_{13}F$	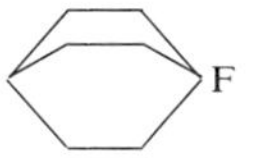	1-Fluorobicyclo (2.2.2)-octane		**2.4**, 81	
$C_8H_{14}O$	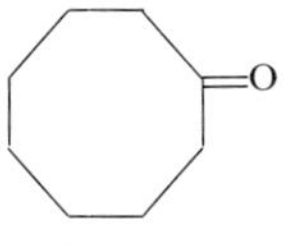	Cyclooctanone	**2.5**, 546		
$C_8H_{14}Si$	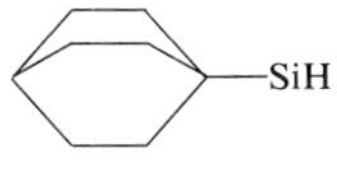	1-Silyl-bicyclo (2.2.2)-octane	**2.4**, 93 a		
C_9					
$C_9H_6CrO_3$	$C_6H_6Cr(CO)_3$	Benzene chromium tricarbonyl			**2.4**, 106
C_9H_8	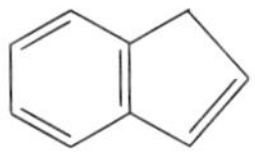	Indene	**2.5**, 662		

*) Including deuterated compounds.

Alphabetic formula	Chemical formula	Name	**Table Nr.**, Substance Nr.		
			II/14	II/6	II/4
C_9H_8O	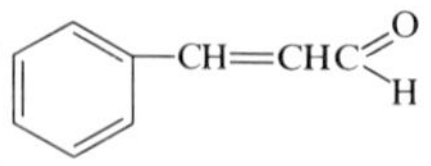	3-Phenylpropenal	**2.5**, 692; **2.9.3**, 49		
$C_9H_{14}O$	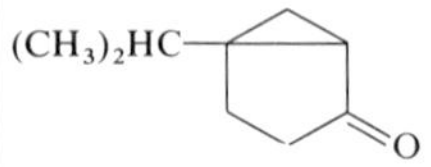	5-(1-Methylethyl) bicyclo (3.1.0) hexan-2-one	**2.5**, 558		
C_{10}					
$C_{10}H_7F$	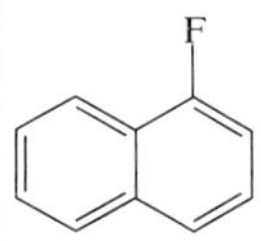	1-Fluoronaphthalene	**2.5**, 729		
$C_{10}H_7F$	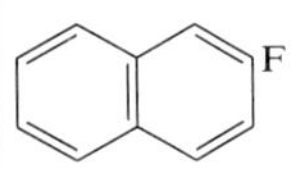	*β*-Fluoronaphthalene			**2.5**, 189
$C_{10}H_8$	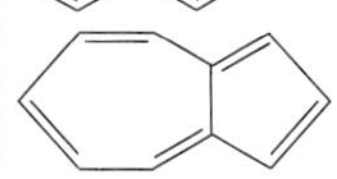	Azulene		**2.5**, 272	**2.5**, 188a
$C_{10}H_{10}$	$C_{10}H_{10}$	Tricyclo (3.3.2.0^{2,8}) deca-3,6,9-triene	**2.4**, 99		
$C_{10}H_{12}O$	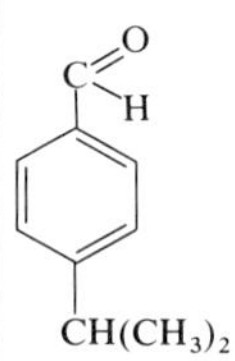	4-Isopropyl-benzaldehyde	**2.5**, 722		
$C_{10}H_{14}O$	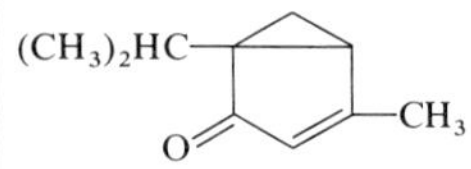	4-Methyl-1-(1-methylethyl) bicyclo (3.1.0)hex-3-en-2-one	**2.5**, 556		
$C_{10}H_{14}O$	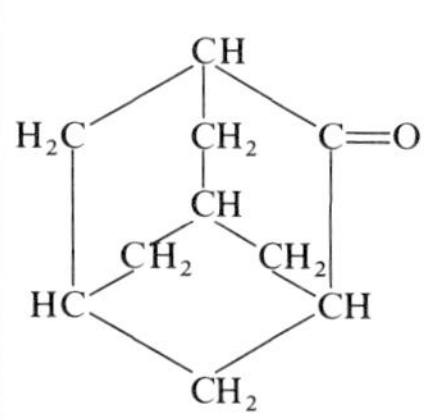	Tricyclo (3.3.1.1.^{3,7}) decan-2-one	**2.5**, 567		
$C_{10}H_{15}Br$	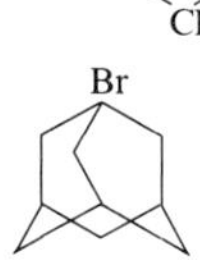	1-Bromo-adamantane		**2.4**, 75	
$C_{10}H_{15}Cl$	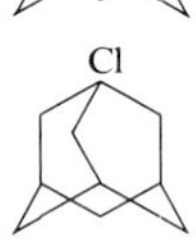	1-Chloro-adamantane			
$C_{10}H_{15}F$	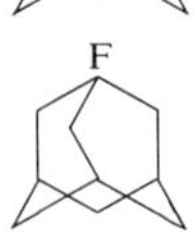	1-Fluoro-adamantane		**2.4**, 73	
$C_{10}H_{15}I$	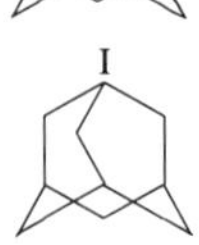	1-Iodoadamantane		**2.4**, 75a	

Alphabetic formula	Chemical formula	Name	Table Nr., Substance Nr.		
			II/14	II/6	II/4
$C_{10}H_{16}$	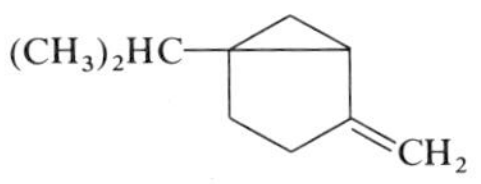	4-Methylene-1-(1-methylethyl) bicyclo (3.1.0) hexane	**2.5**, 557		
$C_{10}H_{16}O$	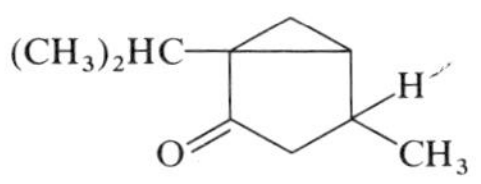	4-Methyl-1-(1-methylethyl) bicyclo (3.1.0) hexan-2-one	**2.5**, 555		
$C_{10}H_{16}O$	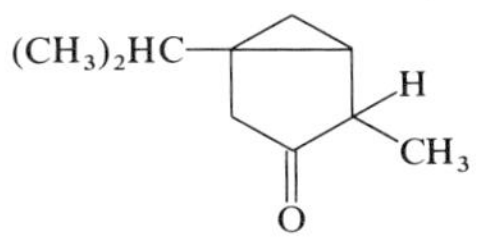	4-Methyl-1-(1-methylethyl) bicyclo (3.1.0) hexan-3-one	**2.5**, 554		
C_{11}					
$C_{11}H_{13}Cl$	$C_{11}H_{13}Cl$	1-Chloropentacyclo $(3.3.3.0^{2,4}.0^{6,8}.0^{9,11})$ undecane	**2.4**, 100		
$C_{11}H_{15}N$	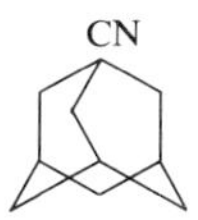	1-Cyanoadamantane		**2.4**, 75b	
C_{13}					
$C_{13}H_{13}Cl$	$C_{13}H_{15}Cl$	1-Chlorotrishomo-bullvalene	**2.4**, 101		
$C_{13}H_{21}O$	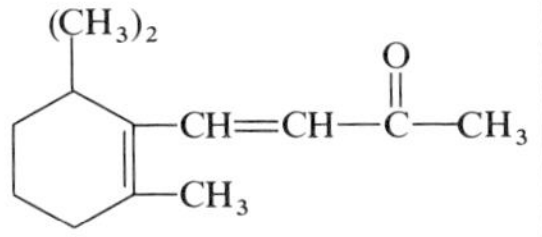	4-(2,6,6-Trimethyl-1-cyclohexen-1-yl)-3-buten-2-one	**2.5**, 654		
$C_{13}H_{21}O$	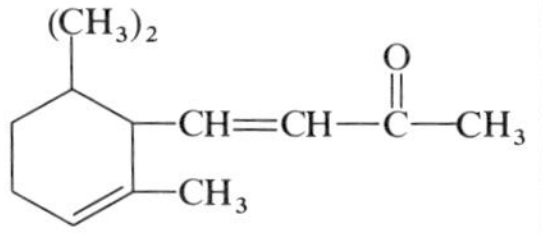	4-(2,6,6,-Trimethyl-2-cyclohexen-1-yl)-3-buten-2-one	**2.5**, 653		
C_{14}					
$C_{14}H_{22}O$	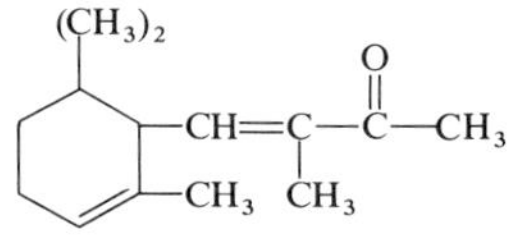	Iso-α-methylionone	**2.5**, 655		
$C_{14}H_{23}O$	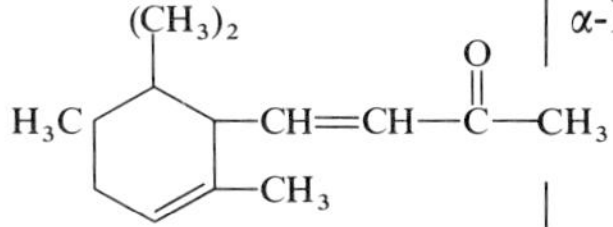	α-Irone	**2.5**, 652		